AF840181

VOLUME NINETY SIX

SEMICONDUCTORS AND SEMIMETALS

III-Nitride Semiconductor Optoelectronics

SERIES EDITORS

CHENNUPATI JAGADISH

Distinguished Professor
Department of Electronic Materials Engineering
Research School of Physics and Engineering
Australian National University
Canberra, ACT2601, Australia

EICKE R. WEBER

Director
Fraunhofer-Institut
für Solare Energiesysteme ISE
Vorsitzender, Fraunhofer-Allianz Energie
Heidenhofstr. 2, 79110
Freiburg, Germany

VOLUME NINETY SIX

SEMICONDUCTORS AND SEMIMETALS

III-Nitride Semiconductor Optoelectronics

Edited by

ZETIAN MI
University of Michigan,
Ann Arbor, MI, United States

CHENNUPATI JAGADISH
Research School of Physics and Engineering,
Australian National University,
Canberra, ACT, Australia

ELSEVIER

ACADEMIC PRESS
An imprint of Elsevier
elsevier.com

Academic Press is an imprint of Elsevier
50 Hampshire Street, 5th Floor, Cambridge, MA 02139, United States
525 B Street, Suite 1800, San Diego, CA 92101-4495, United States
The Boulevard, Langford Lane, Kidlington, Oxford OX5 1GB, United Kingdom
125 London Wall, London, EC2Y 5AS, United Kingdom

First edition 2017

© 2017 Elsevier Inc. All rights reserved

No part of this publication may be reproduced or transmitted in any form or by any means, electronic or mechanical, including photocopying, recording, or any information storage and retrieval system, without permission in writing from the publisher. Details on how to seek permission, further information about the Publisher's permissions policies and our arrangements with organizations such as the Copyright Clearance Center and the Copyright Licensing Agency, can be found at our website: www.elsevier.com/permissions.

This book and the individual contributions contained in it are protected under copyright by the Publisher (other than as may be noted herein).

Notices

Knowledge and best practice in this field are constantly changing. As new research and experience broaden our understanding, changes in research methods, professional practices, or medical treatment may become necessary.

Practitioners and researchers must always rely on their own experience and knowledge in evaluating and using any information, methods, compounds, or experiments described herein. In using such information or methods they should be mindful of their own safety and the safety of others, including parties for whom they have a professional responsibility.

To the fullest extent of the law, neither the Publisher nor the authors, contributors, or editors, assume any liability for any injury and/or damage to persons or property as a matter of products liability, negligence or otherwise, or from any use or operation of any methods, products, instructions, or ideas contained in the material herein.

ISBN: 978-0-12-809584-3
ISSN: 0080-8784

For information on all Academic Press publications visit our website at https://www.elsevier.com/

Working together to grow libraries in developing countries

www.elsevier.com • www.bookaid.org

Publisher: Zoe Kruze
Acquisition Editor: Poppy Garraway
Editorial Project Manager: Shellie Bryant
Production Project Manager: Vignesh Tamil
Cover Designer: Greg Harris

CONTENTS

CONTRIBUTORS

S. Albert
ETSI Telecomunicación, Universidad Politécnica de Madrid, Madrid, Spain. (ch7)

T. Auzelle
Université Grenoble Alpes, INAC-PHELIQS; CEA, INAC-PHELIQS, «Nanophysique et semiconducteurs group», Grenoble, France. (ch6)

C. Bayram
University of Illinois at Urbana–Champaign, Champaign, IL, United States. (ch12)

A.M. Bengoechea-Encabo
ETSI Telecomunicación, Universidad Politécnica de Madrid, Madrid, Spain. (ch7)

P. Bhattacharya
University of Michigan, Ann Arbor, MI, United States. (ch11)

E. Calleja
ETSI Telecomunicación, Universidad Politécnica de Madrid, Madrid, Spain. (ch7)

L.J. Chen
National Tsing-Hua University, Hsinchu, Taiwan. (ch10)

M.H. Crawford
Sandia National Laboratories, Albuquerque, NM, United States. (ch1)

B. Daudin
Université Grenoble Alpes, INAC-PHELIQS; CEA, INAC-PHELIQS, «Nanophysique et semiconducteurs group», Grenoble, France. (ch6)

T. Detchprohm
Center for Compound Semiconductors, School of Electrical and Computer Engineering, Georgia Institute of Technology, Atlanta, GA, United States. (ch4)

A. Dobrinsky
Sensor Electronic Technology, Inc., Columbia, SC, United States. (ch2)

R.D. Dupuis
Center for Compound Semiconductors, School of Electrical and Computer Engineering, Georgia Institute of Technology, Atlanta, GA, United States. (ch4)

T. Frost
University of Michigan, Ann Arbor, MI, United States. (ch11)

W. Guo
University of Massachusetts, Lowell, MA, United States. (ch11)

S. Gwo
National Tsing-Hua University; National Synchrotron Radiation Research Center (NSRRC), Hsinchu, Taiwan. (ch10)

A. Hazari
University of Michigan, Ann Arbor, MI, United States. (ch11)

H. Hirayama
RIKEN, Quantum Optodevice Laboratory, Wako, Saitama, Japan. (ch3)

S. Jahangir
University of Michigan, Ann Arbor, MI, United States. (ch11)

R. Jain
Sensor Electronic Technology, Inc., Columbia, SC, United States. (ch2)

C.T. Kuo
National Tsing-Hua University, Hsinchu, Taiwan. (ch10)

K. Kusakabe
Center for SMART Green Innovation Research, Chiba University, Chiba, Japan. (ch9)

X. Li
Electrical Engineering Program, Computer, Electrical, Mathematical Science and Engineering Division, King Abdullah University of Science and Technology, Thuwal, Saudi Arabia. (ch4)

H.W. Lin
National Tsing-Hua University, Hsinchu, Taiwan. (ch10)

R. Liu
University of Illinois at Urbana–Champaign, Champaign, IL, United States. (ch12)

M.Y. Lu
National Tsing-Hua University, Hsinchu, Taiwan. (ch10)

Y.J. Lu
National Tsing-Hua University, Hsinchu, Taiwan. (ch10)

Z. Mi
University of Michigan, Ann Arbor, MI, United States. (ch5, 8)

M.Á. Sánchez-García
ETSI Telecomunicación, Universidad Politécnica de Madrid, Madrid, Spain. (ch7)

T. Saxena
NXP Semiconductors, Tempe, AZ, United States. (ch2)

M. Shatalov
Sensor Electronic Technology, Inc., Columbia, SC, United States. (ch2)

S.-C. Shen
Center for Compound Semiconductors, School of Electrical and Computer Engineering, Georgia Institute of Technology, Atlanta, GA, United States. (ch4)

M. Shur
Rensselaer Polytechnic Institute, Troy, NY, United States. (ch2)

C.L. Wu
National Cheng-Kung University, Tainan, Taiwan. (ch10)

P.D. Yoder
Center for Compound Semiconductors, School of Electrical and Computer Engineering, Georgia Institute of Technology, Atlanta, GA, United States. (ch4)

A. Yoshikawa
Center for SMART Green Innovation Research, Chiba University, Chiba; Graduate School of Engineering, Kogakuin University, Tokyo, Japan. (ch9)

S. Zhao
McGill University, Montreal, QC, Canada. (ch5, 8)

PREFACE

III-nitride semiconductors have bandgaps that span a very large spectral range, from deep ultraviolet (UV) ($\sim$200 nm) to near infrared ($\sim$1900 nm). Over the past two decades, tremendous progress has been made in GaN-based optoelectronic devices, including light-emitting diodes (LEDs) and lasers operating in the blue and near-UV spectral range. The use of III-nitrides to realize high efficiency LEDs and lasers operating in the deep UV, deep visible, and near-IR spectral range, however, has remained very limited. The underlying causes include the presence of large densities of defects and dislocations, due to the lack of suitable lattice-matched substrates, and large polarization fields and the resulting quantum-confined Start effect. Moreover, it has remained challenging to achieve efficient current conduction in Al-rich AlGaN, due to the large ionization energy (up to 600 meV) for Mg dopant. Similar issues also exist for In-rich InGaN, wherein the realization of efficient p-type conduction has often been limited by the presence of surface electron accumulation.

This book covers the latest breakthrough research and exciting developments in the field of III-nitride compound semiconductors, including quantum well, quantum dot, nanowire, and dot-in-nanowire heterostructures, which have addressed some of the fundamental materials issues of Al-rich AlGaN and In-rich InGaN and have led to optoelectronic devices operating in the deep UV, deep visible, and near-infrared spectral ranges with significantly improved performance. This book includes two parts. Part I is concerned about AlGaN optoelectronic devices, and Part II discusses the emerging III-nitride nanowire heterostructures for application in deep visible and near-infrared optoelectronics.

Part I consists of five chapters. Chapter 1 discusses two of the most significant materials roadblocks to high-performance AlGaN devices: substrates and doping, and further reviews various strategies to mitigate these issues. Chapter 2 discusses key factors currently affecting device performance and reviews progress in development of deep UV LEDs, including high-temperature epitaxy and transparent LED structure design. Chapter 3 presents the epitaxy and performance of AlGaN deep UV LEDs on high-quality AlN on sapphire. The enhancement in device performance by using a transparent p-AlGaN contact layer and by using a low threading dislocation density AlN template is described. Chapter 4 discusses the growth and properties of AlGaN and the simulation, design, processing, and performance of state-of-the-art deep UV lasers and photodetectors. Chapter 5 presents the recent progress made on the growth and characterization of Al(Ga)N nanowires and nanowire deep UV LEDs. The demonstration of electrically pumped semiconductors lasers operating in the UV-B and UV-C bands is also described.

Part II comprises Chapters 6–12. Chapter 6 reviews the self-nucleation process and structural properties of GaN nanowires grown by plasma-assisted molecular beam epitaxy. The crucial issue of GaN crystalline polarity is elucidated. Chapter 7 provides an insight into the selective area epitaxy of InGaN/GaN nanostructures, with a focus on their potential as building blocks for next-generation LEDs operating in the deep visible spectral range. Chapter 8 reviews the recent progress made on the growth, characterization, and device application of InN nanowires. The achievement of intrinsic InN nanowires with the absence of surface electron accumulation and the demonstration of p-type conduction of InN are discussed. Chapter 9 introduces a unique epitaxial process, i.e., dynamic atomic layer epitaxy, and describes the epitaxy and properties of III-N ordered alloys, such as coherent monolayer-InN on/in GaN-matrix nanostructures. Chapter 10 provides a detailed discussion of InGaN nanorod heterostructures, including polarization effects, growth and polarity control, doping and surface properties, heterojunction band alignments, axial heterostructures for full-color and tunable white LEDs, as well as green and full-color core–shell nanorod plasmonic nanolasers. Chapter 11 describes the epitaxial growth and characteristics of edge-emitting electrically pumped GaN/In(Ga)N disk-in-nanowire lasers operating from 533 nm (green) to 1.3 μm. The characteristics of the nanowire heterostructures and the steady state and small-signal modulation characteristics of the lasers are described. Chapter 12 discusses a new method of cubic phase GaN generation: hexagonal-to-cubic phase transition based on novel nanopatterning. The modeling and structural and optical characterization of the novel cubic materials is also discussed.

This book is well suited for students and researchers in the field of semiconductors. It will also be very valuable to researchers and engineers in III-nitrides and optoelectronics. Moreover, the in-depth discussions on the growth and characterization of a broad range of semiconductor nanostructures will benefit students and researchers working on nanomaterials, nanotechnology, and emerging devices.

Zetian Mi,
University of Michigan, Ann Arbor
Chennupati Jagadish,
Australian National University, Canberra
Editors

PART I

AlGaN UV Optoelectronics

CHAPTER ONE

Materials Challenges of AlGaN-Based UV Optoelectronic Devices

M.H. Crawford[1]
Sandia National Laboratories, Albuquerque, NM, United States
[1]Corresponding author: e-mail address: mhcrawf@sandia.gov

Contents

1. INTRODUCTION

AlGaN alloys have emerged as the most promising compound semiconductors for DUV optoelectronics given band gaps that can be tuned over an impressively wide range of the UV spectrum (~363–200 nm). With desirable properties including low size, weight, and operating power, AlGaN-based UV devices have the potential to replace traditional UV sources such as Hg lamps in applications ranging from water purification to fluorescence-based bioagent sensing. However, many AlGaN materials properties are far from ideal. In Fig. 1, we present a generic design for AlGaN-based UV emitters and highlight some of the materials challenges to achieving high-performance light-emitting diodes (LEDs) and laser

Semiconductors and Semimetals, Volume 96
ISSN 0080-8784
http://dx.doi.org/10.1016/bs.semsem.2016.11.001
© 2017 Elsevier Inc. All rights reserved.

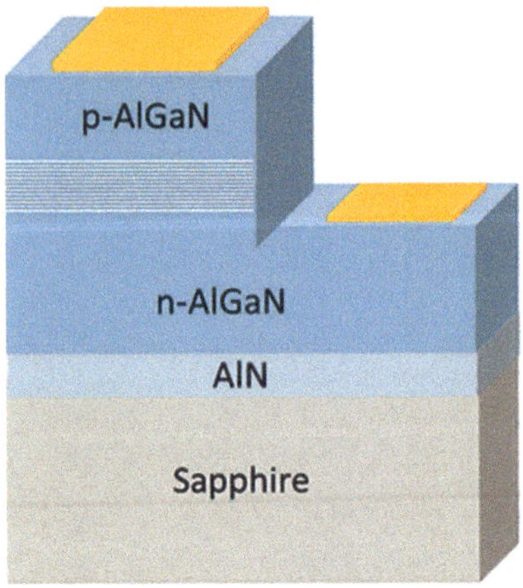

Fig. 1 Schematic of an AlGaN-based deep UV LED and related material challenges.

diodes (LDs). This review focuses on two challenges that are largely responsible for the poor performance of AlGaN UV emitters relative to InGaN visible light emitters: doping and substrate challenges. For each of these material challenges, we present the state of the art and exploratory concepts for overcoming these challenges and enabling higher performance UV devices.

The focus on doping and substrate challenges necessarily omits other topics of interest. In particular, point defects in AlGaN alloys and related impact on radiative efficiency (Chichibu et al., 2011), bandstructure limitations to light extraction (Nam et al., 2004; Wierer et al., 2014), and optical gain (Chow and Kneissl, 2005) are found in the provided references. In addition, an excellent recent review on III-Nitride ultraviolet emitters (Kneissl and Rass, 2016) is recommended.

2. DOPING CHALLENGES OF AlGaN ALLOYS

Typical LEDs and LDs rely upon the ability to achieve both n-type and p-type doping; however, such bipolar doping is challenging for most wide band gap semiconductors (Walukiewicz, 2001; Zunger, 2003). For AlGaN alloys, p-type doping is the major challenge, increasingly so with increasing Al composition and band gap. For these wider band gap alloys, low p-type conductivity limits device performance through Joule heating and poor hole injection, both of which contribute to reduced optical efficiency. P-type GaN is often used for a contact layer in DUV AlGaN LEDs, given higher conductivity than p-type AlGaN, but is absorbing for $\lambda < 363$ nm. Thus, while visible InGaN LEDs have up to 85% light extraction efficiency, DUV AlGaN LEDs have light extraction efficiencies of ~25% or less, largely due to the p-GaN absorption (Shatalov et al., 2017). Solutions are therefore needed for contact layers with both high p-type conductivity and high UV transparency in order to help close the gap between visible and UV LED performance.

Si-doping for n-type AlGaN is effective for Al compositions up to 80% but becomes increasingly less efficient for higher-Al-composition alloys (Borisov et al., 2005; Mehnke et al., 2013; Nakarmi et al., 2004; Taniyasu et al., 2002). Despite both experimental and theoretical studies, consensus on donor activation energies and the dominant mechanism behind the strong increase in n-type resistivity in these high-Al-composition alloys is still lacking. As applications drive AlGaN emitters further into the DUV, greater understanding of these mechanisms, and how to circumvent them, will be critical.

In this section, we review current understanding of the factors behind p-type and n-type doping limitations in AlGaN, potential approaches to overcoming these limitations, and state-of-the-art reports on p-type and n-type AlGaN conductivity. We further emphasize how these advances translate to improved performance of UV emitters.

2.1 P-Type Doping

One of the most formidable materials challenges of DUV AlGaN optoelectronics is ineffective p-type doping. This challenge results from the interplay of several materials properties including increasingly large acceptor ionization energy with Al composition, limitations to dopant solubility, and low formation energies of compensating defects. Density functional theory (DFT) studies identified Mg as a promising acceptor candidate (Van de Walle et al., 1999), and Mg is now the most commonly employed p-type dopant in commercial InGaN-based visible LEDs and LDs. However, Mg activation energies increase from ~160 meV in GaN to as high as 510–630 meV in AlN (Nam et al, 2003; Taniyasu et al., 2006), thereby becoming increasingly problematic for DUV devices with shorter emission wavelengths. Simply increasing Mg dopant concentrations to compensate for this high activation energy has typically not been successful (with exceptions including Gunning et al. (2015), described later). Mg dopant densities $>4 \times 10^{19}$ cm^{-3} have been reported to cause defects such as inversion domains which *decrease* both free carrier concentrations and mobilities (Chakraborty et al., 2007).

These observations point to multiple approaches to improving free hole concentrations in AlGaN-based optoelectronic devices. The first is to discover growth regimes under which high Mg concentrations are achieved *without* creating a high concentration of compensating defects. A second is to explore device architectures where free holes are achieved by mechanisms other than thermal activation from deep acceptors, for example, through polarization engineering approaches. Below, we review the state of the art in Mg doping of AlGaN and present progress in these strategies to achieve higher p-type conductivity in DUV devices.

2.1.1 Optimized Growth Conditions for p-Type Doping of AlGaN

A number of groups have explored the limits of p-type doping of AlGaN through careful optimization of growth conditions. In Table 1, we highlight some of the lowest reported p-type resistivities of AlGaN as a function of Al composition. Several common themes can be gleaned from these reports. First, higher hole concentrations are achieved with the use of high V/III ratio during metalorganic vapor phase epitaxy (MOVPE). This condition is thought to suppress the formation of nitrogen vacancies (V_N), a compensating donor with particularly low formation energies for AlGaN alloys (Van de Walle and Neugebauer, 2004; Van de Walle et al., 1999). Kinoshita et al. (2013) determined an optimal V/III ratio of 1800 by minimizing deep-level photoluminescence (PL) at 4.1 eV; an energy level potentially related to the 3+ charge state of the nitrogen vacancy ($V_N{}^{3+}$) (Nakarmi et al., 2009). This approach to finding optimal V/III ratio yielded a free hole concentration of 1.3×10^{17} cm^{-3} and a resistivity of 47 Ωcm for $Al_{0.70}Ga_{0.30}N$, the lowest p-type resistivity reported for $x > 0.50$ $Al_xGa_{1-x}N$ alloys. Other notable reports include a resistivity of 60 Ωcm for Mg-doped $Al_{0.60}Ga_{0.40}N$ grown by MOVPE in a hot wall reactor using a V/III ratio of ~1560 (Nilsson, 2014).

Another common observation is the need to limit the incorporated Mg concentration to avoid structural degradation that ultimately reduces conductivity, particularly for films grown by MOVPE. Jeon et al. (2005) found that a low resistivity of 10 Ωcm was only achieved for a very narrow window in Mg concentration of around 4×10^{19} cm^{-3} in MOVPE-grown p-$Al_{0.30}Ga_{0.70}N$ films. This window was bound on the low end by the need for sufficiently high Mg concentration to overcome compensating defects and on the high end by the creation of inversion domain boundaries and

Table 1 Highlights of Reported Resistivities, Hole Concentrations, Growth Approach, and Incorporated Mg Concentration of Mg-Doped AlGaN as a Function of Al Composition

Al (%)	Resistivity (Ωcm)	Hole Concentration (cm^{-3})	Growth Approach	Incorporated Mg Concentration (cm^{-3})	References
1	$>1\times10^{8}$	$\sim1\times10^{10}$	MOVPE	2×10^{19}	Taniyasu et al. (2006)
1	$\sim3\times10^{6}$	$\sim2\times10^{11}$	MOVPE	7×10^{18}	Nam et al. (2003)
0.85	7000	$\sim1\times10^{14}$	Hot wall MOVPE	2×10^{19}	Kakanakova-Georgieva et al. (2010)
0.70	47	1.3×10^{17}	MOVPE	3.3×10^{19}	Kinoshita et al. (2013)
0.60	60	—	Hot wall MOVPE	3×10^{19}	Nilsson (2014)
0.45	8	2.7×10^{17}	MOVPE	$\sim4\times10^{19}$	Jeon et al. (2005)
0.40	—	4.75×10^{18}	MOVPE, delta doping, and Indium surfactant	$\sim1\times10^{19}$	Chen et al. (2015)
0.30	10	2.2×10^{17}	MOVPE	3.5×10^{19}	Jeon et al. (2005)
0.27	1.2	2.3×10^{19}	Metal-modulated epitaxy (MBE)	$\sim1\times10^{20}$	Gunning et al. (2015)
0	0.59	1.6×10^{18}	NH_3-MBE, indium surfactant	6×10^{19}	Kyle et al. (2015)
0	0.2	3×10^{18}	MOVPE, low-energy electron beam activation	—	Nakamura et al. (1991)
0	0.19	1.9×10^{19}	Metal-modulated epitaxy (MBE)	$1–2\times10^{20}$	Gunning et al. (2012)

The majority of reported resistivity values were achieved after thermal annealing to reduce hydrogen passivation of Mg acceptors.

conversion to N-polar material. The onset of structural degradation was found to be in a similar range of Mg concentration for $x \sim 0.70$ alloys grown by MOVPE, suggesting little dependence on Al composition (Chakraborty et al., 2007; Kinoshita et al., 2013).

For both MBE and MOVPE growth approaches, employing indium as a surfactant during growth was reported to improve p-type doping of AlGaN (Chen et al., 2015; Kyle et al., 2015). Kyle et al. (2015) found that indium enabled lower NH_3-MBE growth temperatures, beneficial for reduced defect incorporation, while maintaining step-flow growth. They reported 1–2 orders of magnitude reduction of compensating defects and increased free hole concentrations compared to growths without indium. Mg-delta doping by MOVPE, a pulsed-growth approach where metalorganic sources and the Mg source (Cp_2Mg) are turned on and off while NH_3 is flowed continuously, was also reported to reduce compensating defects and increase free hole concentrations in AlGaN (Chen et al., 2015; Nakarmi et al., 2003). Interestingly, these improvements were observed for two distinct cases: when Cp_2Mg was introduced during a growth interruption with metalorganic sources turned off (Nakarmi et al., 2003) or solely when the metalorganic sources are turned on (Chen et al., 2015).

Another approach to enabling high Mg concentrations in p-AlGaN without structural degradation is through a molecular beam epitaxy (MBE) approach called metal-modulated epitaxy (MME) (Gunning et al., 2015). Similar in concept to pulsed-growth MOVPE, MME involves the modulation of group III metal and Mg sources in time, while the N source is held constant. A benefit of MME is that adatom mobility is substantially increased within the metal adlayer, enabling growth at lower temperatures while maintaining crystalline quality. The lower growth temperature is deemed advantageous for p-type doping due to decreased formation energies of compensating defects, such as V_N (Neugebauer and Van de Walle, 1995). Incorporation of Mg $\sim 1 \times 10^{20}$ cm^{-3} into a 100-nm-thick $Al_{0.27}Ga_{0.73}N$ film yielded $p = 2.3 \times 10^{19}$ cm^{-3} and $\mu = 0.2$ cm^2/Vs with no evidence of inversion domains. This free hole concentration is significantly higher than that of any other reported for p-AlGaN, and while the mobility is very low, the corresponding resistivity of 1.2 Ωcm is still among the lowest reported for $x \sim 0.3$ $Al_xGa_{1-x}N$ alloys.

2.1.2 Polarization Engineering Approaches to p-Type Doping

Despite the aforementioned progress, it has proven quite challenging to achieve $>1 \times 10^{18}$ cm^{-3} free hole concentrations in $x > 0.40$ $Al_xGa_{1-x}N$

alloys through optimization of growth conditions alone. Over the past decade, there have been exciting advances in the application of polarization engineering to overcome the challenge of p-type doping in AlGaN. The potential for very large spontaneous polarization in the (0001) crystal orientation arises from the noncentrosymmetric crystal structure of wurzite III-Ns combined with the ionic nature of atomic bonding. For strained layers, spontaneous polarization is augmented by piezoelectric polarization, aligned along a particular crystallographic direction depending on the nature of the strain (e.g., tensile or compressive) (Ambacher et al., 1999). Typical III-N LED heterostructures under equilibrium can have polarization-induced internal fields on the order of MV/cm across quantum wells (QWs) of ~few nm thickness.

The effects of polarization are clearly seen in III-N heterostructures, where the polarization discontinuity leads to a fixed charge at the interface between the two alloys. An exemplar is a GaN/AlGaN high-electron-mobility transistor (HEMT), where fixed charge at the GaN/AlGaN interface is balanced by free electrons from surface donor states (Ibbetson et al., 2000), yielding 2D electron channels with low-temperature mobilities as high as 51,700 cm^2/Vs at a sheet concentration of 2.2×10^{13} cm^{-2} (Smorchkova et al., 1999). As an alternative to abrupt heterointerfaces, various groups have explored distributed polarization doping (DPD), whereby band gap grading results in 3D slabs of free charge (Jena et al., 2002). Among other benefits, DPD avoids heterobarriers of abrupt-interface designs with potential for improved carrier injection. In the following, we review examples of such polarization engineering approaches applied to p-type doping in AlGaN.

2.1.2.1 Mg-Doped Superlattices

Mg-doped AlGaN superlattices are one example of a polarization engineering approach to p-type doping that relies upon abrupt heterointerfaces. As shown in Fig. 2A, these are periodic structures with repeated bilayers of higher- and lower-Al-composition AlGaN, each layer typically being 0.5–10-nm thick. Schubert et al. (1996) first reported the potential to significantly enhance activation of deep acceptors with a heterostructure that is uniformly doped but periodically varying in composition. Assuming a bulk acceptor activation energy $E_a = 200$ meV (similar to that of Mg in GaN) and a valence band offset between the two alloys approximately equal to E_a, their modeling predicted an acceptor activation of 50%, some $10\times$ higher than for a similar material but without band modulation. Effectively,

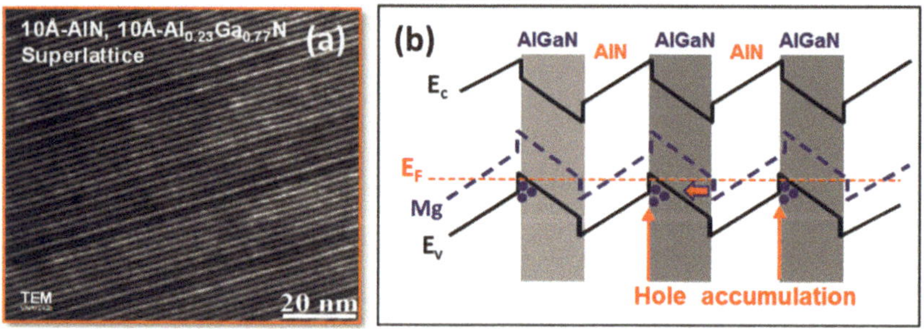

Fig. 2 (A) Cross-sectional transmission electron micrograph of an AlN/$Al_{0.23}Ga_{0.77}N$ Mg-doped superlattice grown by MOVPE. The constituent layers are 10-Å thick. (B) Schematic of the bandstructure of a Mg-doped AlN/AlGaN superlattice. E_C is the conduction band, E_V is the valence band, and E_F is the Fermi level. The schematic highlights the polarization-field-induced band bending that enables hole activation and accumulation in 2D sheets at heterointerfaces. *Panel (A): Reprinted from Allerman, A.A., Crawford, M.H., Miller, M.A., Lee, S.R., 2010. Growth and characterization of Mg-doped AlGaN-AlN short-period superlattices for deep-UV optoelectronic devices. J. Cryst. Growth 312, 756–761 with permission from Elsevier.*

this periodic structure enables ionization of deep acceptors in the wider band gap barrier and accumulation of holes in the narrower band gap well.

III-N superlattices have the added effect of strong polarization fields which lead to a significant modulation of the superlattice bandstructure (Fig. 2B). This enables ionization of Mg acceptors where the band edge is below the Fermi level and accumulation of resulting free holes at the neighboring heterointerface. Such polarization-field-induced activation of holes obviates the need for thermal activation of holes in higher-Al-composition AlGaN, thereby circumventing a primary roadblock to p-type conductivity. As will be discussed later, this enhanced p-type conductivity is in the lateral direction, i.e., perpendicular to the growth direction. In contrast, vertical transport, aligned along the growth direction, is critical for p–n junction devices and can be substantially impeded due to heterobarriers inherent in the superlattice structures.

Over more than a decade, a wide range of AlGaN-based superlattice designs have been explored. In Table 2, we highlight some of the more notable results, organized as a function of average Al composition of the superlattice heterostructures. Earlier studies focused on lower average Al composition superlattices and therefore relatively low Al composition contrast between the layers. Across a range of designs and for both MBE and MOVPE growth techniques, values of $\sim 0.2\ \Omega\text{cm}$ lateral resistivity and $2\text{–}4 \times 10^{18}\ \text{cm}^{-3}$ free hole concentration were commonly achieved.

Table 2 Notable Reported AlGaN-Based Superlattice Designs and Measured Free Hole Concentrations, Lateral Resistivities, and Applied Growth Technique

Material	Thickness (nm)	Average Al	p (cm^{-3})	$\rho_{lateral}$ (Ωcm)	Growth Technique	References
$GaN/Al_{0.2}Ga_{0.80}N$	~7/7	0.10	2.5×10^{18}	0.2	MOVPE	Kozodoy et al. (1999)
$GaN/Al_{0.2}Ga_{0.80}N$	10/10	0.10	3.4×10^{18}	0.2	MBE, modulation doping	Waldron et al. (2001)
$GaN/Al_{0.26}Ga_{0.74}N$	7/7	0.13	4.2×10^{18}	0.19	MOVPE	Yasan and Razeghi (2003)
$GaN/Al_{0.40}Ga_{0.60}N$	5/5	0.20	3×10^{18}	~1.5	MOVPE	Kumakura et al. (2000)
GaN/AlN	5.6/2.4	0.43	$\sim 2 \times 10^{18}$	~0.8	PAMBE	Simon et al. (2010a)
$Al_xGa_{1-x}N/Al_yGa_{1-y}N$	—	0.6	—	9.6	MOVPE	Cheng et al. (2013)
$Al_{0.08}Ga_{0.92}N/AlN$	~0.7/0.7	0.72	1×10^{18}	6	NH_3-MBE	Nikishin et al. (2005)
$Al_{0.23}Ga_{0.77}N/AlN$	0.5/1	0.74	—	6	MOVPE	Allerman et al. (2010)

These values are comparable to some of the best reported for GaN epilayers with lower band gap (Nakamura et al., 1991; Table 1). Assuming relatively little temperature dependence of mobility, temperature-dependent resistivity measurements revealed that the superlattices had substantially reduced acceptor activation energies compared to bulk films of comparable average Al composition (Kozodoy et al., 1999; Saxler et al, 1999). For example, Kozodoy et al. (1999) reported a GaN/$Al_{0.2}Ga_{0.80}N$ superlattice with $E_a \sim 0$ meV for optimized layer thicknesses. In comparison, a reference $Al_{0.1}Ga_{0.9}N$ epilayer had $10\times$ lower hole concentration and a significantly higher acceptor activation energy of 238 meV.

Superlattices with higher-Al-composition alloys for DUV devices employed distinct design elements of ultrathin layers and large Al composition contrast. DUV "digital alloy superlattices (DAS)" using NH_3-MBE with Mg-doped $Al_{0.08}Ga_{0.92}N$ wells (0.5–0.75 nm, i.e., 2–3 monolayers) and AlN barriers (0.75–1.5 nm, i.e., 3–6 monolayers) were developed, demonstrating $p=1\times10^{18}$ cm^{-3} and a lateral resistivity of 6 Ωcm (Nikishin et al., 2005). The DAS had a high effective band gap of 5.1 eV (similar to that of an Al$\sim$0.72 alloy), thereby showing good UV transparency down to 250 nm. Similar in design to the DAS, Mg-doped short-period superlattices (Mg-SPSLs) were developed using the MOVPE growth technique. Employing AlN/$Al_{0.23}Ga_{0.77}N$ Mg-SPSLs with layer thicknesses in the 0.5–1.5 nm range, Allerman et al. (2010) achieved comparable results to the NH_3-MBE-grown DAS structures, including a lateral resistivity of 6 Ωcm for an effective band gap of an $Al_{0.74}Ga_{0.26}N$ alloy. As shown in Fig. 3, Mg-SPSLs with an average Al composition of 61% demonstrated Mg activation energies as low as 18 meV compared to 150 meV for a reference p-GaN epilayer. These Mg-SPSLs also had good transparency down to 260 nm, similar to that of a single Mg-$Al_{0.61}Ga_{0.39}N$ epilayer.

The DASs and Mg-SPSLs represent a significant advance in achieving the desired combination of relatively high p-type conductivity and DUV transparency. However, resistivities reported in Table 2 are measured in the lateral direction, whereas vertical transport is critical for p–n junction devices. A few studies of Mg-DAS and Mg-SPSLs have compared lateral and vertical resistivities, providing important insight into the potential for use in UV devices. DAS LEDs with emission in the 250–290 nm region employed both n-type and p-type DAS's and had a forward turn-on voltage of 5.0 V and a series resistances of ~50 Ω for 350 μm-sized devices. While the individual contribution of the p-DAS was not reported, the group's earlier work on p-DAS's with band gaps ranging from 4.5 to 5.3 eV resulted

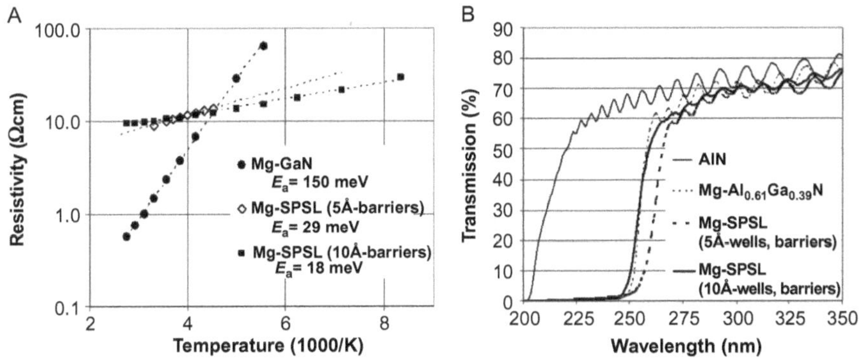

Fig. 3 (A) Resistivity vs inverse of temperature for p-GaN (*circles*) and a Mg-SPSL with 5-Å thick AlN barriers and 5-Å thick (*diamonds*) or 10-Å thick (*squares*) $Al_{0.23}Ga_{0.77}N$ well layers. *Dashed lines* are fits to the data using an Arrhenius model. Activation energies as low as 18 meV were achieved for the Mg-SPSLs, significantly lower than the 150 meV activation energy measured for a lower band gap p-GaN epilayer. (B) Transmission spectra for an AlN-on-sapphire template, a Mg-doped $Al_{0.60}Ga_{0.40}N$ epilayer, and two Mg-SPSL structures. The Mg-SPSLs were found to have similar transmission spectra to a bulk epilayer with the same effective band gap. *Reprinted from Allerman, A.A., Crawford, M.H., Lee, S.R., Clark, B.G., 2014. Low dislocation density AlGaN epilayers by epitaxial overgrowth of patterned templates. J. Cryst. Growth 388, 76–82 with permission from Elsevier.*

in a lateral resistivity of 4 Ωcm and an estimated vertical resistivity of 50 Ωcm (Nikishin et al., 2003). The significantly higher vertical resistivity demonstrates the impact of heterobarriers to vertical transport. Nevertheless, this vertical resistivity compares well to Mg-doped AlGaN epilayer results (Table 1), matching the very lowest reported resistivities at Al = 0.70.

Cheng et al. (2013) developed DUV Mg-SPSLs with a band gap equivalent to an $Al_{0.60}Ga_{0.40}N$ alloy and evaluated both vertical and lateral transport properties. The vertical resistivity of their 200-nm-thick Mg-SPSL under 20 A/cm^2 DC operation was 1.5×10^4 Ωcm, >1000× higher than the lateral resistivity of 9.6 Ωcm. Notably, at high DC current densities of 11 kA/cm^2, the vertical resistivity was dramatically reduced to 10 Ωcm largely due to the effects of Joule heating. While the vertical resistivity at 20 A/cm^2 is very high in the absolute sense, it is 2× lower than that of an equivalent Mg-doped $Al_{0.61}Ga_{0.39}N$ epilayer grown in the same study.

Other studies including vertical transport focused on the potential for Mg-SPSLs as cladding layers in DUV LDs (Martens et al., 2016). An increase of the turn-on voltage of LD heterostructures from 17 to 26 V as the effective alloy composition of the p-SPSL ranged from $x = 37\%$ to 81% was observed and attributed to the increasing heterobarrier between the p-SPSL and p-GaN layer. In contrast, LD series resistances were found to decrease with increasing effective band gap due to increased Joule heating, yielding a differential series resistance of ~40 Ω for a LD employing a 200-nm-thick p-SPSL with effective band gap of $Al_{0.81}Ga_{0.19}N$. While the Mg-SPSL contribution to the series resistance was not delineated, the Mg-SPSLs were reported to have more than 10 × lower vertical resistivity than a 200-nm-thick Mg-doped $Al_{0.81}Ga_{0.19}N$ epilayer also grown in the study. Taken together with the Mg-SPSL development of Cheng et al. (2013), these results demonstrate an increasing benefit of Mg-SPSLs over epilayers for vertical p-type transport in higher effective band gap devices.

2.1.2.2 Distributed Polarization Doping

Mg-SPSL development efforts have achieved impressive advances; however, the vertical resistances at low current densities are still high on an absolute scale. Joule heating at high current densities greatly reduces vertical resistance but adversely impacts device performance, e.g., reduces peak gain in LDs. An alternative polarization engineering strategy uses smoothly graded alloys to create 3D slabs of charge, thereby avoiding heterobarriers of abrupt-interface designs (Jena et al., 2002). Various names have been used to describe this approach and DPD will be used here.

Over the past 5 years or so, DPD applied to the p-type doping challenge of AlGaN has produced compelling results. Unlike for n-type DPD where surface donor states supply free electrons, most studies have found that Mg doping is required as the source of free holes in p-type DPD (Simon et al., 2010b). In one study, Mg-doped AlGaN with Al composition graded from 0% to 30% over 85 nm was grown on N-face GaN substrates by plasma-assisted MBE (PAMBE), yielding high free hole concentrations of $2 \times 10^{18}\ cm^{-3}$ and a hole mobility of $5\ cm^2/Vs$ at room temperature (Fig. 4A, reproduced from Simon et al., 2010b). While Mg doping was employed, the characteristic weak temperature dependence of free hole concentrations confirmed the dominant role of polarization over thermal activation of acceptors (Fig. 4C).

Similar results were achieved in a metal-polar orientation using MOVPE (Zhang et al., 2010). In this study, Mg-doped AlGaN with Al composition graded in the opposite direction, from 30% to 0%, produced $p = 2.6 \times 10^{18}\ cm^{-3}$ and a hole mobility of $\sim 4\ cm^2/Vs$ at room temperature. The free hole concentration is 10 × higher than that of a Mg-doped GaN epilayer grown by the authors. We note that hole concentrations and mobilities as high as $3 \times 10^{18}\ cm^{-3}$ and $9\ cm^2/Vs$, respectively, have

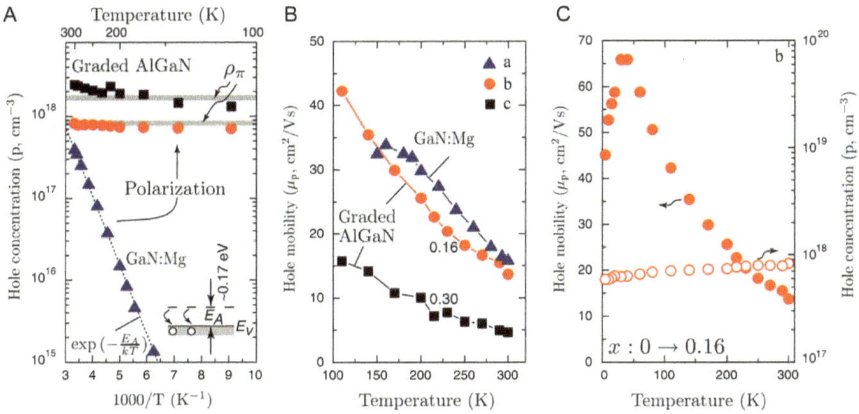

Fig. 4 Hall-effect temperature-dependent (A) hole concentration, (B) hole mobilities, and (C) hole concentration and mobility measured down to $T=4$ K. The polarization-doped graded AlGaN p-type layers show higher hole concentrations and conductivities. Holes in polarization-doped layers are resistant to freeze out at low temperatures, and their mobility and concentration can be measured down to cryogenic temperatures. *Reprinted from Simon, J., Protasenko, V., Lian, C., Xing, H., Jena, D., 2010b. Polarization-induced hole doping in wide band gap uniaxial semiconductor heterostructures. Science 237, 60–64 with permission from The American Association for the Advancement of Science.*

been achieved in other reports of Mg-doped GaN epilayers grown by MOVPE (Nakamura et al., 1991), indicating that resistivity of this DPD structure is approaching that of some of the best reported values for p-GaN.

Both of these DPD examples include GaN in the graded p-layer which would be absorbing for UV wavelengths shorter than ~363 nm. Optimization for DUV devices would require significantly higher Al compositions in the graded region. DPD with such wider band gap AGaN alloys was explored with metal-polar AlGaN grown by MBE on AlN-on-sapphire substrates. Three different DPD structures were explored, grading from AlN to $x=0.9$, 0.8, and 0.7 (Li et al., 2013). Interestingly, the authors explored both Mg and Be as acceptor impurities and noted a striking difference. Be is predicted to have an even lower activation energy than Mg but is less commonly used given a tendency to incorporate interstitially and act as a compensating donor (Van de Walle et al., 1999). Nevertheless, Mg-doped AlGaN structures employing the 100–70% Al composition grade achieved only $p=6\times10^{16}$ cm^{-3}, whereas similar Be-doped structures yielded a very high $p=8\times10^{18}$ cm^{-3} and mobility of ~20 cm^2/Vs at 300 K. For both cases, Hall data showed a weak temperature dependence of hole concentrations, supporting polarization-induced hole activation.

While metal-polar growth is more commonly used in MOVPE, DPD with the N-polar crystal orientation has been reported to have benefits when applied to UV devices (Piprek, 2012; Simon et al., 2010b; Verma et al., 2011). First, the orientation of the polarization field achieves more favorable band profiles for efficient carrier injection and capture into QWs. In addition, grading toward the sample surface proceeds from a lower to a higher band gap AlGaN alloy for p-type doping in the N-polar orientation. This capping layer naturally provides a transparent window for improved extraction of UV emission, in contrast to absorbing p-GaN cap layers used in many DUV LED designs.

Extending DPD to nanowires may offer significant advantages as well. Khokhlev et al. (2013) observed that studies of DPD AlGaN films had been limited to modest concentration gradients of ~0.01 nm^{-1} for n-type structures and ~0.003 nm^{-1} for p-type structures. Whereas large concentration gradients are challenging for DPD AlGaN epilayers due to strain, sufficiently thin nanowires enable grading over the entire AlGaN composition range while avoiding the strain relaxation and defect formation commonly seen in planar films (Glas, 2006). PAMBE was employed to grow AlGaN LED heterostructures on GaN NWs that were nucleated on Si substrates (Carnevale et al., 2012). Metal-polar nanowire LEDs incorporating a 0–100% Al grade for the n-type region, an $A_{0.80}Ga_{0.20}N$/AlN multiple-quantum-disk active region, and a 100–0% Al grade for the p-type layer were realized, yielding DUV emission at 4.4 eV (282 nm) and characteristic diode rectification *without* any Mg doping. The authors hypothesized that for narrow nanowires, the relatively high density of surface states provides sufficient free holes to overcome compensating donors without impurity doping, claiming a distinct advantage over planar films. These nanowire LEDs maintained reasonable IV characteristics and electroluminescence down to 15 K, supporting the role of nonthermal, polarization-induced doping.

2.1.2.3 Tunnel Junctions

Tunnel junctions (TJs) offer a distinct polarization engineering approach to hole injection and have been explored in UV emitters. Based upon original work by Esaki, these structures traditionally involve degenerately doped p–n junctions (Esaki, 1958). Such heavy doping yields very thin depletion regions, allowing for carrier tunneling across the TJ when sufficient reverse bias is applied. This can be seen in NPN LED structures in which forward biasing of the primary diode results in a reverse bias on the TJ

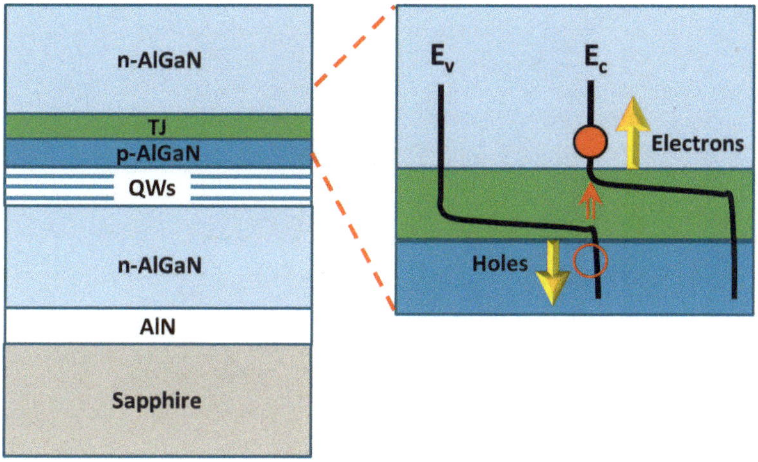

Fig. 5 Schematic of an n–p–n AlGaN-based LED employing a tunnel junction (TJ).

(Fig. 5). This reverse bias creates a favorable band alignment for tunneling of electrons across the TJ, effectively injecting holes into the active layers of the device. While the TJ may not entirely eliminate the need for p-type doping, it reduces series resistance by replacing thick p-type layers with thin p+ layers. In addition, TJ-enabled NPN structures avoid higher-resistance contacts to p-type layers and can eliminate the need for semitransparent contacts given the high conductivity and current spreading of n-type layers. TJs have been widely applied in other III–V semiconductor devices, enabling cascading of multiple junctions of solar cells (King et al., 2000) and intracavity contacts in both vertical-cavity surface-emitting lasers (Boucart et al., 1999) and edge-emitting LDs (Lu et al, 2002; Wierer et al., 1997).

The wide band gaps of III-N alloys, combined with the challenge of achieving very high doping levels without structural degradation, complicate the realization of thin depletion regions and low resistance TJs (Fig. 6A). Early work explored GaN p–n TJs (Jeon et al., 2001) and p-InGaN/n-GaN TJs (Takeuchi et al., 2001) in NPN visible LEDs and showed 0.6–1 V higher voltages than standard LEDs at 16–50 A/cm^2 operating conditions. Here again, leveraging of polarization fields in III-N heterostructures provides an alternative approach. As shown in Fig. 6B, the polarization field resulting from a 4-nm-thick AlN layer in a GaN p–n junction creates sufficient band bending to align neighboring conduction band and valence bands (Schubert, 2010). Thus, polarization fields are predicted to enable thin tunnel barriers without the challenges of very heavy doping (Grundmann and Mishra, 2007; Simon et al., 2009).

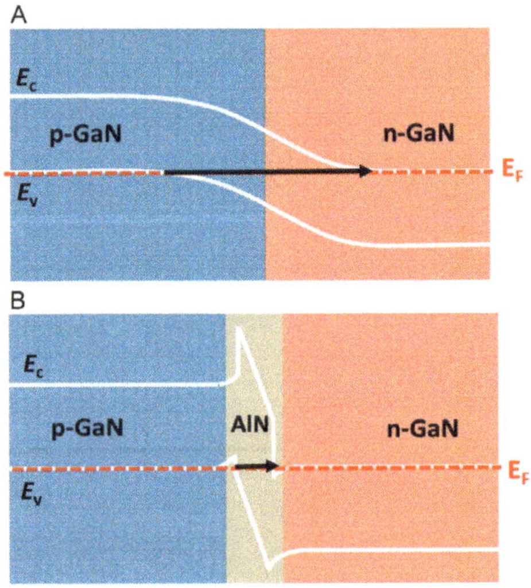

Fig. 6 Band schematic of a (A) GaN p–n junction and (B) GaN p–n junction with AlN tunnel barrier. E_c, E_v, and E_F refer to the conduction band, valence band, and Fermi level, respectively.

To date, a variety of III-N polarization-enhanced TJs have been explored. In Table 3, we highlight some of the best reported results and note several trends. First, as might be expected, lower band gap TJs employing InGaN have yielded lower specific series resistances than higher band gap AlN TJs. In addition, the majority of higher-performing polarization-enhanced TJs are grown by MBE rather than MOVPE, likely due to the ability to achieve higher doping concentrations while maintaining sharp heterointerfaces that produce the largest polarization sheet charge. Nevertheless, we highlight the recent report of Minamikawa et al. (2015) in which an MOVPE-grown p-InGaN/n-GaN TJ achieved 4×10^{-3} Ωcm^2 specific series resistance at lower current densities, decreasing to 4×10^{-4} Ωcm^2 at 5 kA/cm^2 largely due to Joule heating. The same group subsequently reported p-InGaN/n-GaN TJs grown by MOVPE and employing a compositionally graded InGaN tunnel barrier. The graded structures achieved lower specific series resistance, realizing 2.3×10^{-4} Ωcm^2 at a relatively high current density of 3 kA/cm^2 (Takasuka et al., 2016).

Nearly all reported III-N devices employing polarization-enhanced TJs involve InGaN-based active layers and visible emission; however, a few reports of AlGaN-based UV emitters have been made. Zhang et al. (2015) employed a p-$Al_{0.30}Ga_{0.70}N$/4-nm-thick $In_{0.25}Ga_{0.75}N$ tunnel barrier/n-$Al_{0.30}Ga_{0.70}N$ structure in an NPN UV LED with 327 nm

Table 3 Highlights of Reports of III-N Tunnel Junctions (TJs) for Both Polarization-Enhanced and Homojunction Designs

TJ Structure	Specific Resistance (Ωcm^2)	Device	Growth Technique	References
Polarization-enhanced				
p-GaN/InGaN/n-GaN	5×10^{-4}	Blue LED	PAMBE	Krishnamoorthy et al. (2014)
p-GaN/InGaN/n-GaN	1.2×10^{-4}	N/A	PAMBE	Krishnamoorthy et al. (2013)
Graded p-InGaN/n-GaN	2.3×10^{-4}[a]	Blue LED	MOVPE	Takasuka et al. (2016)
p-InGaN/n-GaN	4×10^{-3}[b] 4×10^{-4}[c]	Blue LED	MOVPE	Minamikawa et al. (2015)
p-AlGaN/InGaN/n-AlGaN	5.6×10^{-4}	UV LED	PAMBE	Zhang et al. (2015)
GaN/AlN/GaN	~1[d]	N/A	PAMBE	Simon et al. (2009)
Homojunction				
GaN	1×10^{-5}	N/A	PAMBE	Akyol et al. (2016
GaN	$\leq1.5\times10^{-4}$	Blue LED	MOVPE/NH_3 MBE	Young et al. (2016)
GaN	$\leq3.7\times10^{-4}$	Micro blue LED	NH_3 MBE	Malinverni et al. (2015)
GaN/Al/GaN	$\leq1\times10^{-3}$	Green NW LED	PAMBE	Sadaf et al. (2016)

[a] 3 kA/cm^2.
[b] 26 A/cm^2.
[c] 5 kA/cm^2.
[d] As reported in Krishnamoorthy et al. (2014).
The estimated specific series resistance of the TJs, the device in which the TJ was demonstrated, and the growth technique that was employed are listed.

emission and a low estimated specific TJ resistivity of $5.6\times10^{-4}\ \Omega cm^2$. The LEDs achieved peak wall plug efficiency of 1% which is comparable to some of the best reported for 330 nm AlGaN "standard" (no TJ) LEDs (Kneissl and Rass, 2016; Kneissl et al., 2006). One drawback of this approach is that the InGaN tunnel barrier contributes to absorption of UV emission and is limited in thickness and in composition due to lattice mismatch with AlGaN layers. The use of TJs has also enabled improved performance of

~310 nm-emitting nanowire LEDs (Golam Sarwar et al., 2015). Such nanowires are grown upon p-Si substrates and typically suffer from high forward operating voltages due to a 1.8 eV valence band offset between Si and p-GaN. The use of an InGaN TJ enables growth upon n-Si, taking advantage of the smaller (0.5 eV) conduction band offset and reducing turn-on voltages by ~6 V.

While polarization-enhanced TJs are clearly showing promise, the more traditional p+/n+ homojunction TJ designs for III-N devices have also shown steady performance improvements, primarily due to very high doping levels achieved by MBE. Recent results include n+ (4×10^{20} cm^{-3}) GaN/p+ (5×10^{20} cm^{-3}) GaN TJs grown by PAMBE and showing DC operation up to 150 kA/cm^2 with low differential resistance of 1×10^{-5} Ωcm^2 (Akyol et al., 2016). As a hybrid approach, Young et al. (2016) employed MOVPE for the full LED structure up to the p+ GaN cap layer and completed the NPN LED with degenerately doped n-GaN grown by NH_3 MBE. The differential resistance of the resulting semipolar (20-2-1) blue LEDs, including the TJ contribution, was 1.5×10^{-4} Ωcm^2. We further highlight a distinct TJ design that has been applied to green nanowire LEDs but may ultimately offer advantages for UV LEDs as well. Sadaf et al. (2016) observed that strain relaxation inherent in nanowires reduces the piezoelectric polarization, thereby reducing effectiveness of polarization-engineered TJs. Instead, they employed a 2-nm-thick Al metal as an intermediate layer between p+ and n+ GaN, relying on defects at the Al/GaN interface to enable trap-assisted tunneling. Specific series resistance of the Al-based TJ was estimated to be $<1\times10^{-3}$ Ωcm^2.

These recent advances in p+/n+ homojunction TJ performance have not yet been applied to UV devices but may be a fruitful path for development. Several factors should be evaluated, including absorption loss of degenerately doped GaN at DUV wavelengths compared to InGaN tunnel barriers in polarization-enhanced TJ designs.

2.2 n-Type doping

While n-type doping of $Al_xGa_{1-x}N$ is straightforward for the majority of the alloy composition range, a significant increase in n-type resistivity has been reported for $x>80\%$ (Borisov et al., 2005; Mehnke et al., 2013; Nakarmi et al., 2004; Taniyasu et al., 2002). Achieving a consistent and complete understanding of this increase in resistivity has been frustrated by a number of factors. Difficulties in comparing experimental results arise due to varying levels of impurities such as oxygen which can impact n-type conductivity as a function of Al composition (potentially including DX centers in the case of O_N; Gordon et al., 2014). Consensus on Si activation energies is also still lacking, with experimental reports ranging from ~63 to 570 meV for AlN (Borisov et al., 2005; Collazo et al., 2011; Hermann et al., 2005; Nakarmi et al., 2004; Neuschl et al., 2013; Son et al., 2011; Taniyasu et al., 2002). On the theoretical side, first-principles calculations have predicted dominant mechanisms of dopant compensation by V_{Al}^{3+} (Stampfl and Van de Walle, 2002), DX center formation for both O and Si impurities at high Al compositions (Gordon et al., 2014; Silvestri et al., 2011), and effective doping through low formation energy V_{Al}: 4Si complexes and with no evidence of DX centers (Hevia et al., 2013). Below, we highlight experimental and theoretical studies of n-type doping in high-Al-composition AlGaN and insights into the origins of n-type doping limitations.

Despite large band gaps spanning ~3.4–6.2 eV, n-type conductivity has been achieved in Si-doped AlGaN over the entire composition range. For example, studies of Si-doped AlN grown upon 6H-SiC by PAMBE yielded $n=7.4\times10^{17}$ cm^{-3} and a room temperature resistivity of 1 Ωcm (Ive et al., 2005). Taniyasu et al. (2002) explored Si-doped $Al_xGa_{1-x}N$ ($0.42\leq x\leq 1$) grown by MOVPE and reported a significant decrease in free carrier concentration with increasing Al composition, down to $n=9.5\times10^{16}$ cm^{-3} for AlN. Temperature-dependent Hall measurements reported in that work identified Si activation energies increasing from 8 meV for GaN to 86 meV for AlN. Furthermore, the authors observed that the maximum allowable Si concentration before self-compensation was reduced for higher Al compositions. Thus, both increasing donor activation energy and increasing compensating defect formation were identified as contributing factors to n-type doping limitations with increasing Al composition.

Qualitatively similar conclusions were found by Mehnke et al. (2013) in their study of Si-doped $Al_xGa_{1-x}N$ ($0.81\leq x\leq 0.94$), grown by MOVPE upon reduced dislocation density (~5×10^8 cm^{-2}) AlN epilayers on sapphire substrates. Careful optimization of SiH_4/group III ratio as a function of Al composition achieved a record-low resistivity of 0.026 Ωcm for $Al_{0.81}Ga_{0.19}N$ alloys, with $n=1.5\times10^{19}$ cm^{-3} and $\mu=16.5$ cm^2/Vs. The resistivity increased to 3.35 Ωcm for AlN due to a strong reduction in free electron concentration; in contrast, mobility increased with increasing Al

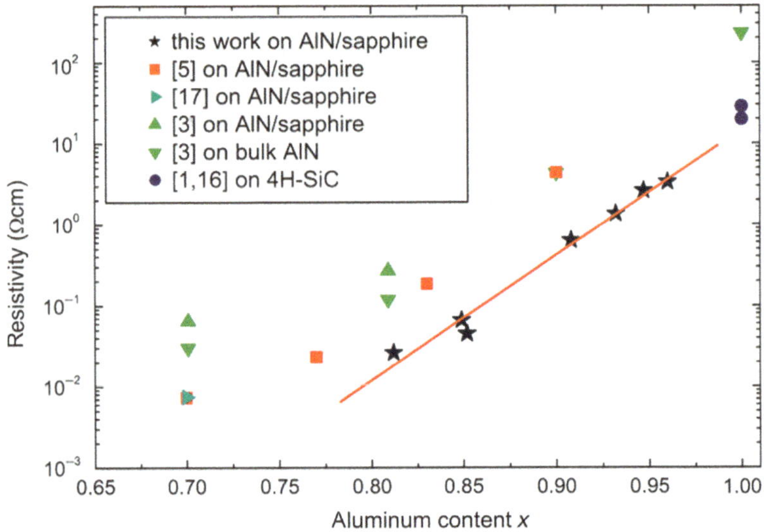

Fig. 7 Comparison of the lowest achieved resistivities of $Al_xGa_{1-x}N$:Si. *Reprinted from Mehnke, F., Wernicke, T., Pingel, H., Kuhn, C., Reich, C., Kueller, V., Knauer, A., Lapeyrade, M., Weyers, M., Kneissl, M., 2013. Highly conductive n-$Al_xGa_{1-x}N$ layers with aluminum mole fractions above 80%. Appl. Phys. Lett. 103, 212109 with permission from AIP Publishing LLC.*

composition in this composition range due to decreased alloy scattering. Collazo et al. (2011) studied n-type doping of high-Al-composition AlGaN grown on low-defect-density bulk AlN substrates which they found to yield lower resistivities than AlGaN films on sapphire. A resistivity of 0.1 Ωcm was achieved for $Al_{0.80}Ga_{0.20}N$ and increasing compensation with increasing Al composition was noted. They further reported a high Si activation energy of 250 meV for AlN. A summary of recent reports of n-type resistivities of $Al_xGa_{1-x}N$, highlighting the strong increase in resistivity in for $x > 0.80$, can be found in the work of Mehnke et al. (2013), reproduced in Fig. 7.

An outstanding issue is the potential role of O and Si DX centers in the observed increase in Si activation energy with increasing Al composition. First-principles calculations employing hybrid functionals identified two Si DX centers with energies fairly close to that of the neutral donor (Silvestri et al., 2011). This close energetic alignment supports the possibility of either shallow donor or DX behavior under nonequilibrium conditions, potentially explaining some of the variability in experimental reports. Other DFT studies employing hybrid functionals predict Si is a DX center for $x > 0.94$ (Gordon et al, 2014). The derived Si activation energy of 150 meV for AlN is consistent with electron paramagnetic resonance

(EPR) experiments of Son et al. (2011), which report the Si DX^- state at 140 meV below the conduction band minimum. This energy is sufficiently low to explain n-type conductivity in AlN, despite DX behavior of the Si impurity.

EPR measurements were also reported for Si-doped $Al_xGa_{1-x}N$ ($0.79 \leq x \leq 1$) films (Trinh et al., 2014). These studies confirmed the presence of a single, metastable Si DX center for Al < 0.84 which closely tracked the activation energy of the neutral donor, leading to shallow-dopant behavior. For Al > 0.84, an additional, stable DX center was observed which had a strongly increasing activation energy with increasing Al. In particular, an activation energy of 240 meV in AlN was determined, similar to the value reported in experimental studies by Collazo et al. (2011). Therefore, despite remaining discrepancies in measured and predicted donor activation energies, there is increasing evidence that Si DX centers are a major factor in the increased donor activation energies of high-Al-composition AlGaN alloys.

3. SUBSTRATES FOR UV OPTOELECTRONICS

3.1 Introduction

Substrates for optoelectronic devices must meet stringent requirements related to crystal structure and orientation, lattice constant, thermal expansion coefficient, dielectric constants, and electrical conductivity. In many of these attributes, substrate technologies employed for AlGaN-based UV optoelectronics are still far from ideal. For example, ideal substrates allow lattice-matching or strain balancing of overlying epitaxially grown device layers, supporting pseudomorphic growth with low defect densities. In devices, such epitaxial layers must span a range of alloy compositions and band gaps for optimal carrier injection, carrier confinement, light extraction, and optical mode confinement. However, unlike $Al_xGa_{1-x}As$ alloys where the in-plane lattice constant changes very little throughout the alloy range, $Al_xGa_{1-x}N$ alloys show a considerable (~2.4%) change in in-plane lattice constant with increasing Al content and band gap. Low-defect growth of AlGaN on commercially available GaN or AlN substrates is therefore limited in both epilayer thicknesses and alloy compositions, making devices in the mid-alloy range (~260–340 nm) particularly challenging. As such, many UV-B to UV-C LEDs are grown on lower-cost, lattice-mismatched sapphire substrates and have threading dislocation densities (TDD) of

$\sim 2 \times 10^9$ to 1×10^{10} cm^{-2}, i.e., >100,000× higher than found in typical III–V optoelectronic devices.

The electrical conductivity of substrates is also critical to device design and performance. The majority of AlGaN LEDs and LDs, including both research and commercial devices, employ electrically insulating sapphire or AlN substrates. Such substrates necessitate n-type and p-type contacts on the same side of the device and device geometries where current is laterally injected into the n-type layer of the device (herein referred to as lateral injection). Compared to structures with bottom n-type contacts and fully vertical injection, lateral-injection devices experience current crowding and local heating and further require more complex and time-consuming device fabrication.

Another critical material parameter for device performance is the crystallographic orientation. For III-N heterostructures, the magnitude of internal polarization fields can be tailored by growth on polar, semipolar, or nonpolar substrates. As detailed in the previous section, engineering of polarization fields enables promising approaches to p-type doping. Conversely, polarization-induced fields may adversely impact light-emitter performance by reducing carrier recombination rates through reduced electron–hole wavefunction overlap in QWs and by blue-shifting of emission wavelength with increasing current due to screening of polarization fields.

The absorption properties of substrates are also a major factor in device performance, potentially the limiting factor to light extraction efficiency of LEDs. While both sapphire and AlN have large band gaps compatible with high transparency throughout the UV-A, UV-B, and UV-C range, bulk AlN substrates have absorption bands in the blue and UV region, likely due to complexes between native vacancies and oxygen impurities (Yan et al., 2014). Si substrates offer a low-cost solution but are strongly absorbing over the entire UV wavelength range. Similarly, 4H and 6H SiC substrates are absorbing in much of the UV range, given band gaps at 3.23 eV (~388 nm) and 3.20 eV (~413 nm), respectively (Goldberg et al., 2001). Finally, GaN substrates are absorbing at UV wavelengths shorter than ~363 nm.

In this section, we examine substrate properties that impact the performance of AlGaN-based optoelectronic devices and confine our discussion to two key substrate challenges. The first challenge is strain management and reduction of extended defects given the relatively large lattice mismatch between typical AlGaN device heterostructures and available substrates. The second challenge is achieving fully vertical-injection-geometry devices given the electrically insulating nature of the most commonly employed substrates. The review will further be focused on polar AlGaN heterostructures and devices. While relatively little has been reported on semipolar and nonpolar AlGaN UV devices compared to InGaN visible devices, recent reports can be found in Brault et al. (2014), Wang and Wu (2012), and Haeger et al. (2012).

3.2 Strain Management and Reduction of Extended Defects

3.2.1 Impact of Extended Defects on Material and Device Properties

Although GaN and AlN bulk substrates offer low TDD, they are used in a minority of commercial III-N optoelectronic devices. GaN substrates with TDD $\leq 1 \times 10^6$ cm^{-2} are used for higher-cost, higher-current-density devices such as blue InGaN LDs, where low TDDs are critical for achieving long operating lifetimes (Nakamura, 1998). However, the majority of commercial III-N LEDs, both InGaN-based visible LEDs and AlGaN-based UV LEDs, are grown on lower-cost, visible/UV transparent sapphire substrates. For both GaN and AlGaN alloys, the large lattice mismatch with these substrates leads to TDDs many orders of magnitude higher than for other III–V optoelectronic materials. Significantly, InGaN-based visible LEDs have demonstrated relatively high efficiencies despite high TDDs. In a study of InGaN LED efficiency vs TDD, Sano et al. (2013) found an internal quantum efficiency (IQE) of ~60% for blue-emitting InGaN LEDs on sapphire substrates, despite TDD of $\sim 1 \times 10^8$ cm^{-2}, and IQE = 90% for LEDs on bulk GaN substrates with TDD $\sim 1 \times 10^6$ cm^{-2}.

For AlGaN-based devices on sapphire substrates, however, TDDs of $\leq 1 \times 10^8$ cm^{-2} are much more difficult to achieve. The growth of both GaN and AlGaN epilayers on sapphire initiates with 3D-island growth and eventual coalescence into a 2D film. For AlGaN, the relatively low mobility of Al on the growth surface, combined with the suppression of alloy decomposition for higher-Al-composition alloys, results in a reduction of lateral mass transport and a higher density of small-area islands (Nitta et al., 2002). As TDs have been shown to originate at the coalescence interfaces of these islands, AlGaN epilayers grown on sapphire substrates typically have higher TDD of $\sim 2 \times 10^9$ to 1×10^{10} cm^{-2}.

Numerous experimental reports confirm the prediction that TDDs $> 1 \times 10^9$ cm^{-2} degrade efficiency of AlGaN-based QWs and devices. The relatively wide range of TDD employed in such studies is made possible by defect reduction techniques, detailed in the following section.

As one example, Iwaya et al. (2001) demonstrated >100 × increase in PL intensity from 352 nm GaN/$Al_{0.08}Ga_{0.92}N$ multi-QWs (MQWs) when TDD was reduced from 1×10^{10} cm^{-2} to 4×10^{7} cm^{-2}. Similarly, Allerman et al. (2014) found a 7 × enhancement of PL and a 15 × enhancement of electroluminescence (EL) from 340 nm AlGaN laser heterostructures, when TDD was reduced from $\sim 5 \times 10^{9}$ cm^{-2} to $\sim 2 \times 10^{8}$ cm^{-2}. The added benefit of improved carrier injection efficiency with reduced TDD was proposed to account for the greater improvement in EL compared to PL.

Beyond reducing efficiency, TDs in III-N devices have been reported to create parasitic current leakage paths that degrade device performance (Kozodoy et al., 1998; Speck and Rosner, 1999). Recently, Moseley et al. (2014, 2015) studied the particularly strong impact of open core screw dislocations (also called nanopipes) on UV LED performance. Similar to what has been found with InGaN LEDs (Lee et al., 2006), nanopipes contributed to both forward and reverse leakage currents in AlGaN LEDs. Moreover, increased densities of nanopipes directly correlated with reduced EL intensity in AlGaN-based 270-nm LEDs, despite being <0.1% of the total TDs in the devices (Moseley et al., 2014).

The reduction of TDD has also been correlated to improved device lifetime, including a 5 × improvement (from 1000 to 5000 h at 20 mA) in 310 nm LED lifetime with TDD reduction from $3.5 \times 10^{9} \times 10$cm^{-2} to 2.8×10^{8} cm^{-2} (Jain et al., 2008). While the dominant degradation mechanism of AlGaN-based LEDs under high currents and temperatures is still under debate, the majority of reliability studies show little change to the optical efficiency of the active region with aging (Moe et al., 2010; Pinos et al., 2010). Instead, an aging-induced increase of nonradiative tunneling currents, likely related to TDs, has been observed. For example, Pinos et al. (2011) proposed that enhanced leakage observed with aging is due to closed-core screw dislocations activated by nitrogen vacancies.

These and other reports of the deleterious effects of TDs have provided significant motivation for developing defect reduction strategies in AlGaN alloys. While GaN and AlN substrates have enabled lowest TDD devices to date, such devices are limited in emission wavelength and design due to lattice mismatch constraints. In the following, we review progress in alternative defect reduction approaches that target the entire AlGaN composition and emission-wavelength range.

3.2.2 Approaches to Strain Management and Defect Reduction

Realization of high-efficiency UV devices requires attention to strain management and mitigation of defects on multiple levels. First and

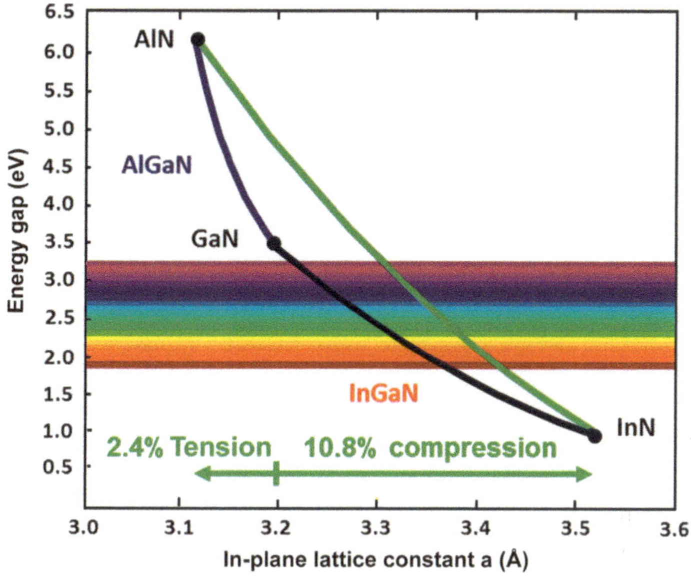

Fig. 8 Band gap vs in-plane lattice constant for III-N alloys.

foremost, the very large lattice mismatch between AlGaN alloys and typical substrates must be mitigated to avoid excessively high TDD. Second, the relatively large change in lattice constant across the AlGaN alloy composition range (Fig. 8) can limit the ability to optimize UV heterostructure designs if strain relaxation and defect formation are to be avoided. In the first case of AlGaN lattice mismatch with common substrates, a critical breakthrough in the III-N field was the discovery that thin AlN or GaN layers, typically grown at lower growth temperatures, enable growth of strain-relaxed GaN films on sapphire substrates with significantly reduced TDD compared to high-temperature growth directly on sapphire (Amano et al., 1986; Nakamura, 1991).

Notably, a similar strategy has been employed for mitigating the second strain management challenge of AlGaN epilayers grown on lattice-mismatched AlGaN epilayers. Kamiyama et al. (2001) employed a 20-nm-thick $Al_xGa_{1-x}N$ interlayer, grown at 500°C by MOVPE, between AlGaN and GaN epilayers grown on a sapphire substrate (Fig. 9A). They determined that crack-free growth of 1.5-μm-thick AlGaN epilayers across the entire AlGaN composition range could be realized, when the AlGaN interlayers had higher Al composition than the overlying AlGaN epilayer. Electrically conductive, Si-doped AlGaN interlayers were also demonstrated, showing compatibility with vertical-injection device geometries.

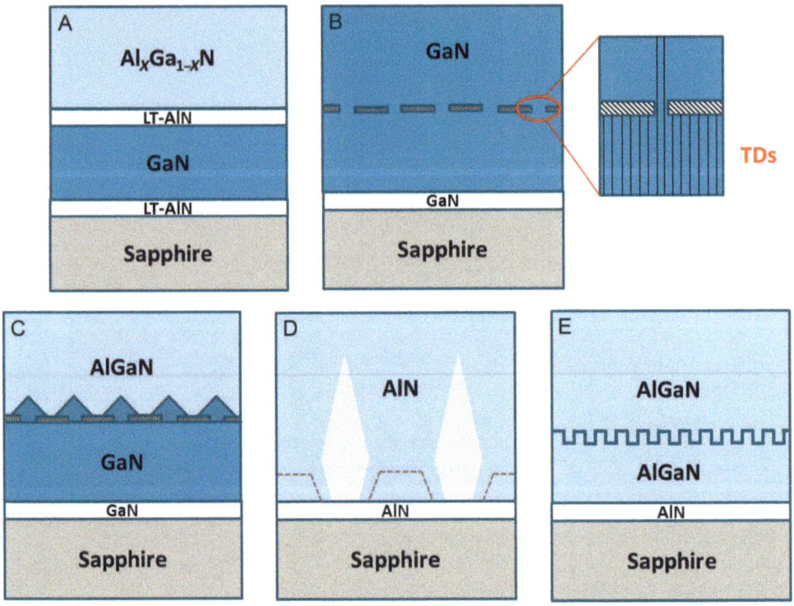

Fig. 9 Schematics of different types of GaN and AlGaN epitaxial overgrowth approaches including (A) epitaxial overgrowth of AlGaN on GaN using low-temperature AlN interlayers, (B) epitaxial lateral overgrowth of GaN on GaN using dielectric stripes showing reduced dislocation density over stripes, (C) facet-controlled epitaxial lateral overgrowth of AlGaN on GaN, (D) overgrowth of AlN on mesa-patterned AlN, and (E) overgrowth of AlGaN on submicron trenches in AlGaN.

However, an undesirably high TDD of $\sim 6 \times 10^9$ cm^{-2} was generated through strain relaxation of the interlayer.

A strategy for growth of thick AlGaN epilayers that successfully *reduces* TDs resulting from substrate lattice mismatch is epitaxial lateral overgrowth (ELO). As described in a comprehensive review article by Beaumont et al. (2001), a standard ELO strategy for GaN-on-sapphire substrates involves deposition of a dielectric layer on top of the high-defect-density GaN epilayer on sapphire substrate, patterning of the dielectric into a periodic series of stripes, and overgrowth of GaN on this patterned template. The dielectric stripes effectively block TDD propagation in the overlying regions, while GaN selectively nucleates on the GaN surface revealed by openings between the stripes. Subsequent lateral growth of GaN over the stripes produces regions of significantly reduced TDD over the masked regions (Fig. 9B).

In contrast to GaN, AlGaN alloys do not show such masked-growth selectivity, typically yielding polycrystalline growth on dielectric materials, and alternative approaches are required. Several groups have employed plasma etching of various micron-scale patterns onto the surface of GaN templates or high-defect-density AlN or AlGaN epilayers, followed by AlGaN overgrowth. As in GaN ELO, the patterned growth surface may either block threading dislocations or facilitate their interaction and annihilation depending on the exact patterning scheme (Imura et al., 2007; Mochizuki et al., 2002; Tsuzuki et al., 2009). Unlike unstrained GaN ELO, however, the strained AlGaN overgrowth may drive misfit dislocation formation (Cherns et al., 2004), cracking and/or compositional variations (Liu et al., 2003), which are also strongly influenced by the surface-patterning scheme.

A number of etched patterns have been explored for epitaxial overgrowth of AlGaN heterostructures and devices. Extensive work has been done on $Al_xGa_{1-x}N$ overgrowth of trenches etched either into substrates, including Si, SiC, and sapphire (Sano et al., 2002) or into AlGaN epilayers on sapphire or SiC substrates (Fig. 9D and E) (Jain et al., 2008; Amano et al., 2009; Kueller et al., 2011, 2013; Allerman et al., 2014). Early demonstration of $Al_{0.13}Ga_{0.87}N$ overgrowth of trenched $Al_{0.13}Ga_{0.87}N$ on SiC substrates employed 5-μm-wide mesas and trenches with widths ranging from 5 to 15 μm wide (Sano et al, 2002). The overgrowth enabled crack-free, 7-μm-thick $Al_{0.13}Ga_{0.87}N$ films with TDD as low as mid-10^6 cm^{-2} over the trenches and 8×10^7 cm^{-2} over the mesas. A potential weakness of trench patterns is that defect reduction is greatest over the trench regions as many TDs originating on the mesa surface propagate through the heterostructure. Facet-controlled epitaxial lateral overgrowth (FACELO), whereby regrowth is performed on pyramids in GaN or AlGaN (Fig. 9C), effectively eliminates the basal plane mesa surface and achieves TDD reduction that is nearly spatially uniform. Kawashima et al. (2004) reported that AlGaN overgrowth of $Al_{0.12}Ga_{0.88}N$ pyramids yielded TDD of 5×10^7 cm^{-2} which was spatially uniform except for a line of higher TDD at the coalescence front of neighboring pyramids.

Another approach to achieving spatially uniform TDD reduction is to perform growth on grooved patterns, where the mesa has submicron dimensions (Figs. 9E and 10A and B). Using this approach, Allerman et al. (2014) demonstrated spatially uniform TDD reduction down to 2–3 $\times 10^8$ cm^{-2} for Al = 0.30 alloys and 4-5 $\times 10^8$ cm^{-2} for Al = 0.60 alloys,

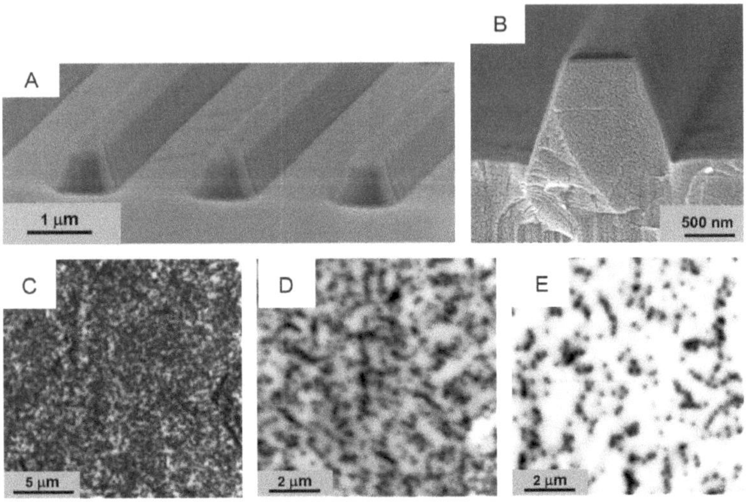

Fig. 10 (A) Scanning electron micrograph of mesas etched into an $Al_{0.30}Ga_{0.70}N$ epilayer on sapphire substrate and (B) cross-sectional image showing submicron mesa widths at both *top* and *bottom* of mesas. *Top-view* cathodoluminescence (CL) images of various AlGaN epilayers and MQWs including (C) a reference nonpatterned 3.4-µm-thick $Al_{0.30}Ga_{0.70}N$ epilayer, (D) AlGaN MQWs on a 6.2-µm-thick, Si-doped $Al_{0.60}Ga_{0.40}N$ epilayer overgrown on a patterned $Al_{0.60}Ga_{0.40}N$ template, and (E) an AlGaN MQW on a patterned $Al_{0.30}Ga_{0.70}N$ template after two patterning and overgrowth cycles. *Dark spots* in CL images indicate threading dislocations. *Reprinted from Allerman, A.A., Crawford, M.H., Miller, M.A., Lee, S.R., 2010. Growth and characterization of Mg-doped AlGaN-AlN short-period superlattices for deep-UV optoelectronic devices. J. Cryst. Growth 312, 756–761 with permission from Elsevier.*

confirming applicability in the challenging mid-alloy range. Conroy et al. (2015) extended the ELO technique to much smaller length scales, developing arrays of AlN nanorods of 1 µm height and 170–330 nm diameter. AlN regrowth on these close-packed structures enabled full coalescence in only ~1.5 µm (compared to 3–20 µm in more standard mesa approaches) and achieved TDD of 3.8×10^8 cm^{-2}.

The strategy of AlGaN overgrowth of patterned templates has directly enabled advancement of LDs to shorter UV wavelengths. Applying AlGaN overgrowth of wide mesas (~20 µm) in GaN, Kamiyama et al. (2005) demonstrated room temperature pulsed lasing at 351 nm, a record short lasing wavelength at that time. The wide mesas were not inhibitive for these LDs since ridge waveguides of ~few micron widths could be carefully aligned to the low TDD regions over the trenches; a strategy not applicable to much larger area LEDs. Crawford et al. (2015) realized reduced TDD AlGaN LDs via AlGaN overgrowth of submicron AlGaN mesa structures and achieved room temperature pulsed-current lasing at 353 nm. Finally,

AlGaN overgrowth of GaN pyramids on GaN-on-sapphire substrates has enabled room temperature pulsed lasing at 336 nm by Yoshida et al. (2008), achieving the shortest wavelength to date for mW-class UV LDs.

While these results represent the state of the art in AlGaN LDs, further reduction in TDD is likely needed to achieve operational lifetimes suitable for commercial applications. Beyond continued optimization of ELO approaches, lower-dimensional structures, including nanowires, nanorods, or nanocolumns, offer a compelling strategy for UV emitters with low TDD. TEM studies of GaN nanowires grown by selective area growth (MBE) on Si (111) substrates found nanowires with diameters <200 nm to be TD-free due to dislocation filtering (Kishino and Ishizawa, 2015). Such nanowires have been successfully used in both UV LEDs (Djavid and Mi, 2016; Kent et al., 2014; Wang et al., 2013; Zhao et al., 2015a) and UV LDs (Li et al., 2015; Zhao et al., 2015b). Notably, AlN-based nanowire LEDs have achieved wavelengths as short as 207 nm, and AlGaN nanowire LDs have achieved room temperature lasing at 289 nm with threshold current densities as low as 300 A/cm^2 (Mi et al., 2016).

3.3 Electrically Conductive Substrates and Alternative Approaches for Vertical-Injection-Geometry Devices

To date, nearly all UV LEDs and LDs are grown upon electrically insulating sapphire and AlN substrates. Such substrates necessitate designs where current is laterally injected into n-type layers (Fig. 11A), leading to degradation of performance and operational lifetime by current crowding

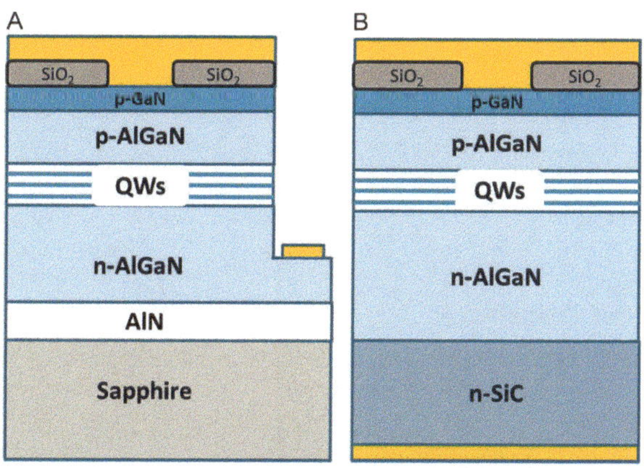

Fig. 11 Schematics of different AlGaN-based device geometries including (A) laterally injected LD structures grown on electrically insulating substrates and (B) vertically injected LD structures grown on electrically conducting substrates.

and local heating (Evstratov et al., 2006; Gong et al., 2006). Efforts to move to full vertical-injection device geometries (Fig. 11B) have focused on n-GaN and n-SiC substrates. These substrates are far from ideal given their large lattice mismatch with AlGaN alloys and absorption at UV wavelengths. This absorption precludes efficient light extraction from DUV LEDs unless substrate removal is pursued. In contrast, the substrate absorption is not prohibitive for vertically injected LD geometries provided cladding layers are sufficiently thick to confine lasing modes away from the absorbing substrate. Below, we describe progress to date in the development of vertical-injection AlGaN light emitters through the use of conductive substrates and substrate removal strategies.

3.3.1 n-Type GaN Substrates

n-GaN substrates have recently been employed to realize the first vertically injected AlGaN-based UV LDs, demonstrating lasing at 356.6 nm (Aoki et al., 2015). In this work, FACELO was applied to a GaN substrate that was periodically masked with SiO_2 stripes, creating GaN pyramids (Hiramatsu et al., 2000). Subsequent overgrowth of AlGaN achieved the strain relaxation needed to realize crack-free laser heterostructures that were several microns thick. While TDD was not reported, similar FACELO efforts by this group resulted in $\sim 10\times$ reduction of TDD down to $3.9 \times 10^8\ \text{cm}^{-2}$ (Yoshida et al., 2007). Ridge waveguide lasers fabricated on these templates had high threshold current densities (56 kA/cm^2) compared to devices on sapphire (8 kA/cm^2; Yoshida et al., 2009), indicating that further optimization is needed.

This vertical-injection design was also leveraged to achieve uniform current injection in 50 μm × 600 μm broad area UV lasers (Taketomi et al., 2016). While threshold current densities of $\sim$39 kA/cm^2 ($I_{th} = 7\text{A}$) were still substantially higher than those of ridge waveguide AlGaN lasers on sapphire, the broad area lasers emitted >1 W peak output power at 339 nm, thereby achieving the highest peak output power reported for any AlGaN-based UV LDs to date.

3.3.2 n-Type SiC Substrates

Growth of high-quality GaN or AlGaN on SiC substrates often involves an electrically insulating AlN nucleation layer, thereby preventing vertical conduction. To overcome this limitation, several groups have developed n-type conducting AlGaN buffer layers for growth on SiC (Moran et al., 2000; Nishikawa et al, 2007; Yoo et al., 2006). A remaining limitation for DUV LEDs on n-SiC is that both the substrate and the commonly employed p-GaN cap layer absorb the DUV emission, greatly reducing light extraction efficiency.

For edge-emitting GaN/AlGaN UV LDs, however, SiC substrates can provide significant advantages over sapphire substrates. Beyond electrical conductivity, n-SiC provides a common (1–100) plane for cleaved facets, closer lattice match to AlGaN alloys and improved thermal conductivity (Kuramata et al., 1998). As early as 1997, vertical-injection InGaN/GaN LDs on n-SiC with conductive n-AlGaN buffer layers were demonstrated in the near-UV region (Doverspike et al., 1997; Kuramata et al., 1997). AlGaN-based UV LDs with conductive n-AlGaN buffer layers on SiC substrates have also been reported (Edmond et al., 2004). These devices achieved CW lasing at 348 nm albeit with operation lifetimes of a few seconds for mW-level operation. Pulsed lasing at 343 nm was also reported, representing a record short wavelength for UV LDs at that time.

3.3.3 Substrate Removal

Removal of the electrically insulating substrate adds process complexity but is another viable path to vertical-injection devices. Sapphire substrate removal by laser lift-off (LLO) is a production-compatible process for thin-film flip-chip visible LEDs, contributing to high light extraction efficiencies up to ~85% (Shchekin et al., 2006). The process involves scanning a high power UV laser (e.g., 248 nm KrF excimer laser) through the transparent sapphire substrate, causing decomposition of GaN at the interface with the sapphire substrate. While much less mature, LLO of AlGaN/AlN has been developed for both laterally injected thin-film flip-chip UV LEDs (Aoshima et al., 2012; Hwang et al., 2011) and vertically injected UV LEDs (Adivarahan et al., 2009; Kawasaki et al., 2006; Takeuchi et al., 2009; Zhou et al., 2006). Notably, LLO using a 100-nm-thick sacrificial GaN layer enabled vertical-injection 322 nm LEDs with 5 × lower resistivity and significantly reduced self-heating compared to lateral-injection LEDs with similar LED active layers (Kawasaki et al., 2006).

The GaN release layer becomes problematic for higher-Al-composition and DUV AlGaN devices due to lattice mismatch and strain limitations. To address this problem, Adivarahan et al. (2009) developed excimer laser-based LLO of 280 nm LEDs grown on standard AlN buffer layers on sapphire. Compared to lateral-injection LEDs fabricated from the same wafer, vertical-injection LEDs had significantly reduced series resistance (7 Ω vs 23 Ω), leading to reduced Joule heating under operation. These vertical-injection 280 nm LEDs achieved 5.5 mW output power and >2000 h operating lifetimes at a current density of 25 A/cm^2.

As a less conventional approach, 2D materials such as graphene and hexagonal boron nitride (hBN) have been explored as release layers between the sapphire substrate and the overlying GaN or AlGaN epilayers. These 2D materials have weak van der Waals bonding between their atomic planes, enabling more facile mechanical separation of the GaN/AlGaN heterostructures from the sapphire substrate. This approach potentially avoids several complications of the LLO process, including the need for very high power UV lasers, time needed to laser scan the entire wafer, and chemical removal of remaining Ga metal on the wafer surface following LLO.

To the author's knowledge, no reports have been made of vertical-injection AlGaN-based UV light emitters with 2D release layers; however, hBN may present a particularly compelling option. Kobayashi et al. (2012) reported that direct nucleation of GaN on hBN/sapphire yielded polycrystalline growth, whereas AlN and AlGaN layers on hBN/sapphire had significantly higher crystalline quality. Despite relatively high TDD of $\sim 9 \times 10^9$ cm^{-2}, a 2-cm-square sample of GaN/$Al_{0.28}Ga_{0.72}N$ HEMTs grown on an AlN epilayer/hBN/sapphire template was separated and transferred to other substrates and maintained 1100 cm^2/Vs mobility and a carrier concentration of 1×10^{13} cm^{-2} (293 K). In the same study, InGaN blue LEDs were grown on an AlGaN epilayer/hBN/sapphire template, mechanically separated from the sapphire substrate, and bonded to alternative substrates with a reflective metal bonding layer. 2 mm-square, 3.4 μm-thick vertically injected LEDs yielded equal or higher EL intensities compared to standard devices on AlN/sapphire, largely due to increased light extraction. Remaining challenges to 2D separation layers include reducing TDD, achieving large-area sample transfers without epilayer cracking, and maintaining high-crystalline-quality growth on these foreign buffer layers.

4. SUMMARY AND OUTLOOK

Over more than 50 years, we have seen the transition from traditional lamps and large-frame lasers to compact, efficient, solid-state sources for many regions of the spectrum. One exception is the UV region (in particular, ≤360 nm) for which mature, high-efficiency solid-state sources are still lacking. In this chapter, we have reviewed outstanding materials challenges of AlGaN alloys which have frustrated the realization of next-generation UV light emitters. The challenge of effective doping has been a particularly large impediment to high-performance AlGaN LED and LDs. Advances in p-type doping with Mg acceptors have been realized through careful optimization of growth conditions, however, achieving $p > 1 \times 10^{18}$ cm^{-3} free holes at high Al compositions by growth optimization alone is extremely challenging. Alternatively, polarization engineering strategies to p-type doping have shown significant promise by circumventing the need for thermal activation of holes and providing new approaches to hole injection. We have detailed advances and remaining challenges in polarization-enabled approaches including Mg-SPSLs, DPD, and TJs, and their performance to date in UV light emitters. Particularly compelling is the implementation of DPD in nanowire structures for which strain accommodation enables grading over the entire AlGaN composition range without materials degradation (Carnevale et al., 2012). n-type doping is straightforward for most of the alloy composition range but has proven challenging in $Al_xGa_{1-x}N$ alloys with $x > 0.8$. This review has examined remaining uncertainties in Si donor activation energies and the role of Si DX centers in the observed high resistivity of high-Al-composition alloys.

The lack of a lattice-matched substrate has created additional challenges, most notably in the creation of high densities of extended defects which reduce radiative efficiency and create current leakage paths in UV emitters. Overgrowth of micropatterned templates has realized $>10\times$ reduction in extended defects and has been the enabling approach to realizing the shortest wavelength mW-class UV LDs to date (Yoshida et al., 2008). AlGaN nanowires offer an approach to even greater reduction of extended defects, and proof-of-concept nanowire UV LEDs and LDs have already been demonstrated (Mi et al., 2016). The performance of AlGaN-based emitters has also been limited by the electrically insulating nature of common substrates, including sapphire and AlN. AlGaN overgrowth of micro-patterned n-type GaN substrates has recently yielded vertically injected broad area UV lasers with the highest peak power to date of >1 W at 338 nm (Taketomi et al., 2016). Given the typical bottom-emitting geometry of UV LEDs, the UV absorption of electrically conducting n-GaN and n-SiC substrates precludes high light extraction efficiency. Beyond LLO strategies, growth of AlGaN LEDs on 2D materials such as hBN is a potential approach to facile mechanical separation of UV-absorbing substrates for greatly enhanced light extraction.

Through this survey, one can see that many of the exciting advances in AlGaN-based emitters are emerging from common themes: leveraging materials and devices at the micro/nanoscale and the polarization properties inherent to polar nitride heterostructures. These themes may well be the enabling concepts for realizing high-performance, solid-state UV light emitters.

ACKNOWLEDGMENTS

The author gratefully acknowledges Andrew Allerman, Andrew Armstrong, and Stephen Lee for suggested improvements to this chapter. Technical contributions of Andrew Allerman, Andrew Armstrong, Jonathan Wierer, Jr., Michael Moseley, Michael Smith, Karen Cross, Mary Miller, Stephen Lee, and Len Alessi are also appreciated. This work was supported by Sandia's Laboratory Directed Research and Development Program. Sandia National Laboratories is a multimission laboratory managed and operated by Sandia Corporation, a wholly owned subsidiary of Lockheed Martin Corporation, for the U.S. Department of Energy's National Nuclear Security Administration under contract DE-AC04-94AL85000.

REFERENCES

Adivarahan, V., Heidari, A., Zhang, B., Fareed, Q., Islam, M., Hwang, S., Balakrishnan, K., Khan, A., 2009. Vertical injection AlGaN multiple-quantum wells active region. Appl. Phys. Express 2, 092102.

Akyol, F., Krishmamoorthy, S., Zhang, Y., Johnson, J., Hwang, J., Rajan, S., 2016. Low-resistance GaN tunnel homojunctions with 150 kA/cm^2 current and repeatable negative differential resistance. Appl. Phys. Lett. 108, 131103.

Allerman, A.A., Crawford, M.H., Miller, M.A., Lee, S.R., 2010. Growth and characterization of Mg-doped AlGaN-AlN short-period superlattices for deep-UV optoelectronic devices. J. Cryst. Growth 312, 756–761.

Allerman, A.A., Crawford, M.H., Lee, S.R., Clark, B.G., 2014. Low dislocation density AlGaN epilayers by epitaxial overgrowth of patterned templates. J. Cryst. Growth 388, 76–82.

Amano, H., Sawaki, H., Akasaki, I., Toyoda, Y., 1986. Metalorganic vapor phase epitaxial growth of a high quality GaN film using an AlN buffer layer. Appl. Phys. Lett. 48, 353–355.

Amano, H., Nagamatsu, K., Takeda, K., Mori, T., Tsuzuki, H., Iwaya, M., Kamiyama, S., Akasaki, I., 2009. Growth and conductivity control of high quality AlGaN and its application to high-performance ultraviolet laser diodes. Proc. SPIE 7216, 72161B.

Ambacher, O., Smart, J., Shealy, J.R., Weimann, N.G., Chu, K., Murphy, M., Schaff, W.J., Eastman, L.F., Dimitrov, R., Wittmer, L., Stutzmann, M., Rieger, W., Hilsenbeck, J., 1999. Two-dimensional electron gases induced by spontaneous and piezoelectric polarization charges in N- and Ga-face AlGaN/GaN heterostructures. J. Appl. Phys. 85, 3222.

Aoki, Y., Kuwabara, M., Yamashita, Y., Takagi, Y., Sugiyama, A., Yoshida, H., 2015. A 350-nm band GaN/AlGaN multiple-quantum-well laser diode on bulk GaN. Appl. Phys. Lett. 107, 151103.

Aoshima, H., Takeda, K., Takehara, K., Ito, S., Mori, M., Iwaya, M., Takeuchi, T., Kamiyama, S., Akasaki, I., Amano, H., 2012. Laser lift-off of AlN/sapphire for UV light-emitting diodes. Phys. Status Solidi C 9, 753–756.

Beaumont, B., Vennegues, Ph., Gibart, P., 2001. Epitaxial lateral overgrowth of GaN. Phys. Status Solidi B 227, 1–43.

Borisov, B., Kuryatkov, V., Kudryavtsec, Yu, Asomoza, R., Nikishin, S., Song, D.Y., Holtz, M., Temkin, H., 2005. Si-doped $Al_xGa_{1-x}N$ ($0.56 \leq x \leq 1$) layers grown by molecular beam epitaxy with ammonia. Appl. Phys. Lett. 87, 132106.

Boucart, J., Starck, C., Gaborit, F., Plais, A., Bouche, N., Derouin, E., Remy, J.C., Bonnet-Gamard, J., Goldstein, L., Fortin, C., Carentier, D., Salet, P., Brillouet, F., Jacquet, J., 1999. Metamorphic DBR and tunnel-junction injection: a CW RT monolithic long-wavelength VCSEL. IEEE J. Sel. Topics Quantum Electron. 5, 520–529.

Brault, J., Rosales, D., Dmilano, B., Leroux, M., Courville, A., Korytov, M., Chenot, S., Vennegues, P., Vinter, B., De Mierry, P., Kahouli, A., Massies, J., Bretangnon, T., Gil, B., 2014. Polar and semipolar $GaN/Al_{0.50}Ga_{0.50}N$ nanostructures for UV light emitters. Semicond. Sci. Technol. 29, 084001.

Carnevale, S.D., Kent, T.F., Phillips, P.J., Mills, M.J., Rajan, S., Meyers, R.C., 2012. Polarization-induced pn diodes in wide-band-gap nanowires with ultraviolet electroluminescence. Nanolett. 12, 915–920.

Chakraborty, A., Moe, C.G., Wu, Y., Mates, T., Keller, S., Speck, J.S., DenBaars, S.P., Mishra, U.K., 2007. Electrical and structural characterization of Mg-doped p-type $Al_{0.69}Ga_{0.31}N$ films on SiC substrate. J. Appl. Phys. 101, 053717.

Chen, Y., Wu, H., Han, E., Yue, G., Chen, Z., Wu, Z., Wang, G., Jiang, H., 2015. High hole concentration in p-type AlGaN by indium-surfactant-assisted Mg-delta doping. Appl. Phys. Lett. 106, 162102.

Cheng, B., Choi, S., Northrup, J.E., Yang, Z., Knollenberg, C., Teepe, M., Wunderer, T., Chua, C.L., Johnson, N.M., 2013. Enhanced vertical and lateral hole transport in high aluminum-containing AlGaN for deep ultraviolet light emitters. Appl. Phys. Lett. 102, 231106.

Cherns, D., Sahonta, S.-L., Will, H.H., Liu, R., Ponce, F.A., Amano, H., Akasaki, I., 2004. The generation of misfit dislocations in facet-controlled growth of AlGaN/GaN films. Appl. Phys. Lett. 85, 4923–4925.

Chichibu, S.F., Hazu, K., Onuma, T., Uedono, A., 2011. Collateral evidence of excellent radiative performance of $Al_xGa_{1-x}N$ alloy films of high AlN mole fractions. Appl. Phys. Lett. 66, 051902.

Chow, W.W., Kneissl, M., 2005. Laser gain properties of AlGaN quantum wells. J. Appl. Phys. 98, 114502.

Collazo, R., Mita, S., Xie, J., Rice, A., Tweedie, J., Dalmau, R., Sitar, Z., 2011. Progress on n-type doping of AlGaN alloys on AlN single crystal substrates for UV optoelectronic applications. Phys. Status Solidi C 8, 2031–2033.

Conroy, M., Zubialevich, V.Z., Li, H., Petkov, N., Holmes, J.D., Parbrook, P.J., 2015. Epitaxial lateral overgrowth of AlN on self-assembled patterned nanorods. J. Mater. Chem. C 3, 431–437.

Crawford, M.H., Allerman, A.A., Armstrong, A.M., Smith, M.L., Cross, K.C., 2015. Laser diodes with 353 nm wavelength enabled by reduced-dislocation-density AlGaN templates. Appl. Phys. Express 8, 112702.

Djavid, M., Mi, Z., 2016. Enhancing the light extraction efficiency of AlGaN deep ultraviolet light emitting diodes by using nanowire structures. Appl. Phys. Lett. 108, 051102.

Doverspike, K., Bulman, G.E., Sheppard, S.T., Kong, H.S., Leonard, M., Dieringer, H., Weeks, T.W., Edmond, J., Brown, J.D., Swindle, J.T., Schetzina, J.F., Song, Y.-K., Kuball, M., Nurmikko, A., 1997. Status of nitride based light emitting and laser diodes on SiC. MRS Proceedings 482, 1169.

Edmond, J., Abare, A., Bergman, M., Bharathan, J., Bunker, K.L., Emerson, D., Haberern, K., Ibbetson, J., Leung, M., Russel, P., Slater, D., 2004. High efficiency GaN-based LEDs and lasers on SiC. J. Cryst. Growth 272, 242–250.

Esaki, L., 1958. New phenomenon in narrow germanium p-n junctions. Phys. Rev. 109, 603–604.

Evstratov, I.Yu., Mymrin, V.F., Karpov, S.Yu., Makarov, Yu.N., 2006. Current crowding effects on blue LED operation. Phys. Status Solidi C 3, 1645–1648.

Glas, F., 2006. Critical dimensions for the plastic relaxation of strained axial heterostructures in free-standing nanowires. Phys. Rev. B 74, 121302(R).

Golam Sarwar, A.T.M., May, B.J., Deitz, J.I., Grassman, T.J., McComb, D.W., Myers, R.C., 2015. Tunnel junction enhanced nanowire ultraviolet light emitting diodes. Appl. Phys. Lett. 107, 101103.

Goldberg, Y., Levinshtein, M.E., Rumyantsev, S.L., 2001. Silicon carbide (SiC). In: Levinshtein, M.E., Rumyantsev, S.L., Shur, M.S. (Eds.), Properties of Advanced Semiconductor Materials: GaN, AIN, InN, BN, SiC, SiGe. John Wiley & Sons, Inc., New York, pp. 93–148.

Gong, Z., Gaevski, M., Adivarahan, V., Sun, W., Shatalov, M., Asif Khan, M., 2006. Optical power degradation mechanisms in AlGaN-based 280 nm deep ultraviolet light-emitting diodes on sapphire. Appl. Phys. Lett. 88, 121106.

Gordon, L., Lyons, J.L., Janotti, A., Van de Walle, C.G., 2014. Hybrid functional calculations of DX centers in AlN and GaN. Phys. Rev. B 89, 085204.

Grundmann, M.J., Mishra, U.K., 2007. Multi-color light emitting diode using polarization-induced tunnel junctions. Phys. Status Solidi C 4, 2830–2833.

Gunning, B., Lowder, J., Moselely, M., Doolittle, W.A., 2012. Negligible carrier freeze-out facilitated by impurity band conduction in highly p-type GaN. Appl. Phys. Lett. 101, 082106.

Gunning, B.P., Fabien, C.A.M., Merola, J.J., Clinton, E.A., Doolittle, W.A., Wang, S., Fischer, A.M., Ponce, F.A., 2015. Comprehensive study of the electronic and optical behavior of highly degenerate p-type Mg-doped GaN and AlGaN. J. Appl. Phys. 117, 045710.

Haeger, D.A., Young, E.C., Chung, R.B., Wu, F., Pfaff, N.A., Tsai, M., Fujito, K., DenBaars, S.P., Speck, J.S., Nakamura, S., Cohen, D.A., 2012. 384 nm laser diode grown on a (20–21) semipolar relaxed AlGaN buffer layer. Appl. Phys. Lett. 100, 161107.

Hermann, M., Furtmayr, F., Bergmaier, A., Dollinger, G., Stutzmann, M., Eickhoff, M., 2005. Highly Si-doped AlN grown by plasma-assisted molecular-beam epitaxy. Appl. Phys. Lett. 86, 192108.

Hevia, D.F., Stampfl, C., Viñes, F., Illas, F., 2013. Microscopi origin of n-type behavior in Si-doped AlN. Phys. Rev. B 88, 085202.

Hiramatsu, K., Nishiyama, K., Onishi, M., Mizutani, H., Narukawa, M., Motogaito, A., Miyake, H., Iyechika, Y., Maeda, T., 2000. Fabrication and characterization of low defect density GaN using facet-controlled epitaxial lateral overgrowth (FACELO). J. Cryst. Growth 221, 316–326.

Hwang, S., Morgan, D., Kesler, A., Lachab, M., Zhang, B., Heidari, A., Nazir, I.A., Dion, J., Fareed, Q., Adivarahan, V., Islam, M., Khan, A., 2011. 276 nm substrate-free flip-chip AlGaN light-emitting diodes. Appl. Phys. Express 4, 032102.

Ibbetson, J.P., Fini, P.T., Ness, K.D., DenBaars, S.P., Speck, J.S., Mishra, U.K., 2000. Polarization effects in AlGaN/GaN heterostructure field effect transistors. Appl. Phys. Lett. 77, 250–252.

Imura, M., Nakano, K., Narita, G., Fujimoto, N., Okada, N., Balakrishnan, K., Iwaya, M., Kamiyama, S., Amano, H., Akasaki, I., Noro, T., Takagi, T., Bandoh, A., 2007. Epitaxial lateral overgrowth of AlN on trench-patterned AlN layers. J. Cryst. Growth 298, 257–260.

Ive, T., Brandt, O., Kostial, H., Friedland, J., Däweritz, L., Ploog, K.H., 2005. Controlled n-type doping of AlN:Si films grown on 6H-SiC (0001) by plasma-assisted molecular beam epitaxy. Appl. Phys. Lett. 86, 024106.

Iwaya, M., Terao, S., Sano, T., Takanami, S., Ukai, T., Nakamura, R., Kamiyama, S., Amano, H., Akasaki, I., 2001. High-efficiency GaN/$Al_xGa_{1-x}N$ multi-quantum-well light emitter grown on low-dislocation density $Al_xGa_{1-x}N$. Phys. Status Solidi A 188, 117–120.

Jain, R., Sun, W., Yang, H., Shatalov, M., Hu, X., Sattu, A., Lunev, A., Deng, J., Shturm, I., Bilenko, Y., Gaska, R., Shur, M.S., 2008. Migration enhanced lateral epitaxial overgrowth of AlN and AlGaN for high reliability deep ultraviolet light emitting diodes. Appl. Phys. Lett. 93, 051113.

Jena, D., Heikman, S., Green, D., Buttari, D., Coffie, R., Xing, H., Keller, S., DenBaars, S., Speck, J.S., Mishra, U.K., 2002. Realization of wide electron slabs by polarization bulk doping in graded III-V nitride semiconductor alloys. Appl. Phys. Lett. 81, 4395–4397.

Jeon, S.-R., Song, Y.-H., Jang, H.-J., Yang, G.M., 2001. Lateral current spreading in GaN-based light-emitting diodes utilizing tunnel junction contacts. Appl. Phys. Lett. 78, 3265–3267.

Jeon, S.-R., Ren, Z., Cui, G., Su, J., Gherasimova, M., Han, J., Cho, H.-K., Zhou, L., 2005. Investigation of Mg doping in high-Al content p-type $Al_xGa_{1-x}N$ ($0.3 < x < 0.5$). Appl. Phys. Lett. 86, 082107.

Kakanakova-Georgieva, A., Nilsson, D., Stattin, M., Forsberg, U., Haglund, A., Larsson, A., Janzen, E., 2010. Mg-doped $Al_{0.85}Ga_{0.15}N$ grown by hot-wall MOCVD with low resistivity at room temperature. Phys. Status Solidi RRL 4, 311–313.

Kamiyama, S., Iwaya, M., Hayashi, N., Takeuchi, T., Amano, H., Akasaki, I., Watanabe, S., Kaneko, Y., Yamada, N., 2001. Low-temperature-deposited AlGaN interlayer for improvement of AlGaN/GaN heterostructure. J. Cryst. Growth 223, 83–91.

Kamiyama, S., Iida, K., Kawashima, T., Kasugai, H., Mishima, S., Honshio, A., Miyake, Y., Iwaya, M., Amano, H., Akasaki, I., 2005. UV laser diode with 350.9 nm lasing wavelength grown by hetero-epitaxial-lateral overgrowth technology. IEEE J. Sel. Top. Quantum Electron. 11, 1069–1073.

Kawasaki, K., Koike, C., Aoyagi, Y., Takeuchi, M., 2006. Vertical AlGaN deep ultraviolet light emitting diode emitting at 322 nm fabricated by the laser lift-off technique. Appl. Phys. Lett. 89, 261114.

Kawashima, T., Iida, K., Miyake, Y., Honshio, A., Kasugai, H., Imura, M., Iwaya, M., Kamiyama, S., Amano, H., Akasaki, I., 2004. High-quality $Al_{0.12}Ga_{0.88}N$ film with low dislocation density grown on facet-controlled $Al_{0.12}Ga_{0.88}N$ by MOVPE. J. Cryst. Growth 272, 377–380.

Kent, T.F., Carnevale, S.D., Sarwar, A.T.M., Phillips, P.J., Klie, R.F., Myers, R.C., 2014. Deep ultraviolet emitting polarization induced nanowire LEDs with $Al_xGa_{1-x}N$ active regions. Nanotechnology 25, 455201.

Khokhlev, K.A., Bulashevich, K.A., Karpov, S.Yu., 2013. Polarization doping for III nitride optoelectronics. Phys. Status Solidi A 210, 1369–1376.

King, R.R., Karam, N.H., Ermer, J.H., Haddad, M., Colter, P., Isshiki, T., Yoon, H., 2000. Next-generation, high efficiency III-V multi-junction solar cells. In: Proceedings of 28th Photovoltaics Specialist Conference, p. 998.

Kinoshita, T., Obata, T., Yanagi, H., Inoue, S.-I., 2013. High p-type conduction in high-Al-content Mg-doped AlGaN. Appl. Phys. Lett. 102, 012105.

Kishino, K., Ishizawa, S., 2015. Selective-area growth of GaN nanocolumns on Si (111) substrates for application to nanocolumn emitters with systematic analysis of dislocation filtering effect of nanocolumns. Nanotechnology 26, 225602.

Kneissl, M., Rass, J. (Eds.), 2016. III-N Ultraviolet Emitters. Springer Series in Materials Science, vol. 227. Springer International Publishing, New York.

Kneissl, M., Yang, Z., Teepe, M., Knollenberg, C., Johnson, N.M., Usikov, A., Dmitriev, V., 2006. Ultraviolet InAlGaN light emitting diodes grown on hydride vapor phase epitaxy AlGaN/sapphire templates. Jpn. J. Appl. Phys. 45, 3905–3908.

Kobayashi, Y., Kumakura, K., Akasaka, T., Makimoto, T., 2012. Layered boron nitride as a release layer for mechanical transfer of GaN-based devices. Nature 484, 223–227.

Kozodoy, P., Ibbetson, J.P., Marchand, H., Fini, P.T., Keller, S., Speck, J.S., DenBaars, S.P., Mishra, U.K., 1998. Electrical characterization of GaN p-n junctions with and without threading dislocations. Appl. Phys. Lett. 73, 975–977.

Kozodoy, P., Hansen, M., DenBaars, S.P., Mishra, U.M., 1999. Enhanced Mg doping efficiency in $Al_{0.2}Ga_{0.8}N$/GaN superlattices. Appl. Phys. Lett. 74, 3681–3683.

Krishnamoorthy, S., Akyol, F., Park, P.S., Rajan, S., 2013. Low resistance GaN/InGaN/GaN tunnel junctions. Appl. Phys. Lett. 102, 113503.

Krishnamoorthy, S., Akyol, F., Rajan, S., 2014. InGaN/GaN tunnel junctions for hole injection in GaN light emitting diodes. Appl. Phys. Lett. 105, 141104.

Kueller, V., Knauer, A., Brunner, F., Zeimer, U., Rodriguez, H., Kneissl, M., Weyers, M., 2011. Growth of AlGaN and AlN on patterned AlN/sapphire templates. J. Cryst. Growth 315, 200–203.

Kueller, V., Knauer, A., Zeimer, U., Kneissl, M., Weyers, M., 2013. Controlled coalescence of MOVPE grown AlN during lateral overgrowth. J. Cryst. Growth 368, 83–86.

Kumakura, K., Makimoto, T., Kobayashi, N., 2000. Enhanced hole generation in Mg-doped AlGaN/GaN superlattices due to piezoelectric field. Jpn. J. Appl. Phys. 39, 2428–2430.

Kuramata, A., Domen, K., Soejima, R., Horino, K., Kubota, S.-I., Tanahashi, T., 1997. InGaN laser diode grown on 6H-SiC substrate using low-pressure metal organic vapor phase epitaxy. Jpn. J. Appl. Phys. 36, L1130–L1132.

Kuramata, A., Horino, K., Domen, K., 1998. GaN-based blue laser diodes grown on SiC substrate as light source for high density optical storage. Fujitsu Sci. Tech. J. 34, 191–203.

Kyle, E.C.H., Kuan, S.W., Young, E.C., Speck, J.S., 2015. Increased p-type conductivity through use of an indium surfactant in the growth of Mg-doped GaN. Appl. Phys. Lett. 106, 222103.

Lee, S.W., Oh, D.C., Goto, H., Ha, J.S., Lee, H.J., Hanada, T., Cho, M.W., Yao, T., Hong, S.K., Lee, H.Y., 2006. Origin of forward leakage current in GaN-based light-emitting devices. Appl. Phys. Lett. 89, 132117.

Li, S., Zhang, T., Wu, J., Yang, Y., Wang, Z., Wu, Z., Chen, Z., Jiang, Y., 2013. Polarization induced hole doping in graded $Al_xGa_{1-x}N$ ($x = 0.7 \sim 1$) layer grown by molecular beam epitaxy. Appl. Phys. Lett. 102, 062198.

Li, K.H., Liu, X., Wang, Q., Zhao, S., Mi, Z., 2015. Ultralow-threshold electrically injected AlGaN nanowire ultraviolet lasers on Si operating at low temperature. Nature Nanotech. 10, 140–144.

Liu, R., Bell, A., Ponce, F.A., Amano, H., Akasaki, I., Cherns, D., 2003. Thick crack-free AlGaN films deposited by facet-controlled epitaxial lateral overgrowth. Phys. Status Solidi C 2136–2140.

Lu, T.C., Shieh, M., Wang, S.C., 2002. Real index-guided InGaAlP red lasers with buried tunnel junctions. Appl. Phys. Lett. 80, 1882–1884.

Malinverni, M., Martin, D., Grandjean, N., 2015. InGaN based micro light emitting diodes featuring a buried GaN tunnel junction. Appl. Phys. Lett. 107, 051107.

Martens, M., Kuhn, C., Ziffer, E., Simoneit, T., Kueller, V., Knauer, A., Rass, J., Wernicke, T., Einfeldt, S., Weyers, M., Kneissl, M., 2016. Low absorption loss p-AlGaN superlattice cladding layer for current-injection deep ultraviolet laser diodes. Appl. Phys. Lett. 108, 151108.

Mehnke, F., Wernicke, T., Pingel, H., Kuhn, C., Reich, C., Kueller, V., Knauer, A., Lapeyrade, M., Weyers, M., Kneissl, M., 2013. Highly conductive n-$Al_xGa_{1-x}N$ layers with aluminum mole fractions above 80%. Appl. Phys. Lett. 103, 212109.

Mi, Z., Zhao, S., Woo, S.Y., Bugnet, M., Djavid, M., Liu, X., Kang, J., Kong, X., Ji, W., Guo, H., Liu, Z., Botton, G.A., 2016. Molecular beam epitaxial growth and characterization of Al(Ga)N nanowire deep ultraviolet light emitting diodes and lasers. J. Phys. D 49, 364006.

Minamikawa, D., Ino, M., Kawai, S., Takeuchi, T., Kamiyama, S., Iwaya, M., Akasaki, I., 2015. GaInN-based tunnel junctions with high InN mole fractions grown by MOVPE. Phys. Status Solidi B 252, 1127–1131.

Mochizuki, S., Detchprohm, T., Sano, S., Nakamura, T., Amano, H., Akasaki, I., 2002. Reduction of threading dislocation density in AlGaN grown on periodically grooved substrates. J. Cryst. Growth 237–239, 1065–1069.

Moe, C.G., Reed, M.L., Garret, G.A., Sampath, A.V., Alexander, T., Shen, H., Wraback, M., Bilenko, Y., Shatalov, M., Yang, J., Sun, W., Deng, J., Gaska, R., 2010. Current-induced degradation of high performance deep ultraviolet light emitting diodes. Appl. Phys. Lett. 96, 213512.

Moran, B., Hansen, M., Craven, M.D., Speck, J.S., DenBaars, S.P., 2000. Growth and characterization of graded AlGaN conductive buffer layers on n+SiC substrates. J. Cryst. Growth 221, 301–304.

Moseley, M., Allerman, A.A., Crawford, M., Wierer, J., Smith, M., Biedermann, L., 2014. Electrical current leakage and open-core threading dislocations in AlGaN-based deep ultraviolet light-emitting diodes. J. Appl. Phys. 117, 095301.

Moseley, M.W., Allerman, A.A., Crawford, M.H., Wierer, J.J., Smith, M.L., Armstrong, A.M., 2015. Detection and modeling of leakage current in AlGaN-based deep ultraviolet light-emitting diodes. J. Appl. Phys. 116, 053104.

Nakamura, S., 1991. GaN growth using GaN buffer layer. Jpn. J. Appl. Phys. 30, L1705–L1707.

Nakamura, S., 1998. High power InGaN-based blue laser diodes with a long lifetime. J. Cryst. Growth 195, 242–247.

Nakamura, S., Senoh, M., Mukai, T., 1991. Highly p-typed Mg-doped GaN films grown with GaN buffer layers. Jpn. J. Appl. Phys. 30, L1708–L1711.

Nakarmi, M.L., Kim, K.H., Li, J., Lin, J.Y., Jiang, H.X., 2003. Enhanced p-type conduction in GaN and AlGaN by Mg-delta-doping. Appl. Phys. Lett. 82, 3041–3043.

Nakarmi, M.L., Kim, K.H., Lin, J.Y., Jiang, H.X., 2004. Transport properties of highly conductive n-type Al-rich $Al_xGa_{1-x}N$ ($x \geq 0.7$). Appl. Phys. Lett. 85, 3769–3771.

Nakarmi, M.L., Nepal, N., Lin, J.Y., Jiang, H.X., 2009. Photoluminescence studies of impurity transitions in Mg-doped AlGaN alloys. Appl. Phys. Lett. 94, 091903.

Nam, K.B., Nakarmi, M.L., Li, J., Lin, J.Y., Jiang, H.X., 2003. Mg acceptor level in AlN probed by deep ultraviolet photoluminescence. Appl. Phys. Lett. 83, 878–880.

Nam, K.B., Li, J., Nakarmi, M.L., Lin, J.Y., Jiang, H.X., 2004. Unique optical properties of AlGaN alloys and related ultraviolet emitters. Appl. Phys. Lett. 84, 5264–5266.

Neugebauer, J., Van de Walle, C.G., 1995. Theory of point defects and complexes in GaN. MRS proceedings 395, 645–656.

Neuschl, B., Thonke, K., Feneberg, M., Goldhahn, R., Wunderer, T., Yang, Z., Johnson, N.M., Xie, J., Mita, S., Rice, A., Collazo, R., Sitar, Z., 2013. Direct determination of the silicon donor ionization energy in homoepitaxial AlN from photoluminescence two-electron transitions. Appl. Phys. Lett. 103, 122105.

Nikishin, S.A., Kuryatkov, V., Chandolu, A., Borisov, B.A., Kipshidze, G.D., Ahmad, I., Holtz, M., Temkin, H., 2003. Deep ultraviolet light emitting diodes based on short period superlattices of AlN/AlGa(In)N. Jpn. J. Appl. Phys. 42, L1362–L1365.

Nikishin, S.A., Holtz, M., Temkin, H., 2005. Digital alloys of AlN/AlGaN for deep UV light emitting diodes. Jpn. J. Appl. Phys. 44, 7221–7226.
Nilsson, D., 2014. Doping of high-Al-content AlGaN grown by MOCVD. Linkoping studies in science and technology dissertation No. 1597.
Nishikawa, A., Kumakura, K., Akasaka, T., Makimoto, T., 2007. Low-resistance graded $Al_xGa_{1-x}N$ buffer layers for vertical conducting devices on n-SiC substrates. J. Cryst. Growth 298, 819–821.
Nitta, S., Yukawa, Y., Watanabe, Y., Kamiyama, S., Amano, H., Akasaki, I., 2002. Mass transport in $Al_xGa_{1-x}N$. Phys. Status Solidi A 194, 485–488.
Pinos, A., Marcinkevicius, S., Yang, J., Gaska, R., Shatalov, M., Shur, M.S., 2010. Optical studies of degradation of AlGaN quantum well based deep ultraviolet light emitting diodes. J. Appl. Phys. 108, 093113.
Pinos, A., Marcinkevicius, S., Shur, M.S., 2011. High current-induced degradation of AlGaN ultraviolet light emitting diodes. J. Appl. Phys. 109, 103108.
Piprek, J., 2012. Ultra-violet light-emitting diodes with quasi acceptor-free AlGaN polarization doping. Opt. Quant. Electron. 44, 67–73.
Sadaf, S.M., Ra, Y.H., Szkopek, T., Mi, Z., 2016. Monolithically integrated metal/semiconductor tunnel junction nanowire light-emitting diodes. Nanolett. 16, 1076–1080.
Sano, S., Detchprohm, T., Yano, M., Nakamura, R., Mochizuki, S., Amano, H., Akasaki, I., 2002. Low-dislocation-density $Al_xGa_{1-x}N$ single crystals grown on grooved substrates. Mater. Sci. Eng. B 93, 197–201.
Sano, T., Doi, T., Inada, S.A., Sugiyama, T., Honda, Y., Amano, H., Yoshino, T., 2013. High internal quantum efficiency blue-green light-emitting diode fabricated on low dislocation density GaN substrate. Jpn. J. Appl. Phys. 52, 08JK09.
Saxler, A., Mitchel, W.C., Kung, P., Razeghi, M., 1999. Aluminum gallium nitride short-period superlattices doped with magnesium. Appl. Phys. Lett. 74, 2023–2025.
Schubert, M., 2010. Interband tunnel junctions for wurzite III-nitride semiconductors based on heterointerface polarization charges. Phys. Rev. B 81, 035303.
Schubert, E.F., Grieshaber, W., Goepfert, I.D., 1996. Enhancement of deep acceptor activation in semiconductors by superlattice doping. Appl. Phys. Lett. 69, 3737–3739.
Shatalov, M., Jain, R., Saxena, T., Dobrinsky, A., Shur, M., 2017. Development of deep UV LEDs and current problems in material and device technology in SEMI: nitride semiconductors. In: Chennupati, J., Mi, Z. (Eds.), Elsevier, pp. 45–84.
Shchekin, O.B., Epler, J.E., Trottier, T.A., Margalith, T., Steigerwald, D.A., Holcomb, M.O., Martin, P.S., Krames, M.R., 2006. High performance thin-film flip-chip InGaN-GaN light-emitting diodes. Appl. Phys. Lett. 89, 071109.
Silvestri, L., Dunn, K., Prawer, S., Ladouceur, F., 2011. Hybrid functional study of Si and O donors in wurzite AlN. Appl. Phys. Lett. 99, 122109.
Simon, J., Zhang, Z., Goodman, K., Xing, H., Kosel, T., Fay, P., Jena, D., 2009. Polarization-induced zener tunnel junctions in wide-band-gap heterostructures. Phys. Rev. Lett. 103, 026801.
Simon, J., Cao, Y., Jena, D., 2010a. Short-period AlN/GaN p-type superlattices: hole transport use in p-n junctions. Phys. Status Solidi C 7, 2386–2389.
Simon, J., Protasenko, V., Lian, C., Xing, H., Jena, D., 2010b. Polarization-induced hole doping in wide band gap uniaxial semiconductor heterostructures. Science 237, 60–64.
Smorchkova, I.P., Elsass, C.R., Ibbetson, J.P., Vetury, R., Heying, B., Fini, P., Haus, E., DenBaars, S.P., Speck, J.S., Mishra, U.K., 1999. Polarization-induced charge and electron mobility in AlGaN/GaN heterostructures grown by plasma-assisted molecular-beam epitaxy. J. Appl. Phys. 86, 4520–4526.
Son, N.T., Bickermann, M., Janzén, E., 2011. Shallow donor and DX states of Si in AlN. Appl. Phys. Lett. 98, 092104.

Speck, J.S., Rosner, S.J., 1999. The role of threading dislocations in the physical properties of GaN and its alloys. Physica B 273–274, 24–32.

Stampfl, C., Van de Walle, C.G., 2002. Theoretical investigation of native defects, impurities, and complexes in aluminum nitride. Phys. Rev. B. 65, 155212.

Takasuka, D., Akatsuka, Y., Ino, M., Koide, N., Takeuchi, T., Iwaya, M., Kamiyama, S., Akasaki, I., 2016. GaInN-based tunnel junctions with graded layers. Appl. Phys. Express 9, 081005.

Taketomi, H., Aoki, Y., Takagi, Y., Sugiyama, A., Kuwabara, M., Yoshida, Y., 2016. Over 1 W record-peak-power operation of a 338 nm AlGaN multiple-quantum well laser diode on a GaN substrate. Jpn. J. Appl. Phys. 55, 05FJ05.

Takeuchi, T., Hasnain, G., Corzine, S., Hueschen, M., Schneider Jr., R.P., Kocot, C., Blomqvist, M., Chang, Y.-l., Lefforge, D., Krames, M.R., 2001. GaN-based light emitting diodes with tunnel junctions. Jpn. J. Appl. Phys. 40, L861–L863.

Takeuchi, M., Maegawa, T., Shimizu, H., Ooishi, S., Ohtsuka, T., Aoyagi, Y., 2009. AlN/AlGaN short-period superlattice sacrificial layers in laser lift-off for vertical-type AlGaN-based deep ultraviolet light emitting diodes. Appl. Phys. Lett. 94, 061117.

Taniyasu, Y., Kasu, M., Kobayashi, N., 2002. Intentional control of n-type conduction for Si-doped AlN and $Al_xGa_{1-x}N$ ($0.42 \leq x < 1$). Appl. Phys. Lett. 81, 1255–1257.

Taniyasu, Y., Kasu, M., Makimoto, T., 2006. An aluminum nitride light-emitting diode with a wavelength of 210 nanometres. Nature 441, 325–328.

Trinh, X.T., Nilsson, D., Ivanov, I.G., Janzen, E., Kakanakova-Georgeva, A., Son, N.T., 2014. Stable and metastable Si negative-U centers in AlGaN and AlN. Appl. Phys. Lett. 105, 162106.

Tsuzuki, H., Mori, F., Takeda, K., Iwaya, M., Kamiyama, S., Amano, H., Akasaki, I., Yoshida, H., Kuwabara, M., Yamashita, Y., Kan, H., 2009. Novel UV devices on high-quality AlGaN using grooved underlying layer. J. Cryst. Growth 311, 2860–2863.

Van de Walle, C.G., Neugebauer, J., 2004. First-principles calculations for defects and impurities: applications to III-nitrides. J. Appl. Phys. 95, 3851–3879.

Van de Walle, C.G., Stampfl, C., Neugebauer, J., McCluskey, M.D., Johnson, N.M., 1999. Doping of AlGaN alloys. MRS Internet J. Nitride Semicond. Res.4S1, G10.4.

Verma, J., Protasenko, V., Kosel, T., Xing, H.G., Jena, D., 2011. N-polar III-nitride quantum well light-emitting diodes with polarization-induced doping. Appl. Phys. Lett. 99, 171104.

Waldron, E., Graff, J., Schubert, E., 2001. Influence of doping profiles on p-type AlGaN/GaN superlattices. Phys. Status Solidi A 188, 889–893.

Walukiewicz, W., 2001. Intrinsic limitations to the doping of wide-band gap semiconductors. Phys. B 302–303, 123–134.

Wang, C.-P., Wu, Y. R., 2012. Study of optical anisotropy in nonpolar and semipolar AlGaN quantum well deep ultraviolet light emission diode. J. Appl. Phys. 112, 033104.

Wang, Q., Connie, A.T., Nguyen, H.P.T., Kibria, M.G., Zhao, S., Sharif, S., Shih, I., Mi, Z., 2013. Highly efficient, spectrally pure 340 nm ultraviolet emission from $Al_xGa_{1-x}N$ nanowire based light emitting diodes. Nanotechnology 24, 345201.

Wierer, J.J., Evans, P.W., Holonyak Jr., N., 1997. Buried tunnel contact junction AlGaAs-GaAs-InGaAs quantum well heterostructures lasers with oxide-confined lateral currents. Appl. Phys. Lett. 71, 2286–2288.

Wierer Jr., J.J., Montano, I., Crawford, M.H., Allerman, A.A., 2014. Effect of thickness and carrier density on the optical polarization of $Al_{0.44}Ga_{0.56}N/Al_{0.55}Ga_{0.45}N$ quantum well layers. J. Appl. Phys. 115, 174501.

Yan, Q., Janotti, A., Scheffler, M., Van de Walle, C.G., 2014. Origins of optical absorption and emission lines in AlN. Appl. Phys. Lett. 105, 111104.

Yasan, A., Razeghi, M., 2003. Very high quality p-type $Al_xGa_{a-x}N$/GaN superlattice. Solid State Electron. 47, 303–306.

Yoo, D., Limb, J., Ryou, J.-H., Lee, W., Dupuis, R.D., 2006. GaN full vertical p-i-n rectifiers employing AlGaN:Si conducting buffer layers on n-SiC substrates. Appl. Phys. Lett. 88, 193503.
Yoshida, H., Takagi, Y., Kuwabara, M., Amano, H., Kan, H., 2007. Entirely crack-free ultraviolet GaN/AlGaN laser diodes grown on 2-in. sapphire substrate. Jpn. J. Appl. Phys. 46, 5782–5784.
Yoshida, H., Yamashita, Y., Kuwabara, M., Kan, H., 2008. Demonstration of an ultraviolet 336 nm AlGaN multiple-quantum-well laser diode. Appl. Phys. Lett. 93, 241106.
Yoshida, H., Kuwabara, M., Yamashita, Y., Takagi, Y., Uchiyama, K., Kan, H., 2009. AlGaN-based laser diodes for short-wavelength ultraviolet region. New J. Phys. 11, 125013.
Young, E.C., Yonkee, B.P., Wu, F., Oh, S.H., DenBaars, S.P., Nakamura, S., Speck, J.S., 2016. Hybrid tunnel junction contacts to III-nitride light-emitting diodes. Appl. Phys. Express 9, 022102.
Zhang, L., Ding, K., Yan, J.C., Wang, J.X., Zeng, Y.P., Wei, T.B., Li, Y., Sun, B.J., Duan, R.F., Li, J.M., 2010. Three-dimensional hole gas induced polarization in (0001)-oriented metal-face III-nitride structure. Appl. Phys. Lett. 97, 062103.
Zhang, Y., Krishnamoorthy, S., Johnson, J., Akyol, F., Allerman, A., Moseley, M.W., Armstrong, A., Hwang, J., Rajan, S., 2015. Interband tunneling for hole injection in III-nitride ultraviolet emitters. Appl. Phys. Lett. 106, 141103.
Zhao, S., Connie, A.T., Dastjerdi, M.H.T., Kong, X.H., Wang, Q., Djavid, M., Sadaf, S., Liu, X.D., Shih, L., Guo, H., Mi, Z., 2015a. Aluminum nitride nanowire light emitting diodes: breaking the fundamental bottleneck of deep ultraviolet light sources. Sci. Rep. 5, 8332.
Zhao, S., Liu, X., Woo, S.Y., Kang, J., Botton, G.A., Mi, Z., 2015b. An electrically injected AlGaN nanowire laser operating in the ultraviolet-C band. Appl. Phys. Lett. 107, 043101.
Zhou, L., Epler, J.E., Krames, M.R., Goetz, W., Gherasimova, M., Ren, Z., Han, J., Kneissl, M., Johnson, N.M., 2006. Vertical injection thin-film AlGaN/AlGaN multiple-quantum well deep ultraviolet light-emitting diodes. Appl. Phys. Lett. 89, 241113.
Zunger, A., 2003. Practical doping principles. Appl. Phys. Lett. 83, 57–59.

CHAPTER TWO

Development of Deep UV LEDs and Current Problems in Material and Device Technology

M. Shatalov*[,1], R. Jain*, T. Saxena[†], A. Dobrinsky*, M. Shur[‡]
*Sensor Electronic Technology, Inc., Columbia, SC, United States
[†]NXP Semiconductors, Tempe, AZ, United States
[‡]Rensselaer Polytechnic Institute, Troy, NY, United States
[1]Corresponding author: e-mail address: shatalov@s-et.com

Contents

1. INTRODUCTION

Advances in ultraviolet light-emitting diodes (UV LEDs) emerged from research of blue LEDs that began with original work of Maruska et al. at Stanford University in 1972, where they investigated electroluminescence from Mg-doped GaN devices. The first functioning GaN:Mg LED was demonstrated in June 1972 (Maruska et al., 1972). In 1983, Khan et al. at Honeywell worked on low-pressure metal–organic chemical vapor

Semiconductors and Semimetals, Volume 96
ISSN 0080-8784
http://dx.doi.org/10.1016/bs.semsem.2016.08.002
© 2017 Elsevier Inc.
All rights reserved.

deposition (MOCVD) of GaN and extended prior work to growth of AlGaN alloys (Khan et al., 1983). In 1986, Amano et al. applied the buffer layer concept, which greatly improved the structural quality of GaN and AlGaN films (Amano et al., 1986) and demonstrated UV/visible p–n junction diode in 1993 (Akasaki et al., 1993).

In the past 15 years UV LEDs progressed rapidly and achieved level of performance comparable with some of visible LEDs. In Fig. 1, we plotted historic increase of the UV-C LED maximum external quantum efficiency (EQE). Difficulty in achieving high crystalline quality of AlGaN layers on sapphire substrates was related to high density of threading dislocations (TDs) and, thus, high nonradiative (NR) recombination rate in the device active region. Early advances in heteroepitaxy of AlN and AlGaN layers led to some suppression of NR recombination by reduction of density of TDs to below 10^{10} cm^{-2} and achieving reasonably good conductivity and contacts for both n- and p-type layers (Han et al., 1998; Kinoshita et al., 2000; Nishida and Kobayashi, 1999; Zhang et al., 2002). Follow-on work was focused on further improvements of AlGaN layer quality and light extraction from LED chip (Hirayama et al., 2014; Kneissl and Rass, 2016; Pernot et al., 2011; Shatalov et al., 2012, 2014) as discussed in the following chapter.

For comparison, we also included in Fig. 1 data for state-of-the-art visible LEDs based on InGaN and AlInGaP (Broell et al., 2014; Hurni et al., 2015; Narukawa et al., 2010; Saito et al., 2013). Maximum reported EQE of UV-C LEDs is close to that of devices operating in the "green gap." High

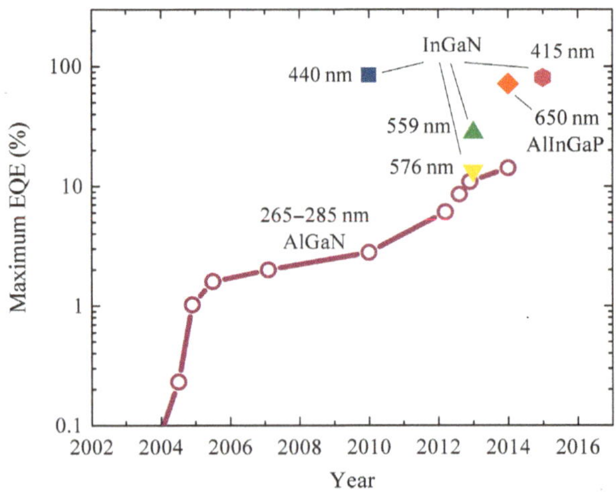

Fig. 1 Historic improvement of UV-C LED efficiency and current state-of-the-art visible LED efficiency.

lattice mismatch between ternary active layers and cladding layer and difficulty of composition control in ternary layers are common factors reducing the EQE outside of peak violet/blue and red regions, where composition of the active layer alloys is closer to binary materials.

UV LEDs are very versatile light sources with emission wavelength spanning from 240 to 360 nm (see Fig. 2). Early technology development was mainly driven by military adopters in biodetection and warning focusing primarily at 270 and 340 nm peak wavelengths for detection of NADH, tryptophan, and tyrosine. Change of the composition in AlGaN active region allows to tune emission from UV-A to UV-C range of the spectrum enabling LED commercial use in UV ink and polymer curing, medical spectroscopy, phototherapy, food growth and preservation, forensic analysis and DNA sequencing, disinfection of water and surface, and gas sensing. Efficiency of commercially available UV LEDs is typically below 5% but due to their good reliability (>10,000 h), small size, fast modulation speed, robustness, and compatibility with driving electronics LED light sources become competitive with compact low-pressure mercury vapor and amalgam bulbs and replace these light sources in a variety of sensing and spectroscopy devices.

Detailed review of UV LED applications was recently published by Kneissl and Rass (2016) and we, therefore, will focus on some aspects of LED performance and problems of material technology. To compete with conventional high power UV bulbs and to create new market opportunities, the LED technology still needs to achieve better performance in output power, operation lifetime, and, most of all, reduced cost. In-depth understanding of AlGaN fundamental material properties is paramount for

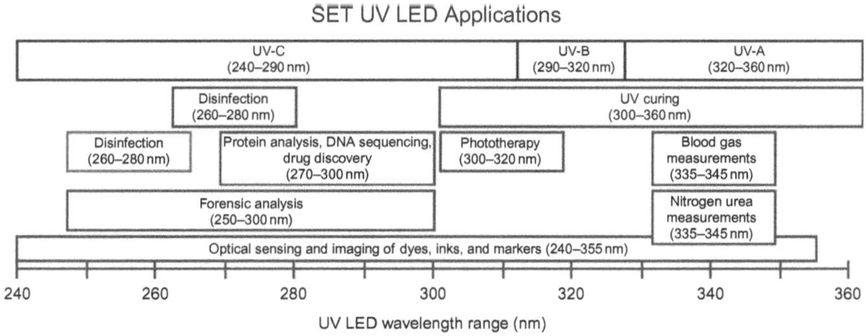

Fig. 2 Examples of UV LED applications enabled by their emission wavelength (applications @ www.s-et.com, n.d.).

improvement of epitaxial technology, new chemistry is invaluable for development of UV packaging materials, and improved manufacturability is essential to reduce cost and achieve widespread adoption of UV LEDs into broad range of consumer applications.

2. EPITAXIAL GROWTH OF AlN AND AlGaN ALLOYS

2.1 High-Temperature MOCVD Growth of AlN on Sapphire

In case of AlN homoepitaxy on bulk substrates, low dislocation density (10^5 cm^{-2} or below) determined by the substrate can be maintained low for epitaxial AlN layers. Thick $Al_xGa_{1-x}N$ layers can be pseudomorphically grown on AlN for $x > 0.7$, while layers with lower composition tend to relax due to strong lattice mismatch causing a significant increase of the threading dislocation density (TDD) to the level of 10^7–10^8 cm^{-2} (Grandusky et al., 2009). In III-nitride heteroepitaxy on foreign substrates treading dislocations introduced by lattice mismatch with the substrate play the most important role in determining device electrical and optical characteristics. Density of TDs in AlN can be significantly reduced from $>10^{10}$ cm^{-2} near the buffer layer to 10^8 cm^{-2} in thick AlN layers grown under optimized conditions.

Quality of the heteroepitaxial AlN films on sapphire strongly depends on growth method and nucleation conditions. Films deposited by MOCVD using low-temperature nucleation layers typically show mosaic structure with very high density of edge-component dislocations (above 5×10^9 cm^{-2}) and slightly lower density of screw-component dislocations (above 1×10^8 cm^{-2}) for a 0.5-μm-thick layer (Wu et al., 2004). Dislocation density, strain, and morphology of the AlN on sapphire are determined by the nucleation on substrate followed by coalescence of the islands, formation of TDs, and relaxation of strain. Multiple methods of AlN quality improvement have been proposed including pulsed precursor flow (Fareed et al., 2007; Hirayama et al., 2007; Khan et al., 1992), high-temperature (HT) growth (Imura et al., 2007b), and modifications of growth mode (Bai et al., 2006; Imura et al., 2007a) to reduce dislocation density and achieve smooth AlN surface morphology. Further reduction of TDD in AlN on sapphire to below 1×10^9 cm^{-2} was achieved by epitaxial lateral overgrowth (Chen et al., 2006; Imura et al., 2006; Jain et al., 2008, 2015). Pulsing of metal–organic precursors and ammonia (see Fig. 3) along with high (above 1200°C) growth temperature enables extreme range of control of vapor

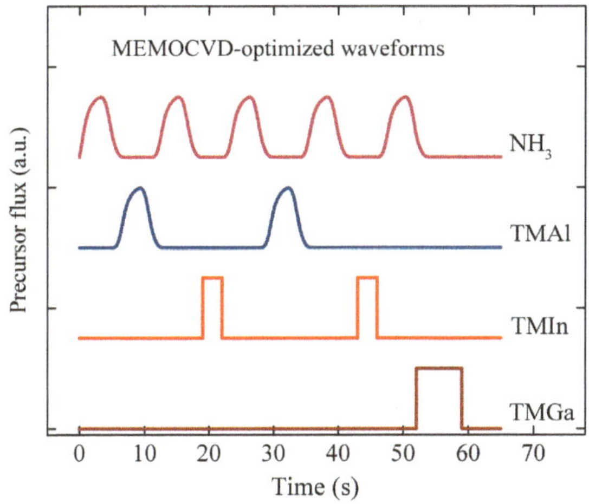

Fig. 3 Typical MEMOCVD precursor waveforms used for AlInGaN growth.

supersaturation (Bryan et al., 2016) and allows to achieve smooth AlN layers with sufficiently low dislocation density.

Migration-enhanced metal–organic chemical vapor deposition (MEMOCVD®) carried out at HT (>1200°C) and low pressure (LP) (<100 Torr) allows to control the epitaxial growth mode during heteroepitaxial growth of AlN on sapphire substrates. Conventionally deposited AlN films at temperatures below 1200°C typically grow in columnar growth mode due to relatively high vapor supersaturation, which leads to high TDD. MEMOCVD conditions drive the supersaturation lower via manipulation of V/III ratio to enhance surface diffusion length of Al adatoms. Duration and waveforms of precursor pulses can be partially overlapped to allow for a continuum of growth technique ranging from atomic layer epitaxy to conventional MOCVD. Such optimized waveforms combined with HT and LP conditions enabled us to increase the growth rates (>3 μm/h) while still maintaining step-flow growth mode.

Using HT-MEMOCVD we were able to achieve crack-free 5- to 10-μm-thick high-quality AlN on sapphire with TDD of 2×10^8 cm^{-2} for all types of dislocations. As shown in Fig. 4, the full width at half maximum (FWHM) of asymmetric (102) X-ray diffraction ω-scans reduces with the thickness of the AlN layer, which is consistent with reduction of the TDD. For example, for a 0.25-μm-thick MOCVD AlN on sapphire very narrow symmetric (002) ω-scan was obtained (FWHM of 11 arcsec), but FWHM of the asymmetric scan was 1070 arcsec. The corresponding TDD of $0.5–1 \times 10^{10}$ cm^{-2} was measured by cross-section transmission

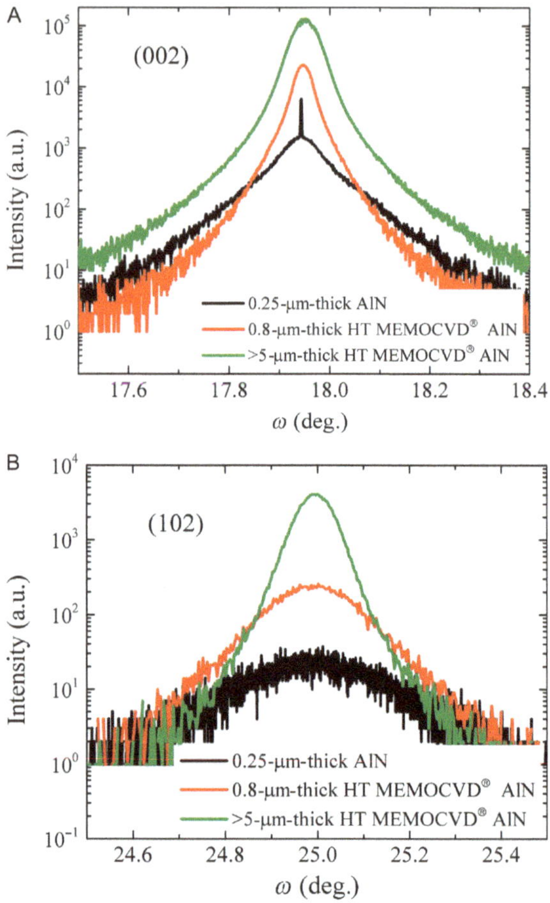

Fig. 4 Symmetric (002) (A) and asymmetric (102) (B) XRD ω-scans of different AlN on sapphire templates.

electron microscope. The FWHM of the asymmetric ω-scan reduces to 720 arcsec and to 300 arcsec for AlN layers grown by HT-MEMOCVD for 0.8 μm and >5 μm layer thickness, respectively. This trend is consistent with reduction of TDD for all types of dislocations from $1–5 \times 10^9$ to $2–3 \times 10^8$ cm^{-2}, which was shown to be sufficient for achieving improvements of the UV LED EQE (Shatalov et al., 2012).

Excellent smooth surface morphology of AlN on sapphire is achieved by HT-MEMOCVD with optimized conditions, as shown in Fig. 5. Atomic-force microscopy (AFM) image of 5×5 μm area shows root-mean-square (RMS) roughness of 0.8 Å confirming step-flow growth mode and high

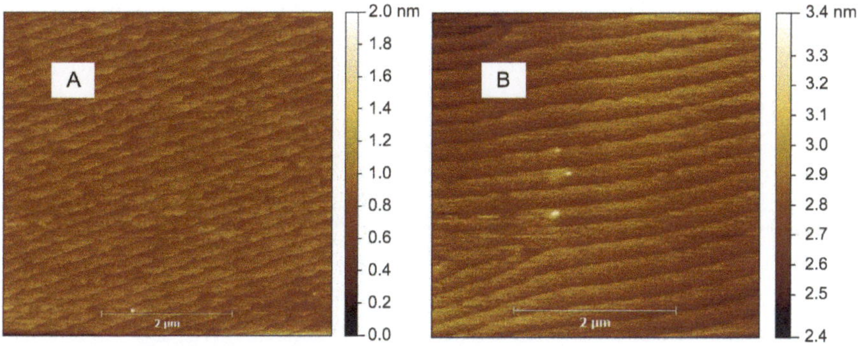

Fig. 5 AFM scan of thick AlN epilayer grown over (A) conventional and (B) patterned sapphire substrates.

surface quality similar to that achieved for homoepitaxial AlN on bulk substrates. AlN films were also deposited over patterned sapphire substrates (PSSs) in an effort to further reduce TDD ($<10^8$ cm^{-2}). Growth over commercially available PSSs typically leads to uncoalesced columnar domains with substantial tilts and twists. PSSs were custom designed with a variety of shapes and patterns to alleviate the coalescence issue. We achieved AlN on PSSs with total TDD of 2×10^8 cm^{-2} using the same HT-MOCVD technique (Jain et al., 2015; Shatalov et al., 2015), as shown in Fig. 5B.

2.2 MOCVD Growth of AlGaN Alloys

Relaxation of compressively strained AlGaN layers on partially strained heteroepitaxial AlN is determined by the AlGaN layer composition, its thickness, and stress of underlying AlN on sapphire. Inclination of the TDs was reported to be a primary relaxation mechanism (Follstaedt et al., 2005). Such relaxation was found to be composition dependent with larger inclination angles corresponding to a larger lattice mismatch (Follstaedt et al., 2009). Such inclination may promote annihilation of dislocations with opposite Burger's vectors in AlGaN layer, but dislocation bending at AlGaN/AlN interface may be accompanied by multiplication of TDs, which degrades AlGaN quality. As shown Fig. 6, stress management allows growing of fully strained $Al_{0.65}Ga_{0.45}N$ layers while maintaining low TDD ($\sim 2 \times 10^8$ cm^{-2}) through the n-AlGaN and multiple quantum well (MQW) active region, as measured by the (104) high-resolution X-ray reciprocal space mapping (RSM).

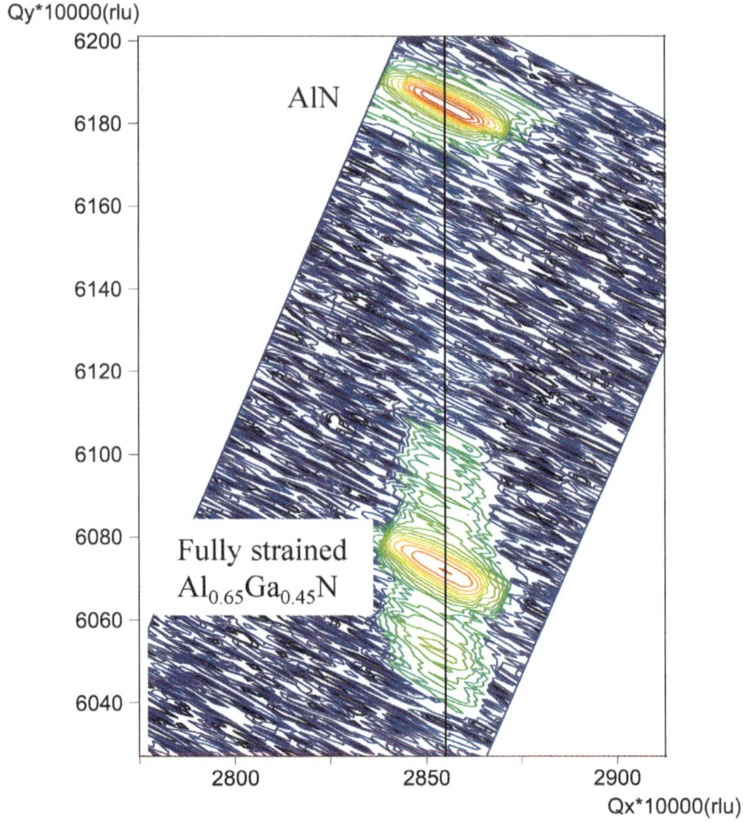

Fig. 6 (104) XRD RSM of fully strained 2-μm-thick n-$Al_{0.65}Ga_{0.35}N$ on >4-μm-thick AlN on sapphire.

The density of dislocations and strain of AlGaN layer on AlN can be controlled but with significant mismatch AlGaN morphology tends to exhibit three-dimensional island growth with significant surface roughness. As discussed later, this effect gives rise to substantial fluctuations of local AlGaN composition and needs to be suppressed by choosing growth conditions to achieve higher recombination rate of band-to-band radiative transitions. As seen from Fig. 7, the RMS roughness of $Al_{0.65}Ga_{0.45}N$ grown at different temperatures and growth rates for the 5×5 μm AFM scans can be reduced significantly from 10.1 to 3.1 Å by controlling growth temperature and gas phase V/III ratio in the range from 1050°C to 1300°C and from 300 to 7000, respectively.

The distribution of nonradiative centers (NRCs) and composition homogeneity can be effectively controlled by composition and AlGaN

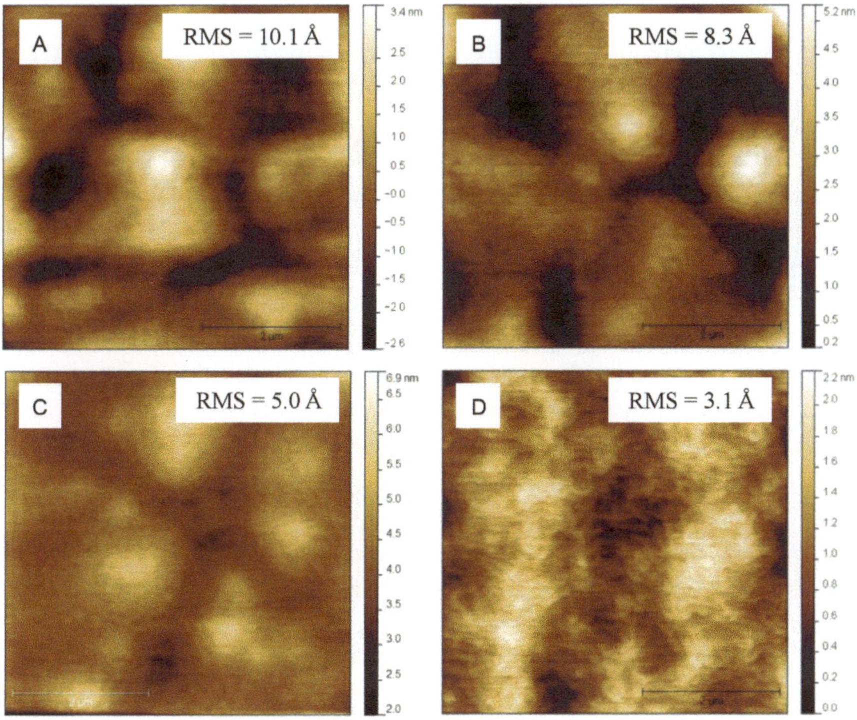

Fig. 7 A 5 × 5 μm AFM scan of $Al_{0.65}Ga_{0.35}N$ layers deposited over AlN template at following temperature and V/III ratio: (A) 1050°C and 3500, (B) 1050°C and 7000, (C) 1300°C and 900, and (D) 1300°C and 400.

growth mode. Thicker, lower composition and partially relaxed AlGaN layers tend to exhibit double-scaled potential fluctuations due to growth domains and random cation distribution (Kazlauskas et al., 2003, 2005). However, layers with higher Al content grown under optimized conditions show random distribution of NRC and PL linewidth primarily determined by homogeneous broadening (Marcinkevičius et al., 2014).

3. OPTICAL PROPERTIES OF AlGaN

As the structural quality of AlGaN alloys has improved and the device designs have evolved, the efficiency of UV LEDs has enhanced considerably (Hirayama et al., 2014; Kneissl and Rass, 2016; Pernot et al., 2011; Shatalov et al., 2012, 2014). Despite a high dislocation density in these alloys due to growth on foreign substrates, the high emission efficiency is ascribed to compositional or quantum well width fluctuations which lead to localization of carriers. The localization of carriers reduces the probability of the carriers to diffuse to centers of nonradiative recombination such as point and extended defects, leading to high emission efficiency (Cho et al., 1998; Collins et al., 2005; Eliseev et al., 1997; Kazlauskas et al., 2003; Kuokstis et al., 2006; Mickevičius et al., 2007). To gain a deeper insight into the overall carrier dynamics determining the efficiency of the system, a clear understanding of the various contributions from the free and localized carriers is essential. In this section, we use the data from time-resolved photoluminescence (TRPL) and light-induced transient grating (LITG) techniques in an attempt to better understand the carrier dynamics and the interplay of free and localized carriers in AlGaN alloys. LITG gives a direct estimation of the free carrier density decay, while TRPL contains contributions from both free and localized carriers. A comparison of the two thus helps reveal the exchange between these states and its importance in the carrier dynamics. We will also present the dependence of carrier lifetime and emission efficiency on temperature and carrier density. This further reveals interesting features about the different carrier dynamics processes that lead to the determination of the emission efficiency in these materials. The spectrally resolved PL characteristics will be discussed to describe the profile of potential fluctuations which explains the features of our results.

The results used here are obtained from two AlGaN epilayers with ~60% Al, grown using a combination of HT-MEMOCVD® and conventional MOCVD approaches resulting in different TD densities in underlying AlN templates and final AlGaN layers. The two samples will be described as LDD (low dislocation density: 2–5 × 10^8 cm^{-2}) and HDD (high dislocation density: 2–5 × 10^9 cm^{-2}). The expected bandgap of AlGaN with 60% Al content is 4.9 eV (Schubert, 2003). To optically excite these samples, the fifth harmonic (213 nm: 5.8 eV) of an Nd:YAG laser was employed. For both TRPL and LITG, a pulsed laser having a pulse width of 30 ps and a repetition rate of 10 Hz was used.

For the TRPL experiments, pulse energy densities from 3 to 300 μJ/cm^2 were used corresponding to the initial carrier densities between 3×10^{17} and 3×10^{19} cm^{-3} (assuming the absorption coefficient of 10^5 cm^{-1}). The highest pulse energy density was chosen such that it was approximately an order of magnitude smaller than the threshold for optical damage in AlGaN (Saxena et al., 2013).

The integrated PL obtained from the TRPL experiments at different excitation levels and different temperatures is shown in Fig. 8. The spectrally integrated PL decay kinetics for the corresponding excitation levels and temperatures are shown in Fig. 9. The LDD sample shows higher PL intensity

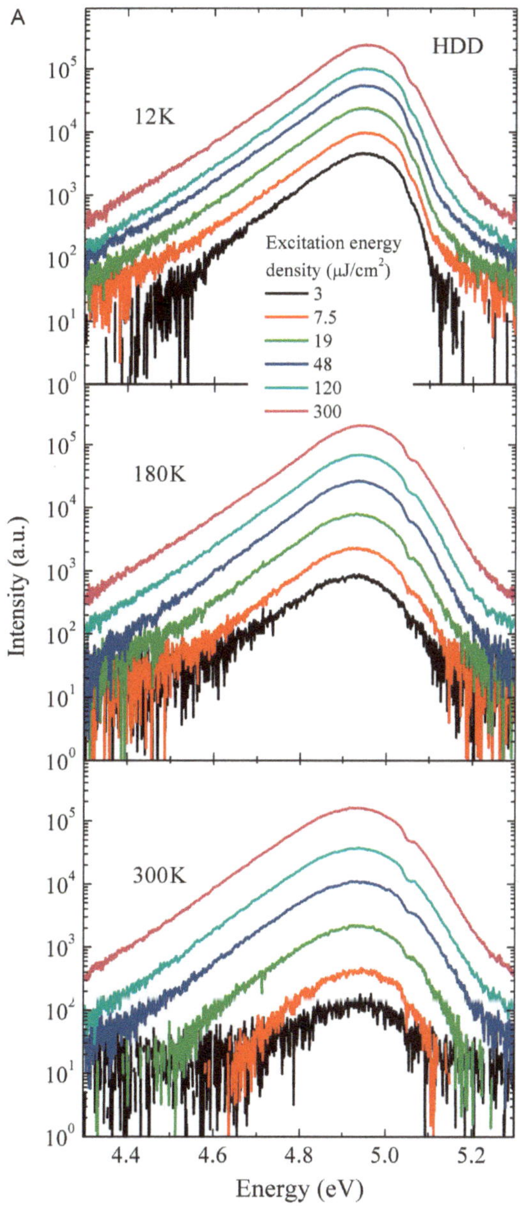

Fig. 8 Integrated spectra obtained from TRPL measurements for HDD (A) and LDD (B) samples at different excitations and temperatures.

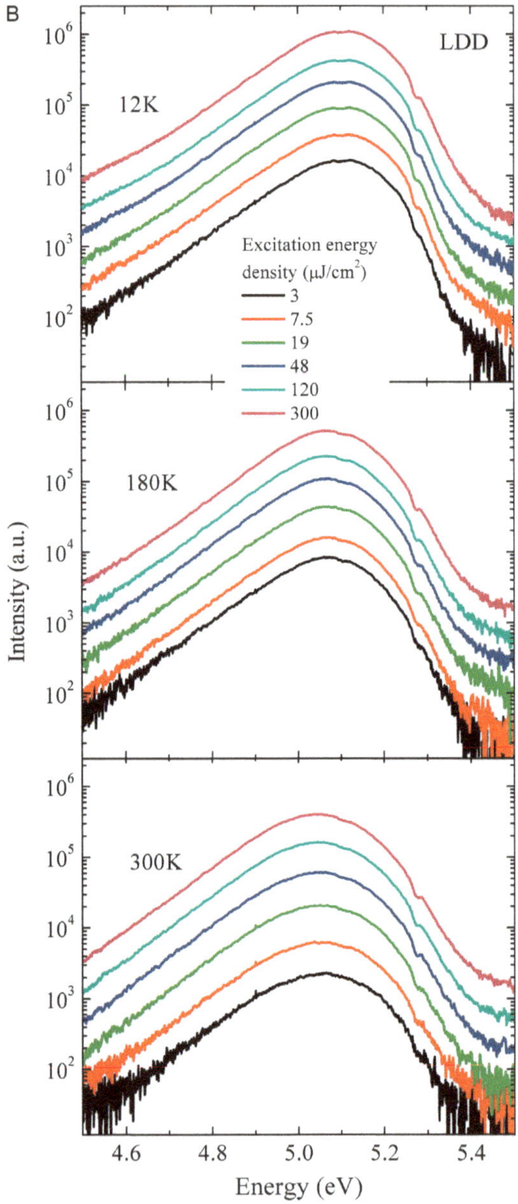
B
LDD
12K
180K
300K
Excitation energy
density (μJ/cm²)
3
7.5
19
48
120
300
Intensity (a.u.)
Energy (eV)
10^6
10^5
10^4
10^3
10^2
4.6
4.8
5.0
5.2
5.4

Fig. 8—Cont'd

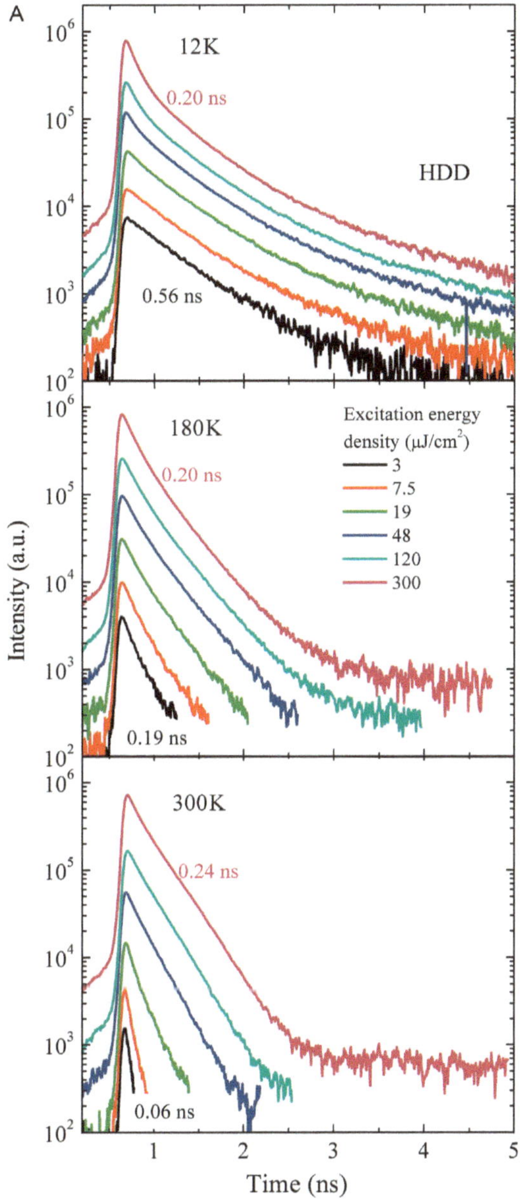

Fig. 9 PL decay kinetics for HDD (A) and LDD (B) samples at different excitations and temperatures. 1/*e* lifetimes for the highest and lowest excitation levels are indicated in each graph.

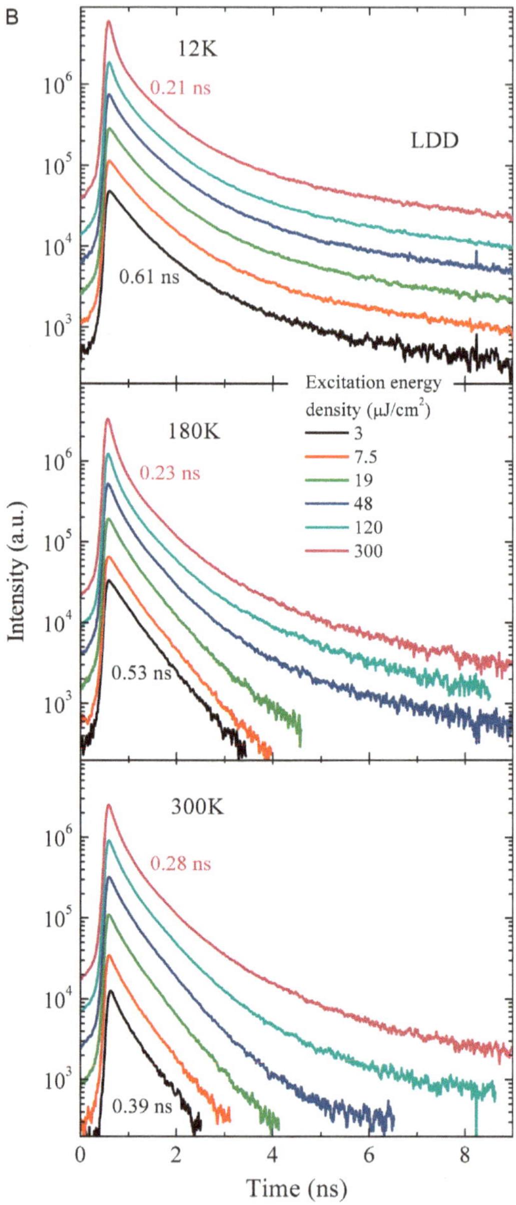
B
12K
0.21 ns
LDD
0.61 ns
180K
0.23 ns
0.53 ns
300K
0.28 ns
0.39 ns
Excitation energy density (μJ/cm²)
3
7.5
19
48
120
300
Intensity (a.u.)
Time (ns)

Fig. 9—Cont'd

than the HDD sample under similar conditions and also shows considerably longer PL decay transients. As the temperature is increased, the decay transients become shorter and the PL intensity also shows a reduction. The interplay between PL decay time and efficiency will be analyzed in more detail in the following subsections.

3.1 Carrier Density and PL Decay Kinetics

The carrier dynamics is described through results from TRPL spectroscopy and LITG measurements (Saxena et al., 2015b). In the LITG experiments, the interference pattern of two 30-ps-long laser pulses at the wavelength of 213 nm ($\hbar\omega = 5.8$ eV) was employed to induce a spatially modulated nonequilibrium carrier distribution leading to modulation of the refractive index according to the Drude–Lorentz model (Miller, 1999):

$$\Delta n(x) = \Delta N(x) n_{eh} = -\frac{e^2}{2n_0\omega^2\varepsilon_0}\Delta N(x)\left[\frac{1}{m_e} + \frac{1}{m_h}\right] \tag{1}$$

where ΔN is the density of the optically excited carriers, ω is the angular frequency of the probe light, n_0 is the refractive index of the unexcited material, ε_0 is the static dielectric permittivity, and n_{eh} is the change in the refractive index due to one electron–hole pair per unit volume.

The diffraction efficiency of the probe beam at 1064 nm, defined as the ratio of its diffracted and transmitted parts, is proportional to the square of the nonequilibrium carrier density integrated over the excited depth as: $\eta \propto \Delta n(t)^2 \propto \left[\int \Delta N(z, t)\mathrm{d}z\right]^2$ (Miller, 1999). The grating decay in these experiments was governed by the recombination lifetime and the grating decay time was proportional to the recombination lifetime.

Since the contribution to the refractive index modulation depends inversely on the effective mass of the carriers, usually the contribution from electrons is higher than the contribution from holes. Similarly, between free and localized carriers, the contribution from localized carriers is significantly lower on account of their higher effective mass. As a consequence, the diffraction efficiency transients from LITG measurements could be used to estimate the "active" carrier density, responsible for refractive index modulation. This carrier density can be extracted from the diffraction efficiency transient by taking its square root. PL intensity, on the other hand, depends on the different mechanisms of radiative recombination and their relative strength: bimolecular radiative recombination of free carriers and unimolecular radiative recombination of localized carriers. The comparison

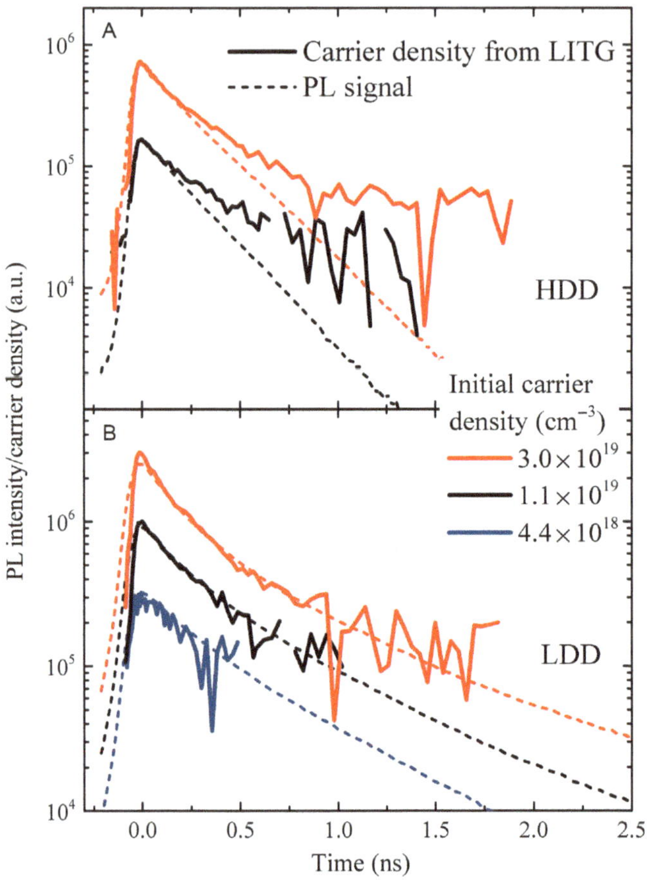

Fig. 10 Decay of PL intensity after short pulse excitation (*dashed lines*) and the carrier density obtained from LITG experiments (*solid lines*) at similar initial carrier densities (indicated) for the HDD (A) and LDD (B) samples.

of the estimated carrier density decay from LITG and the TRPL decay is shown in Fig. 10.

During the initial stages of the decay, the contribution to LITG efficiency is expected to be dominated by the free carriers. However, the PL from free carriers is bimolecular in nature and should decay faster than the free carrier density. As shown Fig. 10, at the beginning of the decay the carrier density estimated from LITG efficiency and the TRPL proceed at the same rate. This is an unexpected result if the PL signal was assumed to have contribution solely from the free carriers. To explain the observation, the influence of an increasing contribution of the PL from localized carriers was proposed (Saxena et al., 2015a,b). The free carriers first relax and populate the localized states and then the localized carriers contribute to the PL signal. Under the appropriate conditions, the PL signal may not decay faster than the free carrier density. At lower carrier density the PL decay starts to be dominated by the localized carriers and should proceed at a rate determined by the overall decay rate of the localized carriers. The slow long tail observed at later stages is likely due to a high density of trapped carriers, which may not contribute to the PL, but may contribute to LITG efficiency after a majority of free carriers have decayed.

Hence, comparison of the LITG efficiency and TRPL decay transients reveals that the localized carriers start contributing to the PL even at the beginning of the transient and effectively lead to the slowing down of the PL decay with respect to the decay of the free carrier density.

3.2 PL Efficiency and Lifetime

The behavior of the luminescence efficiency and PL lifetimes of these AlGaN samples as a function of excitation and temperature underscores the different mechanisms of carrier dynamics that are of utmost importance in gaining a better understanding of these material systems. The PL efficiency usually shows a drop at higher excitation levels. This feature in the efficiency is called efficiency droop and is most commonly reported for the electroluminescence efficiency (Cho et al., 2013; Iveland et al., 2013; Kim et al., 2007; Mickevicius et al., 2013; Piprek, 2010; Saguatti et al., 2012; Tamulaitis et al., 2012). The droop at higher nonequilibrium carrier densities is usually attributed to Auger recombination term in light of the popular "ABC model" (Cho et al., 2013; Delaney et al., 2009; Iveland et al., 2013; Karpov, 2014; Kioupakis et al., 2011, 2013). However, as will be discussed later in this section, this model is insufficient to explain the different observations made from these samples.

The PL decay transients at different excitation energy densities and different temperatures for the two samples are shown in Fig. 9. As can be observed, most of the PL decay transients are strongly nonexponential in nature and, hence, $1/e$ lifetime is taken as the indicator of the PL lifetime. The PL intensity reduces to $1/e$ of its maximum in this time period. The PL efficiency is defined as the ratio of overall integrated PL intensity to the excitation energy density. It is reasonable to expect that most of the PL efficiency arises from the $1/e$ lifetime period. In this section, the relationship between the $1/e$ lifetime and the overall PL efficiency as a function of temperature and excitation energy density would be explained. For brevity, we choose the extreme conditions for temperature, 12K and 300K, while all the intermediate temperatures follow trends based on the discussion of these extreme temperature conditions. Fig. 11 shows the $1/e$ lifetime and PL efficiency for the HDD and LDD samples at 12K and 300K.

At 12K, for both samples, both the $1/e$ lifetime and the PL efficiency show a decrease as the excitation level (or the initial nonequilibrium carrier density) increases. A reduction of lifetime at a higher carrier density is expected if the carriers decay through pathways that have higher order dependence on the carrier density (bimolecular, Auger, etc.). Although the lifetime is not solely determined by the free carriers, as was shown earlier, it is reasonable to expect that the increase in the free carrier concentration is determining the trend of lifetime with excitation level. This is corroborated by observing that the later part of the decay curves at 12K in Fig. 9, which is expected to be determined by the localized carriers, is virtually independent of the excitation level (the decay curves become almost parallel). Thus, the physical mechanisms behind the localized carrier decay are not expected to change significantly with the excitation level. This suggests that the free carriers should be the major contributors in determining the trend in efficiency and lifetime. If the bimolecular recombination enhancement of free carriers

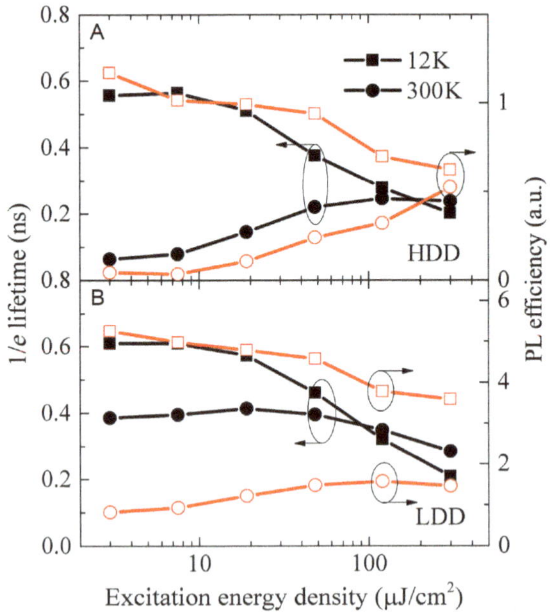

Fig. 11 PL lifetimes (*black*) and efficiency (*red*) as a function of excitation energy density at 12K and 300K for (A) HDD and (B) LDD samples.

was responsible for the decrease in the lifetime with excitation level, it should have simultaneously led to an increase in the PL efficiency, in the light of the simple ABC model. Since this is contrary to the observation, the dominating higher order processes must be NR in nature. Auger recombination may be able to explain these observations; however, Auger recombination loses strength as the temperature is reduced (Hader et al., 2011). Moreover, Auger recombination coefficient for wide bandgap AlGaN is believed to be much smaller than InGaN whose Auger coefficients are already very small $<10^{-30}\ \mathrm{cm}^{-6}/\mathrm{s}$ (Abakumov et al., 1991; Schubert, 2003). Another process causing efficiency reduction could be stimulated emission which might not be collected very efficiently in the front surface configuration and hence appears to have caused the drop in efficiency (Mickevičius et al., 2012a). However, stimulated emission is a fast process depleting the carriers in few tens of picoseconds (Aleksiejūnas et al., 2013) and is expected to have some signature in the PL decay transients. Contribution of the stimulated emission is very unlikely to explain the observed PL efficiency behavior since the transient processes observed are very slow. This leaves density-dependent nonradiative processes, such as the density-activated defect recombination (DADR), first proposed by Hader et al. (2010, 2011) as the possible contenders. The DADR mechanism is temperature independent as it does not need phonon absorption during the nonradiative recombination process and hence can be active at low temperatures as well, leading to shorter lifetime and decreased efficiency, simultaneously. Aleksiejūnas et al. (2013) have also reported similar results on InGaN quantum wells.

At 300K, the dependence of $1/e$ lifetime and efficiency on excitation level is no longer monotonic and is different for the two samples, as shown in Fig. 11. For the HDD AlGaN sample, an increase in both the efficiency and lifetime is simultaneously observed. While the increase in efficiency with increased excitation can be explained by considering the bimolecular recombination of free carriers, the increase in lifetime cannot. The possible mechanism responsible for showing an increase in lifetime and efficiency is the quenching of nonradiative centers as the excitation is increased. At low temperatures, these centers may already have quenched due to temperature and may not be affected by increasing excitation. The quenching effect is expected to shift to lower excitation levels for the LDD sample as the LDD sample is inherently expected to have a lower density of nonradiative centers, which would quench at lower excitations. The lifetime for the LDD sample shows a small increase followed by a decrease as the excitation is increased. The efficiency shows an initial increase and tends to saturate at higher levels. This can be explained by the higher order nonradiative mechanism which is expected to be temperature independent. Compensation of bimolecular radiative enhancement and DADR-induced drop in efficiency may lead to efficiency being less sensitive to excitation level, while the $1/e$ PL lifetime continues to reduce.

Another interesting feature to be noted is that the absolute PL efficiency for the two samples is different, even at 12K. The HDD sample has lower efficiency and lower peak PL emission intensity than the LDD sample. This difference cannot be explained only in terms of the extraction of light from the sample. Therefore, it is reasonable to expect that the sample quality is related to the efficiency and nonradiative recombination is still present at low temperatures. But the Shockley–Read–Hall-type recombination is expected to quench as discussed earlier; hence the only plausible and consistent mechanism is a DADR-like mechanism whose strength depends on the sample quality, but is independent of temperature.

3.3 Spectral Dependence of PL and Carrier Decay

As was discussed earlier, the initial part of the integrated PL decay is contributed by free and localized carriers, while the later slower part of the PL decay is contributed mainly by the localized carriers. The initial fast decay and the later slower decay are observed across the PL spectrum, especially at lower temperatures. For the discussion in this section, results from the HDD sample obtained at the highest excitation level of 300 $\mu J/cm^2$ at three different temperatures are used. The PL decay kinetics at different spectral locations at the different temperatures are shown in Fig. 12. The decay kinetics are described by two characteristic lifetimes, the initial fast decay time and the intermediate decay time. Fig. 13 shows the extracted fast and intermediate PL decay lifetimes as a function of spectral position at different temperatures.

As seen from Fig. 13, the fast decay lifetime has very weak dependence on the spectral position and does not change significantly with temperature. The intermediate decay lifetime has a strong dependence on the spectral position at low temperatures and this dependence weakens as the temperature increases. Moreover, with an increase in temperature, the intermediate decay time tends to reduce.

The intermediate lifetime is almost solely governed by the dynamics of localized carriers. The intermediate lifetime decreases for the either side of

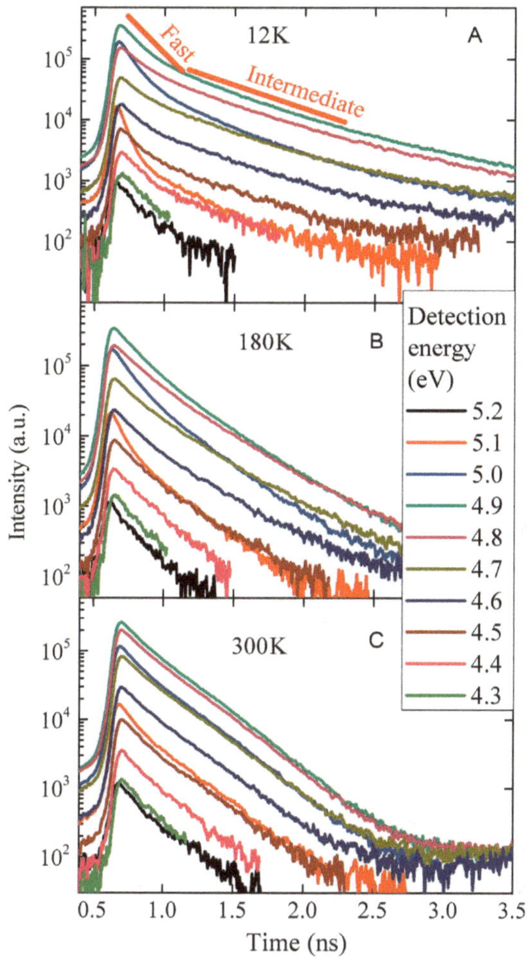

Fig. 12 PL decay kinetics at different spectral positions at 12K (A), 180K (B), and 300K (C).

the PL spectrum and this behavior is the strongest for 12K. There have been many reports showing that the PL decay lifetime increases and becomes constant with decreasing detection energy (Badcock et al., 2014; Morel et al., 2003; Murotani et al., 2008, 2011; Narukawa et al., 1997, 1999; Onuma et al., 2004; Satake et al., 1998). This is ascribed to migration of carriers from higher energy localized states to lower energy localized states within their lifetime through acoustic phonon-assisted hopping (Cohen and Sturge, 1982). However, like a few other reports (Chichibu et al., 2003; Graham et al., 2005; Narukawa et al., 1999; Satake et al., 1998), a decrease in the lifetime at the lower energy end of the PL spectrum is observed for these

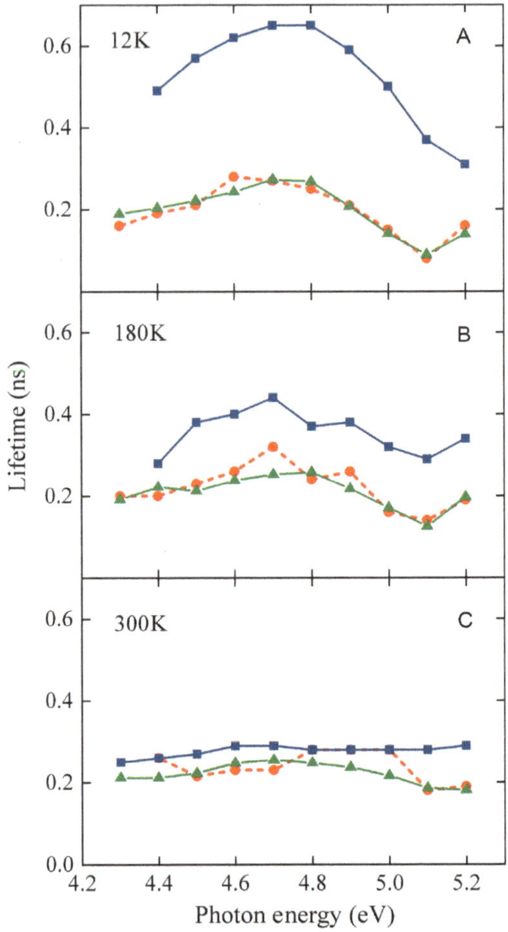

Fig. 13 Dependence of time constants of fast (*red dash line*) and intermediate (*blue solid line*) PL decay components on photon energy at different temperatures: 12K (A), 180K (B), and 300K (C). 1/*e* PL lifetimes (*green solid line*) are also included for comparison.

samples and the model of energy loss hopping cannot be used to explain this feature. To explain the observed dependence of PL decay lifetime on spectral position in the PL band a model of the double-scaled potential profile was proposed (Kazlauskas et al., 2005; Mickevičius et al., 2012a,b) and refined recently (Saxena et al., 2015a,b), as shown schematically in Fig. 14.

The carriers excited deep into the bands by pulsed optical excitation can quickly thermalize and reach the corresponding position on this potential profile. This should be followed by the decay of these free carriers and their

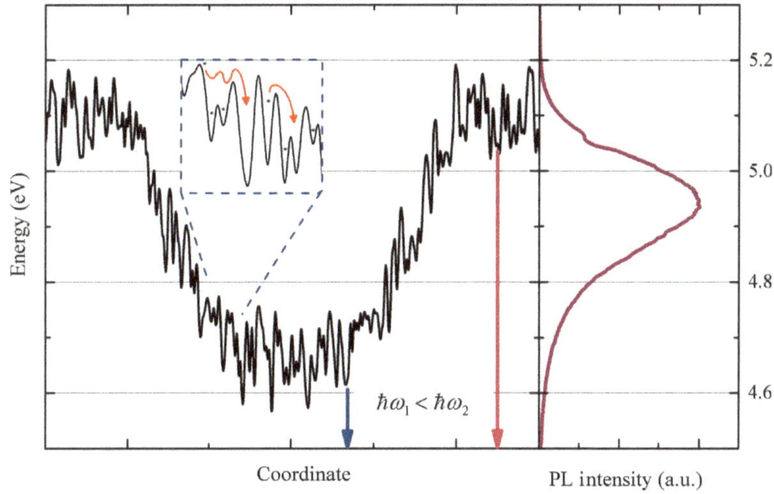

Fig. 14 Schematic of energy potential fluctuations in real space and PL spectrum.

relaxation to the localized states. Once the contribution from the free carriers is over, the intermediate lifetime is solely governed by the decay of the localized carriers at any given energy. The localized carriers can be replenished by the carriers from higher energy states if the higher energy carriers have enough time to hop/relax to the lower energy states. This is quite possible at low temperatures where the lifetimes of localized carriers are expected to be long. While it could be assumed that the radiative decay of localized excitons does not depend on the location of the carrier on the energy slope, the processes leading to replenishment of these carriers do depend on the location.

As described by the Miller–Abrahams expression, the probability of hopping or the hopping rate between localized states decreases with distance between the states (Miller and Abrahams, 1960):

$$\nu_{i\to j} = \nu_0 \exp\left(-\frac{2r_{ij}}{\alpha} - \frac{E_j - E_i + |E_j - E_i|}{2kT}\right), \tag{2}$$

where ν_0 is the attempt to escape frequency, α is the decay length of exciton wave function, E_i and E_j are energies of the initial and final states, and r_{ij} is the spatial separation between the initial and final states.

In the model depicted in Fig. 14, due to a smaller slope in the energy profile, the distance between localized states becomes larger at the lower part of the low-energy region and hence the probability of hopping to the bottom of the low-energy region decreases. This may lead to accumulation of carriers on the slope. Hence, the effective lifetime is lower. Thus, the central part of the band which corresponds to the peak of the PL sees an efficient replenishment from the higher energy carriers and corresponds to the longest lifetimes. As the temperature rises, the nonradiative rates begin to increase and can lead to a reduced replenishment of carriers in the middle of the band and hence the lifetime spectrum becomes flatter.

The fast initial decay is present across the spectrum with a weak dependence on the spectral position. The presence of this fast component is feasible according to the model described here. The initial part of the PL decay is governed by the free carrier dynamics at any spectral position and the compensation by carriers localized at the same energy (but a different location in real space), as discussed earlier. In the light of this model, the free carriers are expected to be created and their dynamics are expected to be similar at all spectral positions. Any dependence of the fast PL lifetime on the spectral position, thus, should bear the signature of compensation through the localized carriers. As a result, the spectral dependence of the fast decay component bears a resemblance to that of the intermediate component arising solely from the localized carriers. However, this dependence is weaker than that of the intermediate component.

The insensitivity of the fast decay lifetime to temperature could be explained by considering the different processes which tend to compensate each other; for example, with rising temperature, the bimolecular radiative rate decreases, while the nonradiative rate increases, thus affecting the lifetime in opposite ways.

4. UV LED DEVICE DESIGN AND PERFORMANCE

4.1 Efficiency of UV LED Devices

As discussed at the beginning of this chapter, the maximum EQE of current state-of-the-art UV-B and UV-C LEDs does not exceed 15%. In Fig. 15, we compare the components contributing to the peak EQE for state-of-the-art AlGaN UV-C (Shatalov et al., 2014) and InGaN/GaN violet (Hurni et al., 2015) LEDs.

At 20 mA drive current the electrical efficiency (EE) of UV LEDs is about twice as low as that of visible violet LED due to significant Ohmic losses in the semitransparent p-layers of the LED structure and across contacts to n-AlGaN layer, whereas at low current density the $EE > 1$ was demonstrated for 415 nm emission violet LED devices. Higher EE can be achieved for UV-C devices by optimization of doping conditions for high

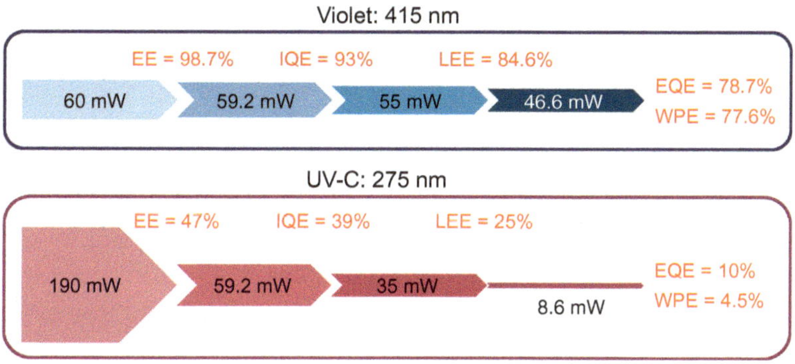

Fig. 15 Efficiency of state-of-the-art visible (violet) and UV LEDs.

Al-content AlGaN layers and contact formation. The internal quantum efficiency (IQE) is substantially lower for the heteroepitaxially grown UV-C LED structure due to significantly higher density of dislocations. At high density of NRC lower AlGaN alloy homogeneity promotes radiative transitions through recombination of localized carriers. However, at low NRC density (for AlGaN alloys grown coherently on bulk substrates under close to stoichiometric conditions) enhanced alloy disorder may cause reduction of the effective radiative recombination rate reducing IQE. For devices grown on bulk substrates (for example, laser structures), higher homogeneity of AlGaN composition is beneficial.

Furthermore, the IQE of LED is affected by high stress in the semiconductor structure, the presence of polarization field that is dependent on composition of semiconductor layers. In UV-C LED the electrical (hole) injection into the active layer is also significantly lower due to lower free hole concentration and low diffusion coefficient in Al-rich AlGaN blocking and contact layers. However, the major factor effecting EQE is the substantial single-pass light absorption inside LED structure that is currently significantly higher in UV-C structures with fully absorbing p-GaN or semitransparent p-AlGaN or SPSL (short-period superlattice) layers than in nearly fully transparent InGaN violet (David et al., 2014) and blue/green LEDs. Visible LEDs utilize p-type reflective contacts such as Ni/Ag contacts (Horng et al., 2014), Ag/Zn/Ag (Yum et al., 2012), Ru/Ni/ITO, or Ni/Ag/Ru/Ni/Au (Jang et al., 2004). The reflectance of such contacts in the visible spectral range exceeds 90%. However, in UV range, especially in UV-C, the reflectivity of most commonly used metals is relatively low. Ni/Al-based contacts (Hirayama et al., 2014) offer >90% reflectance in the UV range but may exhibit stability problems at high current density, whereas Ni/Rh or Ni/Rh/Al contacts (Gaska et al., 2014; Lunev et al., 2015) were much more stable at high current but their reflectance is typically 70–80% in UV-C range. Furthermore, visible LEDs can be readily encapsulated resulting in significant gains in the light extraction (David et al., 2014), whereas UV stable and transparent encapsulation for UV LEDs, and in particular for UV-C LEDs, remains to be a challenge. While all these issues prevented efficient light extraction from UV LEDs, in recent years, there has been a significant progress in understanding and improving light extraction from UV LEDs.

4.2 UV LED Chip Design

As discussed before, maximum of the EQE is typically achieved at low current density of approximately 10 A/cm^2. Thus, majority of low-power (<5 mW CW) UV LED chips are designed to have small junction area to achieve maximum efficiency at 20 mA. For many applications, especially those related to disinfection, higher output powers are required, thereby leading to operation at higher currents and likely at higher current densities. In addition to nonthermal reduction of the EQE at higher current densities (droop), the heat dissipation becomes an important issue for UV LED devices operating at high input power levels since the operating voltage is approximately factor of 2 higher for UV devices than for their visible counterparts. Thermal management requires the use of large chip geometries that inevitably causes reduction of light extraction efficiency (LEE) in flip-chip devices. The challenges in improving efficiency of UV LEDs are schematically illustrated in Fig. 16.

Fig. 16 shows the semiconductor layers of a typical UV LED epitaxially grown on a substrate. For majority of commercial UV LEDs sapphire is frequently used as a UV transparent substrate, mostly due to its low cost and very good UV transparency. Devices fabricated on sapphire substrate are typically flip chip mounted to submount or circuit board. This approach allows to extract light through the substrate, while improving heat transfer from the n-type and p-type contacts toward the circuit board. Fig. 16 shows that performance of the device can be improved by addressing four major challenges: improvement of light extraction from the device, improvement of IQE through advances in epitaxial growth and bandgap engineering, reduction of electrical losses, and improvements of heat dissipation from the chip and package.

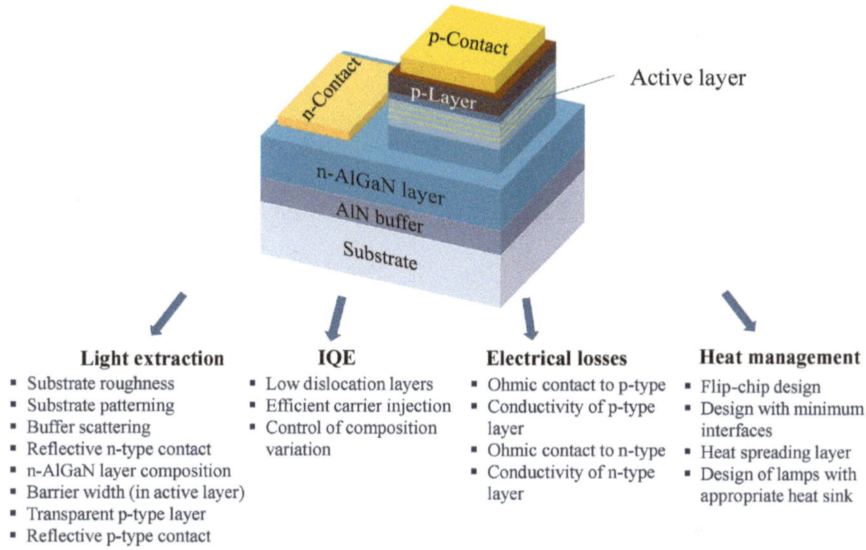

Fig. 16 Factors affecting LED efficiency.

Improvement of LEE is one of the key goals for widespread adoption of UV LED technology for a multitude of applications. Also, design and fabrication of devices operating at high current density with extended lifetime is equally important aspect of UV LED performance. The improvements in reliability can be addressed by maturing of the epitaxial and device fabrication technology. While detailed description of various failure mechanisms is outside the scope of this chapter, we believe it is important to recognize that this is an increasingly important topic in UV LED device research.

4.3 Improvements in Light Extraction from UV LED Devices

One of the key improvements in light extraction in UV LED is reduction of the single-pass photon absorption in the LED device. Several approaches have been proposed to reduce internal absorption in LED structures with p-GaN contact layer, such as reduction of p-GaN layer thickness (Akiba et al., 2012) or partial removal of p-GaN layer (Inazu et al., 2011). In both approaches, conventional AlGaN:Mg was used as electron-blocking layer giving rise to substantial forward voltage increase and reduction of device reliability. The AlGaN/GaN (Goepfert et al., 2000; Kozodoy et al., 1999), AlN/AlGa(In)N (Kipshidze et al., 2003), and AlN/AlGaN (Allerman et al., 2010) SPSL structures were proposed to improve ionization

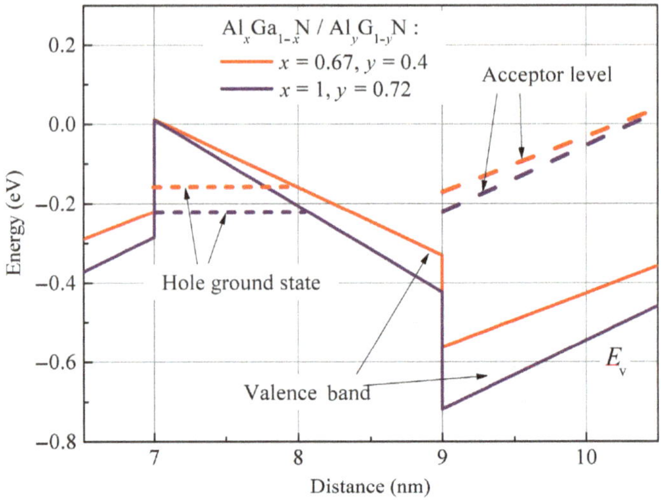

Fig. 17 The simulated band diagram of $Al_xGa_{1-x}N/Al_yGa_{1-y}N$ SPSLs. Acceptor level in SPSL barriers and hole ground state in wells are schematically shown.

of Mg acceptors and increase free hole concentration in p-type layers with high average Al content. Major drawback of SPSL structure is that the improvement of conductivity was significant in lateral direction (in plane of SPSL), whereas vertical conductivity across superlattice Al-rich layers was still too high as compared to p-GaN. Use of $Al_xGa_{1-x}N/Al_yGa_{1-y}N$ SPSL (Shatalov et al., 2012) allows to tune the AlN molar ratios x and y in well and barrier to provide the best transparency while still maintaining appropriate levels of vertical conductivity (see Fig. 17).

As seen from Fig. 17, the changes in molar values of quantum wells and barriers affect both the hole quantized energy levels in quantum wells and the position of acceptor energy level. In general, selecting a composition of the SPSL that allows alignment of hole ground state in quantum well and acceptor level in quantum barrier can lead to improved ionization and increased p-type conductivity (Shatalov et al., 2013). Taking into account pronounced fluctuations of the alloy composition in ternary AlGaN layers the SPSL structure may be modulated in lateral direction by the fluctuations of composition providing additional path for increased activation of acceptors, enhanced vertical conductivity, and high average transparency (Shur et al., 2014), as schematically shown in Fig. 18. Similar concept of 3D SPSL structures with lateral conductivity of 0.7 Ω cm at room temperature was discussed recently (Zheng et al., 2016). Single-pass normal

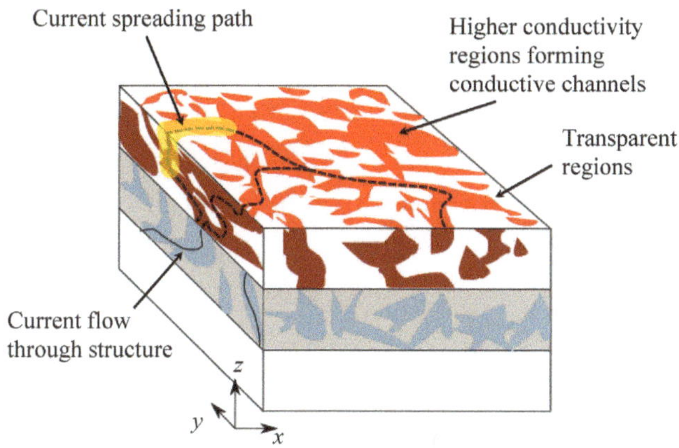

Fig. 18 Schematic representation of current flow in AlGaN layers with fluctuations of alloy composition.

incidence transmission was measured to increase from 7% for the LED heterostructure with approximately 0.2-μm-thick p-type GaN layer to 70% for the LED structure with SPSL (Shatalov et al., 2014). Corresponding effective absorption coefficient was estimated to reduce from 10,000 to below 500 cm^{-1}, which is still too high to enable the use of thick p-layer for volumetric light extraction but leads to significant improvement in LEE.

Utilizing transparent SPSL is necessary, but not sufficient for improving light extraction for UV LEDs. Fig. 19 demonstrates major problems for light extraction from UV LED devices. First the light is lost at the substrate-to-ambient interface due to light trapping from total internal reflection (TIR) and Fresnel reflection. Typical substrate can be sapphire which has an index or refraction of about 1.82 for wavelength of 280 nm resulting in TIR at angles of larger than 33 degrees to surface normal. As shown in Fig. 19, only the light directed within the set of escape cones can exit the substrate, while the remaining of light is reflected back into the structure and largely absorbed by metallic contacts or semitransparent p-type semiconductor layer. Based on geometrical ray-tracing simulations, only about half of all the UV light entering substrate structure can exit the substrate structure to the ambient. It has to be understood that these numbers are an approximation and largely depend on all elements of the device, i.e., wavelength of analysis, type and absorption of the contacts, and absorption of all of the semiconductor layers.

Fig. 19 illustrates that sapphire-ambient interface (here referred to as first TIR interface) can lead to significant light trapping. Due to changes in the

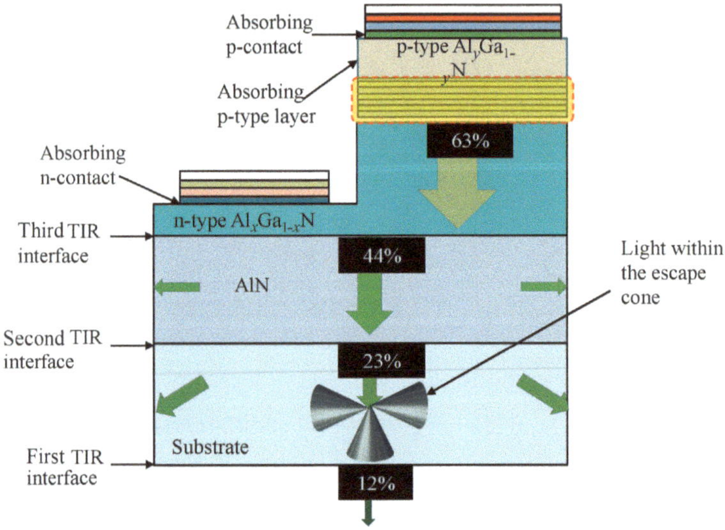

Fig. 19 Schematic of light loss at interfaces and in the layers of the LED structure.

index of refraction between sapphire, AlN, and $Al_xGa_{1-x}N$ layers, more TIR interfaces lead to additional light trapping. Furthermore, absorption in p-type and n-type metal contact and residual absorption of the p-type layer result in additional reduction of LEE. It needs to be understood that addressing one issue within the device (say improving transparency of the p-type layer) does not result in significant gains of the extraction efficiency, as the light will still be absorbed at the contacts. Only improving light extraction at all stages will lead to a gain in extraction efficiency. Fig. 20 shows the simulated LEE based on ray tracing for an LED device with the heterostructure shown in Fig. 19. The improvement in transmission of the p-type layer alone does not lead to significant gains in efficiency unless it is combined with high reflectivity p-contact metal. Similarly, increase of the LEE due to improved reflective p-contact is not significant if absorption is too high. Only when combining both of the approaches, LEE can be improved by a factor of 2 or higher.

Semiconductor heterostructure with low optical losses enables multiple pass light extraction in flip-chip geometry when p-type metal contact has high reflectivity in the 250–350 nm spectral range. Metal contacts commonly used for visible LEDs, such as Ag, are highly absorbing in the UV range. Aluminum has high reflectivity in the UV range, but it does not form an ohmic contact with p-type semiconductor layer. The suitable contact

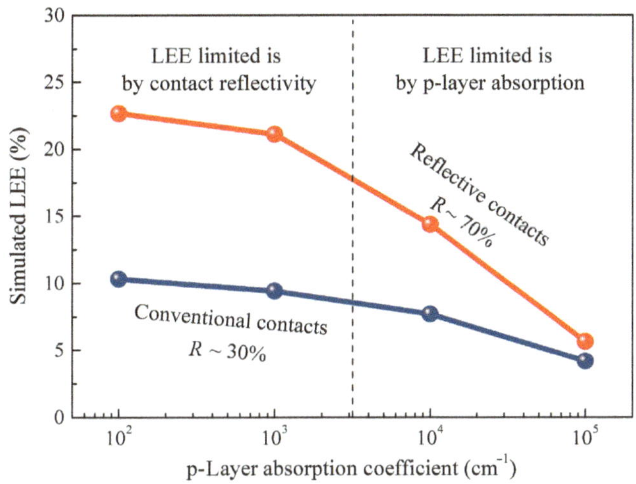

Fig. 20 Simulated LEE for reflective and conventional p-metal as a function of p-layer absorption.

needs to provide low voltage drop and have good adhesion to the semiconductor surface. Contacts comprising a multistack of metallic layers can provide an ohmic contact with satisfactory reflectivity (Gaska et al., 2014; Lunev et al., 2015). The metal stack is formed from the first thin layer being an ohmic layer, such as Ni for example, with the thickness of the layer chosen to be sufficiently transparent to UV light. The layer is then followed by protective diffusion barrier layer that is also sufficiently thin and does not significantly absorb UV light. Finally, the stack is followed by the reflective contact layer, where aluminum can be used. As shown in Fig. 21, the normal incidence reflectivity spectrum of the multilayer metal stack with thick ohmic interlayer deposited on sapphire can reach 70% in the UV-C region.

As seen from ray-tracing simulations, a multistage approach in improving light extraction from a UV LED utilizing surface roughness, improvements of p-type layer transparency, reflective ohmic contacts, and use of encapsulation may increase LEE of UV-C LEDs from typical 7–8% to level of 25% and above. This approach combined with optimization of the MQW active region design and reduction of NRC density resulted in record high power and efficiency of UV-C devices, as shown in Fig. 22.

Large chip UV-C LEDs show a peak EQE of 8%, which is below that of small chip design and limited by residual absorption in the large chip. At drive current of 300 mA, large chip UV-C LEDs are capable of delivering as much as 100 mW per single die enabling UV sources with power density similar or surpassing traditional UV light sources.

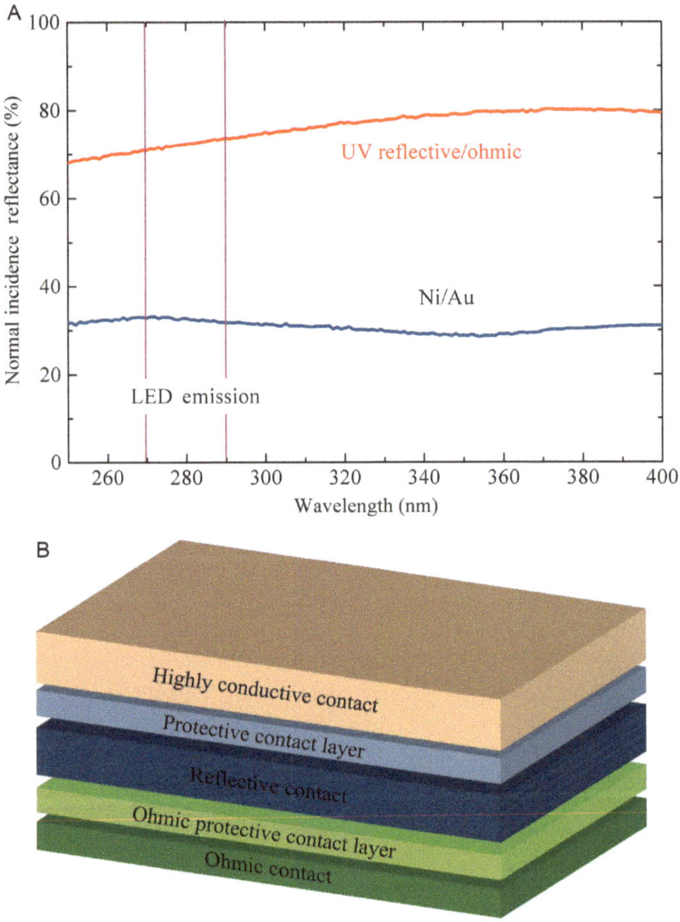

Fig. 21 (A) Reflectivity of the complex UV-reflective/ohmic metal contact; (B) schematics of the contact layer stack.

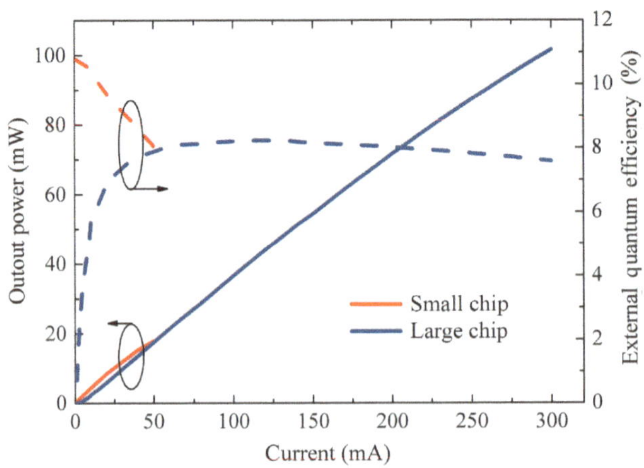

Fig. 22 L–I (*solid lines*) and EQE vs current (*dashed lines*) of small and large chip state-of-the-art UV-C LED.

5. CONCLUSIONS

To conclude, we discussed the development of AlN and AlGaN layer growth on sapphire substrates and importance of dislocation density reduction on the efficiency of deep ultraviolet (DUV) LEDs. It is shown that TDD of 10^8 cm^{-2} is sufficiently low to achieve the EQE above 10% for LEDs emitting at 275–280 nm.

The importance of localized carriers in the overall PL of high aluminum content AlGaN epitaxial layers was underscored through comparison of LITG and TRPL decay kinetics. The temperature and excitation dependence of efficiency and PL decay lifetime revealed different features including existence of nonradiative recombination of higher order at low temperatures and saturation of nonradiative centers at HTs. It was also observed that the structural quality of the samples determines the low-temperature efficiency and the IQE cannot be assumed to be unity at low temperatures. A new model for potential fluctuations featuring regions with the average potential lower than that in the surrounding areas was discussed as an improvement to the previous model of double-scaled potential fluctuations. This model was used to describe the spectral dependence of PL decay times and revealed the exchange between localized states at different energy positions on the profile.

Further improvement of the transparent p-layer vertical conductivity and reduction of alloy fluctuations in the AlGaN MQW region is needed for development of DUV LEDs to accelerate their use in power demanding applications, such as disinfection, decontamination, and curing.

ACKNOWLEDGMENTS

The authors would like to thank our colleagues at the Sensor Electronic Technology, Inc., Rensselaer Polytechnic Institute, US Army Research Laboratory, Vilnius University, and KTH Royal Institute of Technology.

REFERENCES

Abakumov, V.N., Perel, V.I., Yassievich, I.N., 1991. Nonradiative Recombination in Semiconductors. Elsevier Science, Amsterdam.

Akasaki, I., Amano, H., Murakami, H., Sassa, M., Kato, H., Manabe, K., 1993. Growth of GaN and AlGaN for UV/blue p-n junction diodes. J. Cryst. Growth 128, 379–383. http://dx.doi.org/10.1016/0022-0248(93)90352-W.

Akiba, M., Hirayama, H., Tomita, Y., Tsukada, Y., Maeda, N., Kamata, N., 2012. Growth of flat p-GaN contact layer by pulse flow method for high light-extraction AlGaN deep-

UV LEDs with Al-based electrode. Phys. Status Solidi 9, 806–809. http://dx.doi.org/10.1002/pssc.201100370.

Aleksiejūnas, R., Ščajev, P., Nargelas, S., Malinauskas, T., Kadys, A., Jarašiūnas, K., 2013. Impact of diffusivity to carrier recombination rate in nitride semiconductors: from bulk GaN to (In,Ga)N quantum wells. Jpn. J. Appl. Phys. 52, 08JK01. http://dx.doi.org/10.7567/JJAP.52.08JK01.

Allerman, A.A., Crawford, M.H., Miller, M.A., Lee, S.R., 2010. Growth and characterization of Mg-doped AlGaN–AlN short-period superlattices for deep-UV optoelectronic devices. J. Cryst. Growth 312, 756–761. http://dx.doi.org/10.1016/j.jcrysgro.2009.12.047.

Amano, H., Sawaki, N., Akasaki, I., Toyoda, Y., 1986. Metalorganic vapor phase epitaxial growth of a high quality GaN film using an AlN buffer layer. Appl. Phys. Lett. 48, 353. http://dx.doi.org/10.1063/1.96549.

applications @ www.s-et.com [WWW Document], n.d. http://www.s-et.com/applications.html.

Badcock, T.J., Dawson, P., Davies, M.J., Kappers, M.J., Massabuau, F.C.P., Oehler, F., Oliver, R.A., Humphreys, C.J., 2014. Low temperature carrier redistribution dynamics in InGaN/GaN quantum wells. J. Appl. Phys. 115, 8–14. http://dx.doi.org/10.1063/1.4868628.

Bai, J., Dudley, M., Sun, W.H., Wang, H.M., Khan, M.A., 2006. Reduction of threading dislocation densities in AlN/sapphire epilayers driven by growth mode modification. Appl. Phys. Lett. 88, 1–3. http://dx.doi.org/10.1063/1.2170407.

Broell, M., Sundgren, P., Rudolph, A., Schmid, W., Vogl, A., Behringer, M., 2014. New developments on high-efficiency infrared and InGaAlP light-emitting diodes at OSRAM opto semiconductors. In: Streubel, K.P., Jeon, H., Tu, L.-W., Strassburg, M. (Eds.), Light-Emitting Diodes: Materials, Devices, and Applications for Solid State Lighting XVIII, 90030L. Proc. SPIE., Volume 9003. San Francisco, CA. http://dx.doi.org/10.1117/12.2039078.

Bryan, I., Bryan, Z., Mita, S., Rice, A., Tweedie, J., Collazo, R., Sitar, Z., 2016. Surface kinetics in AlN growth: a universal model for the control of surface morphology in III-nitrides. J. Cryst. Growth 438, 81–89. http://dx.doi.org/10.1016/j.jcrysgro.2015.12.022.

Chen, Z., Qhalid Fareed, R.S., Gaevski, M., Adivarahan, V., Yang, J.W., Khan, A., Mei, J., Ponce, F.A., 2006. Pulsed lateral epitaxial overgrowth of aluminum nitride on sapphire substrates. Appl. Phys. Lett. 89, 081905. http://dx.doi.org/10.1063/1.2245436.

Chichibu, S., Onuma, T., Sota, T., DenBaars, S.P., Nakamura, S., Kitamura, T., Ishida, Y., Okumura, H., 2003. Influence of InN mole fraction on the recombination processes of localized excitons in strained cubic $In_xGa_{1-x}N$/GaN multiple quantum wells. J. Appl. Phys. 93, 2051. http://dx.doi.org/10.1063/1.1535746.

Cho, Y.-H., Gainer, G.H., Fischer, A.J., Song, J.J., Keller, S., Mishra, U.K., DenBaars, S.P., 1998. "S-shaped" temperature-dependent emission shift and carrier dynamics in InGaN/GaN multiple quantum wells. Appl. Phys. Lett. 73, 1370. http://dx.doi.org/10.1063/1.122164.

Cho, J., Schubert, E.F., Kim, J.K., 2013. Efficiency droop in light-emitting diodes: challenges and countermeasures. Laser Photon. Rev. 7, 408–421. http://dx.doi.org/10.1002/lpor.201200025.

Cohen, E., Sturge, M.D., 1982. Fluorescence line narrowing, localized exciton states, and spectral diffusion in the mixed semiconductor CdS_x Se_{1-x}. Phys. Rev. B 25, 3828–3840. http://dx.doi.org/10.1103/PhysRevB.25.3828.

Collins, C.J., Sampath, A.V., Garrett, G.A., Sarney, W.L., Shen, H., Wraback, M., Nikiforov, A.Y., Cargill, G.S., Dierolf, V., 2005. Enhanced room-temperature luminescence efficiency through carrier localization in $Al_xGa_{1-x}N$ alloys. Appl. Phys. Lett. 86, 031916. http://dx.doi.org/10.1063/1.1856702.

David, A., Hurni, C.A., Aldaz, R.I., Cich, M.J., Ellis, B., Huang, K., Steranka, F.M., Krames, M.R., 2014. High light extraction efficiency in bulk-GaN based volumetric violet light-emitting diodes. Appl. Phys. Lett. 105, 231111. http://dx.doi.org/10.1063/1.4903297.

Delaney, K.T., Rinke, P., Van de Walle, C.G., 2009. Auger recombination rates in nitrides from first principles. Appl. Phys. Lett. 94, 191109. http://dx.doi.org/10.1063/1.3133359.

Eliseev, P.G., Perlin, P., Lee, J., Osiński, M., 1997. "Blue" temperature-induced shift and band-tail emission in InGaN-based light sources. Appl. Phys. Lett. 71, 569. http://dx.doi.org/10.1063/1.119797.

Fareed, Q., Gaska, R., Shur, M., 2007. Methods of growing nitride-based film using varying pulses. US Patent 7,192,849.

Follstaedt, D.M., Lee, S.R., Provencio, P.P., Allerman, A.A., Floro, J.A., Crawford, M.H., 2005. Relaxation of compressively-strained AlGaN by inclined threading dislocations. Appl. Phys. Lett. 87, 121112. http://dx.doi.org/10.1063/1.2056582.

Follstaedt, D.M., Lee, S.R., Allerman, A.A., Floro, J.A., 2009. Strain relaxation in AlGaN multilayer structures by inclined dislocations. J. Appl. Phys. 105, 083507. http://dx.doi.org/10.1063/1.3087515.

Gaska, R., Shatalov, M., Shur, M., 2014. Deep ultraviolet light emitting diode. US Patent US8907322B2, issued Dec 9, 2014.

Goepfert, I.D., Schubert, E.F., Osinsky, A., Norris, P.E., Faleev, N.N., 2000. Experimental and theoretical study of acceptor activation and transport properties in p-type AlxGa1-xN/GaN superlattices. J. Appl. Phys. 88, 2030–2038. http://dx.doi.org/10.1063/1.1305842.

Graham, D.M., Soltani-Vala, A., Dawson, P., Godfrey, M.J., Smeeton, T.M., Barnard, J.S., Kappers, M.J., Humphreys, C.J., Thrush, E.J., 2005. Optical and microstructural studies of InGaN/GaN single-quantum-well structures. J. Appl. Phys. 97, 103508. http://dx.doi.org/10.1063/1.1897070.

Grandusky, J.R., Smart, J.A., Mendrick, M.C., Schowalter, L.J., Chen, K.X., Schubert, E.F., 2009. Pseudomorphic growth of thick n-type AlxGa1-xN layers on low-defect-density bulk AlN substrates for UV LED applications. J. Cryst. Growth 311, 2864–2866. http://dx.doi.org/10.1016/j.jcrysgro.2009.01.101.

Hader, J., Moloney, J.V., Koch, S.W., 2010. Density-activated defect recombination as a possible explanation for the efficiency droop in GaN-based diodes. Appl. Phys. Lett. 96, 221106. http://dx.doi.org/10.1063/1.3446889.

Hader, J., Moloney, J.V., Koch, S.W., 2011. Temperature-dependence of the internal efficiency droop in GaN-based diodes. Appl. Phys. Lett. 99, 181127. http://dx.doi.org/10.1063/1.3658031.

Han, J., Crawford, M.H., Shul, R.J., Figiel, J.J., Banas, M., Zhang, L., Song, Y.K., Zhou, H., Nurmikko, A.V., 1998. AlGaN/GaN quantum well ultraviolet light emitting diodes. Appl. Phys. Lett. 73, 1688. http://dx.doi.org/10.1063/1.122246.

Hirayama, H., Yatabe, T., Noguchi, N., Ohashi, T., Kamata, N., 2007. 231–261 nm AlGaN deep-ultraviolet light-emitting diodes fabricated on AlN multilayer buffers grown by ammonia pulse-flow method on sapphire. Appl. Phys. Lett. 91, 071901. http://dx.doi.org/10.1063/1.2770662.

Hirayama, H., Maeda, N., Fujikawa, S., Toyoda, S., Kamata, N., 2014. Recent progress and future prospects of AlGaN-based high-efficiency deep-ultraviolet light-emitting diodes. Jpn. J. Appl. Phys. 53, 100209.

Horng, R.-H., Chuang, S.-H., Tien, C.-H., Lin, S.-C., Wuu, D.-S., 2014. High performance GaN-based flip-chip LEDs with different electrode patterns. Opt. Express 22 (Suppl. 3), A941–A946. http://dx.doi.org/10.1364/OE.22.00A941.

Hurni, C.A., David, A., Cich, M.J., Aldaz, R.I., Ellis, B., Huang, K., Tyagi, A., Delille, R.A., Craven, M.D., Steranka, F.M., Krames, M.R., 2015. Bulk GaN flip-chip violet light-emitting diodes with optimized efficiency for high-power operation. Appl. Phys. Lett. 106, 031101. http://dx.doi.org/10.1063/1.4905873.

Imura, M., Nakano, K., Kitano, T., Fujimoto, N., Narita, G., Okada, N., Balakrishnan, K., Iwaya, M., Kamiyama, S., Amano, H., Akasaki, I., Shimono, K., Noro, T., Takagi, T., Bandoh, A., 2006. Microstructure of epitaxial lateral overgrown AlN on trench-patterned AlN template by high-temperature metal-organic vapor phase epitaxy. Appl. Phys. Lett. 89, 221901. http://dx.doi.org/10.1063/1.2364460.

Imura, M., Fujimoto, N., Okada, N., Balakrishnan, K., Iwaya, M., Kamiyama, S., Amano, H., Akasaki, I., Noro, T., Takagi, T., 2007a. Annihilation mechanism of threading dislocations in AlN grown by growth form modification method using V/III ratio. J. Cryst. Growth 300, 136–140. http://dx.doi.org/10.1016/j.jcrysgro.2006.11.013.

Imura, M., Nakano, K., Fujimoto, N., Okada, N., Balakrishnan, K., Iwaya, M., Kamiyama, S., Amano, H., Akasaki, I., Noro, T., Takagi, T., Bandoh, A., 2007b. Dislocations in AlN epilayers grown on sapphire substrate by high-temperature metal-organic vapor phase epitaxy. Jpn. J. Appl. Phys. 46, 1458–1462. http://dx.doi.org/10.1143/JJAP.46.1458.

Inazu, T., Fukahori, S., Pernot, C., Kim, M.H., Fujita, T., Nagasawa, Y., Hirano, A., Ippommatsu, M., Iwaya, M., Takeuchi, T., Kamiyama, S., Yamaguchi, M., Honda, Y., Amano, H., Akasaki, I., 2011. Improvement of light extraction efficiency for AlGaN-based deep ultraviolet light-emitting diodes. Jpn. J. Appl. Phys. 50, 122101. http://dx.doi.org/10.1143/JJAP.50.122101.

Iveland, J., Martinelli, L., Peretti, J., Speck, J.S., Weisbuch, C., 2013. Direct measurement of Auger electrons emitted from a semiconductor light-emitting diode under electrical injection: identification of the dominant mechanism for efficiency droop. Phys. Rev. Lett. 110, 177406. http://dx.doi.org/10.1103/PhysRevLett.110.177406.

Jain, R., Sun, W., Yang, J., Shatalov, M., Hu, X., Sattu, A., Lunev, A., Deng, J., Shturm, I., Bilenko, Y., Gaska, R., Shur, M.S., 2008. Migration enhanced lateral epitaxial overgrowth of AlN and AlGaN for high reliability deep ultraviolet light emitting diodes. Appl. Phys. Lett. 93, 1–3. http://dx.doi.org/10.1063/1.2969402.

Jain, R., Sun, W., Yang, J., Shatalov, M., Dobrinsky, A., Shur, M., Gaska, R., 2015. Patterned layer design for group III nitride layer growth. Patent number: 9105792.

Jang, H.W., Kim, J.K., Kim, S.Y., Yu, H.K., Lee, J.L., 2004. Ohmic contacts for high power LEDs. Phys. Status Solidi C Conf. 1, 2831–2836. http://dx.doi.org/10.1002/pssa.200405111.

Karpov, S., 2014. ABC-model for interpretation of internal quantum efficiency and its droop in III-nitride LEDs. In: 14th International Conference on Numerical Simulation of Optoelectronic Devices. IEEE, pp. 17–18. http://dx.doi.org/10.1109/NUSOD.2014.6935334. http://ieeexplore.ieee.org/document/6935334/.

Kazlauskas, K., Tamulaitis, G., Žukauskas, A., Khan, M.A., Yang, J.W., Zhang, J., Simin, G., Shur, M.S., Gaska, R., 2003. Double-scaled potential profile in a group-III nitride alloy revealed by Monte Carlo simulation of exciton hopping. Appl. Phys. Lett. 83, 3722. http://dx.doi.org/10.1063/1.1625111.

Kazlauskas, K., Žukauskas, A., Tamulaitis, G., Mickevičius, J., Shur, M.S., Fareed, R.S.Q., Zhang, J.P., Gaska, R., 2005. Exciton hopping and nonradiative decay in AlGaN epilayers. Appl. Phys. Lett. 87, 172102. http://dx.doi.org/10.1063/1.2112169.

Khan, M.A., Skogman, R.A., Schulze, R.G., Gershenzon, M., 1983. Electrical properties and ion implantation of epitaxial GaN, grown by low pressure metalorganic chemical vapor deposition. Appl. Phys. Lett. 42, 430–432. http://dx.doi.org/10.1063/1.93953.

Khan, M.A., Kuznia, J.N., Skogman, R.A., Olson, D.T., 1992. Low pressure metalorganic chemical vapor deposition of AIN over sapphire substrates. Appl. Phys. Lett. 61, 2539–2541.

Kim, M.-H., Schubert, M.F., Dai, Q., Kim, J.K., Schubert, E.F., Piprek, J., Park, Y., 2007. Origin of efficiency droop in GaN-based light-emitting diodes. Appl. Phys. Lett. 91, 183507. http://dx.doi.org/10.1063/1.2800290.

Kinoshita, A., Hirayama, H., Ainoya, M., Aoyagi, Y., 2000. Room-temperature operation at 333 nm of Al0.03Ga0.97N/Al0.25Ga0.75N quantum-well light-emitting diodes with Mg-doped superlattice layers. Appl. Phys. Lett. 77, 175–177.

Kioupakis, E., Rinke, P., Delaney, K.T., Van de Walle, C.G., 2011. Indirect Auger recombination as a cause of efficiency droop in nitride light-emitting diodes. Appl. Phys. Lett. 98, 161107. http://dx.doi.org/10.1063/1.3570656.

Kioupakis, E., Yan, Q., Steiauf, D., Van De Walle, C.G., 2013. Temperature and carrier-density dependence of Auger and radiative recombination in nitride optoelectronic devices. New J. Phys. 15, 125006. http://dx.doi.org/10.1088/1367-2630/15/12/125006.

Kipshidze, G., Kuryatkov, V., Zhu, K., Borisov, B., Holtz, M., Nikishin, S., Temkin, H., 2003. AlN/AlGaInN superlattice light-emitting diodes at 280 nm. J. Appl. Phys. 93, 1363. http://dx.doi.org/10.1063/1.1535255.

Kneissl, M., Rass, J., 2016. III-Nitride ultraviolet emitters. In: Kneissl, M., Rass, J. (Eds.), Springer Series in Materials Science. Springer International Publishing, Cham. http://dx.doi.org/10.1007/978-3-319-24100-5.

Kozodoy, P., Hansen, M., DenBaars, S.P., Mishra, U.K., 1999. Enhanced Mg doping efficiency in Al[sub 0.2]Ga[sub 0.8]N/GaN superlattices. Appl. Phys. Lett. 74, 3681. http://dx.doi.org/10.1063/1.123220.

Kuokstis, E., Sun, W.H., Shatalov, M., Yang, J.W., Asif Khan, M., 2006. Role of alloy fluctuations in photoluminescence dynamics of AlGaN epilayers. Appl. Phys. Lett. 88, 261905. http://dx.doi.org/10.1063/1.2218275.

Lunev, A., Dobrinsky, A., Shatalov, M.S., Gaska, R., Shur, M., 2015. Ultraviolet reflective contact. US9184346B2.

Marcinkevičius, S., Jain, R., Shatalov, M., Yang, J., Shur, M., Gaska, R., 2014. High spectral uniformity of AlGaN with a high Al content evidenced by scanning near-field photoluminescence spectroscopy. Appl. Phys. Lett. 105, 241108. http://dx.doi.org/10.1063/1.4904710.

Maruska, H.P., Rhines, W.C., Stevenson, D.A., 1972. Preparation of Mg-doped GaN diodes exhibiting violet electroluminescence. Mater. Res. Bull. 7, 777–781. http://dx.doi.org/10.1016/0025-5408(72)90127-4.

Mickevičius, J., Tamulaitis, G., Kuokštis, E., Liu, K., Shur, M.S., Zhang, J.P., Gaska, R., 2007. Well-width-dependent carrier lifetime in AlGaN/AlGaN quantum wells. Appl. Phys. Lett. 90, 131907. http://dx.doi.org/10.1063/1.2717145.

Mickevičius, J., Jurkevičius, J., Shur, M.S., Yang, J., Gaska, R., Tamulaitis, G., 2012a. Photoluminescence efficiency droop and stimulated recombination in GaN epilayers. Opt. Express 20, 25195–25200. http://dx.doi.org/10.1364/OE.20.025195.

Mickevičius, J., Tamulaitis, G., Shur, M., Shatalov, M., Yang, J., Gaska, R., 2012b. Internal quantum efficiency in AlGaN with strong carrier localization. Appl. Phys. Lett. 101, 211902. http://dx.doi.org/10.1063/1.4767657.

Mickevicius, J., Tamulaitis, G., Shur, M., Shatalov, M., Yang, J., Gaska, R., 2013. Correlation between carrier localization and efficiency droop in AlGaN epilayers. Appl. Phys. Lett. 103, 011906. http://dx.doi.org/10.1063/1.4813259.

Miller, A., 1999. Transient grating studies of carrier diffusion and mobility in semiconductors. In: Garmire, E., Kost, A. (Eds.), Nonlinear Optics in Semiconductors II. Semiconductors and Semimetals. Academic Press, pp. 287–312. https://www.elsevier.com/books/nonlinear-optics-in-semiconductors-ii/author/978-0-12-752168-8.

Miller, A., Abrahams, E., 1960. Impurity conduction at low concentrations. Phys. Rev. 120, 745–755. http://dx.doi.org/10.1103/PhysRev.120.745.

Morel, A., Lefebvre, P., Kalliakos, S., Taliercio, T., Bretagnon, T., Gil, B., 2003. Donor-acceptor-like behavior of electron-hole pair recombinations in low-dimensional (Ga,In)N/GaN systems. Phys. Rev. B 68, 1–6. http://dx.doi.org/10.1103/PhysRevB.68.045331.

Murotani, H., Yamada, Y., Taguchi, T., Ishibashi, A., Kawaguchi, Y., Yokogawa, T., 2008. Temperature dependence of localized exciton transitions in AlGaN ternary alloy epitaxial layers. J. Appl. Phys. 104, 053514. http://dx.doi.org/10.1063/1.2975970.

Murotani, H., Kittaka, R., Kurai, S., Yamada, Y., Miyake, H., Hiramatsu, K., 2011. Recombination dynamics of localized excitons in AlxGa1-xN (0.37 < x < 0.81) ternary alloys. Phys. Status Solidi 8, 2133–2135. http://dx.doi.org/10.1002/pssc.201000899.

Narukawa, Y., Kawakami, Y., Fujita, S., Fujita, S., Nakamura, S., 1997. Recombination dynamics of localized excitons in In0.20Ga0.80N-In0.05Ga0.95N multiple quantum wells. Phys. Rev. B 55, R1938–R1941. http://dx.doi.org/10.1103/PhysRevB.55.R1938.

Narukawa, Y., Kawakami, Y., Fujita, S., Nakamura, S., 1999. Dimensionality of excitons in laser-diode structures composed of $In_xGa_{1-x}N$ multiple quantum wells. Phys. Rev. B 59, 10283–10288. http://dx.doi.org/10.1103/PhysRevB.59.10283.

Narukawa, Y., Ichikawa, M., Sanga, D., Sano, M., Mukai, T., 2010. White light emitting diodes with super-high luminous efficacy. J. Phys. D Appl. Phys. 43, 354002. http://dx.doi.org/10.1088/0022-3727/43/35/354002.

Nishida, T., Kobayashi, N., 1999. 346 nm emission from AlGaN multi-quantum-well light emitting diode. Phys. Status Solidi 176, 45–48. http://dx.doi.org/10.1002/(SICI)1521-396X(199911)176:1<45::AID-PSSA45>3.0.CO;2-0.

Onuma, T., Chichibu, S.F., Uedono, A., Sota, T., Cantu, P., Katona, T.M., Keading, J.F., Keller, S., Mishra, U.K., Nakamura, S., DenBaars, S.P., 2004. Radiative and nonradiative processes in strain-free $Al_xGa_{1-x}N$ films studied by time-resolved photoluminescence and positron annihilation techniques. J. Appl. Phys. 95, 2495. http://dx.doi.org/10.1063/1.1644041.

Pernot, C., Fukahori, S., Inazu, T., Fujita, T., Kim, M., Nagasawa, Y., Hirano, A., Ippommatsu, M., Iwaya, M., Kamiyama, S., Akasaki, I., Amano, H., 2011. Development of high efficiency 255–355 nm AlGaN-based light-emitting diodes. Phys. Status Solidi 208, 1594–1596. http://dx.doi.org/10.1002/pssa.201001037.

Piprek, J., 2010. Efficiency droop in nitride-based light-emitting diodes. Phys. Status Solidi 207, 2217–2225. http://dx.doi.org/10.1002/pssa.201026149.

Saguatti, D., Bidinelli, L., Verzellesi, G., Meneghini, M., Meneghesso, G., Zanoni, E., Butendeich, R., Hahn, B., 2012. Investigation of efficiency-droop mechanisms in multi-quantum-well InGaN/GaN blue light-emitting diodes. IEEE Trans. Electron Devices 59, 1402–1409. http://dx.doi.org/10.1109/TED.2012.2186579.

Saito, S., Hashimoto, R., Hwang, J., Nunoue, S., 2013. InGaN light-emitting diodes on c-face sapphire substrates in green gap spectral range. Appl. Phys. Express 6, 111004. http://dx.doi.org/10.7567/APEX.6.111004.

Satake, A., Masumoto, Y., Miyajima, T., Asatsuma, T., Nakamura, F., Ikeda, M., 1998. Localized exciton and its stimulated emission in surface mode from single-layer $In_xGa_{1-x}N$. Phys. Rev. B 57, R2041–R2044. http://dx.doi.org/10.1103/PhysRevB.57.R2041.

Saxena, T., Tamulaitis, G., Shatalov, M., Yang, J., Gaska, R., Shur, M.S., 2013. Low threshold for optical damage in AlGaN epilayers and heterostructures. J. Appl. Phys. 114, 203103. http://dx.doi.org/10.1063/1.4834520.

Saxena, T., Nargelas, S., Mickevičius, J., Kravcov, O., Tamulaitis, G., Shur, M., Shatalov, M., Yang, J., Gaska, R., 2015a. Spectral dependence of carrier lifetime in high aluminum content AlGaN epitaxial layers. J. Appl. Phys. 118, 085705. http://dx.doi.org/10.1063/1.4929499.

Saxena, T., Shur, M., Nargelas, S., Podlipskas, Ž., Aleksiejūnas, R., Tamulaitis, G., Shatalov, M., Yang, J., Gaska, R., 2015b. Dynamics of nonequilibrium carrier decay in AlGaN epitaxial layers with high aluminum content. Opt. Express 23, 19646. http://dx.doi.org/10.1364/OE.23.019646.

Schubert, E.F., 2003. Light-Emitting Diodes. Cambridge University Press, New York.
Shatalov, M., Sun, W., Lunev, A., Hu, X., Dobrinsky, A., Bilenko, Y., Yang, J., Shur, M., Gaska, R., Moe, C., Garrett, G., Wraback, M., 2012. AlGaN deep-ultraviolet light-emitting diodes with external quantum efficiency above 10%. Appl. Phys. Express 5, 082101. http://dx.doi.org/10.1143/APEX.5.082101.
Shatalov, M.S., Gaska, R., Yang, J., Shur, M., 2013. Semiconductor material doping based on target valence band discontinuity. US8426225B2. doi:10.1126/science.Liquids.
Shatalov, M., Sun, W., Jain, R., Lunev, A., Hu, X., Dobrinsky, A., Bilenko, Y., Yang, J., Garrett, G.A., Rodak, L.E., Wraback, M., Shur, M., Gaska, R., 2014. High power AlGaN ultraviolet light emitters. Semicond. Sci. Technol. 29, 084007. http://dx.doi.org/10.1088/0268-1242/29/8/084007.
Shatalov, M., Jain, R., Yang, J., Shur, M., Gaska, R., 2015. Patterned substrate design for layer growth. US8981403B2.
Shur, M., Shatalov, M.S., Dobrinsky, A., Gaska, R., Yang, J., 2014. Emitting device with compositional and doping inhomogeneities in semiconductor layers. US8787418B2.
Tamulaitis, G., Mickevičius, J., Dobrovolskas, D., Kuokštis, E., Shur, M.S., Shatalov, M., Yang, J., Gaska, R., 2012. Carrier dynamics and efficiency droop in AlGaN epilayers with different Al content. Phys. Status Solidi 9, 1677–1679. http://dx.doi.org/10.1002/pssc.201100550.
Wu, Y., Hanlon, A., Kaeding, J.F., Sharma, R., Fini, P.T., Nakamura, S., Speck, J.S., 2004. Effect of nitridation on polarity, microstructure, and morphology of AlN films. Appl. Phys. Lett. 84, 912. http://dx.doi.org/10.1063/1.1646222.
Yum, W.-S., Jeon, J.-W., Sung, J.-S., Seong, T.-Y., 2012. Highly reliable Ag/Zn/Ag ohmic reflector for high-power GaN-based vertical light-emitting diode. Opt. Express 20, 19194. http://dx.doi.org/10.1364/OE.20.019194.
Zhang, J.P., Chitnis, A., Adivarahan, V., Wu, S., Mandavilli, V., Pachipulusu, R., Shatalov, M., Simin, G., Yang, J.W., Khan, M.A., 2002. Milliwatt power deep ultraviolet light-emitting diodes over sapphire with emission at 278 nm. Appl. Phys. Lett. 81, 4910. http://dx.doi.org/10.1063/1.1531835.
Zheng, T.C., Lin, W., Liu, R., Cai, D.J., Li, J.C., Li, S.P., Kang, J.Y., 2016. Improved p-type conductivity in Al-rich AlGaN using multidimensional Mg-doped superlattices. Sci. Rep. 6, 21897. http://dx.doi.org/10.1038/srep21897.

CHAPTER THREE

Growth of High-Quality AlN on Sapphire and Development of AlGaN-Based Deep-Ultraviolet Light-Emitting Diodes

H. Hirayama[1]
RIKEN, Quantum Optodevice Laboratory, Wako, Saitama, Japan
[1]Corresponding author: e-mail address: hirayama@riken.jp

Contents

1. INTRODUCTION

Growth techniques of AlN/AlGaN semiconductors and recent advances in AlGaN-based deep-ultraviolet (DUV) light-emitting diodes (LEDs) are demonstrated. 220–350 nm-band DUV LEDs have been realized by developing crystal growth techniques for wide-bandgap AlN and AlGaN semiconductors. Significant increases in internal quantum efficiency (IQE) have been achieved for AlGaN DUV emissions by developing low-threading-dislocation-density (TDD) AlN buffer layers grown on sapphire substrates. The electron injection efficiency (EIE) of the LEDs was also significantly increased by introducing a multiquantum barrier (MQB). We also discuss light extraction efficiency (LEE), which is the most important parameter for achieving high-efficiency DUV LEDs. We succeeded in improving

Semiconductors and Semimetals, Volume 96
ISSN 0080-8784
http://dx.doi.org/10.1016/bs.semsem.2016.11.002
© 2017 Elsevier Inc.
All rights reserved.

LEE by developing a transparent p-AlGaN contact layer. The maximum external quantum efficiency (EQE) obtained was 7% for a 279-nm DUV LED. EQE could be increased by up to several tens of percent through the improvement of LEE by utilizing transparent contact layers and photonic nanostructures.

In Section 2, the research background including device applications, history, and the current status of DUV LEDs is described. We will describe the development of crystal growth techniques for obtaining high-quality AlN and AlGaN crystals in Section 3. We will describe the achievement of high IQE for AlGaN DUV emission and the realization of DUV LEDs in Sections 4 and 5, respectively. We will go on to discuss several issues for increasing the efficiencies of DUV LEDs, i.e., EIE and LEE, in Section 6. Finally, the future prospects of DUV LEDs are discussed in Section 7.

2. RESEARCH BACKGROUND OF DUV LEDs

The development of semiconductor light sources operating in the DUV region, such as DUV LEDs and laser diodes (LDs), is quite an important subject because they are required for a wide variety of applications. Fig. 1 shows the image of application fields of DUA (200–280 nm), UVB (280–315 nm), and UVC (315–400 nm) LEDs and LDs. UV LEDs are expected to be used in applications such as sterilization, water purification, medicine and biochemistry, agriculture, white light illumination, light

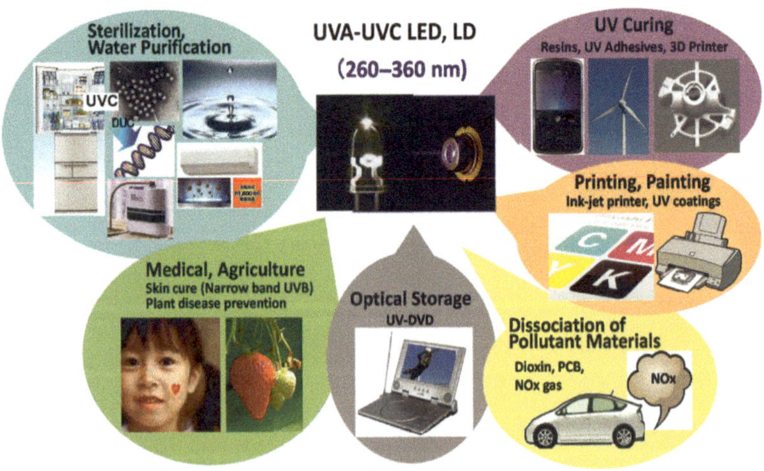

Fig. 1 Potential applications of DUA-UVC LEDs and LDs.

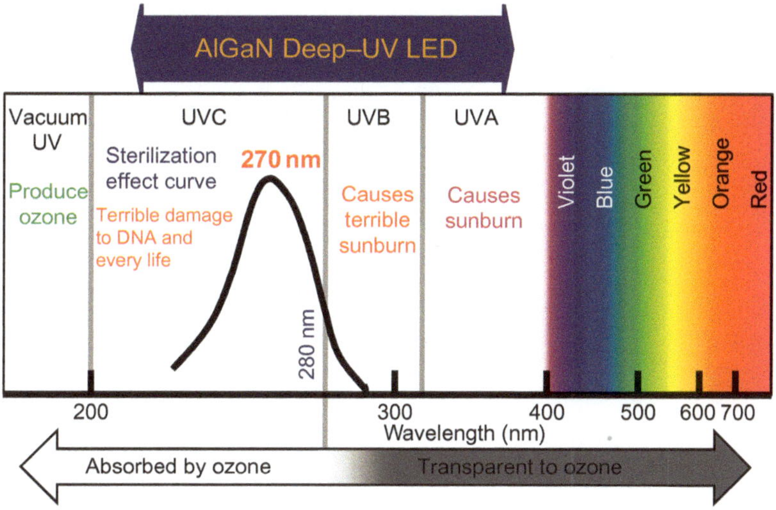

Fig. 2 Classification of UV light and the wavelength range achieved by AlGaN DUV LEDs.

source of high-density optical memory, fluorescence analytical systems, and related information sensing fields. They are also very important for UV curing, three-dimensional (3D) printer, UV adhesives, printing, painting, or courting (Hirayama, 2005; Zukauskas et al., 2002).

Fig. 2 shows the classification of UV light and the wavelength range achieved by AlGaN DUV LEDs. For applications involving sterilization or water purification with direct UV-light treatment, UVC light with the wavelength range between 260 and 280 nm is most suitable. For the UV purification process using a titanium-oxide (TiO_2) catalyst, the wavelength range between 320 and 380 nm is also useful. For application to white LED illumination produced by a mixture of red–green–blue phosphors excited by a UV LED, wavelengths around 340 nm in UVA are considered to be most suitable, taking into account the efficient absorption by phosphors (<350 nm). The wavelength range achieved by AlGaN LED covers from UVA to UVC as shown in Fig. 2.

Because of their wide direct transition energy range in the UV, covering the region between 6.2 eV (AlN) and 3.4 eV, AlGaN is attracting considerable attention as candidate materials for the realization of DUV LEDs and LDs (Hirayama, 2005). Fig. 3 shows the relationship between the direct transition bandgap energy and the lattice constant of the wurtzite (WZ) InAlGaN material system and the lasing wavelengths of various gas lasers. The main advantages of using AlGaN for DUV light sources are (1) the

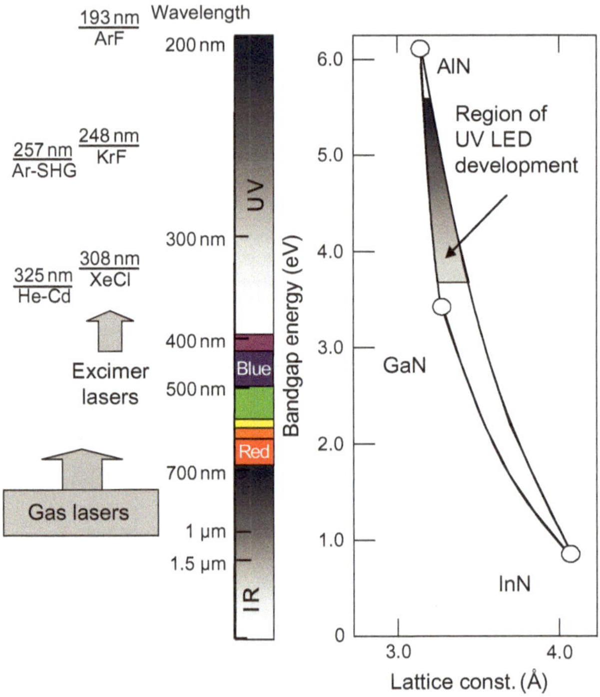

Fig. 3 Relationship between the direct transition bandgap energy and the lattice constant of the wurtzite (WZ) InAlGaN material system and the lasing wavelengths of various gas lasers.

possibility of obtaining high-efficiency optical emission from quantum wells (QWs), (2) the possibility of producing both p- and n-type semiconductors in the wide-bandgap spectral region, (3) their physical properties, i.e., nitrides are mechanically hard and the devices have long life-times, and (4) the fact that the materials are free from harmful arsenic, mercury, and lead (Hirayama, 2005).

Research into AlGaN-based UV LEDs for wavelengths shorter than 360 nm was initiated by several research groups between 1996 and 1999 (Han et al., 1998; Kinoshita et al., 2000; Nishida et al., 2001). In the United States, the effort directed at DUV light sources was driven by DARPA's Semiconductor Ultraviolet Optical Sources (SUVOS) program. A group at the University of South Carolina reported the first 250–280 nm AlGaN-based DUV LEDs between 2002 and 2006 (Adivarahan et al., 2004a,b; Sun et al., 2004). A group at NTT reported the shortest wavelength (210 nm) LED using an AlN-emitting layer in 2006 (Taniyasu et al., 2006). We started research into AlGaN-based DUV LEDs in 1997 and reported the first efficient

DUV (230 nm) photoluminescence (PL) from AlGaN/AlN QWs (Hirayama et al., 2002a), and a 333-nm AlGaN QW UV LED on SiC in 1999 (Kinoshita et al., 2000). We have also developed high-efficiency UV LEDs using In-incorporation effects into AlGaN (Hirayama, 2005; Hirayama et al., 2002b,c). We have demonstrated several mW CW operation from 340 to 350 nm InAlGaN QW UV LEDs on both GaN single-crystal substrates (Hirayama et al., 2004) and sapphire substrates (Fujikawa et al., 2008).

The development of 280 nm band AlGaN DUV LEDs to achieve high-efficiency and high-power operation has become extremely competitive recently because they are expected to have a huge market for sterilization applications. We developed a growth method for low TDD AlN templates on sapphire substrates in 2007 (Hirayama et al., 2007) and achieved high IQE (>60%) for AlGaN and quaternary InAlGaN QWs in the DUV region (Hirayama et al., 2008a, 2009). We also achieved high EIE by introducing an MQB design as an electron-blocking layer (EBL) (Hirayama et al., 2010a) and demonstrated AlGaN and quaternary InAlGaN-based UV LEDs with a wide emission range (222–351 nm) (Fujikawa et al., 2012; Hirayama et al., 2008b, 2009, 2010a,b). We have also improved the LEE of DUV LED by developing a transparent p-AlGaN contact layer and a highly reflective p-type electrode (Hirayama et al., 2014a; Maeda and Hirayama, 2014) and have achieved an EQE of 7% (Hirayama et al., 2014b). RIKEN and Panasonic have developed commercially available DUV LED modules for sterilization use in 2014 (270-nm, 10-mW LED module with lifetime longer than 10,000 h, EQE of 2–3%) (Mino et al., 2012, 2013).

High-efficiency DUV LEDs with wavelength between 260 and 280 nm have developed by several companies (Fujioka et al., 2010; Grandusky et al., 2011, 2013; Hwang et al., 2011; Kinoshita et al., 2012, 2013; Mickevičius et al., 2013; Pernot et al., 2010; Shatalov et al., 2010, 2012). Sensor Electronic Technology has developed commercially valuable UV LEDs with wavelengths ranging between 240 and 360 nm (Mickevičius et al., 2013; Shatalov et al., 2010, 2012), and they have reported a maximum EQE of 11% for 278 nm LED (Shatalov et al., 2012). Nikkiso has also reported over 10% EQEs (Pernot et al., 2010). Crystal IS and Tokuyama have developed efficient DUV LED on single-crystal AlN substrates fabricated by a sublimation method (Grandusky et al., 2011, 2013) and hydride vapor phase epitaxy (Kinoshita et al., 2012, 2013), respectively. Nitek and Nichia have also developed high-efficiency DUV LEDs (Fujioka et al., 2010; Hwang et al., 2011).

The next targets in UV device research are to develop EQEs of several tens of percent for 220–350 nm LEDs and to achieve 250–330 nm LDs.

However, the realization of high EQE UV LEDs with wavelengths below 360 nm is still challenging owing to some major problems. The sudden drop in the efficiency of UV LEDs below 360 nm is mainly due to the following three factors:

1. IQE of AlGaN is more sensitive to TDD than that of InGaN.
2. The hole concentration of p-AlGaN is low, resulting in low injection efficiency.
3. LEE is low because of the absorption of UV light in p-GaN contact layers.

The development of low TDD AlN templates is most important because IQE of AlGaN QWs is as low as 1% if we use conventional templates with high TDD. To obtain high IQE of more than 60%, the reduction of TDD to below 5×10^8 cm^{-2} is required (Hirayama et al., 2008a, 2009). To fabricate such a low TDD AlN template on sapphire, it is necessary to introduce some special growth conditions. Low TDD AlN single-crystal wafers have advantages for high IQE (Grandusky et al., 2011, 2013; Kinoshita et al., 2012, 2013), although they are expensive for use as commercially available DUV LEDs. We are using the "ammonia (NH_3) pulsed-flow multilayer (ML) growth" method to fabricate AlN templates on sapphire and have obtained IQE of over 60% from AlGaN QWs (Hirayama et al., 2007, 2008a, 2009). Also, for the realization of high IQE DUV emission, the use of quaternary InAlGaN with a few percent of indium (In) is effective (Hirayama, 2005; Hirayama et al., 2009).

The device properties of AlGaN DUV LEDs strongly depend on the properties of the p-AlGaN. The hole concentration of p-AlGaN with high Al content (Al > 60%) is low (as low as 10^{14} cm^{-3}) owing to its deep acceptor level, i.e., 240 (GaN)–590 meV (AlN). EIE of a DUV LED is reduced owing to the leakage of electrons to the p-side layers. The high series resistance of p-type layers also becomes a problem for the device properties.

Owing to the lack of high-hole-density p-type AlGaN, we must use p-GaN contact layers. The use of a p-GaN contact layer results in a significant reduction in LEE owing to the strong absorption of DUV light. LEE of a DUV LED is typically below 8%. Transparent p-AlGaN contact layers and highly reflective p-type electrodes are desirable for realizing high LEE devices.

The current EQE of 270 nm DUV LEDs in our group is approximately 7%, which is determined by 60% IQE, 80% EIE, and 15% LEE. Further improvements in EQE are expected as we start the production of commercially available DUV LEDs. Techniques for increasing each of these efficiencies are described in the following sections.

3. GROWTH OF HIGH-QUALITY AlN ON SAPPHIRE SUBSTRATE

In order to realize high-efficiency DUV LEDs, it is necessary to develop a low TDD AlGaN/AlN template. The TDD of a conventional AlN buffer layer on a sapphire substrate, which was fabricated using a low-temperature (LT) AlN buffer, was greater than 2×10^{10} cm^{-2}. On the other hand, TDD of 10^8–10^9 cm^{-2} is required in order to obtain high IQE values of several tens % from AlGaN QWs. Several fabrication methods have been reported for obtaining high-quality AlN buffers, for example, the use of AlN/AlGaN superlattices (SLs) grown with alternating gas feeds (Sun et al., 2004), AlGaN buffer layers deposited by epitaxial lateral overgrowth (ELO) (Iida et al., 2004), and a combination of GaN/AlN SLs and AlGaN produced by alternate source-feeding epitaxy on SiC (Takano et al., 2004).

In the former experiments, we grew AlN layers directly onto sapphire substrates at high growth temperature (HT) after an initial nitridation treatment using NH_3. The growth temperature was around 1300°C, and the V/III ratio was a relatively low value. As the nitridation time was increased from 5 to 10 min, the full width at half-maximum (FWHM) of X-ray diffraction (10–12) and (0002) ω-scan rocking curves (XRCs) was reduced to below 600 arcsec. The value of the FWHM of (10–12) XRC corresponds to the edge-type TDD. We found that larger AlN nuclei are formed in the initial stages of the growth process by introducing longer nitridation times, and that edge dislocations are reduced by embedding them in a thick AlN layer. However, heavy nitridation on sapphire becomes the cause of a polarity inversion from Al to N polarity, which leads to the generation of abnormal nuclei on the AlN surface. We also found that a long nitridation time leads to cracks on the AlN surface.

It is necessary to satisfy several conditions to achieve high-quality AlGaN/AlN templates that are applicable to DUV emitters, i.e., low TDD, crack-free, atomically flat surfaces and stable Al (+*c*) polarity. To obtain all of the conditions mentioned earlier, we have introduced an "NH_3 pulsed-flow ML growth" method for fabricating AlN layers on sapphire (Hirayama et al., 2007). Fig. 4 shows the typical gas flow sequence and a schematic view of the growth control method using pulsed- and continuous-flow gas feeding growth that is used for the NH_3 pulsed-flow ML AlN growth.

The samples were grown on sapphire (0001) substrates by low-pressure metal-organic chemical vapor deposition (LP-MOCVD). First, an AlN

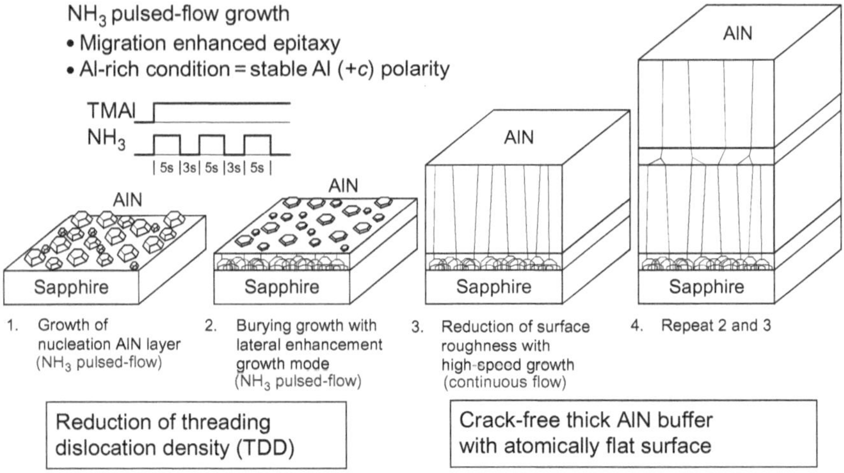

Fig. 4 Gas flow sequence and schematic view of the growth control method used for "an NH_3 pulsed-flow multilayer (ML) AlN growth technique."

nucleation layer and a "burying" AlN layer were deposited, both by NH_3-pulsed-flow growth. The trimethylaluminum (TMAl) flow was continuous during the NH_3-pulsed-flow sequence, as shown in Fig. 4. Low TDD AlN can be achieved by promoting the coalescence of the AlN nucleation layer. After the growth of the first AlN layer, the surface is still rough because of the low growth rate by the pulsed-flow mode growth. We introduced a high-growth-rate continuous-flow mode to reduce the surface roughness. By repeating the pulsed- and continuous-flow modes, we can obtain crack-free, thick AlN layers with atomically flat surfaces. NH_3 pulsed-flow growth is effective for obtaining high-quality AlN because of the enhancement of precursor migration. Furthermore, it is effective for obtaining stable Al (+*c*) polarity, which is necessary for suppressing polarity inversion from Al to N by maintaining Al-rich growth conditions.

As described earlier, we used three different growth conditions in this method, i.e., the initial deposition in order to fabricate an AlN nucleation layer, migration enhancement epitaxy for decreasing TDD, and a high growth rate using a conventional continuous-flow mode. The detailed growth conditions were described in Hirayama et al. (2007, 2008b). The typical growth rates in the pulsed- and continuous-flow modes were approximately 0.6 and 6 $\mu m\ h^{-1}$, respectively. We found recently that a low V/III ratio and a higher growth temperature (~1400°C) are more suitable for obtaining a low TDD AlN growth on sapphire.

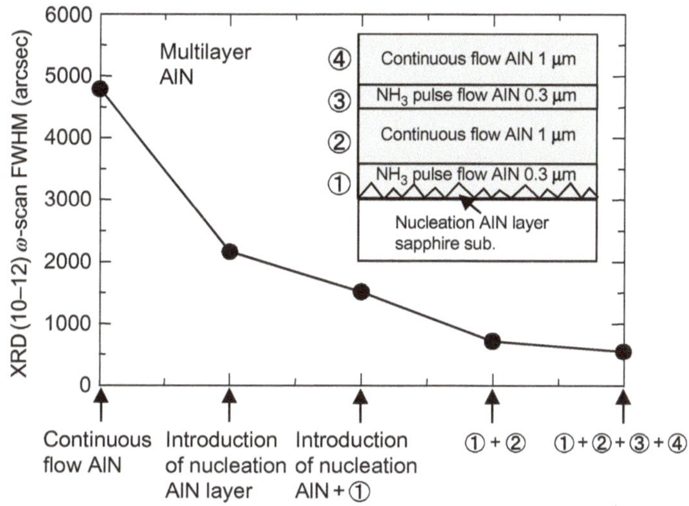

Fig. 5 Reduction of FWHM of X-ray diffraction (10–12) ω-scan rocking curve (XRC) for various stages of ML AlN growth.

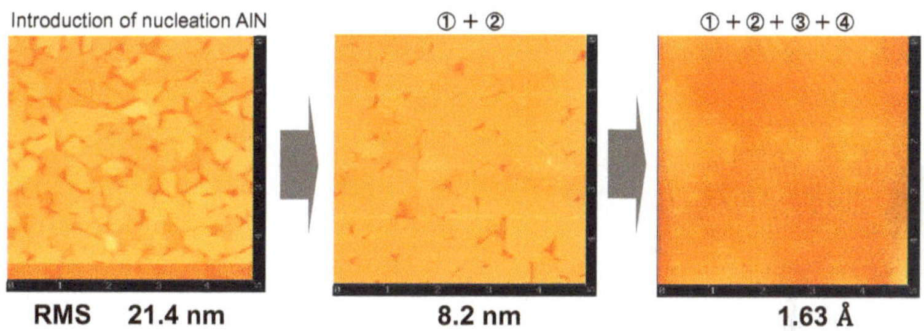

Fig. 6 AFM images of the surface of ML AlN with an area of 5 × 5 μm^2 for various stages of growth.

The advantage of using ML AlN for the DUV LED is that low TDD AlN can be obtained without the need for AlGaN layers, yielding a device structure with minimal DUV absorption. An AlGaN-free buffer is believed to be important for realizing sub-250 nm band high-efficiency LEDs.

Fig. 5 shows the FWHM of the X-ray diffraction (10–12) ω-scan rocking curves (XRCs) for various stages of the ML AlN growth. The FWHM of XRC (10–12) for AlN was reduced from 2160 to 550 arcsec by introducing "two-times repetition" of the NH_3 pulsed-flow ML AlN growth. Fig. 6 shows atomic-force microscope (AFM) images of the surface of ML AlN on sapphire at various stages of the ML AlN growth. We can observe that the surface was improved by growing the multilayers of AlN, and we can finally confirm an atomically flat surface, as demonstrated in Fig. 6. The root-mean-square value of the surface roughness of the ML AlN obtained from the AFM image was 0.16 nm.

Fig. 7 shows (A) a schematic structure and (B) a cross-sectional transmission electron microscope (TEM) image of an AlGaN/AlN template with a 5-step ML AlN buffer layer grown on a sapphire substrate. The total thickness of the ML AlN buffer was typically 4 μm. The typical FWHMs of (10–12) and (0002) XRCs of the ML AlN were approximately 370 and 180 arcsec, respectively, which were achieved by a highly uniform 3×2-in. reactor MOCVD (Mino et al., 2012). The minimum FWHMs obtained for a 1×2-in. reactor MOCVD were approximately 290 and 180 arcsec, respectively. The minimum edge- and screw-type dislocation densities of the ML AlN were below 5×10^8 and $4 \times 10^7\ cm^{-2}$, respectively, as observed by a cross-sectional TEM image.

For further reduction of TDD of the AlN layer on sapphire, we were introducing AlN ELO on patterned sapphire substrate (PSS). Also recently, we are planning to introduce high-temperature annealing for the reduction of TDD. These methods are effective to reduce TDD to be the order of $10^7\ cm^{-2}$ and will be useful to obtain higher IQE.

4. INCREASE IN IQE

We observed a remarkable enhancement of the DUV emission of AlGaN QWs by fabricating them on low TDD AlN templates (Hirayama et al., 2008a, 2009). Fig. 8 shows a cross-sectional TEM image of an AlGaN multi-(M) QW DUV LED with an emission wavelength of 227 nm fabricated on an ML AlN buffer. We used a thin QW in order to obtain a high IQE by suppressing the effects of the polarization field spontaneously applied in the well. This is believed to be particularly important for obtaining the atomically smooth heterointerfaces that are necessary in order to achieve a high IQE from such a thin QW. The atomically flat heterointerfaces of the 1.3-nm-thick three-layer QWs are confirmed as observed in the cross-sectional TEM image shown in Fig. 8.

Fig. 9 shows PL spectra of AlGaN QWs fabricated on ML AlN templates with various values of XRC (10–12) FWHM, as measured at room temperature (RT). The peak emission wavelengths of the QWs were around 255 nm. The QWs were excited with a 244-nm Ar-ion second harmonics generation (SHG) laser. The excitation power density was fixed at

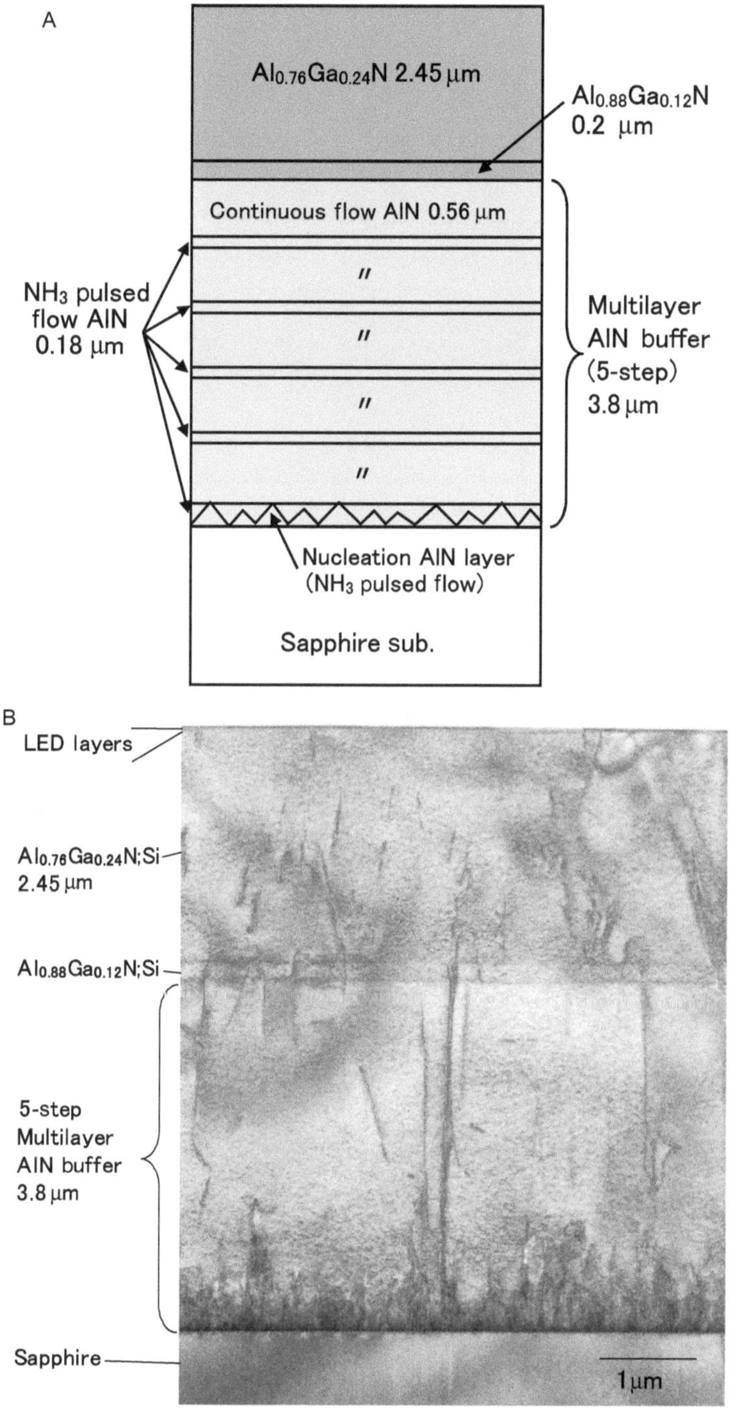

Fig. 7 (A) Schematic structure and (B) cross-sectional TEM image of an AlGaN/AlN template including a 5-step ML AlN buffer layer grown on a sapphire substrate.

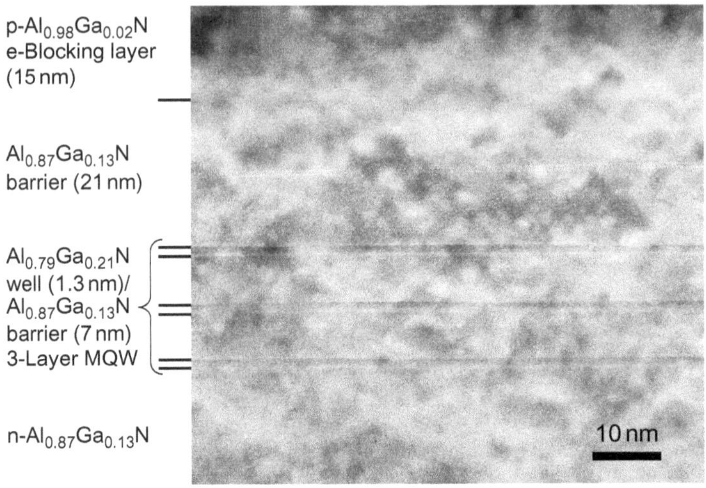

Fig. 8 Cross-sectional TEM image of the quantum well region of an AlGaN MQW DUV LED.

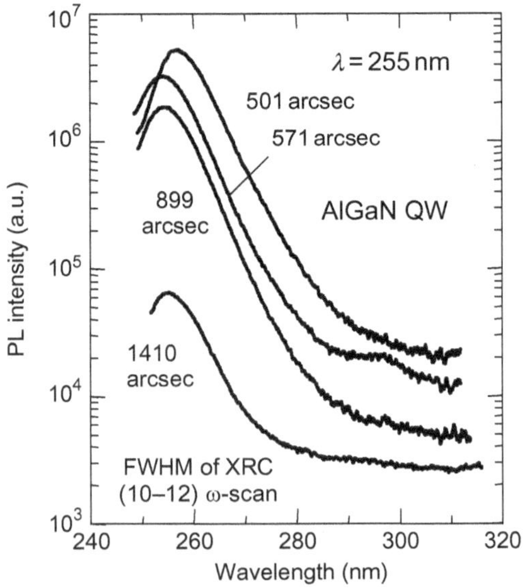

Fig. 9 Photoluminescence (PL) spectra of AlGaN QWs on multilayer (ML) AlN templates with various values of XRC (10–12) FWHM measured at room temperature.

200 W cm^{-2}. The PL emission intensity of the AlGaN QW was significantly increased by improving the XRC (10–12) FWHM, as shown in Fig. 9. We can see from Fig. 9 that the emission efficiency of AlGaN depends strongly on the edge-type TDD.

Fig. 10 shows the PL peak intensity as measured at RT for 255-nm-emission AlGaN QWs as a function of XRC (10–12) FWHM. The PL intensity was increased by approximately 80 times by reducing the XRC (10–12) FWHM from 1400 to 500 arcsec. The PL intensity increased rapidly when the FWHM of the XRC was reduced to 500–800 arcsec. The rapid increase in PL intensity can be explained by a reduction of a non-radiative recombination rate as the distance between TDs becomes greater compared with the carrier diffusion length in the QW. We obtained similar enhancement of the emission from AlGaN QWs with various wavelength QWs. The relationships between IQE and TDD in DUV emission AlGaN QWs were also investigated in Shatalov et al. (2010) and Ban et al. (2011).

The quaternary alloy InAlGaN is attracting considerable attention as a candidate material for realizing DUV LEDs, since efficient UV emission as well as higher hole concentrations can be realized due to In-incorporation effects. The incorporation of a few percent of In into AlGaN is considered to be quite effective for obtaining high IQE because an efficient DUV emission can be obtained due to the In-segregation effect, which has already been investigated for the ternary InGaN alloy. We have described the advantages of the use of the quaternary InAlGaN alloy in Hirayama (2005) and Hirayama et al. (2002b,c, 2009).

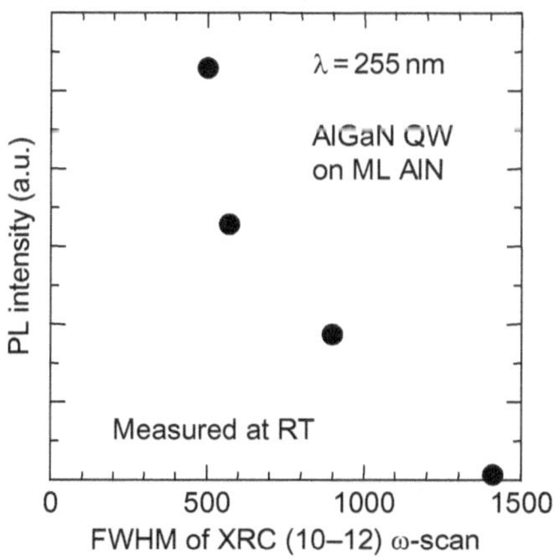

Fig. 10 PL intensity of AlGaN QWs as a function of XRC (10–12) FWHM of AlGaN buffers measured at room temperature.

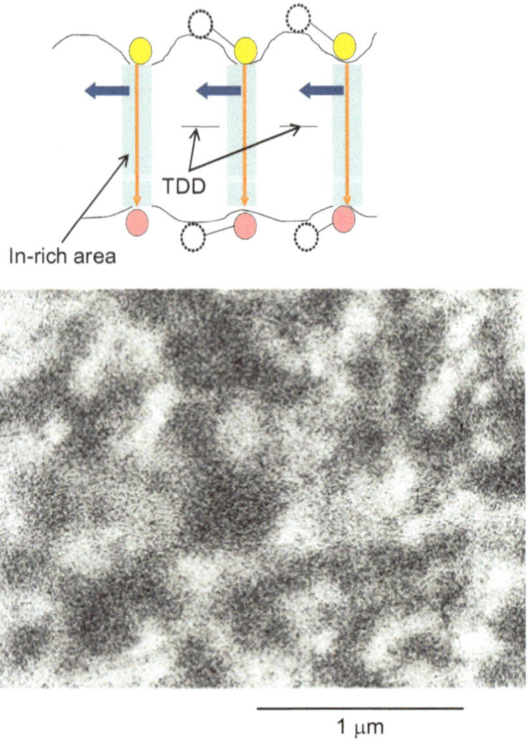

Fig. 11 Cathodeluminescence (CL) image obtained from a quaternary InAlGaN layer and a schematic image showing carrier recombination in an InAlGaN alloy.

Fig. 11 shows a cathodeluminescence (CL) image obtained from a quaternary InAlGaN layer (Hirayama, 2005) and a schematic image of carrier recombination in an InAlGaN alloy. Emission fluctuations in the submicron region were clearly observed in the CL image. The emission fluctuation is considered to be due to carrier localization in the In-segregation area. The CL images obtained for quaternary InAlGaN were very similar to those obtained for InGaN films. Electron–hole pairs localized in the low-potential valley emit before they are trapped in nonradiative centers induced by dislocations. Therefore, the advantage of the In incorporation is that the emission efficiency is less sensitive to TDD.

Fig. 12 shows the temperature dependence of the integrated PL intensity measured for an InAlGaN/InAlGaN MQW with an emission wavelength of 338 nm fabricated on a high-temperature (HT) AlN buffer on sapphire. The TDD of the HT AlN was approximately 2×10^{10} cm^{-2}. The IQE can be roughly estimated from the temperature dependence of the integrated PL

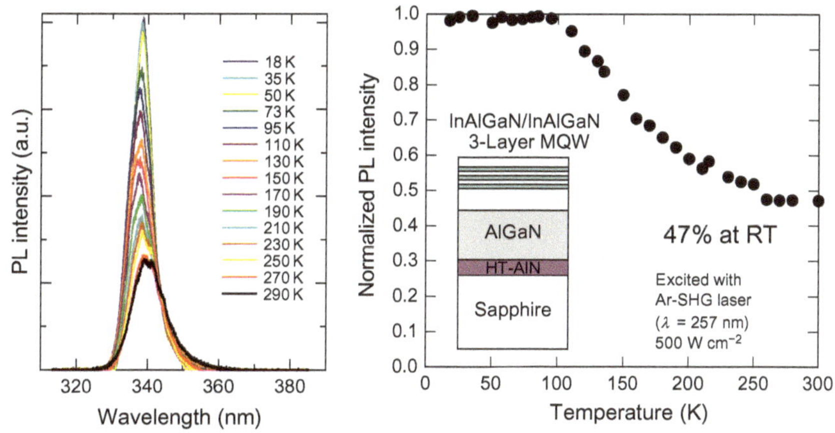

Fig. 12 Temperature dependence of integrated PL intensity measured for an InAlGaN/InAlGaN MQW with an emission wavelength of 338 nm fabricated on a high-temperature (HT) AlN buffer on sapphire.

intensity if we assume that the nonradiative recombination rate is quite low at low temperature. The estimated IQE was approximately 47% at RT, from Fig. 12. We found that high IQE can be obtained for InAlGaN QWs in the wavelength range between 310 and 380 nm, even when using a high TDD template (Hirayama, 2005; Hirayama et al., 2002b,c).

Then we took up the challenge of developing crystal growth for high-quality InAlGaN alloys emitting at the "sterilization" wavelength (280 nm) (Hirayama et al., 2009). The crystal growth of high-Al-composition quaternary InAlGaN is relatively difficult because In incorporation becomes more difficult with increasing growth temperature, which is required in order to maintain the crystal quality of high-Al-content AlGaN. We achieved high-quality quaternary InAlGaN layers with high Al content (>45%) by using relatively low growth rate epitaxy, i.e., 0.03 μm h^{-1}. The emission intensity of a 280-nm-band quaternary InAlGaN QW at RT was increased by five times by reducing the growth rate from 0.05 to 0.03 μm h^{-1}. Fig. 13 shows the PL spectra of a quaternary InAlGaN QW measured at 77K and at RT. We obtained extremely high-intensity PL emission at RT. The ratio of the integrated intensity of the RT-PL against the 77K-PL was 86%. Thus, high IQE was obtained from the quaternary InAlGaN QW at RT.

Fig. 14 summarizes the wavelength dependence of the ratio of the integrated PL intensity [PL measured at RT against PL measured at low temperature (usually below 20K)] investigated in 2008 (Hirayama et al., 2009),

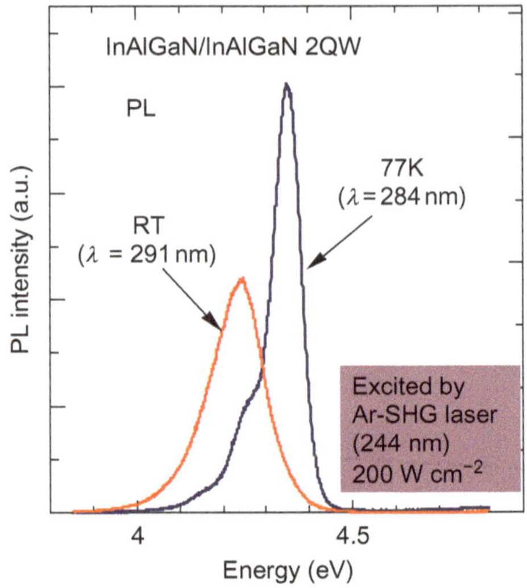

Fig. 13 PL spectra of a quaternary InAlGaN/InAlGaN QW emitting at 282 nm measured at 77K and at RT.

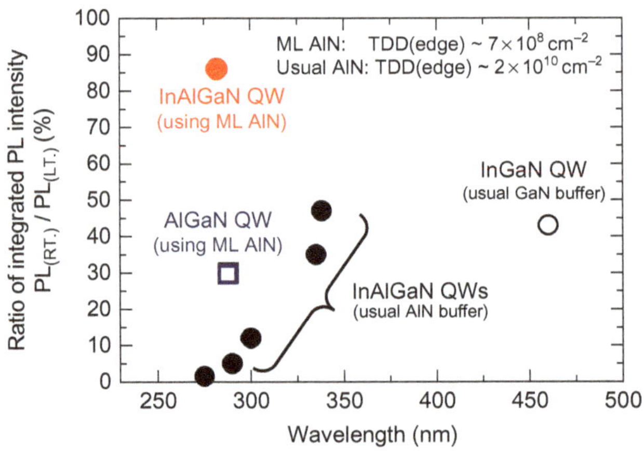

Fig. 14 Wavelength dependence of the ratio of the integrated PL intensity (PL measured at RT against PL measured at low temperature) for AlGaN and quaternary InAlGaN QWs fabricated on conventional high TDD AlN and on low TDD ML AlN templates on sapphire substrates.

which is related to IQE. IQE of 340 nm InAlGaN QW was estimated to be 30–50%, even when we use a high TDD template (TDD$\sim 2\times 10^{10}$ cm^{-2}). However, IQE was reduced to below 2% for short-wavelength (280 nm) QWs; even we use quaternary InAlGaN QWs. On the other hand, we achieved high IQE by introducing low TDD ML AlN templates. The ratios of the integrated PL intensity obtained for 280 nm QWs were approximately 30% and 86% for an AlGaN QW and an InAlGaN QW, respectively, when we use low TDD ML AlN templates (TDD $\sim 7 \times 10^8$ cm^{-2}). The IQE at RT can also be estimated from the PL intensity observed at RT. We observed higher IQE values (50–60%) for AlGaN QWs by realizing a further reduction of the TDD and by optimizing the AlGaN QW growth conditions.

5. 222–351 NM AlGaN AND InAlGaN DUV LEDs

AlGaN and quaternary InAlGaN MQW DUV LEDs were fabricated on low TDD ML AlN templates (Fujikawa et al., 2012; Hirayama et al., 2007, 2008a,b, 2009, 2010a,b, 2014a,b; Maeda and Hirayama, 2014; Mino et al., 2012). Fig. 15 shows a schematic of the structure of an AlGaN-based DUV LED fabricated on a sapphire substrate. Table 1 shows the typical design values of the Al compositions (x) in the $Al_xGa_{1-x}N$ wells, the buffer and barrier layers, and the EBLs that were used for the 222–273 nm AlGaN MQW LEDs. High-Al-composition AlGaN layers were used in order to obtain short-wavelength DUV emissions, as shown in Table 1. A typical LED structure consisted of an approximately 4-μm-thick undoped ML

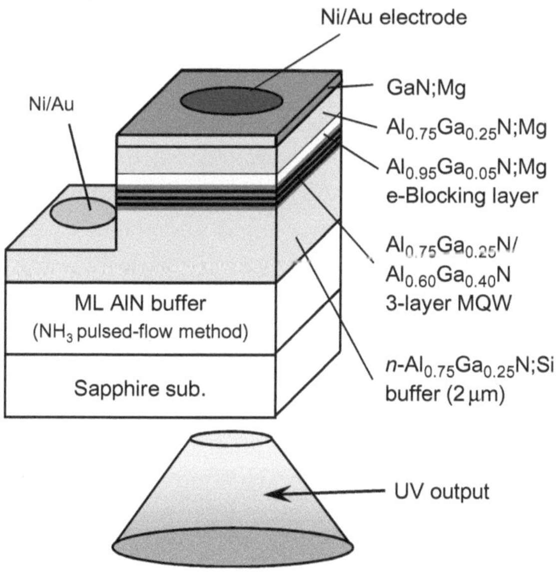

Fig. 15 Schematic structure of a typical AlGaN-based DUV LED fabricated on a sapphire substrate.

Table 1 Typical Design Values of Al Compositions (x) in $Al_xGa_{1-x}N$ Wells, Buffer and Barrier Layers, and Electron-Blocking Layers (EBLs) Used for 222–273 nm AlGaN MQW LEDs

Wavelength (nm)	Well	Barrier and Buffer	Electron-Blocking Layer
222	0.83	0.89	0.98
227	0.79	0.87	0.98
234	0.74	0.84	0.97
248	0.64	0.78	0.96
255	0.60	0.75	0.95
261	0.55	0.72	0.94
273	0.47	0.67	0.93

AlN buffer layer grown on sapphire, a 2-μm-thick Si-doped AlGaN buffer layer, followed by a three-layer undoped MQW region consisting of approximately 1.5-nm-thick AlGaN wells and 7-nm-thick AlGaN barriers, an approximately 20-nm-thick undoped AlGaN barrier, a 15-nm-thick Mg-doped AlGaN EBL, a 10-nm-thick Mg-doped AlGaN p-layer, and an approximately 20-nm-thick Mg-doped GaN contact layer. The quantum well thickness was varied within the range between 1.3 and 2 nm. Thin quantum wells are preferable for AlGaN QWs in order to suppress the effects of large piezoelectric fields in the well. Ni/Au electrodes were used for both n-type and p-type electrodes. The typical size of the p-type electrode was $300 \times 300\ \mu m^2$. The output power that radiated into the back of the LED was measured using a Si photodetector located behind the LED sample, which was calibrated to measure the luminous flux so that the output power of a fabricated flip-chip LED device gives an accurate value. The output power of the flip-chip LED was measured precisely using an integrated sphere system (Mino et al., 2012). The LEDs were measured under "bare wafer" or "flip-chip" conditions. The forward voltages (Vf) of the bare wafer and the flip-chip samples with an injection current of 20 mA were approximately 15 and 8.3 V, respectively.

Fig. 16 shows the electroluminescence (EL) spectra of the fabricated AlGaN and InAlGaN MQW LEDs with emission wavelengths between 222 and 351 nm, all measured at RT with an injection current of approximately 50 mA. As can be seen in Fig. 16, single-peaked operation was obtained for every sample. The deep-level emissions were two orders of magnitude smaller than that of the main peak.

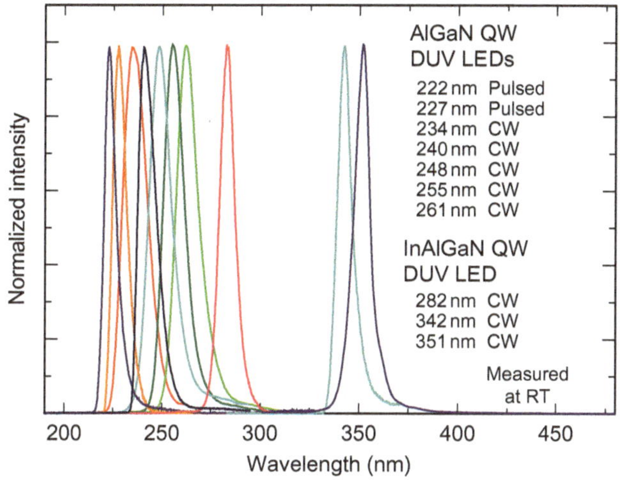

Fig. 16 Electroluminescence (EL) spectra of fabricated AlGaN and quaternary InAlGaN MQW LEDs with emission wavelengths between 222 and 351 nm, all measured at room temperature (RT) with injection currents of around 50 mA.

Fig. 17 shows the EL spectra of a 227-nm AlGaN LED on a log scale (Hirayama et al., 2008b). We obtained single-peaked EL spectra, even for sub-230 nm wavelength LEDs. The deep-level emissions with wavelengths at around 255 and 330–450 nm were more than two orders of magnitude smaller than the main peak. These peaks may correspond to deep-level emissions associated with Mg acceptors or other impurities. The output power of the 227-nm LED was 0.15 mW at an injection current of 30 mA, and the maximum EQE was 0.2% under RT pulsed operation. The pulse width and the repetition frequency were 3 μs and 10 kHz, respectively.

Fig. 18 shows (A) the EL spectra for various injection currents and (B) current vs output power (*I–L*) and EQE (η_{ext}) characteristics for a 222-nm AlGaN MQW LED measured under RT pulsed operation (Hirayama et al., 2010b). Single-peaked operation of a 222-nm DUV AlGaN MQW LED was realized, which is the shortest record wavelength ever reported for a QW LED. The output power of the 222 nm LED was 0.14 μW at an injection current of 80 mA, and the maximum EQE was 0.003% under RT pulsed operation.

It has been reported that "normal" *c*-axis direction emission (vertical emission) is difficult to obtain from an AlN (0001) or a high-Al-content AlGaN surface because the optical transition between the conduction band and the top of the valence band is mainly only allowed for light that has its

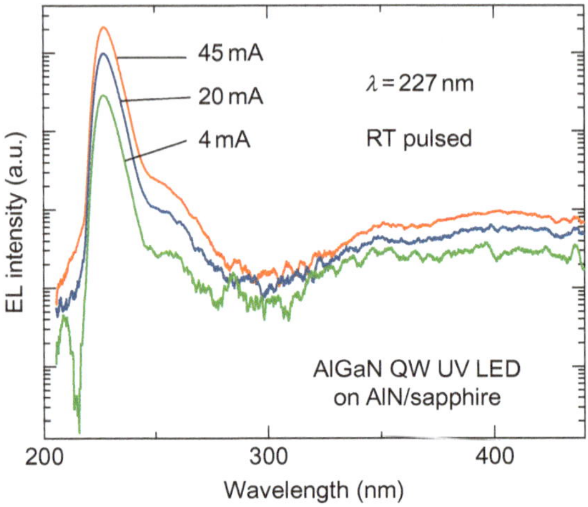

Fig. 17 EL spectra on a log scale of a 227-nm AlGaN DUV LED for various injection currents.

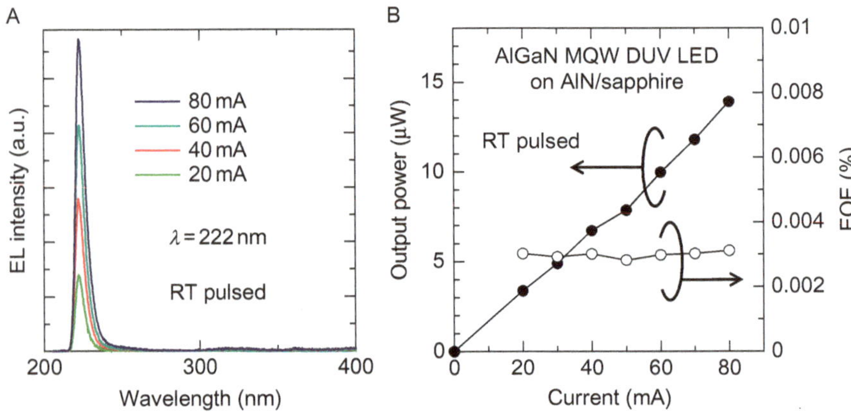

Fig. 18 (A) EL spectra for various injection currents and (B) current vs output power (*I*–*L*) and EQE (η_{ext}) characteristics for a 222-nm AlGaN MQW LED measured under RT pulsed operation.

electric field parallel to the *c*-axis direction of AlN ($E||c$) (Taniyasu et al., 2006). The suppression of the vertical emission is a quite severe problem for AlGaN-based DUV LEDs because it results in a significant reduction in the LEE. Several groups have reported that vertical *c*-axis emission is suppressed for high-Al-content AlGaN QWs (Banal et al., 2009; Kawanishi et al., 2006). Banal et al. reported that the critical Al composition for "polarization switching" could be expanded to approximately 0.82 by

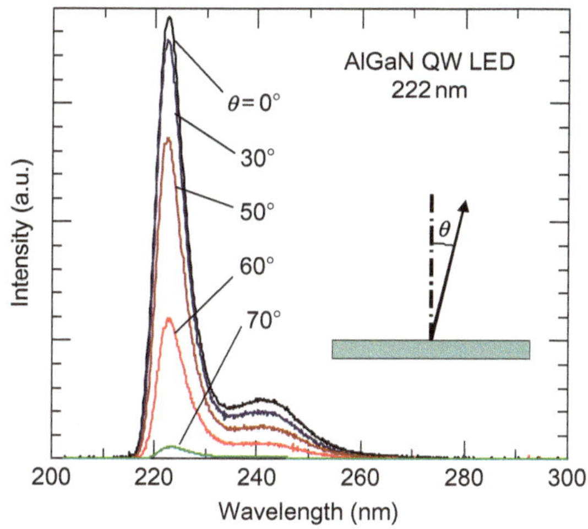

Fig. 19 Radiation angle dependence of emission spectra of a 222-nm AlGaN QW DUV LED.

using a very thin (1.3 nm) QW, when AlGaN QW was fabricated on an AlN/sapphire template (Banal et al., 2009).

Fig. 19 shows the radiation angle dependence of emission spectra of a 222-nm AlGaN QW LED. We demonstrated that "normal" *c*-axis-direction emission can be obtained, even for short-wavelength (222 nm) LEDs with a high-Al-composition AlGaN QW, as shown in Fig. 19 (Hirayama et al., 2010b). It was found that vertical *c*-axis emission can be obtained for an AlGaN QW LED on AlN/sapphire, even when the Al composition range of the AlGaN QW is as high as 83%.

We fabricated quaternary InAlGaN-based DUV LEDs in order to increase IQE and EIE of the DUV LEDs. Fig. 20 shows a schematic structure and a cross-sectional TEM image of an InAlGaN QW DUV LED. We confirmed that the surface roughness of the InAlGaN layer was significantly improved by introducing a Si-doped InAlGaN buffer layer. The InAlGaN-based DUV LED is considered to be attractive for achieving high EQE due to the higher IQE and higher hole concentration obtained by In-segregation effects. Fig. 21 shows (A) an EL spectrum and (B) the current vs output power (*I–L*) and EQE characteristics of an InAlGaN-based QW DUV LED with an emission wavelength of 282 nm. The maximum output power and EQE were 10.6 mW and 1.2%, respectively, under RT CW operation. From these results, we found that quaternary InAlGaN QWs and p-type InAlGaN are quite useful for achieving high-efficiency DUV LEDs (Hirayama et al., 2009).

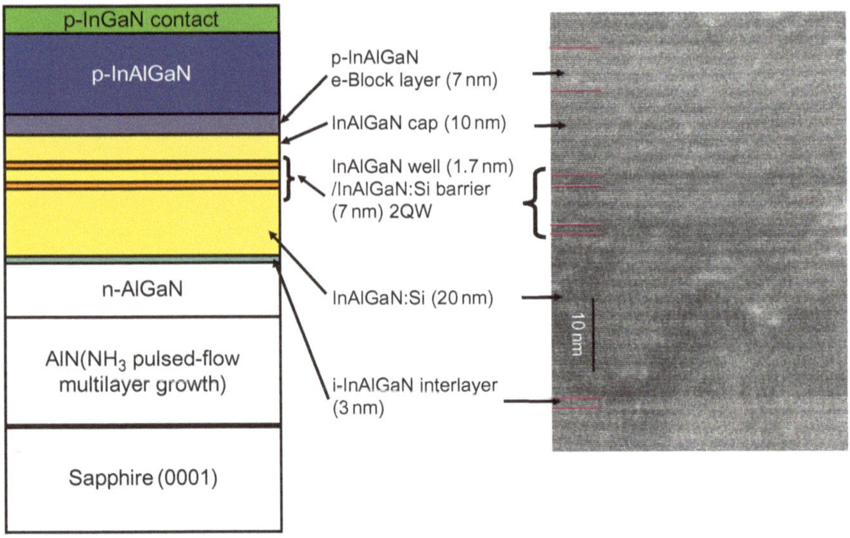

Fig. 20 Schematic structure and cross-sectional TEM image of a quaternary InAlGaN QW DUV LED with emission wavelength at 282 nm.

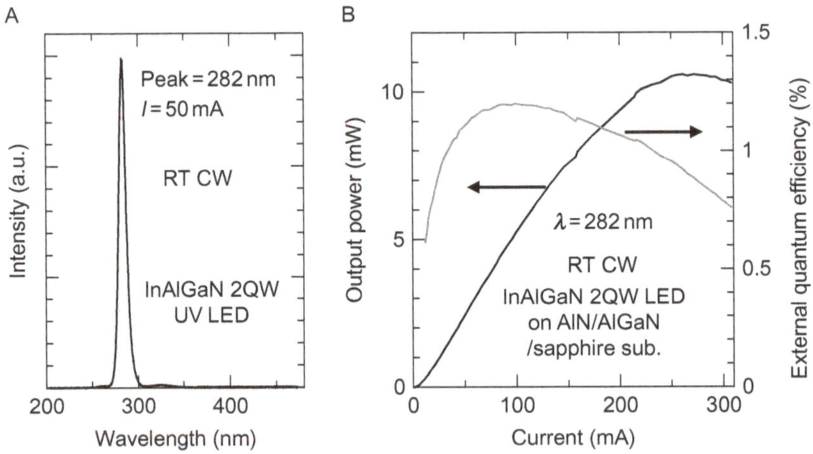

Fig. 21 (A) EL spectrum and (B) current vs output power (*I–L*) and EQE characteristics of the quaternary InAlGaN QW DUV LED with emission wavelength at 282 nm.

6. INCREASE IN EIE BY MQB

Despite achieving high IQE in DUV as mentioned earlier, EQE of the LED was still as low as 1–2% (Hirayama et al., 2009). The low EQE figures for AlGaN DUV LEDs compared with those for InGaN blue LEDs are a result of low EIE into the QW due to electron leakage caused by low hole concentrations in the p-type AlGaN layers, as well as inferior LEEs (<8%) due to strong UV absorption in the p-GaN contact layer and the p-side electrode. The values of EIE for 250–280 nm band AlGaN-based DUV LEDs were roughly estimated to be 10–30% (Hirayama et al., 2009). We have introduced an MQB as an EBL in an AlGaN QW LED and have consequently achieved a marked increase in EIE (Hirayama et al., 2010a).

Fig. 22 shows schematic images of the electron flow for AlGaN DUV LEDs with (A) an MQB EBL and (B) a conventional single-barrier EBL. A large barrier height is required for the EBL to obtain a sufficiently high EIE for the suppression of overflow electrons above the QW into the p-type AlGaN layers. We have tried using AlN or high-Al-composition (Al > 0.95)

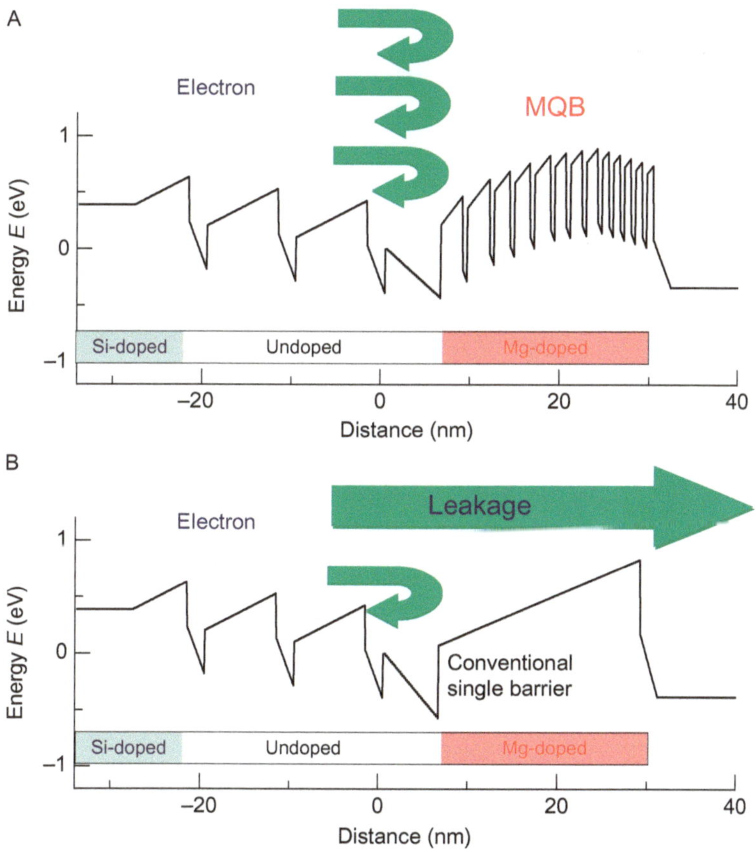

Fig. 22 Schematic images of the electron flow for AlGaN DUV LEDs with (A) an MQB EBL and (B) a conventional single-barrier EBL.

AlGaN layers for the EBL (Hirayama et al., 2008b, 2009, 2010b); however, the barrier heights of these EBLs were still not sufficient high to obtain the desired high EIE. Indeed, EIE is estimated to be particularly low (<20%) for short-wavelength AlGaN LEDs (<250 nm).

Such a material limitation can be overcome by enhancing the "effective" barrier height through the introduction of an MQB, which causes multi-reflection effects in the wavefunctions. The MQB was predicted theoretically by Iga et al. (1986), and the effects were demonstrated experimentally in GaInP/AlInP red LDs (Kishino et al., 1991). It has been reported that the effective electron barrier height of an MQB in comparison with a bulk potential barrier is increased by as much as 30% for GaAs/AlAs and by 50% for GaInAs/InP MQBs. It is believed that the use of AlN/AlGaN or AlGaN/AlGaN MQBs would be quite effective for increasing the "effective" barrier height of an EBL and, as a result, would contribute to the realization of high EQE AlGaN DUV LEDs.

Fig. 23 shows electron transmittance through an AlGaN/AlGaN MQB (red line) and a conventional single-barrier EBL (black line) calculated for a 250-nm AlGaN QW LED. The multireflection effects in the heterostructures are analyzed by a transfer-matrix method. It was shown that the "effective" electron barrier height of an MQB in comparison with a conventional single-barrier EBL is increased by maximally two times for an AlGaN/AlGaN MQB by using barriers with thickness modulation.

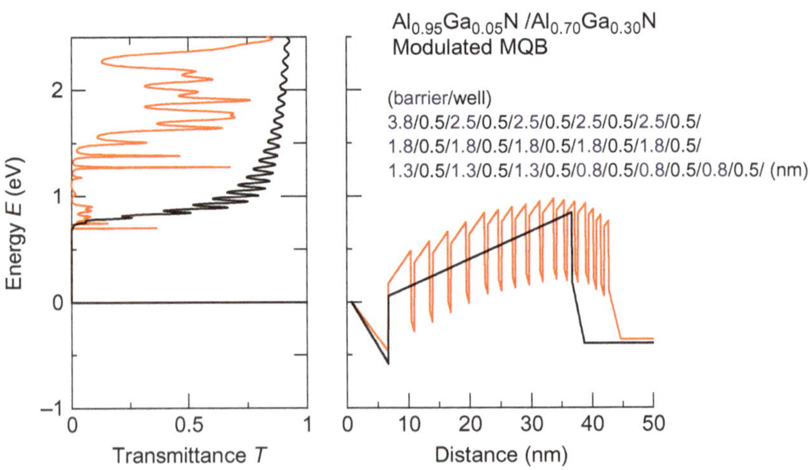

Fig. 23 Electron transmittance through AlGaN/AlGaN MQB (*red line*) and conventional single-barrier EBL (*black line*) calculated for a 250-nm-band AlGaN QW LED.

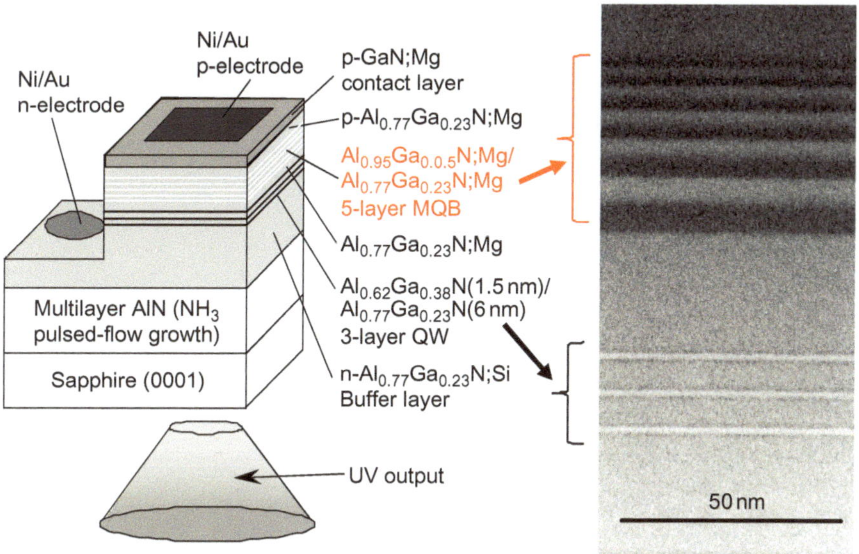

Fig. 24 Schematic structure and cross-sectional TEM image of a 250-nm AlGaN QW DUV LED with an MQB.

Fig. 24 shows a schematic structure and a cross-sectional TEM image of a fabricated 250-nm AlGaN QW DUV LED with an MQB EBL. We investigated an appropriate MQB structure experimentally for use with 250 nm DUV LEDs. We found that the insertion of an initial thick barrier is important for obtaining the reflection of lower energy electrons. We also found that thin barriers contribute to the reflection of higher-energy electrons. The optimized MQB structure for a 250-nm AlGaN QW LED was a five-layer $Al_{0.95}Ga_{0.05}N/Al_{0.77}Ga_{0.23}N$ MQB with thicknesses of **7**/4/**5.5**/4/**4**/2.5/**4**/2.5/**4** nm, in which the bold letters are for barriers and the normal letters are for valleys. We should design the total thickness of the MQB to be within 40 nm because a coherent length exists for obtaining the multireflection effect of an MQB.

Fig. 25 shows (A) a current vs output power (*I–L*) and (B) a current vs EQE (η_{ext}) characteristics for a 250-nm AlGaN MQW LED with an MQB and with a single-barrier EBL, both measured under RT CW operation. Significant increases in output power and EQE were observed when the single EBL was replaced by the MQB. The maximum output powers of the 250-nm LED with the MQB and with the single-barrier EBL were 15 and 2.2 mW, respectively. The EQE of a 250-nm LED was increased by approximately four times by the introduction of the MQB. From

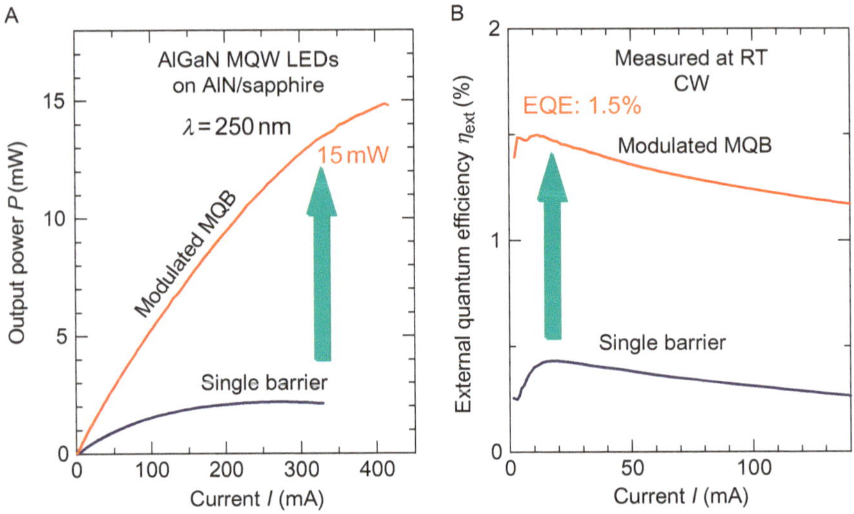

Fig. 25 (A) Current vs output power (*I–L*) and (B) current vs EQE (η_{ext}) characteristics for the 250 nm AlGaN MQW LEDs with an MQB and with a single EBL.

Fig. 25, we have estimated that the EIE of the 250-nm LED was improved from approximately 25% to more than 80% by introducing the MQB.

Fig. 26 shows current vs output power (*I–L*) characteristic for a 237-nm AlGaN MQW LED with an MQB and with a single-barrier EBL, both measured under RT CW operation. The enhancement of EIE when using the MQW was found to be extremely high for short-wavelength DUV LEDs, as seen in Fig. 26. The output power of the 234-nm LED was increased by approximately 12 times by replacing a single-barrier EBL by an MQB.

Fig. 27 summarizes the wavelength dependence of EQE of AlGaN DUV LEDs with MQBs and single-barrier EBLs. The enhancement factors of the value of EQE by introducing the MQB are approximately 10, 4, and 3 times for 235, 250, and 270 nm AlGaN LEDs, respectively. We demonstrated high-output-power 270 nm AlGaN MQW LED with an MQB measured under RT CW operation. A CW output power of 33 mW was obtained for a bare-chip sample. Higher output power may be obtained by performing heat dissipation by using a flip-chip geometry. The value of the EQE for a 270-nm AlGaN DUV LED with an MQB was 3.8% without using any structures for increasing LEE (Fujikawa et al., 2012).

Fig. 28 summarizes the maximum output powers of AlGaN- and InAlGaN-based DUV LEDs fabricated on low TDD ML AlN templates that were achieved by the RIKEN between 2007 and 2011 (Fujikawa et al.,

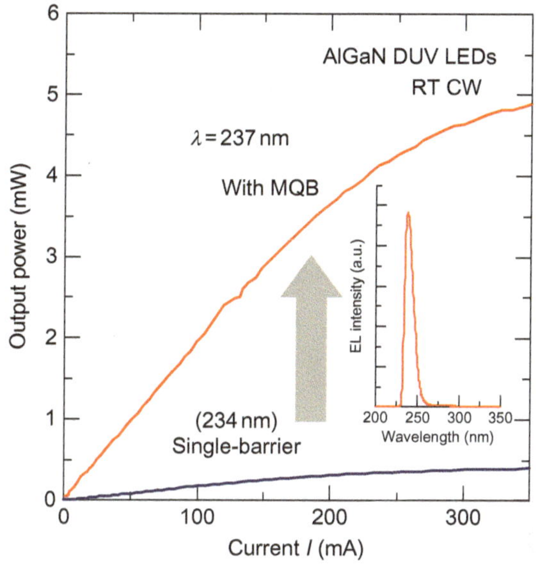

Fig. 26 Current vs output power (*I–L*) characteristics for the 237 nm AlGaN MQW LEDs with an MQB and with a single EBL.

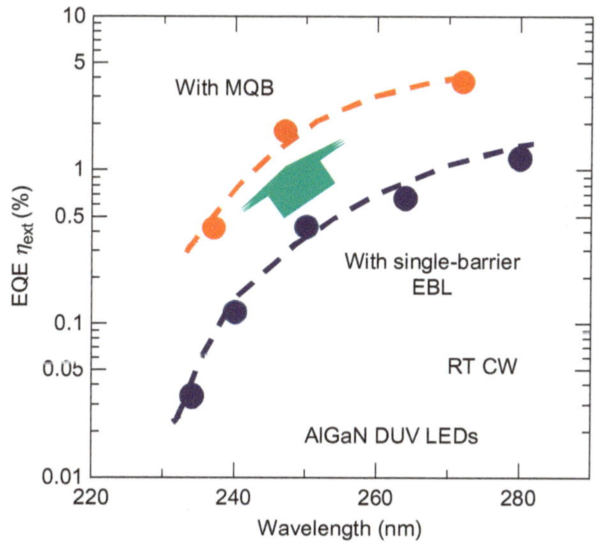

Fig. 27 Wavelength dependence of EQE of AlGaN DUV LEDs with MQBs and single EBLs.

2012; Hirayama et al., 2008a,b, 2009, 2010a,b; Maeda and Hirayama, 2014). We have achieved significant increases in the EQE and output powers of AlGaN-based DUV LEDs by introducing low TDD AlN templates and MQB EBLs. The maximum output powers obtained were 15–33 mW

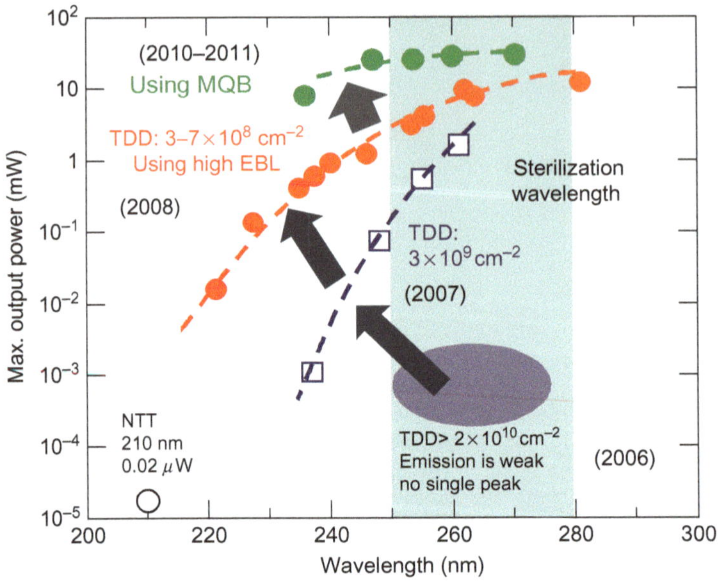

Fig. 28 Maximum output powers of AlGaN and InAlGaN-based DUV LEDs fabricated on low TDD ML AlN templates that were achieved by RIKEN between 2007 and 2011 (Fujikawa et al., 2012; Hirayama et al., 2008a,b, 2009, 2010a,b; Maeda and Hirayama, 2014).

for 245–270 nm single-chip LEDs. These achievements will contribute to accelerating the practical application of DUV LED and to expanding them to a wide range of applications.

RIKEN and Panasonic have developed commercially available DUV LED modules for sterilization use. For developing a commercially available device, the reproducibility and the uniformity of an AlN template and an AlGaN LED layer structure are required for obtaining a constant high EQE and a long device lifetime. To obtain the reproducibility for an AlN and an AlGaN layer is particularly difficult because the growth condition is very sensitive for a vapor reaction between NH_3 and TMAl, which are induced by a high growth temperature (1200–1400°C). We achieved a highly uniform AlN template grown by a NH_3 pulsed-flow ML growth method on sapphire using a 3 × 2-in. wafer product type MOCVD. The fluctuation of the FWHM of XRCs in an AlN template on sapphire was within 5%. We also constantly obtain XRC (10–12) FWHM of 340 arcsec for AlN templates on sapphire. Technique to obtain such a highly uniform and low TDD template wafer is suitable for a product of commercial DUV LEDs. Fig. 29 shows (A) a bird's eye view and (B) operating properties of a

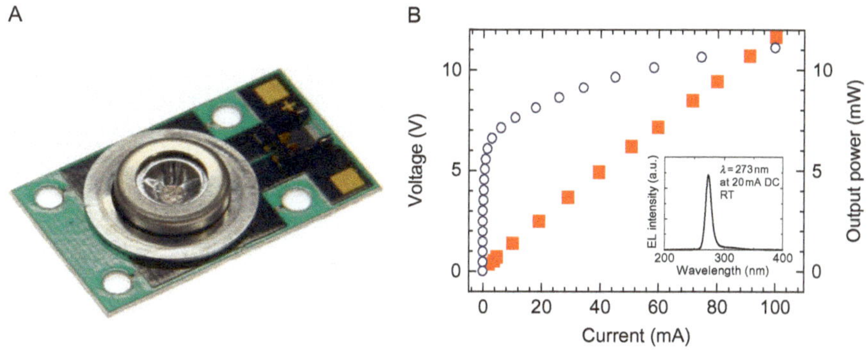

Fig. 29 (A) A bird's eye view and (B) operating properties of a commercially available 270-nm 10-mW DUV LED module for an application to sterilization developed by RIKEN and Panasonic.

270-nm 10-mW DUV LED module for an application to sterilization. We integrated six chips for the DUV LED module. A lifetime longer than 10,000 h has been already achieved for devices with EQE of 2–3% (Mino et al., 2012, 2013).

7. FUTURE LED DESIGN FOR HIGH LEE

The improvement in LEE is a particularly important for a research of an AlGaN DUV LED because LEE of an AlGaN DUV LED is quite low. The increase in LEE is not so easy because of the lack of suitable transparent p-type contact layers, transparent p-type electrodes, highly reflective p-type electrodes useful in a DUV wavelength range.

Fig. 30 shows schematic images of the improvement in LEE of an AlGaN DUV LED that can be achieved by replacing the conventional p-GaN contact-layer structure by the novel LED structure proposed in this work (Hirayama et al., 2014b). In a conventional DUV LED, the QW emission in the upward direction is completely absorbed by the p-GaN contact layer. The DUV emission in the downward direction is reflected by the sapphire/air interface (<16% of the downward light is extracted and >84% of the downward light is reflected mainly by a total reflection). As a result, LEE of a conventional DUV LED is reduced to be smaller than 8%. Although we use an encapsulating technique, photonic nanostructures on the surface of sapphire substrates, or a growth geometry on a PSS, the LEE improvement is not so high (maximum LEE is expected to around 20%), because of a heavy absorption of a p-GaN contact layer.

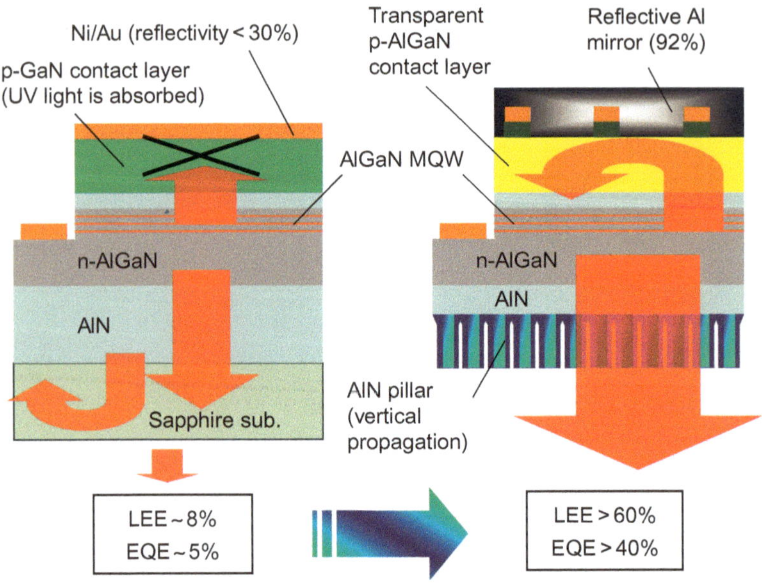

Fig. 30 Schematic images of the improvement in LEE of an AlGaN DUV LED by introducing a transparent p-AlGaN contact layer, a highly reflective p-type electrode, and a vertical-light-propagation AlN pillar array.

To realize high LEE, the combination of a transparent contact layer, a highly reflective p-type electrode, and a vertical light-propagation photonic structure is desirable. Recently, we have fabricated a DUV LED with a high-Al-content p-AlGaN contact layer. We found that the p-AlGaN can act as a highly transparent p-type contact layer. The reflectance of a conventional Ni/Au p-electrode is low (~25%) and is not useful as a highly reflective mirror. The reflectance of Al in DUV is 92%, but Ohmic contacts are difficult to obtain. The introduction of a mesh-type electrode with Al mirror windows is one of the solutions for a highly reflective p-type electrode, as shown on the right of Fig. 30. If the reflectivity of the p-type electrode is not sufficiently high (~80%), a photonic structure with a vertical light-propagation property is required to extract the light with the minimum number of reflections. In other words, an efficient light coupling into a photonic structure such as a pillar array is important for obtaining an efficient light extraction.

LEE of an AlGaN DUV LED similar to we proposed was investigated by Ryu et al. (2013), based on the basis of finite-difference time-domain (FDTD) simulations. They also concluded from the calculation that the combination of a vertical LED geometry with a photonic nanostructure and a transparent p-AlGaN contact layer is important for obtaining high

LEE in AlGaN DUV LEDs. They demonstrated that LEE of the vertical AlGaN DUV LED is improved by about 10 times, when the thickness of the p-GaN contact layer is reduced from 25 nm to zero. They also demonstrated from the simulation that a maximum LEE as high as 72% can be obtained for the transverse-electric mode, if a transparent p-AlGaN contact layer is introduced in the vertical LEDs. Such results match well with our proposal.

We will demonstrate a very simple estimation of LEE for a proposed DUV LED. We assume that the equivalent reflectance for the upward-direction light is 80%, which is determined by the reflectivity of a p-electrode and the absorption through a p-AlGaN contact layer. We also assume that the equivalent coupling coefficient of the downward light into a pillar array is 40%. In other words, 40% of the downward light can be extracted through the pillar array. Of course, they are average values taking into account all propagating light to every direction. At first, 50% light is emitted from the QW toward the downward direction and 40% light is reflected and returned by the p-electrode, and then 90% light reaches in front of a pillar array. The amount of the first-time extraction light through the pillar array is calculated to be $90\% \times 0.4 = 36\%$. Then, 54% light is reflected by the pillar array. The amount of the second-time extraction light through the pillar array is obtained as $54\% \times 0.8 \times 0.4 = 17.3\%$. By the same manner, the amounts of the third- and fourth-time extraction light are calculated as 8.3% and 4%, respectively. By integrating the amounts of the first- to the third-time extraction light, we can obtain high LEE of more than 61.6%. The equivalent coupling coefficient of 40% is considered to be reasonable by comparing with FDTD results. Through the above estimations, we can obtain high LEE with the minimum number of reflections.

We demonstrated a DUV LED with a highly transparent p-AlGaN contact layer. Fig. 31 shows the transparency of a high-Al-content Mg-doped p-AlGaN layer, grown on an AlGaN/AlN/sapphire template usually used for a DUV LED. We confirmed that the transparency of a 120-nm-thick p-AlGaN layer with an Al composition of 60% is higher than 94% for 279 nm DUV light, as measured using a spectrophotometer. The transparency of the p-AlGaN contact layer used for an actual DUV LED was estimated to be approximately 97% taking into account the contact layer thickness (70 nm).

As a p-type electrode, we replaced the conventional Ni(25 nm)/Au(150 nm) with a highly reflective Ni(1 nm)/Al(150 nm). Current injection was made possible by inserting a very thin Ni layer (<1 nm) between

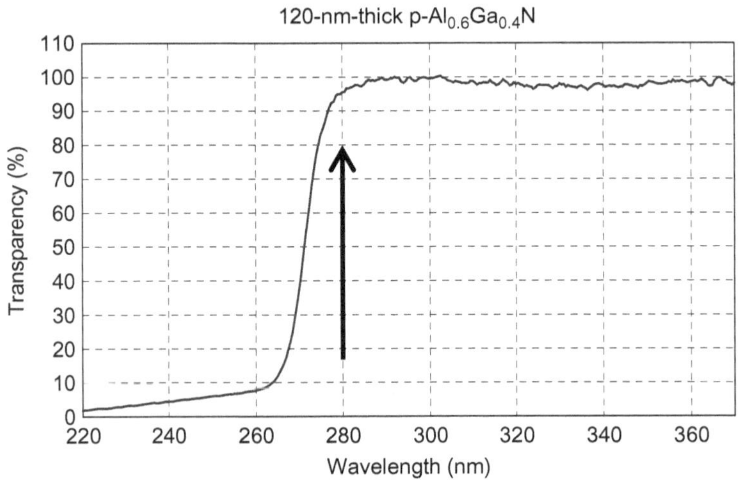

Fig. 31 Transparency of a high-Al-content Mg-doped p-AlGaN layer, grown on an AlGaN/AlN/sapphire template.

the Al layer and the p-AlGaN contact layer. Fig. 32 shows the wavelength dependent of the reflectivity for various type p-type electrodes used for AlGaN DUV LEDs. In the actual 270 nm DUV LED devices, the reflectivity of the p-type electrode was increased from 30% to 64% by replacing the conventional Ni(25 nm)/Au(150 nm) electrode with a Ni(1 nm)/Al(150 nm) electrode.

Fig. 33 shows the current vs EQE (η_{ext}) characteristics of 279 nm AlGaN MQW DUV LEDs with transparent p-AlGaN contact layers, comparing them with conventional Ni(25 nm)/Au(150 nm) and highly reflective Ni(1 nm)/Al(150 nm) p-electrodes measured under RT CW operation. EQE was significantly increased from 4% to 7% (by 1.7 times) owing to the increase in LEE induced by replacing the conventional Ni/Au p-electrode with a highly reflective Ni/Al electrode. We can estimate that the transparency of the p-AlGaN contact layer is higher than 95% from the enhancement factor of LEE. The LEE enhancement factors that were achieved in 265–279 nm DUV LEDs by replacing the Ni/Au p-electrode with the highly reflective Ni/Al electrode were approximately 1.3–1.7.

We observed an increase in the Vf when p-GaN was replaced with a transparent p-AlGaN contact layer. Vf at an injection current of 5 mA increased from 11 to 17 V for a bare wafer DUV LED when the contact layer was changed from p-GaN to p-AlGaN. Therefore, the high resistivity of p-contacts caused by the low hole concentration is still a problem for the

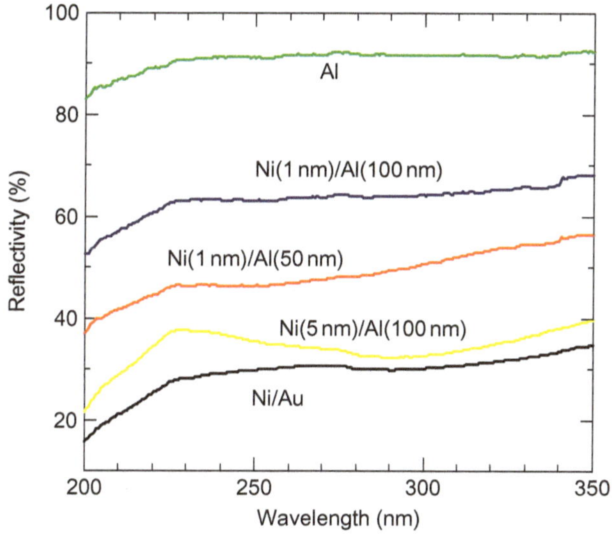

Fig. 32 Wavelength dependent on the reflectivity for various type p-type electrodes for AlGaN DUV LEDs.

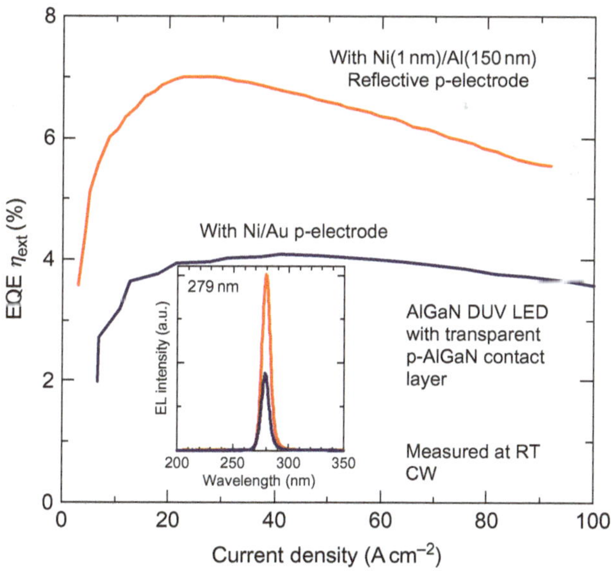

Fig. 33 Current vs EQE (η_{ext}) characteristics of 279 nm AlGaN MQW DUV LEDs with transparent p-AlGaN contact layers in comparison with conventional Ni/Au and highly reflective Ni/Al p-electrodes, measured under RT CW operation.

use of p-AlGaN contact layers. The increase in Vf causes a reduction in the wall-plug efficiency of the device. We aim to improve Vf by using mesh-type p-electrodes.

We are planning to realize the proposed structure as shown in Fig. 30 by introducing pillar AlN structure grown on PSS and laser lift-off techniques, with the aim of achieving a marked improvement in LEE for UVC LEDs. We hope that the application of such devices to fields involving sterilization will be widely expanded through the achievement of high-efficiency UVC LEDs in the near future.

8. SUMMARY

We demonstrated the recent progress in AlGaN-based DUV LEDs, which has been achieved by the development of crystal growth techniques for wide-bandgap AlN and AlGaN. Significant increases in IQE have been achieved for AlGaN QW DUV emissions by using low TDD AlN on sapphire grown by an NH_3 pulsed-flow multilayer growth method. 222–351 nm DUV LEDs have been achieved using the high IQE emission layers. EIE of the DUV LED was significantly increased by controlling the electron flow using an MQB. We also demonstrated the improvement of LEE by using a transparent p-AlGaN contact layer and a highly reflective p-electrode. The maximum EQE obtained was 7% for a 279-nm DUV LED. EQE could be increased up to several tens of percent in the near future by improving LEE by utilizing transparent contact layers and pillar array buffer layers.

REFERENCES

Adivarahan, V., Wu, S., Zhang, J.P., Chitnis, A., Shatalov, M., Madavilli, V., Gaska, R., Khan, M.A., 2004a. High-efficiency 269 nm emission deep ultraviolet light-emitting diodes. Appl. Phys. Lett. 84, 4762.

Adivarahan, V., Sun, W.H., Chitnis, A., Shatalov, M., Wu, S., Maruska, H.P., Asif Khan, M., 2004b. 250 nm AlGaN light-emitting diodes. Appl. Phys. Lett. 85, 2175.

Ban, K., Yamamoto, J., Takeda, K., Ide, K., Iwaya, M., Takeuchi, T., Kamiyama, S., Akasaki, I., Amano, H., 2011. Internal quantum efficiency of whole-composition-range AlGaN multiquantum wells. Appl. Phys. Expr. 4 (5), 052101.

Banal, R.G., Funato, M., Kawakami, Y., 2009. Optical anisotropy in [0001]-oriented $Al_xGa_{1-x}N$/AlN quantum wells ($x>0.69$). Phys. Rev. B 79, 121308(R).

Fujikawa, S., Takano, T., Kondo, Y., Hirayama, H., 2008. Realization of 340-nm-band high-output-power (7 mW) InAlGaN quantum well ultraviolet light-emitting diode with p-type InAlGaN. Jpn. J. Appl. Phys. 47, 2941.

Fujikawa, S., Hirayama, H., Maeda, N., 2012. High-efficiency AlGaN deep-UV LEDs fabricated on a- and m-axis oriented c-plane sapphire substrates. Phys. Status Solidi C 9 (3–4), 790–793.

Fujioka, A., Misaki, T., Murayama, T., Narukawa, Y., Mukai, T., 2010. Improvement in output power of 280-nm deep ultraviolet light-emitting diode by using AlGaN multi quantum wells. Appl. Phys. Expr. 3 (4), 041001.

Grandusky, J.R., Gibb, S.R., Mendrick, M.C., Moe, C., Wraback, M., Schowalter, L.J., 2011. High output power from 260 nm pseudomorphic ultraviolet light-emitting diodes with improved thermal performance. Appl. Phys. Expr. 4 (8), 082101.

Grandusky, J.R., Chen, J., Gibb, S.R., Mendrick, M.C., Moe, C.G., Rodak, L., Garrett, G.A., Wraback, M., Schowalter, L.J., 2013. 270 nm pseudomorphic ultraviolet light-emitting diodes with over 60 mW continuous wave output power. Appl. Phys. Expr. 6 (3), 032101.

Han, J., Crawford, M.H., Shul, R.J., Figiel, J.J., Banas, M., Zhang, L., Song, Y.K., Zhou, H., Nurmikko, A.V., 1998. AlGaN/GaN quantum well ultraviolet light emitting diodes. Appl. Phys. Lett. 73, 1688.

Hirayama, H., 2005. Quaternary InAlGaN-based high-efficiency ultraviolet light-emitting diodes. J. Appl. Phys. 97, 091101.

Hirayama, H., Enomoto, Y., Kinoshita, A., Hirata, A., Aoyagi, Y., 2002a. Efficient 230–280 nm emission from high-Al-content AlGaN-based multi-quantum wells. Appl. Phys. Lett. 80, 37.

Hirayama, H., Kinoshita, A., Yamabi, T., Enomoto, Y., Hirata, A., Araki, T., Nanishi, Y., Aoyagi, Y., 2002b. Marked enhancement of 320–360 nm UV emission in quaternary $In_xAl_yGa_{1-x-y}N$ with In-segregation effect. Appl. Phys. Lett. 80, 207.

Hirayama, H., Enomoto, Y., Kinoshita, A., Hirata, A., Aoyagi, Y., 2002c. Room-temperature intense 320 nm-band UV emission from quaternary InAlGaN-based multi-quantum wells. Appl. Phys. Lett. 80, 1589.

Hirayama, H., Akita, K., Kyono, T., Nakamura, T., Ishibashi, K., 2004. High-efficiency 352 nm quaternary InAlGaN-based ultraviolet light-emitting diodes grown on GaN substrates. Jpn. J. Appl. Phys. 43, L1241.

Hirayama, H., Yatabe, T., Noguchi, N., Ohashi, T., Kamata, N., 2007. 231-261 nm AlGaN deep-ultraviolet light-emitting diodes fabricated on AlN multilayer buffers grown by ammonia pulse-flow method on sapphire. Appl. Phys. Lett. 91, 071901.

Hirayama, H., Yatabe, T., Ohashi, T., Kamata, N., 2008a. Remarkable enhancement of 254–280 nm deep ultraviolet emission from AlGaN quantum wells by using high-quality AlN buffer on sapphire. Phys. Status Solidi C 5, 2283.

Hirayama, H., Noguchi, N., Yatabe, T., Kamata, N., 2008b. 227 nm AlGaN light-emitting diode with 0.15 mW output power realized using thin quantum well and AlN buffer with reduced threading dislocation density. Appl. Phys. Expr. 1, 051101.

Hirayama, H., Noguchi, N., Fujikawa, S., Norimatsu, J., Takano, T., Tsubaki, K., Kamata, N., 2009. 222-282 nm AlGaN and InAlGaN based high-efficiency deep-UV-LEDs fabricated on high-quality AlN on sapphire. Phys. Status Solidi A 206, 1176.

Hirayama, H., Tsukada, Y., Maeda, T., Kamata, N., 2010a. Marked enhancement in the efficiency of deep-ultraviolet AlGaN light-emitting diodes by using a multiquantum-barrier electron blocking layer. Appl. Phys. Expr. 3, 031002.

Hirayama, H., Noguchi, N., Kamata, N., 2010b. 222 nm deep-ultraviolet AlGaN quantum well light-emitting diode with vertical emission properties. Appl. Phys. Expr. 3, 032102.

Hirayama, H., Maeda, N., Fujikawa, S., Toyoda, S., Kamata, N., 2014a. Improvement light-extraction efficiency of AlGaN Deep-UV LEDs by using transparent contact layers. Optronics 2, 58.

Hirayama, H., Maeda, N., Fujikawa, S., Toyota, S., Kamata, N., 2014b. Recent progress and future prospects of AlGaN-based high-efficiency deep-ultraviolet light-emitting diodes. Jpn. J. Appl. Phys. 53 (10), 100209.

Hwang, S., Morgan, D., Kesler, A., Lachab, M., Zhang, B., Heidari, A., Nazir, H., Ahmad, I., Dion, J., Fareed, Q., Adivarahan, V., Islam, M., Khan, A., 2011. 276 nm substrate-free flip-chip AlGaN light-emitting diodes. Appl. Phys. Expr. 4 (3), 032102.

Iga, K., Uenohara, H., Koyama, F., 1986. Electron reflectance of multiquantum barrier (MQB). Electron. Lett. 22, 1008.
Iida, K., Kawashima, T., Miyazaki, A., Kasugai, H., Mishima, A., Honshio, A., Miyake, Y., Iwaya, M., Kamiyama, S., Amano, H., Akasaki, I., 2004. 350.9 nm UV laser diode grown on Low-dislocation-density AlGaN. Jpn. J. Appl. Phys. 43 (No. 4A), L499–L500.
Kawanishi, H., Senuma, M., Yamamoto, M., Niikura, E., Nukui, T., 2006. Extremely weak surface emission from (0001) *c*-plane AlGaN multiple quantum well structure in deep-ultraviolet spectral region. Appl. Phys. Lett. 89, 081121.
Kinoshita, A., Hirayama, H., Ainoya, M., Hirata, A., Aoyagi, Y., 2000. Room-temperature operation at 333 nm of $Al_{0.03}Ga_{0.97}N/Al_{0.25}Ga_{0.75}N$ quantum-well light emitting diodes with Mg-doped superlattice layers. Appl. Phys. Lett. 77, 175.
Kinoshita, T., Hironaka, K., Obata, T., Nagashima, T., Dalmau, R., Schlesser, R., Moody, B., Xie, J., Inoue, S., Kumagai, Y., Koukitu, A., Sitar, Z., 2012. Deep-ultraviolet light-emitting diodes fabricated on AlN substrates prepared by hydride vapor phase epitaxy. Appl. Phys. Expr. 5 (12), 122101.
Kinoshita, T., Obata, T., Nagashima, T., Yanagi, H., Moody, B., Mita, S., Inoue, S., Kumagai, Y., Koukitu, A., Sitar, Z., 2013. Performance and reliability of deep-ultraviolet light-emitting diodes fabricated on AlN substrates prepared by hydride vapor phase epitaxy. Appl. Phys. Expr. 6 (9), 092103.
Kishino, K., Kikuchi, A., Kaneko, Y., Nomura, I., 1991. Enhanced carrier confinement effect by the multiquantum barrier in 660 nm GaInP/AlInP visible lasers. Appl. Phys. Lett. 58 (17), 1822.
Maeda, N., Hirayama, H., 2014. Realization of high-efficiency deep-UV LEDs using transparent p-AlGaN contact layer. Phys. Status Solidi C 10, 1521.
Mickevičius, J., Tamulaitis, G., Shur, M., Shatalov, M., Yang, J., Gaska, R., 2013. Correlation between carrier localization and efficiency droop in AlGaN epilayers. Appl. Phys. Lett. 103, 011906.
Mino, T., Hirayama, H., Takano, T., Noguchi, N., Tsubaki, K., 2012. Highly-uniform 260 nm-band AlGaN-based deep-ultraviolet light-emitting diodes developed by 2-inch×3 MOVPE system. Phys. Status Solidi C 9, 749.
Mino, T., Hirayama, H., Takano, T., Tsubaki, K., Sugiyama, M., 2013. Development of 260 nm band deep-ultraviolet light emitting diodes on Si substrate. Proc. SPIE 8625, 59.
Nishida, T., Saito, H., Kobayashi, N., 2001. Efficient and high-power AlGaN-based ultraviolet light-emitting diode grown on bulk GaN. Appl. Phys. Lett. 78, 711.
Pernot, C., Kim, M., Fukahori, S., Inazu, T., Fujita, T., Nagasawa, Y., Hirano, A., Ippommatsu, M., Iwaya, M., Kamiyama, S., Akasaki, I., Amano, H., 2010. Improved efficiency of 255–280 nm AlGaN-based light-emitting diodes. Appl. Phys. Expr. 3 (6), 061004.
Ryu, H.Y., Choi, I.G., Choi, H.S., Shim, J.I., 2013. Investigation of light extraction efficiency in AlGaN deep-ultraviolet light-emitting diodes. Appl. Phys. Expr. 6 (062101).
Shatalov, M., Sun, W., Bilenko, Y., Sattu, A., Hu, X., Deng, J., Yang, J., Shur, M., Moe, C., Wraback, M., Gaska, R., 2010. Large chip high power deep ultraviolet light-emitting diodes. Appl. Phys. Expr. 3, 062101.
Shatalov, M., Sun, W., Lunev, A., Hu, X., Dobrinsky, A., Bilenko, Y., Yang, J., 2012. AlGaN deep-ultraviolet light-emitting diodes with external quantum efficiency above 10%. Appl. Phys. Expr. 5 (8), 082101.
Sun, W.H., Adivarahan, V., Shatalov, M., Lee, Y., Wu, S., Yang, J.W., Zhang, J.P., Khan, M.A., 2004. Continuous wave milliwatt power AlGaN light emitting diodes at 280 nm. Jpn. J. Appl. Phys. 43, L1419.
Takano, T., Narita, Y., Horiuchi, A., Kawanishi, H., 2004. Room-temperature deep-ultraviolet lasing at 241.5 nm of AlGaN multiple-quantum-well laser. Appl. Phys. Lett. 84, 3567.
Taniyasu, Y., Kasu, M., Makimoto, T., 2006. An aluminum nitride light-emitting diode with a wavelength of 210 nanometers. Nature 444, 325.
Zukauskas, A., Shue, M.S., Gaska, R., 2002. Introduction to Solid-State Lighting. Wiley, New York, ISBN: 978-0-471-21574-5.

CHAPTER FOUR

III-N Wide Bandgap Deep-Ultraviolet Lasers and Photodetectors

T. Detchprohm*, X. Li†, S.-C. Shen*, P.D. Yoder*, R.D. Dupuis*,[1]
*Center for Compound Semiconductors, School of Electrical and Computer Engineering, Georgia Institute of Technology, Atlanta, GA, United States
†Electrical Engineering Program, Computer, Electrical, Mathematical Science and Engineering Division, King Abdullah University of Science and Technology, Thuwal, Saudi Arabia
[1]Corresponding author: e-mail address: dupuis@gatech.edu

Contents

Semiconductors and Semimetals, Volume 96
ISSN 0080-8784
http://dx.doi.org/10.1016/bs.semsem.2016.09.001
© 2017 Elsevier Inc.
All rights reserved.

ABBREVIATIONS

AFM atomic force microscopy
APD avalanche photodiode
CH split-off (valence band)
DBR distributed Bragg reflector
DD dislocation density
DUV deep ultraviolet
ELO epitaxial lateral overgrowth
FS free-standing
GM Geiger-mode
HH heavy hole
HVPE hydride vapor-phase epitaxy
ICP inductively coupled plasma
IQE internal quantum efficiency
ITO indium-tin-oxide
LD laser diode
LED light-emitting diode
LP-MOCVD low-pressure MOCVD
LT low-temperature
MBE molecular-beam epitaxy
MOCVD metalorganic chemical vapor deposition
MQW multiple quantum well
M–S metal–semiconductor
PALE pulsed atomic layer epitaxy
PD photodiode
PIN *p*-type/intrinsic/*n*-type junction
PL photoluminescence
PV photovoltaic
QW quantum well
SAM separate absorption and multiplication
SB solar-blind
SE stimulated emission
SL superlattice
SPE spontaneous emission
SPSL short-period superlattice
TDD threading dislocation density
TE transverse electric
TEM transmission electron microscopy
TM transverse magnetic
TMAl trimethylaluminum
UV ultraviolet
VCSEL vertical-cavity surface-emitting laser
VPE vapor-phase epitaxy

1. INTRODUCTION

The metalorganic chemical vapor deposition (MOCVD) epitaxial growth technology was first reported in the scientific literature by Manasevit (1968). Similar processes and experimental results were previously described in the patent literature by other workers, e.g., Scott et al. (1957), Miederer et al. (1965), and Ruehrwein (1965, 1967), prior to 1968; however, no reports of successful MOCVD growth were published. Manasevit was primarily interested in technologies for the heteroepitaxial growth of III–V compound semiconductors on insulating oxide substrates, the analog of the silicon on insulator and silicon on sapphire technology that he had also pioneered earlier (Manasevit and Simpson, 1964). Manasevit's early work on MOCVD growth of compound semiconductors, particularly his work over the period 1968–75, established that MOCVD could be used to grow a wide variety of III–V (as well as II–VI and IV–VI) heteroepitaxial single-crystal semiconductor films on various insulating substrates, including sapphire (Al_2O_3), BeO, diamond, and spinel ($MgAl_2O_4$). In particular, Manasevit et al. (1971) expanded the research on MOCVD growth of III–Vs to include the heteroepitaxial growth of the wide-bandgap III-Ns including GaN and AlN on Al_2O_3 and $MgAl_2O_4$ using trimethylgallium, trimethylaluminum (TMAl), and ammonia (NH_3) at growth temperatures of 925–975°C for GaN and 1150–1250°C for AlN heteroepitaxial films. However, the results reported by Manasevit and several other researchers who became interested in MOCVD growth of III–V compound semiconductors in that time period did not create much enthusiasm for this materials growth technology due to the limited quality of the semiconductor films produced and the lack of any demonstration of device performance data comparable to that reported for semiconductor devices grown by other more established III–V epitaxial materials technologies, in particular, by liquid-phase epitaxy and hydride and halide vapor-phase epitaxy (VPE) and by the recently demonstrated molecular-beam epitaxy (MBE) technology (Cho, 1970).

Consequently, very little work was reported on MOCVD growth of III–V epitaxial materials in the early 1970s, and in particular, for III-N films. Duffy et al. (1973) reported related work on MOCVD growth of AlN and

GaN on (0001) and (11–20) sapphire and (111) silicon substrates. Morita et al. (1981a) also reported on MOCVD-grown AlN films on sapphire at temperatures in the range 1000–1200°C. In addition, Morita et al. (1981b) reported the MOCVD growth of AlN metal–insulator–semiconductor structures on (111) Si substrates at 1200°C. Khan et al. (1983a) reported the low-pressure MOCVD (LP-MOCVD) growth of GaN on sapphire and the fabrication of Schottky barrier diodes on Be^+ and N^+ ion-implanted GaN layers. Khan et al. (1983b) also reported LP-MOCVD growth of single-crystal $Al_xGa_{1-x}N$ alloys on sapphire over the entire alloy composition range. This was the first report of the growth of $Al_xGa_{1-x}N$ alloys by MOCVD. Hashimoto et al. (1984) reported the properties of Zn-doped GaN films on sapphire grown by MOCVD using both an N_2 and an N_2+H_2 ambient.

The next big improvement in MOCVD-grown III-N films was reported in 1986 when Amano et al. (1986) reported the atmospheric-pressure MOCVD growth of GaN on an AlN intermediate template layer grown on sapphire. This AlN intermediate or buffer layer was deposited at lower temperature (900–1000°C) and then annealed at higher temperature (950–1060°C) before the GaN layer was deposited at this same high temperature. This was the first report of the use of a "lower-temperature" AlN buffer layer for MOCVD growth of GaN, which resulted in greatly improved crack-free GaN/sapphire heteroepitaxial films with good crystallinity and surface morphology. Further work on MOCVD growth of AlGaN on *c*-plane sapphire and (111) Si substrates was reported by Koide et al. (1986) who described the growth of single-crystal $Al_xGa_{1-x}N$ films on sapphire with AlN mole fractions as large as $x=0.40$ and $Al_xGa_{1-x}N$ films up to $x=0.80$. In this work, they reported that AlGaN alloy thin films grown by MOCVD follow Vegard's law. In addition, Khan et al. (1986) reported the MOCVD growth of $Al_xGa_{1-x}N$ ($0<x<0.24$) using this low-temperature AlN buffer layer approach that resulted in the first observation of band-edge photoluminescence emission from AlGaN alloys.

Using a modification of this low-temperature AlN buffer layer approach, Amano et al. (1989) reported the first *p*-type GaN films grown by any process. They used MOCVD to grow Mg-doped GaN films on sapphire with a thin low-temperature ($T_g=600°C$) 50 nm thick AlN buffer layer and a high-temperature ($T_g=1040°C$) GaN film doped with Mg using bis-cyclopentadienyl Mg as a source and the activation of Mg acceptors using postgrowth low-energy electron-beam irradiation. They also reported the creation of the first GaN-based *p-n* junction light-emitting diodes (LEDs) in this paper. The first growth of high-quality InGaN films by MOCVD was reported by Yoshimoto et al. (1991). Akasaki et al. (1993) reported the growth of AlGaN and GaN for ultraviolet (UV) and blue *p-n* junction LEDs. The first MOCVD growth and photoluminescence (PL) characterization of UV-emitting $Al_xGa_{1-x}N$-GaN quantum well (QW) heterostructures having $0.06<x<0.13$ was reported by Krishnankutty et al. (1992) who described the effects of strain on the low-temperature (77 K) PL emission of the QWs.

This body of work on MOCVD growth of GaN, InGaN, and AlGaN created intense interest in the III-N materials and resulted in a rapid expansion of the research in the wider-bandgap AlGaN alloys including the development of AlGaN-based UV LEDs and UV avalanche photodiodes (APDs). Generally, the UV spectral region is divided into UV-A (315–400 nm), UV-B (280–315 nm), and UV-C (100–280 nm) regions. This chapter will review the recent work on the development of III-N DUV (DUV or UV-C, i.e., $\lambda<280$ nm) lasers and photodetectors. Extensive review of the current performance of III-N near-UV LEDs (UV-A, UV-B) and other related device and materials issues are covered in other chapters in this volume and in Kneissl and Rass (2016).

2. MOCVD GROWTH OF III-N DUV MATERIALS AND HETEROSTRUCTURES

2.1 Substrate Selection Issues

The materials in the AlInGaN alloy system are generally grown by MOCVD as heteroepitaxial films with the most stable wurtzite structure. However, this alloy system has a large degree of in-plane lattice mismatch and thermal strain associated with heteroepitaxial growth. Fig. 1 shows the in-plane (*c*-plane) lattice mismatch for AlInGaN alloys calculated for growth on AlN substrates. For DUV devices, the growth generally begins on AlN heteroepitaxial films grown on *c*-plane sapphire substrates. However, recently, AlN bulk substrates have emerged as a promising substitute for sapphire for DUV devices because of low dislocation density (DD) of $\sim10^4$ cm^{-2} as well as having a similar lattice constant and thermal expansion coefficient to those of Al-rich AlGaN (Wunderer et al., 2011). This ensures a relatively low DD in the AlGaN heterostructures. Despite these recognized benefits, current AlN substrates have some drawbacks. The use of bulk AlN is constrained by the limited

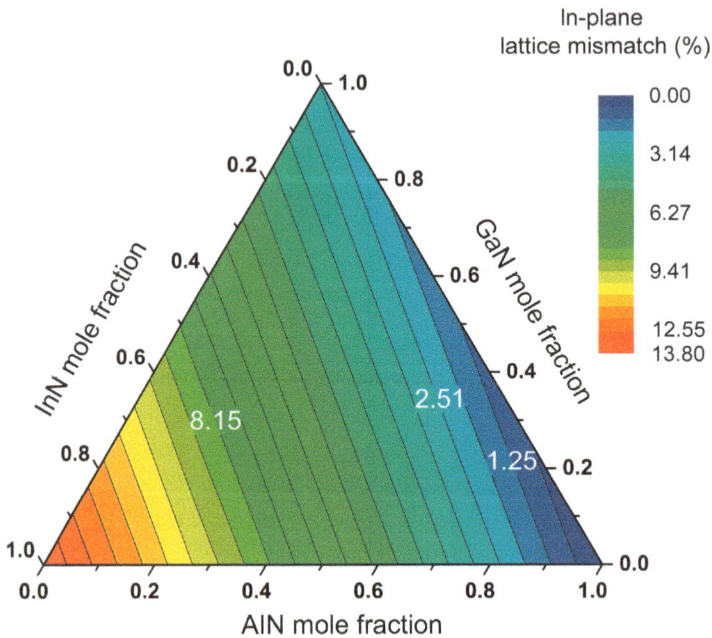

Fig. 1 An in-plane lattice mismatch map of AlInGaN alloys grown on AlN. Numbers on the map indicate the mismatch values for $In_{0.5}Ga_{0.5}N$, GaN, and $Al_{0.5}Ga_{0.5}N$ (from *left* to *right*) (Detchprohm, 2015).

supply, high cost, and smaller wafer size. In addition, the current manufacturing process introduces carbon impurities in the AlN crystal that is absorptive in the DUV region (Collazo et al., 2012). Similarly, the relatively high cost constrains the closely lattice-matched *c*-plane SiC substrates from wide-spread use in scalable applications. Furthermore, SiC is absorbing in the DUV. On the contrary, sapphire substrates do not have these issues. Therefore, most of the III-N DUV materials and heterostructures have been grown on sapphire substrates which are relatively inexpensive, DUV transparent, and readily available in large in size (up to 300 mm or 12 in. dia.).

2.2 Growth of High-Quality AlN on Sapphire Templates

The MOCVD growth of III-N DUV heterostructures generally begins with a heteroepitaxial AlN layer grown on *c*-plane sapphire, since the AlN is optically transparent to the DUV emission and closely lattice-matched to III-N

DUV heterostructures. However, the large lattice and thermal expansion mismatch between AlN and sapphire typically leads to a high threading dislocation density (TDD) over 10^{10} cm^{-2} unless special growth procedures or conditions are employed. This is undesirable for the performance of DUV emitters as the internal quantum efficiency (IQE) is generally inversely proportional to the density of dislocation-related nonradiative recombination centers (Ban et al., 2011). Hence, it is crucial to reduce the DD of the AlN layers.

A common approach to reducing the TDD is the use of epitaxial lateral overgrowth (ELO). AlN layers are regrown on patterned seeding AlN layers (Zeimer et al., 2013). However, because the ELO approach involves fabrication-like etching as well as a regrowth process of the many-μm thick AlN secondary layer to coalesce over the patterned AlN layer or sapphire substrate, this process is associated with higher cost and longer processing time, uneven surfaces, and growth complexity. Another approach, the pulsed atomic layer epitaxy (PALE) process, where the N source is supplied in a pulsed mode to allow Al atoms additional time to mobilize on the epitaxial surface, has been used (Paduano and Weyburne, 2005). In some studies, the ELO and PALE were collectively employed to expedite the coalescence (Hirayama et al., 2009).

In addition to the ELO and PALE, high-temperature growth above 1200°C has been employed independently or collectively with the ELO and PALE to achieve low TDD and smooth surface morphology by MOCVD, where the mobility of Al atoms on the epitaxial surface is enhanced at high temperatures (Imura et al., 2006). However, there are concerns regarding the high-temperature growth approach. Not only does it require a special reactor configuration and/or reactor parts to reach and maintain high temperatures, but it can also cause considerable thermal stress in the heteroepitaxial layer due to the large thermal expansion mismatch between the AlN layers and sapphire (Hearne et al., 1999). In addition, the serious wafer bowing at high temperatures can deteriorate wafer uniformity such as the layer thickness and the composition of layers grown on the AlN layers (Hoffmann et al., 2011). To address these issues, there have been some attempts to grow AlN layers at temperatures below 1200°C on sapphire (Kakanakova-Georgieva et al., 2012) and SiC (Zhang et al., 2003). However, the surfaces of these AlN layers were found to suffer from a high density of defects. There were few successful studies of growing high-quality planar AlN layers grown on sapphire substrates below 1200°C.

Recently, Li et al. (2015b,c) reported a three-step method to grow high-quality AlN layers on sapphire substrates at relatively low temperatures by MOCVD without the use of ELO or the PALE method. The three-layer AlN structure comprised a 15-nm thick buffer layer, a 50-nm thick intermediate layer, and a 3.4-μm thick AlN layer grown at 930, 1130, and 1100°C sequentially on a *c*-plane sapphire substrate. The resulting AlN layer had smooth surfaces with well-defined terraces and low RMS roughness of 0.07 nm for $1 \times 1\ \mu m^2$ atomic force microscopy (AFM) scan and the total TDD was $2 \times 10^9\ cm^{-2}$ as determined by transmission electron microscopy (TEM). Band-edge emission from AlN films was observed at 208 nm by 300 K PL measurements. This level of DD can lead to a relatively high IQE of ~50% for DUV emitters, lessening the necessity of using high-cost AlN and SiC substrates. The residual impurity concentrations were comparable to those of AlN layers grown at higher temperatures, i.e., at 1200–1600°C. A high growth efficiency of 2280 μm/mol was achieved, indicating reduced parasitic reactions between TMAl and NH_3. This study demonstrates that relatively high-quality AlN layers on sapphire substrates can be grown at temperatures achievable for most modern MOCVD systems.

2.3 Strain Effects

Our discussion is primarily limited to devices fabricated from III-N epitaxial materials having DUV energy gaps and currently achievable with relatively high quality and is thus mainly focused on *c*-plane $Al_xGa_{1-x}N$ alloys with high AlN mole fractions, x, of at least $x \sim 0.45$ for device applications in the UV-C spectral wavelength region, i.e., 200–280 nm. Epitaxial films of AlGaN alloys for this purpose are primarily grown on AlGaN/sapphire templates with at least the same or higher AlN mole fraction or on an AlN substrate. With such substrates, a heteroepitaxial AlGaN layer is subject to strain induced by thermal expansion difference, and lattice mismatch during the growth while the thermal expansion difference generally affects the material during temperature ramping. For AlN and high-AlN-mole-fraction $Al_xGa_{1-x}N$ ($x > \sim 0.9$) grown on sapphire substrates, there exists a tensile stress at the interface with the substrate at the growth temperature (typically >1000°C) (Brunner et al., 2013) due the thermal expansion coefficient different between the layers and the substrate. For the case of bulk/quasi bulk substrates, experimental studies of the lattice parameters of *c*-plane bulk AlN, and HVPE grown free-standing (FS)-GaN substrates as a function of temperature were carried out by

Figge et al. (2009) and Roder et al. (2005). After data processing with temperature-corrected refractive indices, they estimated thermal expansion coefficients as a function of substrate temperature and compared these values between the two binary crystals. The maximum difference in the thermal mismatch was reported to be at ~700 K (Figge et al., 2009). Though the thermal expansion coefficients for high-AlN-mole-fraction AlGaN alloys have not yet been reported due to the lack of bulk or quasi-bulk materials, thermal expansion mismatch effects, e.g., cracking, can be minimized by utilizing careful temperature ramps in an MOCVD process.

For lattice-mismatch-induced biaxial strain in $Al_xGa_{1-x}N$ grown on AlN, an in-plane strain (ε_{xx}) is defined as $\varepsilon_{xx}(x) = (a_{measured} - a_0(x))/a_0(x)$ where $a_{measured}$ is the a-plane lattice constant of $Al_xGa_{1-x}N$ derived by X-ray diffraction, and $a_0(x)$ is a strain-free a-lattice constant at the AlN mole fraction of x, while out-of-plane strain ($\varepsilon_{zz}(x)$) is defined as $\varepsilon_{zz}(x) = -2(C_{13}(x)/C_{33}(x))^* \varepsilon_{xx}(x)$, where $C_{13}(x)$ and $C_{33}(x)$ are the elastic constants of $Al_xGa_{1-x}N$ alloy, determined by assuming a linear interpolation between the values for GaN and AlN. For DUV applications, growing $Al_{0.45}Ga_{0.55}N$ directly on AlN result in an in-plane lattice mismatch of 1.36%, a compressive in-plane strain of −0.013, and a tensile c-axis strain of 0.026 when the layer undergoes fully biaxial strain on AlN. These values get smaller as the mole fraction increases. Any $Al_xIn_yGa_{1-x-y}N$ layer coherently grown on AlN is bound to be in compressive in-plane strain. A relaxed AlGaN template can be acquired on various foreign substrates as explained in Section 2.1. An AlGaN layer grown on either a sapphire substrate with a low-temperature deposited AlN buffer or thin AlN template on sapphire substrate was employed as a platform for developing solar-blind (SB) photodetectors, and DUV LEDs since the late 1990s. Cantu et al. (2003a,b) reported relaxation of Si-doped $Al_{0.49}Ga_{0.51}N$ grown on $Al_{0.62}Ga_{0.38}N$ template on sapphire degraded the layer surface morphology as inclined dislocations were developed within the grown layer, even though this heterointerface system was reported to have compressive in-plane strain of as little as −0.003. Romanov and Speck (2003) suggested that edge dislocations contributed to the misfit stress relaxation by inclining their line direction that corresponded to their effective climb. Such inclination was accelerated by Si doping that caused surface roughening due to the dopant antisurfactant effect. Follstaedt et al. (2005) observed similar relaxation with inclination of edge dislocations in a larger composition-contrast heterostructures composed of undoped $Al_{0.61}Ga_{0.39}N$ on AlN/sapphire templates; however, there was no sign of surface roughening associated with such relaxation. These direct-growth studies were performed on a compromised quality of available AlGaN and AlN templates, i.e., wafers with TDD in the range of upper 10^9–10^{10} cm^{-2}. With recently improved quality of AlN/sapphire templates whose TDD is typically in the range of mid-10^8 to lower 10^9 cm^{-2} (after Imura et al., 2006, 2007b), AlGaN layers can be grown with the same quality as that of the AlN template. For example, Shimahara et al. (2011) demonstrated such quality of AlGaN even with Si doping for AlN mole fractions greater than 0.61 as long as the layer remained highly strained to the template. A linear dependence of the free-electron concentration up to $n = 2 \times 10^{18}$ cm^{-3} was confirmed with a donor activation rate close to 1 for $Al_{0.65}Ga_{0.35}N$. This implies that a direct growth of AlGaN on AlN bulk substrate in preserved pseudomorphic mode is favorable for device applications that require low material defects such as laser diodes (LDs) and SB avalanche photodiodes.

Another approach employed for growing AlGaN on AlN is inserting single or multiple superlattices (SLs) generating step-graded AlGaN heterostructures as strain-management layers. The use of five-period $Al_{0.20}Ga_{0.80}N$/AlN superlattices was initiated as threading dislocation filter for 1.5 μm thick $Al_{0.20}Ga_{0.80}N$ grown on AlN templates (Wang et al., 2002; Zhang et al., 2002). This group later applied this approach to grow fully relaxed $Al_{0.55}Ga_{0.45}N$ layer (Sun et al., 2005). With 40 pairs of an $Al_{0.85}Ga_{0.15}N$/AlN SL, the electron mobility was improved to 120–130 cm^{-2} V s from 50 to 70 cm^{-2} V s in the five-pair case, while the screw DD was reduced to 7×10^7 cm^{-2}; however, it had little effect on edge dislocations in the AlGaN layer on top. Besides, since all of these layers were grown via PALE, each individual AlGaN layer in the superlattice was naturally formed in a short-period superlattice of $Al_xGa_{1-x}N/Al_yGa_{1-y}N$ with a period of ~1.55 nm (6 monolayers). Ren et al. (2007) investigated the crystallographic quality of $Al_{0.50}Ga_{0.50}N$ grown on three sets of 10-period of 15 nm $Al_xGa_{1-x}N$/15 nm $Al_yGa_{1-y}N$ with x/y in an order of 1.00/0.80, 0.80/0.65, and 0.65/0.50, as a function relaxation degree and reported a 100% relaxed AlGaN had a rough surface morphology and broader (10–12) rocking curve linewidth indicating growing number of edge and/or mixed threading dislocation compared to those of almost pseudomorphic AlGaN. These results point out that it is important to preserve AlGaN under a fully strained condition in order to have the AlGaN film resemble the quality of the AlN substrate, in particular when a bulk AlN substrate is utilized for device applications sensitive to material quality. Considering growing a single layer of AlGaN on AlN, the layer gradually relaxes as it grows thicker. The critical thickness for this relaxation primarily depends on the AlGaN alloy composition. Grandusky et al. reported that $Al_{0.60}Ga_{0.40}N$ and $Al_{0.70}Ga_{0.30}N$ layers grown directly on AlN bulk substrates remained pseudomorphically strained up to thickness of 0.5 and 1.0 μm, respectively. Manning et al. (2009) performed an in-situ monitoring stress evolution in 500 nm thick Si-doped $Al_{0.61}Ga_{0.39}N$ grown on top of AlN templates on 6H-SiC and found out that strain in both the undoped and moderately doped AlGaN layers, i.e., $[Si]=3.2\times10^{18}$ cm^{-3}, switched from a compressive layer to a tensile one when the such layer reached thickness of approximately 0.6 and 0.4 μm, respectively. Such transition occurred as early as ~120 nm in the AlGaN layer with [Si] of 2.5×10^{19} cm^{-3}. These turning points of strain condition could be interpreted as starting points of strain relaxation. Since the TDD in this case was as high as $\sim10^{10}$ cm^{-2}, the actual critical thickness for AlGaN on AlN substrates requires further investigation. Besides, there are several strain-induced effects in a heterostructure of this type that impact device performance such as piezoelectric polarization and optical polarization; however, these effects are beyond the scope discussed in this section.

2.4 Doping Issues

Two significant factors affecting the carrier concentration in high-AlN-mole-fraction AlGaN are (1) the activation energy of donor or acceptor and (2) the density of the compensating point defects and impurities such as oxygen and carbon impurities. Generally, Si and Mg are commonly utilized as dopants for *p*- and *n*-type alloy materials, respectively. Some alternative dopants are Ge and C though only C was reported as a *p*-type dopant (Kawanishi and Tomizawa, 2012). For the shallow donors, the activation energy rapidly increases when the AlN mole fraction is 0.8 or greater (Collazo et al., 2012). The activation energy values for Si for $x=0.81$, 0.9, and 1.0 were 30, 60, and 250 meV, respectively (Collazo et al., 2011; Taniyasu et al., 2006). With such activation energy distribution, Mehnke et al. (2013) achieved free-electron concentrations in the range of 1.5×10^{19} cm^{-3} at 300 K for Si-doped $Al_{0.81}Ga_{0.19}N$ grown on an ELO-AlN/sapphire template.

For the shallow Mg acceptor, the activation energy values were much higher. Analyzing the optical properties of Mg-doped AlGaN, Imura et al. (2007a) reported that the acceptor activation energy increased with the AlN mole fraction and was in the range of 400–1000 meV for $0.5\leq x\leq1$. This suggests that the free hole concentration is very low at room temperature. More than an order of magnitude larger dopant concentration is required to reach a desirable free hole concentration that is typically $p \sim 1 \times 10^{18}$ cm^{-3}. The most useful indicator to justify the electrical properties of p-AlGaN would likely be its bulk resistivity instead of the free hole concentration. To improve the p-type conductivity, several groups utilized superlattice structures and achieved reasonable lateral carrier transport with low effective acceptor activation energies and high free hole concentrations (Allerman et al., 2010; Cheng et al., 2013; Zheng et al., 2016). Recently Zheng et al. reported a low bulk resistivity of 0.7 Ω cm for a multidimensional Mg-doped short-period superlattice (SPSL) of $Al_{0.51}Ga_{0.49}N/Al_{0.63}Ga_{0.37}N$. The properties of n-type and p-type AlGaN material are summarized in Tables 1 and 2, respectively.

3. III-N DEVICE DESIGN AND SIMULATION

III-N compounds pose unique challenges for both device design and theoretical modeling. Although considerable progress has been made in the epitaxial growth and processing of wurtzite III-N materials over the past 25 years, this remains in many respects an immature material system. Typical TDDs for binary GaN range from 5×10^4 cm^{-2} (on bulk substrates) to 5×10^9 cm^{-2} (on nonnative substrates), introducing fixed mid-gap electronic states that may degrade charge carrier mobility through a Coulomb interaction. Intrinsic material grown by MOCVD typically exhibits an unintentional n-type doping on the order of 10^{16} cm^{-3}, nearly an order of magnitude higher than that of other common compound semiconductors. Passivation of exposed III-N surfaces, both vertical and horizontal, is also less efficient than in other common III–V material systems, leading to enhanced surface recombination and leakage currents. Moreover, wurtzite III-nitrides are displacive ferroelectrics and exhibit electrostatically significant spontaneous polarization charge at hetero-interfaces, that is, augmented by an interfacial piezoelectric polarization of like or greater magnitude under tensile or compressive strain. Interfacial polarization charge is well known to affect the localized quantization of bound electrons and holes in QWs (Ryou et al., 2009), as well as the transport of free electrons and holes nonlocally. These and other considerations complicate the direct measurement of both electrical and optical properties of wurtzite III-N materials and influence the reliability of the values documented in the archival literature.

Table 1 Summary of Electrical Properties of Si-Doped *n*-Type AlGaN

References	*x*	Bulk Resistivity (Ω cm)	*n* (cm^{-3})	μ_n ($cm^2\ v^{-1}\ s^{-1}$)	Substrate
Cantu et al. (2003a,b)	0.62	0.0620	1.3×10^{17}		Sapphire
Nam et al. (2002)	0.65	0.1500	2.1×10^{18}	20	AlN/Sapphire template
Cantu et al. (2003a,b)	0.65	0.0001[a]	2.5×10^{19}	22	Sapphire
Nakarmi et al. (2005)	0.70	0.0075	3.3×10^{19}		AlN/Sapphire template
Al tahtamouni et al. (2008)	0.75	0.0440	5.6×10^{18}	26	AlN/SiC template
	0.75	0.0380	7.3×10^{18}	24	
	0.75	0.0320	8.1×10^{18}	23.3	
	0.75	0.0270	9.5×10^{18}	21.1	
Kakanakova-Georgieva et al. (2013)	0.77	<0.05	Low 10^{18}	80	4H-SiC
Collazo et al. (2011)	0.80	0.1000	1.0×10^{18}	40	AlN bulk
Mehnke et al. (2013)	0.81	0.0260	1.5×10^{19}	16.5	ELO-AlN/sapphire template
	0.86	0.0450			
	0.91	0.6300			
	0.95	2.6200			
	0.96	3.3500			
Taniyasu et al. (2006)	1.00	N/A	1.75×10^{15}	125	4H-SiC

[a]Indicates isoelectronic doping with In.

3.1 Simulation of Basic Materials Properties

Seminal electronic structure calculations of GaN and AlN were performed from first-principles density-functional theory by Rubio et al. (1993), from which many empirical theories for electronic structure and optical properties

Table 2 Summary of Electrical Properties of *p*-type AlGaN

References	x	Bulk Resistivity (Ω cm)	p (cm^{-3})	μ_n (cm^2 v^{-1} s^{-1})	Dopant	Template
Yu et al. (2006)	0.35	3.5			Mg	AlN/Sapphire template
Jeon et al. (2005)	0.45	8	2.7×10^{17}	1.4	Mg	AlN/Sapphire template
	0.5	10	2.2×10^{17}	2.7	Mg	AlN/Sapphire template
Ji et al. (2016)	0.5	3.31			Mg	AlN/Sapphire template
Chakraborty et al. (2007)	0.69	10 at 670 K			Mg	SiC substrate
Nakarmi et al. (2005)	0.7	100,000 at RT			Mg	AlN/Sapphire template
	0.7	40 at 800 K			Mg	AlN/Sapphire template
Kinoshita et al. (2013)	0.7	47			Mg	AlN/Sapphire template
Kakanakova-Georgieva et al. (2010)	0.85	7000	1.0×10^{15}	2	Mg	SiC substrate
Allerman et al. (2010)	0.45[a]	5			Mg	*p*-SPSL AlGaN
	0.74[a]	6			Mg	*p*-SPSL AlGaN
Zheng et al. (2016)	0.51/0.63	0.7	3.5×10^{18}		Mg	*p*-SPSL AlGaN
Kawanishi and Tomizawa (2012)	0.27		5.0×10^{18}		C	AlGaN/sapphire

[a]Average AlN mole fraction.

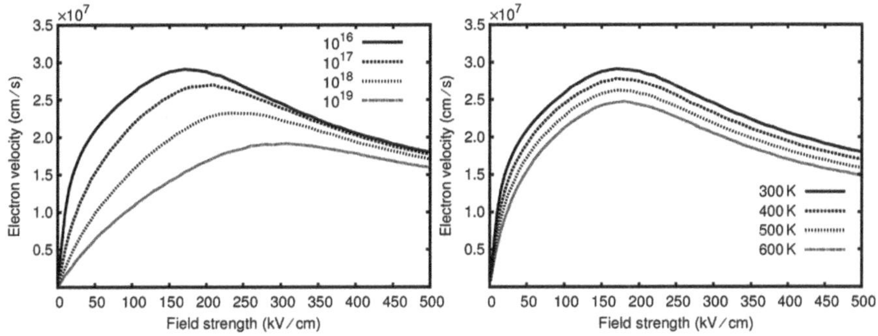

Fig. 2 Dependency of electron drift velocity in binary GaN on doping concentration and temperature.

have been developed, including the empirical pseudopotential and *k* dot *p* methods. From these latter approaches, fundamental material properties such as effective mass and optical absorption (and gain) are more readily extracted. Among the most fundamental electrical properties of any semiconductor material is the dependence of charge carrier mobility on doping level and temperature, which depend on both electronic structure and charge transport. Yet even for binary GaN, the authors are aware of no comprehensive, documented experimental study of carrier mobility. The first systematic theoretical study of the temperature- and doping dependence of electron mobility in bulk GaN was performed by Sridharan and Yoder (2008) based on full-band ensemble Monte Carlo simulation calibrated to Hall measurements documented in the archival literature. These results are reproduced in Fig. 2 and provide valuable insight into low- and high-field carrier dynamics encountered in realistic devices under realistic operating conditions.

3.2 Comparison of Simulation Techniques

The numerical analysis of UV wurtzite III-N LDs is considerably more complicated than that of infrared LDs made from common III–V cubic semiconductors. Due to the larger bandgap energies, the dynamic range of free carrier densities is several orders of magnitude higher. This problem is exacerbated at III-N heterojunctions, where band discontinuity energies are typically larger, and may lead to large gradients in free carrier concentration when thermionic emission boundary conditions are applied. Due to the large interfacial polarization charges, simple analytic models for carrier confinement in QWs fail, even under flatband conditions. Indeed, polarization fields are so high that carrier confinement is achieved by virtue of both QW and quantum well barrier material. In addition, the difficulties associated with making Ohmic contacts to Al-rich *p*-type AlGaN necessitate the use of a region of inverse compositional gradient (Satter et al., 2014), within which volumetrically distributed polarization charge of negative sign draws holes electrostatically from the narrower-gap *p*-type Ohmic contact region, through the compositionally graded region, facilitating their efficient electrical injection into the active region.

Our approach to LD simulation (Satter et al., 2012, 2014) involves the coupled system of (1) charge transport equations for free electrons and holes, (2) rate equations for quantum-confined electrons and holes, (3) models for an exhaustive set of radiative and nonradiative generation/recombination mechanisms, (4) Poisson equation for electrostatic self-consistency, (5) *k* dot *p* bandstructure calculations, (6) lattice heat equation for thermal transport, (7) optical mode calculations, and (8) photon rate equations (accounting for both intrinsic and diffractive losses). Optical mode calculations are performed according to the vector Helmholtz equation supported by a comprehensive model we have developed and calibrated for the anisotropic complex dielectric function for each of the III-nitride materials used for UV emitters, spanning a broad range of frequency, as well as tensile and compressive strain conditions.

In contrast to LDs, there exists a wealth of analytical and semianalytical models to describe the time- and frequency-dependent operation of photodiodes (Yoder and Flynn, 2006); a common approximation in such models is the neglect of diffusion, the validity of which depends on the application and must be evaluated on a case-by-case basis. Of equal concern for wurtzite III–N photodiodes is the analytic models' neglect of the field-dependence of the saturated drift velocity (see Fig. 2), and the related overestimation of transit time within the active region. For the case of unity-gain UV photodiodes, moment-based methods such as drift diffusion and energy balance fully resolve the first approximation, and partially mitigate the second, though they are all susceptible to the anomalous velocity overshoot effect. It is well known from decades of work on silicon technologies that the drift-diffusion and energy balance methods are unable to correctly predict the breakdown voltage of simple one-dimensional homojunction *p-n* diodes, even if the latter is equipped with realistic energy relaxation time parameters. One must therefore go beyond the ubiquitous moment-based method for reliable linear- and Geiger-mode (GM) analysis of UV III–N avalanche photodiodes.

The complex operation of UV wurtzite III-N APDs poses exceptional challenges for predictive modeling and design, including (1) transient operation involving large voltage and current swings and the associated Joule heating, (2) nonlocal and nonstationary charge transport at high fields, (3) nonlinear avalanche generation, and (4) interactions with an external circuit. We have addressed these challenges through the development of a state of the art multiscale full-band ensemble electrothermal Monte Carlo device simulation tool (Sridharan et al., 2009). Coupled self-consistently to thermal transport equations and models for biasing and load circuitry, the Monte Carlo model provides reliable insights into device operation and serves as a highly accurate guide to device design. The Monte Carlo model itself considers all relevant scattering mechanisms, including polar optic and deformation potential electron–phonon interactions, hole- and electron-initiated impact ionization, piezoelectric scattering, extended defect scattering, and ionized impurity scattering. Wave-vector-dependent atomic pseudopotentials have been generated to facilitate calculations of GaN and AlGaN bandstructure for arbitrary AlN mole fraction. Treatment of electron dynamics and kinematics are in all respects fully consistent with the nonlocal empirical pseudopotential bandstructure. The technique of electrothermal Monte Carlo simulation, in which electron energy loss to the phonon bath serves as a source term for solution of the lattice heat equation, was pioneered by the authors (Yoder and Fichtner, 1998). The resulting lattice temperature profiles are then fed back to the calculation of spatially dependent electron–phonon scattering rates for self-consistency. Electrical self-consistency is achieved via repeated solution of the Poisson equation, using the calculated electron and hole densities and fixed polarization and ionized charge as source terms.

As an example, we have applied our model to the theoretical analysis of a simple GM separate absorption and multiplication (SAM) APD structure featuring a 200-nm *n*-GaN buffer layer, followed by an intrinsic GaN absorption layer of between 500 and 1000 nm, followed by a 50-nm *n*-AlGaN graded translation layer, a 300- to 500-nm thick intrinsic AlGaN multiplication layer, a heavily doped 50 nm layer of *p*-GaN, followed by a 20-nm *p*++ GaN cap layer, grown on a bulk GaN substrate. Polarization charge as well as intentional heavy doping of the AlGaN graded transition layer contributes to a large discontinuity in electric field strength between the absorption and multiplication regions. The electric field profile is depicted in Fig. 3. With light incident from the top of the structure, electron–hole pairs are generated within both the absorption and

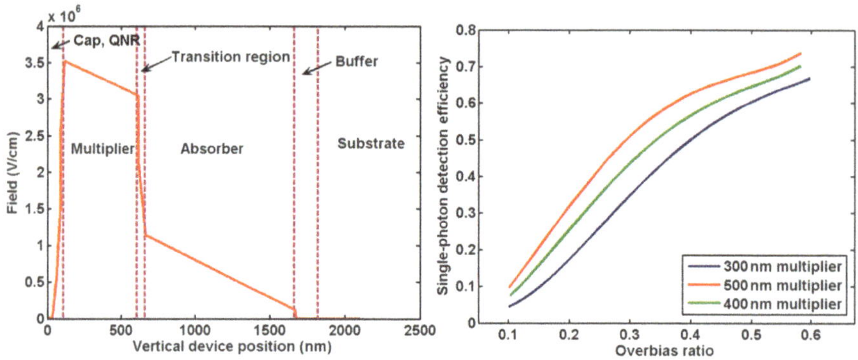

Fig. 3 *Left*: Electric field profile within a Geiger-mode SAM APD. *Right*: Single-photon detection efficiency as a function of overbias ratio.

multiplication region at a rate that decays exponentially with depth into the device. We find that electron–hole pairs photogenerated furthest from the *p*-side of the depletion region have the greatest likelihood of triggering avalanche breakdown, due to a higher effective hole ionization coefficient. As a consequence, APD structures with the multiplier above the absorber, such as the one considered here, will favor the injection of holes into the multiplication region and exhibit higher single-photon detection efficiency. With this design, thicker multiplication regions increase the number of electron–hole pairs generated above the absorber, but this is more than compensated for by the enhanced multiplication feedback of the thicker absorbers, with a net result of an enhanced single-photon detection efficiency, as shown in Fig. 3.

4. PROCESSING OF III-N DUV EMITTERS AND PHOTODETECTORS

The fabrication of III-N semiconductor devices shares many similarities to conventional III–V compound semiconductor devices. Precision mesa etching and high-aspect-ratio device geometry are commonly employed. The ion implantation of III-N materials is typically used for device isolation and not for dopant incorporation purposes. The sintering process for Ohmic contacts on wide-bandgap materials requires a higher temperature compared to GaAs- and InP-based devices. Nonalloyed metal–semiconductor (M–S) junctions usually form Schottky contacts. The significant annealing temperature difference between the *p*-type and *n*-type Ohmic contact processes mandate separate annealing steps in the fabrication of III-N bipolar devices. The etched mesa sidewalls usually induce additional leakage paths when the devices are under electrical stress, resulting in high dark current in photodetectors and reduced quantum efficiency in emitters. Proper surface treatment and device passivation are key elements in enabling high-performance optoelectronic devices. Common practice in the fabrication of UV lasers also employs dielectric mirrors to form an enhanced resonant cavity.

4.1 Ohmic Contacts

4.1.1 n-Type Contacts

Successful formation of Ohmic contacts of III-N materials requires the unpinning of the interface states at the M-S junction. An *n*-type III-N contact can be achieved by choosing a low-work function (χ) contacting material with a proper sintering process to create transition layers to enhance the electron conduction across the interface. Commonly used *n*-type III-N contacts are titanium-based multilayer stacks such as Ti/Al/Ti/Au and Ti/Al/Ni/Au. The typical annealing temperature is between 700 and 950°C. After the annealing step, the Ti layer on top of the GaN can form a TiN ($\chi = 3.74$ eV) and a $TiAl_3$ complex at the interface. The Ti or Ni layer adjacent to the gold is used as the diffusion barrier and also prevents the undesired oxidation of the underlying layers during the annealing process. Refractory metals such as molybdenum (Mo) are also commonly used as the first layer to replace titanium for low-contact resistance performance.

As the bandgap energy of the III-N film increases, Ti/Al-based contacts may not be a suitable choice. Ti/Al-based metal stacks are reportedly incapable of forming Ohmic contacts for $n\text{-}Al_xGa_{1-x}N$ ($x > \sim 0.6$). Vanadium/aluminum-based contacts were studied, and this system exhibited better properties for high-Al-content *n*-type AlGaN layers (Schweitz et al., 2002). As a comparison of the contact resistance (ρ_c) for V-based and Ti-based Ohmic contact on an $Al_{0.55}Ga_{0.45}N$ film, annealed V/Ti contacts show an optimal ρ_c of $2 \times 10^{-5}\ \Omega\ cm^2$ at 775°C, while the Ti/Al contacts exhibited higher ρ_c of $3 \times 10^{-4}\ \Omega\ cm^2$ at a higher annealing temperature of 825°C (Kao et al., 2016). For V-based contacts on $n\text{-}Al_{0.06}Ga_{0.94}N$ films, ρ_c can be as low as $6.6 \times 10^{-6}\ \Omega\ cm^2$. As the AlN mole fraction increases in AlGaN from 6% to 73%, ρ_c increases from $6.6 \times 10^{-6}\ \Omega\ cm^2$ to $4.4 \times 10^{-3}\ \Omega\ cm^2$, and R_{sh} increases from $5 \times 10^{-3}\ \Omega$ cm to 5.6 Ω cm. The increase in ρ_c and R_{sh} can be attributed to the lower free-carrier concentration and large bandgap energy of $n\text{-}Al_xGa_{1-x}N$ films as the AlN mole fraction increases. It is also observed that Ti-based metal stacks cannot form Ohmic contacts to AlGaN when the AlN mole fraction is greater than 60%.

4.1.2 p-Type Contacts

Due to the high activation energy of Mg in III-N layers, the *p*-type III-N Ohmic contact to devices is usually achieved by using a heavily magnesium-doped GaN layer as a capping layer to get around the lack of sufficient free holes in Mg-doped high-aluminum-content III-N layers. Consequently, typical *p*-type contacts to III-N layers in UV emitters and detectors are formed by using either Ni/Au, Ni/Ag, or indium-tin-oxide (ITO). The annealing temperature of a *p*-type contact is below 600°C with typical ρ_c between 10^{-3} and 10^{-4} Ω cm^2. The choice of metal stack, the thicknesses of each layer, and the annealing conditions are dependent on the UV optical properties of these films for specific optoelectronic devices of interest. Ni/Au and ITO are known as the preferred transparent contact for visible-blue wavelengths. However, the absorption coefficient for these materials increases dramatically in the UV wavelengths. Careful design of the device structure must be considered to minimize the undesired UV absorption in the *p*-contact layers. Approaches, such as excluding the *p*-contact layers from the path of the photon flux, or minimizing the optical field by enforcing a node in the desired optical resonant modes at these layers, are among a few optoelectronic performance enhancement experiments that have been studied so far.

4.2 Etching of III-N Materials

The etching of III-N materials usually serves as two purposes. The first is to provide a mesa-type device topology to expose the underlying layers for Ohmic contacts, electric field engineering, device isolation, or waveguide formation. The second is to remove specific semiconductor layers through selective etching or surface treatment. Dry etching of the AlInGaN materials uses plasma tools such as inductively coupled plasma (ICP), reactive ion etching, or chemically assisted ion-beam etching. Commonly used chemicals in dry etching are chlorine-based species. As the aluminum content increases in III-N materials, a mixture of boron tetrachloride (BCl_3) and chlorine (Cl_2) along with carrier gases such as argon or helium can be used to achieve desired etching rate. Selective etching of GaN over AlGaN can also be achieved using Cl_2 in plasma etchers. Extensive research on the dry etching of III-N materials has been reported (e.g., Pearton et al., 2006). Wet etching of III-N materials was also studied (Zhuang and Edgar, 2005).

Dry etching is usually preferred to wet etching for the III-N mesa etching step because dry etching can achieve a smooth surface morphology even with a material with high density of as-grown defects. On the other hand, wet-chemical etching of III-N materials has preferential orientation-dependent etching characteristics that are suitable for surface roughening or for defect-site revealing (Youtsey et al., 1997). They can also be used for surface modification in III-N device processes. UV-photon-assisted wet etching in potassium hydroxide-based solutions is the commonly used approach. Additional electrodes can be introduced in the etching solution to facilitate photon-assisted electrochemical (PEC) etching. A simplification of PEC etching can be implemented using a mixture of strong oxidant (e.g., potassium persulfate) and KOH to achieve an electrode-less etching processing (Bardwell et al., 2001). Various surface treatment techniques have employed optimized $KOH/K_2S_2O_8$ electrode-less PEC etching process. For example, a GaN mesa was first etched using a Cl_2/Ar mixture in an ICP that showed a rather rough sidewall surface that was subsequently treated in $KOH/K_2S_2O_8$ solution, and a smooth sidewall was obtained (Shen et al., 2007). This smooth sidewall morphology helped drastically to reduce the reverse-bias leakage current in devices, which was validated in the current–voltage measurement of fabricated mesa III-N diodes, and helped to realize high-performance AlGaN optoelectronic and electronic devices.

4.3 Passivation of III-N Devices

Proper device passivation is also a key to high-performance III-N UV optoelectronics as the leakage current directly impacts the quantum efficiency and the noise performance of the devices. Typical III-N device passivation methods include plasma-enhanced chemical vapor deposition-grown silicon nitride (SiN_x) or SiO_2, in-situ SiN_x, benzocyclobutene or spin-on glass. Each method has demonstrated effective reduction in the leakage current in III-N devices.

5. PERFORMANCE OF III-N DUV LASERS AND PHOTODETECTORS

5.1 Overview of DUV Lasers

Semiconductor DUV lasers can enable compact solutions to important applications including Raman spectroscopy and non-line-of-sight communication. The III-N semiconductors are promising materials for compact, reliable, low-cost, and efficient DUV lasers because of a proper direct bandgap range as well as high chemical and mechanical toughness. In particular, the wavelength of the III-N direct band-edge transition can be as short as 210 nm at 300 K. The commercial success of MOCVD-grown III-nitride blue LEDs and LDs has encouraged researchers to dream about and work toward a similar device performance for DUV LEDs and LDs. Unfortunately, the wall-plug efficiency of most of the commercial DUV LEDs is still in the low single-digit range. Moreover, researchers have not yet demonstrated a DUV LD.

Several paramount challenges exist for the demonstration of the first DUV LD. First, the highly mature blue-emitting InGaN materials system can no longer be used due to its relatively small bandgap energy. Blue-shifting the emission to the DUV range requires significant addition of Al to GaN. This increases the in-plane lattice mismatch between the III-N and the most common sapphire or GaN substrates, degrading the material quality. Thus, tremendous research is needed to create high-quality AlGaN materials for high quantum efficiency. Second, the optical confinement structure has to be optimized given the small refractive index variation as a function of Al composition in AlGaN. Third, the theoretically predicted optical polarization switching from transverse electric (TE) ($E_{TE} \perp c$-axis) to transverse magnetic (TM) ($E_{TM} \parallel c$-axis) of the stimulated emission for various AlGaN active regions needs to be considered. This is important for the design of LD structures as the TM-polarized light will leak deeper into the absorptive *p*-cladding region due to its broader beam profile. Fourth, the activation energy of Mg acceptors in Al-rich AlGaN is considerably larger than that in GaN. This leads to an insufficient concentration of free holes in the active region to support stimulated emission under forward bias.

It is difficult to solve all the issues simultaneously. A strategy adopted by many researchers is to focus on the first three issues, and especially, the material quality, to produce optically pumped DUV lasers preferably with short wavelengths and low thresholds prior to addressing the *p*-type doping issue. Takano et al. (2004) reported the first optically pumped AlGaN multiple quantum well (MQW) DUV laser that was grown on a SiC substrate and emitted at 241.5 nm in spite of having a large threshold of 1200 kW cm^{-2}. After that, Wunderer et al. (2011) demonstrated an optically pumped AlGaN MQW laser at 267 nm with a significantly reduced threshold of 126 kW cm^{-2} grown on a bulk AlN substrate. Subsequently, different groups managed to gradually push the wavelengths of the optically pumped AlGaN lasers grown on bulk AlN substrates down to 237 nm while maintaining low thresholds (Bryan et al., 2015; Kao et al., 2013).

5.2 Optically Pumped DUV Lasers on Sapphire

As discussed earlier, sapphire substrates are more practical for DUV LDs than AlN and SiC substrates because of lower cost, high availability, and larger area. For example, we have grown pseudomorphic AlGaN MQW heterostructures for optical pumping experiments (Li et al., 2014). An AlGaN grading waveguide layer, with a five-period AlGaN MQW active region designed for laser emission at ~250 nm and an AlGaN cap layer for surface passivation were grown sequentially by MOCVD on AlN/sapphire templates. The composition and thickness of these AlGaN layers were optimized to enhance the optical confinement and thus reduce the laser threshold. Subsequent to the growth, the wafer was cleaved into Fabry–Perot laser bars after being scribed by laser or hand. The laser scribing process led to smoother facets than the hand scribing. No high-reflectivity coating was applied to either facet and thus the facets retained a reflectance of ~0.2 in the DUV region at the wavelength of operation ~250 nm. By optical pumping, Li et al. (2014, 2015a) demonstrated a plurality of edge-emitting lasers at 237–256 nm, as shown by some examples in Fig. 4. As shown in Fig. 5, the lasers possessed similar or lower thresholds than those of the reported lasers on the AlN substrates at similar wavelengths, indicating excellent optical properties. In particular, the lowest threshold is 61 kW cm^{-2} for a laser emitting at ~256 nm, the lowest reported value in the vicinity of the wavelength.

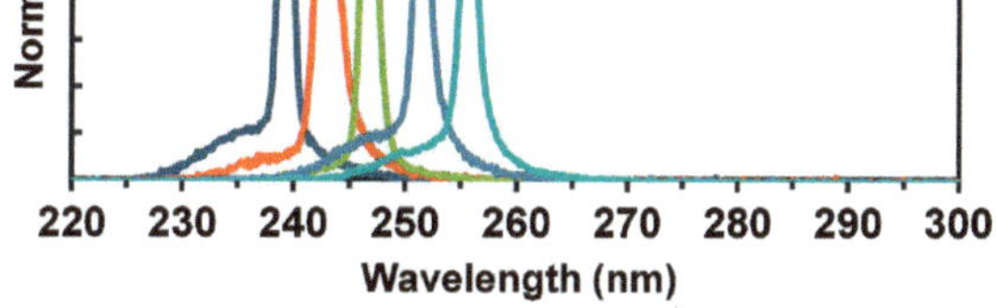

Fig. 4 Stimulated-emission spectra of the optically pumped AlGaN MQW DUV lasers grown on (0001) sapphire substrates with emission at 239–256 nm at excitation power densities above the respective threshold.

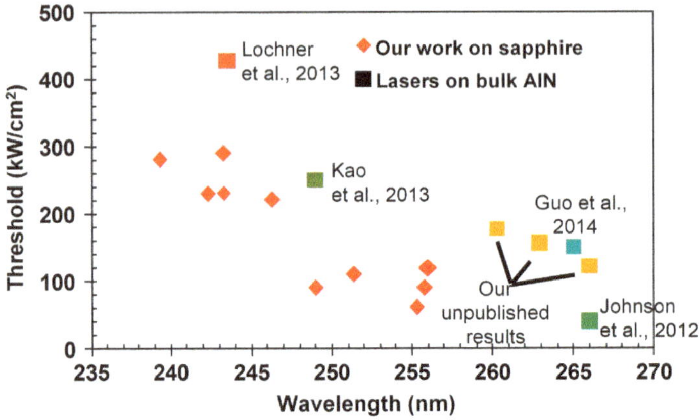

Fig. 5 Summary of thresholds of the reported state of the art optically pumped AlGaN MQW DUV lasers grown on sapphire substrates and AlN substrates (Guo et al., 2014; Johnson et al., 2012).

To facilitate the design of DUV lasers operating at shorter wavelengths, it is important to know the wavelength range where the TE-dominant lasing switches to TM-dominant lasing. The TE–TM switching is related to the valence-band structure of AlGaN. When the topmost valence band is the heavy hole (HH) band, the dominant band transition is between the conduction band and HH band that leads to TE-dominant emission. With an increased Al composition and thus a shorter emission wavelength, the split-off hole (CH) band moves closer to the conduction band relative to the HH band that triggers the switching from TE- to TM-polarized emission when the CH band crosses over the HH band and thus becomes the topmost band. The polarization degree, defined as $\rho=(I_{TE}-I_{TM})/(I_{TE}+I_{TM})$, can be calculated wherein I_{TE} and I_{TM} represent the intensity of TE- and TM-polarized emission, respectively. Fig. 6 shows summary of the above-threshold polarization degrees of the lasers demonstrated in our studies. Both TE- and TM-dominant DUV-stimulated emission from lasers grown on sapphire have been demonstrated. As indicated by the dashed line in Fig. 6, the rapid variation between TE- and TM-dominance with respect to the change in lasing wavelength from 243 to 249 nm is distinct from the previous studies, wherein the spontaneous emission (SPE) from AlGaN structures made a similar extent of polarization switch at a considerably longer wavelength span (Kolbe et al., 2010; Banal et al., 2009). This can be attributed to the dramatic change in the ratio of TE-to-TM gain coefficients for the DUV AlGaN MQW lasers in the vicinity of TE–TM switch.

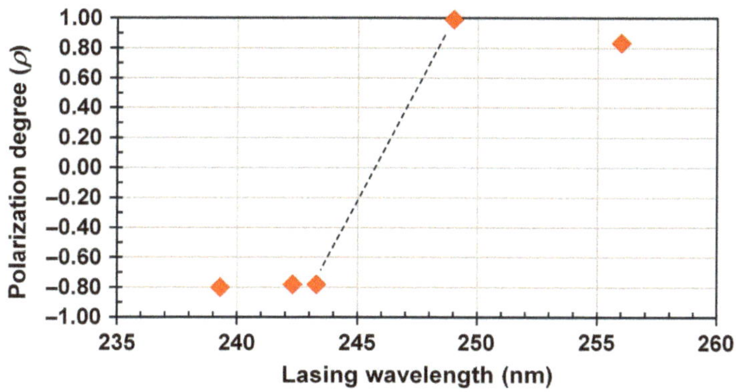

Fig. 6 Summary of the above-threshold polarization degree of DUV lasers grown on sapphire substrates in our studies.

Although the earlier discussion focuses on edge-emitting lasers, vertical-cavity surface-emitting lasers (VCSELs) possess advantages including high-speed modulation, good beam quality, and easy control of the device production process. Despite good progress for the development of III-N edge-emitting lasers in the near-UV-to-visible range (i.e., longer than ~390 nm), the development of surface-emitting III-N lasers has been much slower, especially for DUV lasers. We demonstrated the onset of DUV surface-stimulated emission from *c*-plane AlGaN MQW heterostructures grown on sapphire substrates by optical pumping at 300 K (Li et al., 2015a). As shown in Fig. 7, the onset of stimulated emission (SE) became observable at a pumping power density of 630 kW cm^{-2}. Spectral deconvolution reveals superposition of a linearly amplified SPE peak at $\lambda \sim 257.0$ nm with a FWHM of ~12 nm and a superlinearly amplified SE peak at $\lambda \sim 260$ nm with a narrow FWHM of less than 2 nm. In particular, the wavelength of 260 nm is the shortest wavelength of surface SE from III-nitride MQW heterostructures reported to date. AFM and scanning TEM measurements were employed to investigate the material and structural quality of the AlGaN heterostructures, showing smooth surface and sharp layer interfaces.

5.3 Fabry–Perot Injection Laser Limits

For an electrically driven III-N LD, a Fabry–Perot (FP) injection LD is the most common device geometry that employs the confinement of photons emitted in the active region within *n*- and *p*-type waveguiding layers along a transverse direction by utilizing low-refractive-index *n*- and *p*-type cladding layers. Typically, two parallel cleaved crystallographic planes

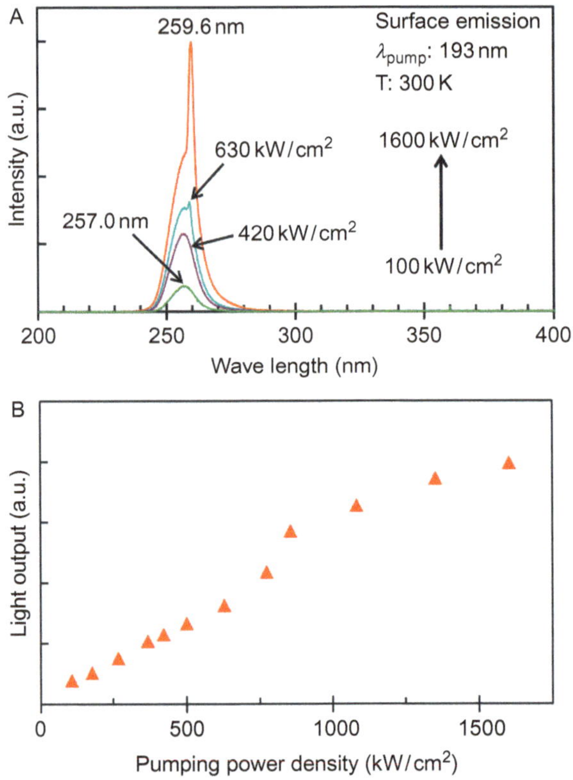

Fig. 7 (A) Surface emission spectra under power-dependent optical pumping and (B) light output intensity of surface emission as a function of pumping power density.

perpendicular to the waveguiding layers are formed as optical feedback mirrors at a designed resonator length, e.g., 500–1500 μm. This type of LD utilizes InGaN/GaN active regions for coherent emission in the near-UV to green spectral regions, and InGaN/AlGaN or GaN/AlGaN or $Al_xGa_{1-x}N/Al_yGa_{1-y}N$ ($x \neq y$) active regions for emission in the UV-A region. The shortest wavelength electrical injection LD to date is 336 nm from a structure consisting of $Al_{0.06}Ga_{0.94}N/Al_{0.16}Ga_{0.84}N/$ $Al_{0.16}Ga_{0.84}N/$ $Al_{0.30}Ga_{0.70}N$ for quantum well/quantum barrier/waveguide/cladding layers (Yoshida et al., 2008). The IQE of AlGaN quantum wells is known to dominantly depend on the TDD in the material; however, with the improvement of the MOCVD and III-N native substrate technologies, the defect density can be reduced.

As mentioned in Sections 5.1 and 5.2, the optically stimulated emission of III-N UV-C lasers has been reported by several groups (e.g., Bryan et al., 2015; Lochner et al., 2013; Martens et al., 2014; Wunderer et al., 2011; Xie et al., 2013). The current technical challenges for UV-C FP-LD are primarily (1) limited free hole concentration and low hole mobility in *p*-type Mg-doped AlGaN due high Mg acceptor activation energy, (2) low refractive index contrast between the waveguide and cladding layers, (3) low hole injection efficiency, and (4) TM contributions to the optical polarization of the stimulated emission peak. The first three issues are solely related to the unavailability of highly conducting *p*-type AlGaN alloys with high AlN mole fractions. Cheng et al. (2013) demonstrated that Mg-doped AlGaN SPSLs with an average AlN mole fraction of 0.6 in their 295 nm separated confinement heterostructure LD were able to operate at current densities up to 11 and 21 kA cm^{-2} in DC and pulse current mode, respectively. However, such a *p*-SPSL was subject to large band discontinuities for holes, and this caused a large diode turn-on voltage, leading to excessive Joule heating and subsequent optical gain suppression. Satter et al. (2014) suggested an inverse-taper AlGaN cladding layer design that utilized composition graded *p*-layers to generate a polarization field that effectively drove holes into the active region, and this concept was later demonstrated by Liu et al. (2015). In such inverted-taper designs, the fabricated 290 nm MQW DH emitter was able to sustain a DC current of at least 500 mA and a pulsed current of at least 1.07 A that corresponds to a current density of 10 and 18 kA cm^{-2} at a maximum measured voltage of 15 and 20 V with the measured series resistance of 15 and 11 Ω cm, respectively. Hole transport in AlGaN is still a major concern, and further studies are needed toward the development of the DUV LDs. For instance, the limited *p*-type conductivity in high-AlN-mole-fraction AlGaN available by either the *p*-SPSL or *p*-inverse-taper technique still limits the maximum refractive index contrast to be formed by the *p*-$Al_{0.45}Ga_{0.55}N$ waveguide and *p*-$Al_{0.60}Ga_{0.40}N$ cladding for the shortest possible emission wavelength of ~280 nm.

Another major concern for DUV LDs relates to the optical polarization since TE-polarized waveguide modes are preferred as these modes do not expand as deeply into the *p*-region as do TM modes. Thus, TE-mode operation reduces the intrinsic losses that are caused by UV absorption in the heavily doped *p*-region and the *p*-contact metal. Depending on strain condition of the quantum wells, the interband transitions can yield either TE- or TM-polarized light (Northrup et al., 2012). The shortest stimulated emission with high TE polarization was at 253 nm from an AlGaN MQW grown on an AlN bulk substrate as demonstrated by

Kolbe et al. (2010). With the above material property-based technological barriers, it is still quite challenging to achieve an electrically driven LD employing the currently reported properties of MOCVD- or MBE-grown high-AlN-mole-fraction AlGaN materials.

5.4 III-N UVVCSEL Issues and Distributed Bragg Reflector Mirrors

For III-N VCSELs, continuous-wave operation has been achieved only for InGaN-based active region VCSELs in which the cavity was formed between two sets of dielectric distributed Bragg reflectors (DBRs), and these reported devices had emitting wavelengths longer than 380 nm (Onishi et al., 2012). To fabricate a DUV VCSEL (e.g., at $\lambda \leq 280$ nm) by using at least one AlGaN-based semiconductor DBR, all III-N layers must have an absorption band-edge energy greater than the emission energy as the photons ideally make many round trips between the DBR mirrors through the layers inside the cavity without any optical absorption loss.

With this requirement in mind, two major technical challenges are inevitable: (1) achieving high *p*-type conductivity of the high-AlN-mole-fraction ($x > 0.45$) AlGaN and (2) creating highly reflective DBR mirrors. For the former challenge, the situation is similar to that in the previous discussion of edge-emitting DUV-LDs. Until a better hole-transport mechanism can be discovered, an electrically driven DUV VCSEL is expected to employ *p*-AlGaN with AlN mole fractions close to where sufficient free hole concentrations can be achieved. For the latter challenge, there is a limited choice of AlGaN to be utilized for a DUV-DBR. Due to lattice mismatch, thermal expansion coefficient mismatch, in-plane composition variations, and low refractive-index contrast, the quality, and reflectivity of DUV DBRs has often been compromised (Moe et al., 2006).

At this time, there are few reports attempting to produce AlGaN-based DUV DBRs. Recently, some strain management approaches such as employing a thick AlGaN buffer layer (Moe et al., 2006) and low-temperature (LT) AlN (Franke et al., 2016) have been applied to suppress cracking in the DBR stacks grown on AlN templates. Moe et al. (2006) reported that the crystallographic quality of $Al_{0.58}Ga_{0.42}N$/AlN DBRs grown on an AlN/6H-SiC template was abruptly improved by introducing a thick $Al_{0.83}Ga_{0.17}N$ strain-relief layer. A maximum reflectivity of 82.8% at 278 nm was achieved with a stopband of ~10 nm for the 21-period DBR; however, cracking still became an issue when the pair number reached 25. In the latter case, two LT-AlN interlayers were introduced in the growth 25.5 pairs of the AlN/$Al_{0.65}Ga_{0.35}N$ DBR that demonstrated a peak reflectivity of 97.7 with the center wavelength of 270 nm and stopband of 8 nm (Franke et al., 2016). Berger et al. (2015) pointed out that thermal mismatch between AlN and sapphire was the dominant cause of cracking and exploited thin AlN/sapphire template ($d_{AlN}\sim$few hundred nm) as a platform for the growth of a 50-pair $Al_{0.7}Ga_{0.3}N$/AlN DBR stack without cracking. Such DBR mirrors achieved reflectivity of 98% at 273 nm. In all these reports, the refractive-index contrast was merely 6% or less, and thus large number of AlGaN/AlN pairs was necessary in order to reach a reflectivity of 99% or greater typically required for VCSEL operation.

However, owing to the direct-band transition of AlGaN ternary semiconductors over the entire alloy composition range, an excitonic resonance is observed for the real part of the dielectric function (ε_1) near the band-edge energies in this AlGaN ternary alloy system even at high AlN mole fractions, including for AlN (after Brunner et al., 1997; Feneberg et al., 2014; Wagner et al., 2001). Due to such excitonic effects, the AlGaN refractive indices increase rapidly near the bandgap energies before optical absorption becomes dominant. For instance, such enhanced refractive index contrast with negligible absorption of Mg-doped $Al_{0.733}Ga_{0.267}N$ and AlN is found to cover a photon energy range of ~320 meV (equivalent to ~11 nm in spectral wavelength range) from data reported by Feneberg et al. (2014). For this reason, Detchprohm et al. (2016) tuned the AlGaN band edge close to the desired VCSEL emission energy in order to benefit from such enhanced refractive-index contrast in an AlGaN/AlN DBR structure for the 220–250 nm DUV region. The AlGaN/AlN DBR structures were grown on 1.5 μm-thick AlN/sapphire templates with TDDs in the lower 10^9 cm^{-2} range. The AlGaN layers were grown as a SPSL structures of AlGaN and AlN. No cracking was observed even for the total pair number of 50. Reflectivity spectra of a 30.5-pair (SPSL-$Al_{0.87}Ga_{0.13}N$)/AlN DBR, and a (SPSL-$Al_{0.73}Ga_{0.27}N$)/AlN DBR are exhibited in Fig. 8B and D together with transmission spectra of a 78-nm thick SPSL-$Al_{0.87}Ga_{0.13}N$ (Fig. 8A), and a 72-nm thick SPSL-$Al_{0.73}Ga_{0.27}N$ grown on an AlN template (Fig. 8C). In both cases, the reflectivity peaks were located just before the absorption from the AlGaN layers became dominant. The peak reflectivity values were 96.9% at $\lambda_{center}=226$ nm and 95.7% at $\lambda_{center}=247$ nm for a SPSL-$Al_{0.87}Ga_{0.13}N$/AlN DBR, and SPSL-$Al_{0.73}Ga_{0.27}N$/AlN DBR, respectively. This approach to the growth of high-reflectivity DUV DBRs may provide a pathway to the realization of a practical DUV electrically driven VCSEL.

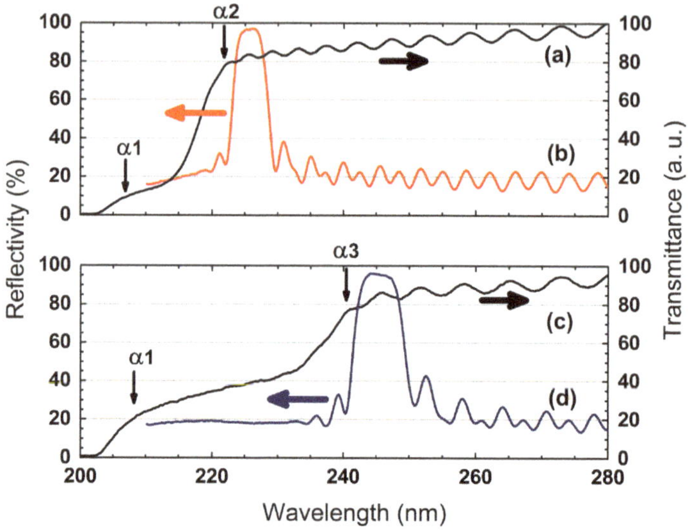

Fig. 8 Optical transmission spectra (*black*) of 78 nm thick SPSL-$Al_{0.87}Ga_{0.13}N$ on AlN template (A), and 72 nm thick SPSL-$Al_{0.73}Ga_{0.27}N$ on AlN template (C), and reflectivity spectra (*red, blue*) of (B) 30.5-pair (SPSL-$Al_{0.87}Ga_{0.13}N$)/AlN DBR for $\lambda_{center}=226$ nm, and (D) 30.5-pairs (SPSL-$Al_{0.73}Ga_{0.27}N$)/AlN DBR for $\lambda_{center}=247$ nm. α1, α2, and α3 indicate the absorption onset of AlN template, SPSL-$Al_{0.87}Ga_{0.13}N$, and SPSL-$Al_{0.73}Ga_{0.27}N$, respectively. *After Detchprohm, T., Liu, Y.-S., Mehta, K., Wang, S., Xie, H., Kao, T.-T., Shen, S.-C., Yoder, P.D., Ponce, F.A., Dupuis, R.D., 2016. Sub 250 nm Deep-UV AlGaN/AlN distributed Bragg reflectors. Appl. Phys. Lett. (submitted for publication).*

6. III-N DUV PHOTODETECTORS

III-N-based DUV photodetectors can replace conventional types of DUV photodetectors in a wide range of applications such as combustion engine control, missile plume detection, corona discharge detection, flame detection, UV astronomy, and chemical/biological battlefield reagent detection. For III-N semiconductors, such wide-bandgap materials suitable for the visible-blind or SB UV photon detections can be achieved in various combinations of epitaxial layers including AlGaN, AlInN, and AlInGaN. However, with the currently limited quality of achievable alloys, we confine our discussion to AlGaN. $Al_xGa_{1-x}N$ ternary alloys have a direct energy bandgap and high absorption coefficient ($\alpha > 10^5$ cm^{-2}) above the bandgap energy for the whole alloy composition range. The intrinsic band-edge absorption of $Al_xGa_{1-x}N$ can be engineered to cover a cut-off wavelength from 365 nm for $x=0$ to 210 nm for $x=1$ by simply altering the alloy composition. To utilize these materials as a photon absorber in SB-UV photodiode, requires at least $x \geq 0.45$ for a cut-off wavelength of 280 nm. Using an AlGaN absorber layer with an AlN mole fraction of $0.45 < x < 1$, an AlGaN-based photodetector can cover the whole UV-C range, i.e., 200–280 nm.

6.1 DUVPIN Photodiodes

SB III-N DUV PDs have been studied by many groups for several years (e.g., Campbell et al., 2003; Dupuis and Campbell, 2002). A UV AlGaN *p-i-n* photodiode (PD) often operates under low reverse bias with a stable and relatively constant electric field ($|E| < 1$MV cm^{-1}) across the entire *i*-AlGaN layer where a space-charge region is formed. The photon absorption efficiency and high-frequency detectability is much improved through the use of a wider depletion region as compared to that of a typical MSM detector. For example, GaN UV PIN photodetectors showed a 300-K noise-equivalent power (NEP) of 4.27×10^{-17} W Hz$^{-0.5}$ and a detectivity (D^*) of 1.66×10^{14} cm Hz$^{0.5}$ W^{-1} at 20 V reverse bias and $\lambda = 360$ nm (Zhang et al., 2009). In the late 1990s, the early III-N UV PDs were formed on GaN/sapphire templates largely due to limited epitaxial quality of high-AlN-mole-fraction AlGaN layers as well as the relatively poor *p*-type conductivity of Mg-doped AlGaN. Parish et al. (1999) and Tarsa et al. (2000) employed *i*-$Al_{0.33}Ga_{0.67}N$ and *i*-$Al_{0.30}Ga_{0.70}N$ cladded by *n*- and *p*-GaN layers to demonstrate UV-PDs with a cut-off wavelength of ~295 nm and external quantum efficiency $\eta_{ex} = 21.7\%$ and 34.8% at a peak absorption wavelength of ~285 nm under zero external bias, respectively. This group also attempted an improved optical performance by introducing a UV-transparent "window" of *n*-$Al_xGa_{1-x}N$ ($x > 0.30$) grown on sapphire instead of the standard GaN/sapphire template for back-side illumination. This device, however, had a lower $\eta_{ex} = 14.9\%$ at 275 nm under zero external bias as its performance was subject to the compromised quality of the AlGaN window layer which subsequently affected the *i*-AlGaN quality. Depending on the AlN mole fraction of that optical window layer, the device then exhibited a cut-on wavelength around 260 nm narrowing the spectral detection range down to approximately 35 nm. Pernot et al. (2000) reported an SB DUV *p-i-n* PD utilizing low TDD (~mid-10^9 cm^{-2}) $Al_{0.44}Ga_{0.56}N$:Si ($n = 1 \times 10^{18}$ cm^{-3}) and undoped $Al_{0.44}Ga_{0.56}N$ grown GaN/sapphire template via AlN interlayer as *n*- and *i*-layers, while *p*-GaN cap layer was used having a free hole carrier concentration of $p \sim 1 \times 10^{18}$ cm^{-3}. The UV-PD cut-off wavelength, peak absorption wavelength, and η_{ex} without external bias were 280 nm, 270 nm, and 5.4%, respectively. The front-illumination photoresponse was measured through a meshed contact on the p-GaN, and the device was demonstrated to detect the UV-C signature of a natural gas flame, i.e., 250–280 nm, at intensities as low as hundreds of nW cm^{-2} under room-light ambient (Hirano et al., 2001). However, the performance in these early SB-p-i-n PDs was limited due to either an absorbing p-GaN in front-illumination mode or a compromised crystallographic quality of the AlGaN window and absorber layer in the back-illumination mode. Lambert et al. (2000) grew an epitaxial structure utilizing an AlGaN template on sapphire using an all AlGaN p-i-n device except for a thin p++ GaN for low electrical contact resistance purposes. The device structure incorporated several compositionally graded layers of AlGaN as transparent windows on sapphire. For example, a p-i-n structure of n^+-$Al_{0.57}Ga_{0.43}N$/i-$Al_{0.48}Ga_{0.52}N$/p-$Al_{0.48}Ga_{0.52}N$ was successfully grown on lightly doped n-$Al_{0.57}Ga_{0.43}N$ template on sapphire as shown in Fig. 9. The η_{ex} was 42% and 48% at 269 nm under zero and −10 V bias, respectively. This type of device was used in fabricating a full 256 × 256 SB imaging arrays (Reine et al., 2006), and the best device performance was reported as η_{ex} = 58.1% and 64.5% at ~275 nm under zero and −5 V bias, respectively. These high-performance

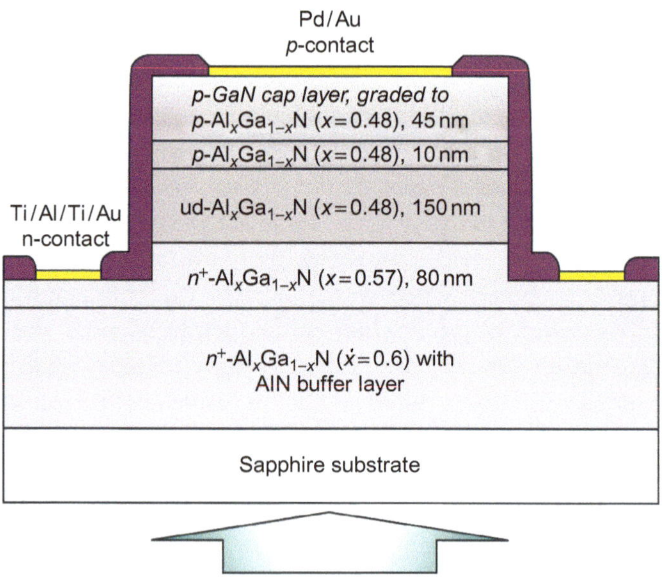

Fig. 9 Schematic cross section of an all AlGaN SB-p-i-n PD design for back illumination. *After Collins, C.J., Chowdhury, U., Wong, M.M., Yang, B., Beck, A.L., Dupuis, R.D., Campbell, J.C., 2002. Improved solar-blind detectivity using an* $Al_xGa_{1-x}N$ *heterojunction* p–i–n *photodiode. Appl. Phys. Lett. 80, 3754–3756.*

SB AlGaN/sapphire PIN photovoltaic (PV) photodetector structures were used in fabricating the first large-area 256 × 256 SB imaging arrays (Reine et al., 2006).

Recently, MOCVD has been further developed for high-AlN-mole-fraction AlGaN and AlN growth on sapphire to create films having a relatively low TDD of $\sim 10^9$ cm^{-3} or less as the layer growth temperature was raised to 1200–1500°C. Cicek et al. (2013b) utilized a PALE process with this high-temperature MOCVD scheme to develop SB-n^+-$Al_{0.55}Ga_{0.45}N$/i-$Al_{0.40}Ga_{0.60}N$/p-$Al_{0.38}Ga_{0.62}N$ PDs grown on an AlN/sapphire template. The APDs were measured under back-side illumination, and these devices reached $\eta_{ex} = 80\%$ and 89% at ~275 nm under zero and −5 V bias, respectively.

Hybrid SB-n-i-p PDs utilizing n-$Al_{0.80}Ga_{0.20}N$/i-AlN/i-SiC/p-SiC (4H SiC polytype with epitaxial growth on the Si face) have also been reported (Rodak et al., 2013). Though illumination was directed through the III-N epitaxial materials, the absorption mainly took place in the i-SiC. The selectivity for the UV-C absorption was introduced by the polarization electric field across the AlN layer creating a barrier for transport of photogenerated electrons from the M valence-band valley of the SiC but allowing the transport of photogenerated electrons from the SiC Γ and L conduction-band valleys to be collected at the n-$Al_{0.80}Ga_{0.20}N$. At zero external bias, η_{ex} was 20% at 226 nm with cut-off wavelength of ~235 nm; however, the peak absorption wavelength and cut-off wavelength were red-shifted to 242 and 260 nm under reverse bias of 40 V, respectively, while the device reached its maximum η_{ex} of 76%.

The development status of III-N-based SB-p-i-n PDs is summarized in Table 3. The reported D^* values are in the range of 10^{12}–10^{14} cm Hz$^{1/2}$ W^{-1}. Future work on the design of SAM PIN PDs and the use of native III-N substrates will undoubtedly result in improved performance.

6.2 III-N UV Avalanche Photodiodes (APDs)

Semiconductor APDs can offer high photocurrent gain comparable to photomultiplier tubes, combined with the benefits of small size, high reliability, high speed, low operation voltage, low power consumption, low cost, and all-solid-state integration. Although UV-enhanced Si single-photon detectors are commercially available and SiC-based APDs have demonstrated impressive GM operation, III-N APDs possess unique bandgap engineering capabilities and a direct bandgap that are important to achieve a SB operation with high quantum efficiencies.

Table 3 Summary of Performance of SB *p-i-n* PDs in an Chronological Order

					Unbiased Condition				Under Reverse Bias					
Substrate	*n*	*i*	*p*	Illumination Direction	λ_{peak} (nm)	$\lambda_{cut\text{-}on}$ (nm)	$\lambda_{cut\text{-}off}$ (nm)	peak η_x (%)	λ_{peak} (nm)	Peak η_x (%)	Bias (V)	J_{dark} (A cm^2)	D^a (cm Hz$^{1/2}$ W^{-1}) (Unbiased)	References
Sapphire	GaN:Si	$Al_{0.33}Ga_{0.67}N$	GaN:Mg	Front	286		295	21.7				1×10^{-8} at -5 V		Parish et al. (1999)
Sapphire	$Al_{0.44}Ga_{0.56}N$:Si	$Al_{0.44}Ga_{0.56}N$	GaN:Mg	Front	270		280	5.4 ($\eta_i = 50\%$)				$<1 \times 10^{-11}$ at 0 V	1.2×10^{13}	Pernot et al. (2000)
Sapphire	GaN:Si	$Al_{0.30}Ga_{0.70}N$	GaN:Mg	Front	285		295	34.8						Tarsa et al. (2000)
	$Al_xGa_{1-x}N$:Si ($x > 0.3$)	$Al_{0.30}Ga_{0.70}N$	GaN:Mg	Back	285		295	14.9						
Sapphire	$Al_{0.40}Ga_{0.60}N$:Si	$Al_{0.40}Ga_{0.60}N$	$Al_{0.40}Ga_{0.60}N$:Mg	Back	277			12		35	-60			Lambert et al. (2000)
Sapphire	$Al_{0.47}Ga_{0.53}N$:Si	$Al_{0.39}Ga_{0.61}N$	$Al_{0.47}Ga_{0.53}N$:Mg	Back	279			26		31	-5			Wong et al. (2001)

Sapphire	$Al_{0.57}Ga_{0.43}N$:Si	$Al_{0.48}Ga_{0.52}N$	$Al_{0.48}Ga_{0.57}N$:Mg	Back	269		282	42	270	48	−10	$<8.2 \times 10^{-11}$ at −5 V	2×10^{14} at 269 nm	Collins et al. (2002)
Sapphire	$Al_{0.54}Ga_{0.46}N$:Si	$Al_{0.46}Ga_{0.54}N$	$Al_{0.46}Ga_{0.54}N$:Mg	Back	270	262	282	58.1		64.5	−5			Reine et al., 2006[a]
Sapphire	$Al_{0.45}Ga_{0.55}N$:Si	$Al_{0.40}Ga_{0.60}N$	$Al_{0.38}Ga_{0.62}N$:Mg	Back	279			38		57	−5		8.7×10^{12} at 272 nm	Cicek et al. (2013a)[a]
Sapphire	$Al_{0.45}Ga_{0.55}N$:Si	$Al_{0.40}Ga_{0.60}N$	$Al_{0.38}Ga_{0.62}N$:Mg	Back	275			80		89	−5	$<2 \times 10^{-9}$ at −10 V		Cicek et al. (2013b)
Si-face 4H-SiC	$Al_{0.80}Ga_{0.20}N$	AlN/*i*-SiC	*p*+SiC	Front	226		250	20	241	78	−40			Rodak et al. (2013)[b]
Sapphire	$Al_{0.45}Ga_{0.55}N$:Si	$Al_{0.40}Ga_{0.60}N$	$Al_{0.38}Ga_{0.62}N$:Mg	Back	278		290	49		66	−5	$<2 \times 10^{-9}$ at −5 V		Cicek et al. (2014)[a]

[a]Indicate a focal plane array of SB *p-i-n* PDs.
[b]Hybrid *n-i-p* design.

The primary approaches for III-N UV-A APDs use GaN as the absorption layer. In 2001, the first GM GaN *p-i-n* APD was demonstrated for a structure grown by VPE on a sapphire substrate and was tested at 325 nm (Verghese et al., 2001). However, for low-leakage current operation, low-defect-density FS-GaN or bulk GaN substrates are the preferred platform. Commercially available FS-GaN substrates have DDs of $<10^5$ cm^{-2}. GaN APDs grown FS-GaN substrates have successfully demonstrated high-gain "visible-blind" UV-A APDs with avalanche gain of $>10^4$ (Shen et al., 2007) and GM-APDs (Choi et al., 2009). Improvements in GaN FS substrates have led to even better performance devices.

The inherently SB III-N APDs, based upon wide-bandgap AlGaN alloys, have not been able to achieve this level of performance and GM operation. At this time, there are no reports of true SB DUV AlGaN APDs. However, visible-blind GaN-based III-N UV APDs have been studied by many groups (e.g., Butun et al., 2008; Carrano et al., 2000; Pau et al., 2007, 2008; Shen et al., 2007). As described earlier, most early work on AlGaN-based PIN photodetectors was focused on the "PV mode" of operation near zero bias (e.g., Reine et al., 2006). All of the work reported so far on DUV AlGaN PIN APDs has been based on heteroepitaxial films grown on (0001) sapphire substrates and, consequently, the devices have had a high density of dislocations $\sim 10^9$–10^{10} cm^{-2} (Limb et al., 2008). The use of sapphire, while allowing the growth of PIN structures that are compatible with back-side UV illumination, leads to a high concentration of defects in the device and correspondingly, high dark currents at reverse bias voltages well below the avalanche breakdown.

State of the art $Al_xGa_{1-x}N$ ($x=0.05$)-based PIN APDs in the UV spectral range can be grown on FS-GaN substrates (Kim et al., 2015). The current density vs voltage (J–V) characteristics for one of these top-illuminated AlGaN PIN APDs with device mesa diameters of 30–70 μm in the dark and under SB illumination is shown in Fig. 10 and in the photocurrent density and gain vs mesa area are shown in Fig. 11 (using $\lambda = 280$ nm illumination). The breakdown voltage (V_{BR}) and the dark-current densities were derived from the onset of dark-current multiplication and the averaged dark-current density at the reverse biases between 0 and 60 V, respectively. Even though the averaged dark-current density increases with the mesa area, the values are maintained at $<10^{-5}$ A cm^{-2} for all the mesa areas. These devices exhibit similar properties to the homoepitaxial GaN PIN APDs grown on FS-GaN substrates with very low dark current densities (instrument limited) of $\sim 5\times10^{-9}$ A cm^{-2} over the voltage range $0 < V_R < 50$ V and avalanche gains $>10^6$.

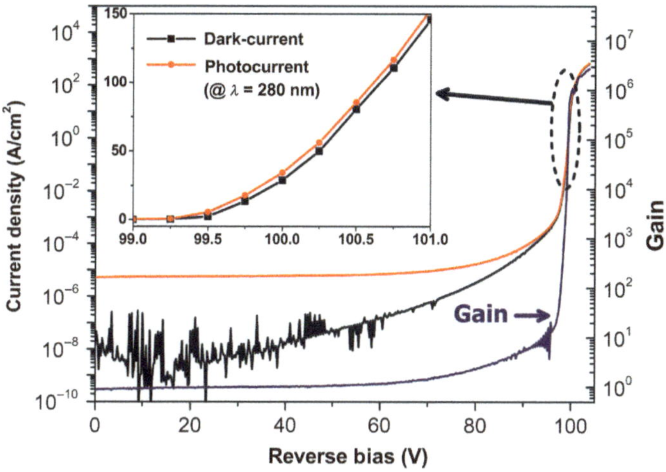

Fig. 10 Reverse bias *J–V* characteristics of an $Al_{0.05}Ga_{0.95}N$ UV-APD grown on free-standing GaN substrates with a circular mesa diameter of 30 μm with and without UV light illumination at $\lambda = 280$ nm. The gain of the APD is also shown on the right-hand axis. *After Kim, J., Ji, M.-H., Detchprohm, T., Dupuis, R.D., Sood, A., Dhar, N.K., 2014. Growth and characterization of GaN avalanche photodiodes grown on free-standing GaN substrates by metalorganic chemical vapor deposition. International Workshop on III-Nitrides 2014, 24–29 August 2014, Wroclaw, Poland.*

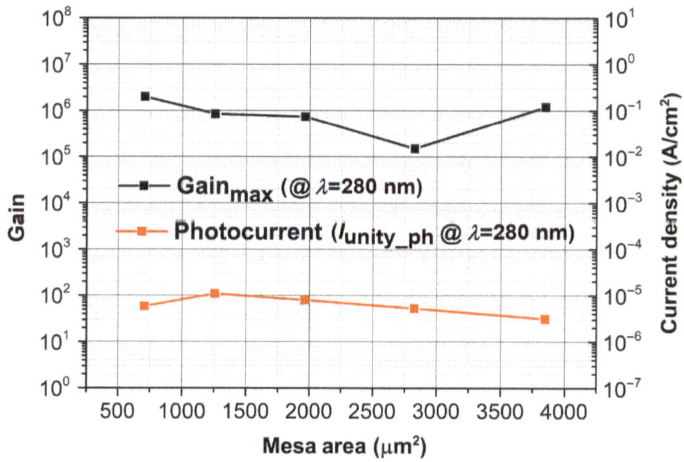

Fig. 11 Photocurrent densities and avalanche gains of $Al_{0.05}Ga_{0.95}N$ UV-APDs under 280-nm UV illumination for various area devices showing low leakage current densities for all device areas measured. *After Kim, J., Ji, M.-H., Detchprohm, T., Dupuis, R.D., Sood, A., Dhar, N.K., 2014. Growth and characterization of GaN avalanche photodiodes grown on free-standing GaN substrates by metalorganic chemical vapor deposition. International Workshop on III-Nitrides 2014, 24–29 August 2014, Wroclaw, Poland.*

While there are no reports of truly SB AlGaN GM DUV APDs, APDs with high Al-content AlGaN layers and advanced APD structure designs such as AlGaN-based SAM devices have been reported (Huang et al., 2012; Wang et al., 2014). An attractive alternative choice of substrate platform for DUV III-N GM APDs would be DUV-transparent bulk AlN substrates, which are only becoming available in small areas at the time of this writing. In fact, III-N DUV APDs grown on bulk AlN substrates have not yet been reported.

7. CONCLUSIONS

The MOCVD growth of III-N DUV materials and devices is a dynamic and rapidly advancing field. We have tried to provide a short summary of the status of research in this area that is focused on two important device applications: the DUV laser and the DUV photodetector. The materials challenges in this alloy composition range are significant, and critical device design trade-offs must be made with the benefit of comprehensive device design and simulation that include the most accurate experimental materials properties. The use of high-quality substrates will provide a platform for the advancement of device performance, as it has had for all other similar III–V devices. The efforts of many past and current researchers have paved the way for a very bright future for DUV devices which will have a significant impact benefitting all of humanity.

ACKNOWLEDGMENTS

The work at Georgia Institute of Technology was supported over several years in part by DARPA, NSF, and the US Army Research Office. We thank the School of ECE and the College of Engineering at Georgia Institute of Technology for additional support, and RDD acknowledges the continued support of the Steve W. Chaddick Endowed Chair in Electro-Optics and the Georgia Research Alliance.

REFERENCES

Akasaki, I., Amano, H., Murakami, H., Sassa, M., Kato, H., Manabe, K., 1993. Growth of GaN and AlGaN for UV/blue p-n junction diodes. J. Cryst. Growth 128, 379–383.

Al tahtamouni, T.M., Sedhain, A., Lin, J.Y., Jiang, H.X., 2008. Si-doped high Al-content AlGaN epilayers with improved quality and conductivity using indium as a surfactant. Appl. Phys. Lett. 92, 092105-1–092105-3.

Allerman, A.A., Crawford, M.H., Miller, M.A., Lee, S.R., 2010. Growth and characterization of Mg-doped AlGaN–AlN short-period superlattices for deep-UV optoelectronic devices. J. Cryst. Growth 312, 756–761.

Amano, H., Sawaki, I., Akasaki, I., Toyoda, Y., 1986. Metalorganic vapor phase epitaxial growth of a high quality GaN film using an AlN buffer layer. Appl. Phys. Lett. 48, 353–355.

Amano, H., Kito, M., Hiramatsu, K., Akasaki, I., 1989. P-type conduction in Mg-doped GaN treated with low-energy electron irradiation (LEEBI). Jpn. J. Appl. Phys. 28, L2112–L2114.

Ban, K., Yamamoto, J.I., Takeda, K., Ide, K., Iwaya, M., Takeuchi, T., Kamiyama, S., Akasaki, I., Amano, H., 2011. Internal quantum efficiency of whole-composition-range AlGaN multiquantum wells. Appl. Phys. Express 4, 052101-1–052101-3.

Banal, R.G., Funato, M., Kawakami, Y., 2009. Optical anisotropy in [0001]-oriented $Al_xGa_{1-x}N/AlN$ quantum wells ($x > 0.69$). Phys. Rev. B 79, 121308.

Bardwell, J.A., Webb, J.B., Tang, H., Fraser, J., Moisa, S., 2001. Ultraviolet photoenhanced wet etching of GaN in $K_2S_2O_8$ solution. J. Appl. Phys. 89, 4142–4149.

Berger, C., Franke, A., Dadgar, A., Schmidt, G., Bläsing, J., Goldhahn, R., Krost, A., Hoffmann, A., Christen, J., Strittmatter, A., 2015. High-reflectivity distributed Bragg reflectors for DUV to visible spectral range. In: Paper presented at the 6th International Symposium on Growth of III-Nitrides. November 8–13, Hamamatsu, Japan.

Brunner, D., Angerer, H., Bustarret, E., Freudenberg, F., Höpler, R., Dimitrov, R., Ambacher, O., Stutzmann, M., 1997. Optical constants of epitaxial AlGaN films and their temperature dependence. J. Appl. Phys. 82, 5090–5096.

Brunner, F., Mogilatenko, A., Kueller, V., Knauer, A., Weyers, M., 2013. Stress evolution during $Al_xGa_{1-x}N/AlN$ growth on sapphire. J. Cryst. Growth 376, 54–58.

Bryan, Z., Bryan, I., Mita, S., Tweedie, J., Sitar, Z., Collazo, R., 2015. Strain dependence on polarization properties of AlGaN and AlGaN-based ultraviolet lasers grown on AlN substrates. Appl. Phys. Lett. 106, 232101-1–232101-5.

Butun, B., Tut, T., Ulker, E., Yelboga, T., Ozbay, E., 2008. High-performance visible-blind GaN-based p-i-n photodetectors. Appl. Phys. Lett. 92, 033507-1–033507-3.

Campbell, J.C., Wang, S., Zheng, X., Li, X., Li, N., Ma, F., Sun, X., Collins, C.J., Beck, A.L., Yang, B., Hurst, J.B., Sidhu, R., Holmes, A.L., Chowdhury, U., Wong, M.M., Dupuis, R.D., Huntington, A., Coldren, L.A., Chen, Z., Kim, E.-T., Madhukar, A., 2003. Photodetectors: UV to IR. SPIE 4999, 410–422.

Cantu, P., Keller, S., Mishra, U.K., DenBaars, S.P., 2003a. Metalorganic chemical vapor deposition of highly conductive $Al_{0.65}Ga_{0.35}N$ films. Appl. Phys. Lett. 82, 3683–3685.

Cantu, P., Wu, F., Waltereit, P., Keller, S., Romanov, A.E., Mishra, U.K., DenBaars, S.P., Speck, J.S., 2003b. Si doping effect on strain reduction in compressively strained Al0.49Ga0.51N thin films. Appl. Phys. Lett. 83, 674–676.

Carrano, J.C., Lambert, D.J.H., Eiting, C.J., Li, T., Wang, S., Yang, B., Beck, A.L., Dupuis, R.D., Campbell, J.C., 2000. GaN avalanche photodiodes. Appl. Phys. Lett. 76, 924–926.

Chakraborty, A., Moe, C.G., Wu, Y., Mates, T., Keller, S., Speck, J.S., DenBaars, S.P., Mishra, U.K., 2007. Electrical and structural characterization of Mg-doped p-type Al0.69Ga0.31N films on SiC substrate. J. Appl. Phys. 101, 053717-1–053717-6.

Cheng, B., Choi, S., Northrup, J.E., Yang, Z., Knollenberg, C., Teepe, M., Wunderer, T., Chua, C.L., Johnson, N.M., 2013. Enhanced vertical and lateral hole transport in high aluminum-containing AlGaN for deep ultraviolet light emitters. Appl. Phys. Lett. 102, 231106-1–231106-4.

Cho, A.Y., 1970. Morphology of epitaxial growth of GaAs by a molecular beam method: observation of surface structures. J. Appl. Phys. 41, 2780–2786.

Choi, S., Kim, H.-J., Zhang, Y., Bai, X., Yoo, D., Limb, J., Ryou, J.-H., Shen, S.-C., Yoder, P.D., Dupuis, R.D., 2009. Geiger-mode operation of GaN avalanche photodiodes grown on GaN substrates. IEEE Photonic Technol. Lett. 21, 1526–1528.

Cicek, E., McClintock, R., Vashaei, Z., Zhang, Y., Gautier, S., Cho, C.Y., Razeghi, M., 2013a. Crack-free AlGaN for solar-blind focal plane arrays through reduced area epitaxy. Appl. Phys. Lett. 102, 051102.

Cicek, E., McClintock, R., Cho, C.Y., Rahnema, B., Razeghi, M., 2013b. $Al_xGa_{1-x}N$-based back-illuminated solar-blind photodetectors with external quantum efficiency of 89%. Appl. Phys. Lett. 103, 191108-1–191108-4.

Cicek, E., McClintock, R., Haddadi, A., Gaviria Rojas, W.A., Razeghi, M., 2014. High performance solar-blind ultraviolet 320 × 256 focal plane arrays based on $Al_xGa_{1-x}N$. IEEE J. Quantum Electron. 50, 593–597.

Collazo, R., Mita, S., Xie, J., Rice, A., Tweedie, J., Dalmau, R., Sitar, Z., 2011. Progress on n-type doping of AlGaN alloys on AlN single crystal substrates for UV optoelectronic applications. Phys. Status Solidi C 8, 2031–2033.

Collazo, R., Xie, J., Gaddy, B.E., Bryan, Z., Kirste, R., Hoffmann, M., Dalmau, R., Moody, B., Kumagai, Y., Nagashima, T., Kubota, Y., 2012. On the origin of the 265 nm absorption band in AlN bulk crystals. Appl. Phys. Lett. 100, 191914-1–191914-5.

Collins, C.J., Chowdhury, U., Wong, M.M., Yang, B., Beck, A.L., Dupuis, R.D., Campbell, J.C., 2002. Improved solar-blind detectivity using an $Al_xGa_{1-x}N$ heterojunction *p–i–n* photodiode. Appl. Phys. Lett. 80, 3754–3756.

Detchprohm, T., 2015. Calculations of lattice mismatch in the AlInGaN alloy system (unpublished).

Detchprohm, T., Liu, Y.-S., Mehta, K., Wang, S., Xie, H., Kao, T.-T., Shen, S.-C., Yoder, P.D., Ponce, F.A. and Dupuis, R.D., 2016 Sub 250 nm deep-UV AlGaN/AlN distributed Bragg reflectors, *Appl. Phys. Lett.*, (submitted for publication).

Duffy, M.T., Wang, C.C., O'Clock Jr., G.D., McFarlane, S.H., Zanzucchi, P.J., 1973. Epitaxial growth and piezoelectric properties of AlN, GaN and GaAs on sapphire or spinel. J. Electron. Mater. 2, 359–372.

Dupuis, R.D., Campbell, J.C., 2002. Ultraviolet photodetectors based upon III-N materials. In: Ren, F., Zolper, J. (Eds.), Wide Energy Bandgap Electronic Devices. World Scientific Publishing Co. Inc., Singapore.

Feneberg, M., Osterburg, S., Romero, M.F., Garke, B., Goldhahn, R., Neumann, M.D., Esser, N., Yan, J., Zeng, J., Wang, J., Li, J., 2014. Optical properties of magnesium doped $Al_xGa_{1-x}N$ ($0.61 \leq x \leq 0.73$). J. Appl. Phys. 116, 143103-1–143103-7.

Figge, S., Kröncke, H., Hommel, D., Epelbaum, B.M., Figge, S., Kröncke, H., Hommel, D., Epelbaum, B.M., 2009. Temperature dependence of the thermal expansion of AlN. Appl. Phys. Lett. 94, 101915-1–101915-3.

Follstaedt, D.M., Lee, S.R., Provencio, P.P., Allerman, A.A., Floro, J.A., Crawford, M.H., 2005. Relaxation of compressively-strained AlGaN by inclined threading dislocations. Appl. Phys. Lett. 87, 121112-1–121112-3.

Franke, A., Hoffmann, M.P., Hernandez-Balderrama, L., Kaess, F., Bryan, I., Washiyama, S., Bobea, M., Tweedie, J., Kirste, R., Gerhold, M., Collazo, R., Sitar, Z., 2016. Strain engineered high reflectivity DBRs in the deep UV. SPIE 9748, 97481G.

Guo, W., Bryan, Z., Xie, J., Kirste, R., Mita, S., Bryan, I., Hussey, L., Bobea, M., Haidet, B., Gerhold, M., Collazo, R., Sitar, Z., 2014. Stimulated emission and optical gain in AlGaN heterostructures grown on bulk AlN substrates. J. Appl. Phys. 115, 10.

Hashimoto, M., Amano, M., Sawaki, N., Akasaki, I., 1984. Effects of hydrogen in an ambient on the crystal growth of GaN using $Ga(CH_3)_3$ and NH_3. J. Cryst. Growth 68, 163–168.

Hearne, S., Chason, E., Han, J., Floro, J.A., Figiel, J., Hunter, J., Amano, H., Tsong, I.S.T., 1999. Stress evolution during metalorganic chemical vapor deposition of GaN. Appl. Phys. Lett. 74, 356–358.

Hirano, A., Pernot, C., Iwaya, M., Detchprohm, T., Amano, H., Akasaki, I., 2001. Demonstration of flame detection in room light background by solar-blind AlGaN PIN photodiode. Phys. Status Solidi A 188, 293–296.

Hirayama, H., Fujikawa, S., Norimatsu, J., Takano, T., Tsubaki, K., Kamata, N., 2009. Fabrication of a low threading dislocation density ELO-AlN template for application to deep-UV LEDs. Phys. Status Solidi C 6 (S2), S356–S359.

Hoffmann, V., Knauer, A., Brunner, C., Einfeldt, S., Weyers, M., Tränkle, G., Haberland, K., Zettler, J.T., Kneissl, M., 2011. Uniformity of the wafer surface temperature during MOVPE growth of GaN-based laser diode structures on GaN and sapphire substrate. J. Cryst. Growth 315, 5–9.

Huang, Y., Chen, D.J., Lu, H., Dong, K.X., Zhang, R., Zheng, Y.D., Li, L., Li, Z.H., 2012. Back-illuminated separate absorption and multiplication AlGaN solar-blind avalanche photodiodes. Appl. Phys. Lett. 101, 253516-1–253516-4.

Imura, M., Nakano, K., Fujimoto, N., Okada, N., Balakrishnan, K., Iwaya, M., Kamiyama, S., Amano, H., Akasaki, I., Noro, T., Takagi, T., Bandoh, A., 2006. High-temperature metal-organic vapor phase epitaxial growth of AlN on sapphire by multi transition growth mode method varying V/III ratio. Jpn. J. Appl. Phys. 45, 8639–8643.

Imura, M., Kato, N., Okada, N., Balakrishnan, K., Iwaya, M., Kamiyama, S., Amano, H., Akasaki, I., Noro, T., Takagi, T., Bandoh, A., 2007a. Mg-doped high-quality $Al_xGa_{1-x}N$ ($x=0$-1) grown by high-temperature metal-organic vapor phase epitaxy. Phys. Status Solidi C 4, 2502–2505.

Imura, M., Nakano, K., Fujimoto, N., Okada, N., Balakrishnan, K., Iwaya, M., Kamiyama, S., Amano, H., Akasaki, I., Noro, T., Takagi, T., Bandoh, A., 2007b. Dislocations in AlN epilayers grown on sapphire substrate by high-temperature metal-organic vapor phase epitaxy. Jpn. J. Appl. Phys. 46, 1458–1462.

Jeon, S.-R., Ren, Z., Cui, G., Su, J., Gherasimova, M., Han, J., Cho, H.-K., Zhou, L., 2005. Investigation of Mg doping in high-Al content p-type $Al_xGa_{1-x}N(0.3<x<0.5)$. Appl. Phys. Lett. 86, 082107-1–082107-3.

Ji, M.-H., Kim, J., Detchprohm, T., Dupuis, R.D., 2016. (unpublished).

Johnson, N.M., Cheng, B., Choi, S., Chua, C.L., Knollenberg, C., Northrup, J.E., Teepe, M.R., Wunderer, T., Yang, Z., 2012. The 9th International Symposium on Semiconductor Light Emitting Devices, Berlin, Germany.

Kakanakova-Georgieva, A., Nilsson, D., Stattin, M., Forsberg, U., Haglund, Å., Larsson, A., Janzén, E., 2010. Mg-doped $Al_{0.85}Ga_{0.15}N$ layers grown by hot-wall MOCVD with low resistivity at room temperature. Phys. Status Solidi (RRL)—Rapid Res. Lett. 4, 311–313.

Kakanakova-Georgieva, A., Nilsson, D., Janzén, E., 2012. High-quality AlN layers grown by hot-wall MOCVD at reduced temperatures. J. Cryst. Growth 338, 52–56.

Kakanakova-Georgieva, A., Nilsson, D., Trinh, X.T., Forsberg, U., Son, N.T., Janzén, E., 2013. The complex impact of silicon and oxygen on the n-type conductivity of high-Al-content AlGaN. Appl. Phys. Lett. 102, 132113-1–132113-4.

Kao, T.T., Liu, Y.S., Satter, M.M., Li, X.H., Lochner, Z., Yoder, P.D., Detchprohm, T., Dupuis, R.D., Shen, S.C., Ryou, J.H., Fischer, A.M., 2013. Sub-250 nm low-threshold deep-ultraviolet AlGaN-based heterostructure laser employing HfO_2/SiO_2 dielectric mirrors. Appl. Phys. Lett. 103, 211103-1–211103-4.

Kao, T., Liu, Y., Detechprohm, T., Dupuis, R.D., Shen, S.-C., 2016. Vanadium-based Ohmic contact for aluminum-rich n-AlGaN. In: Technical Digest, CSMANTECH, May 17–19, 2016, Miami, FL, USA.

Kawanishi, H., Tomizawa, T., 2012. Carbon-doped p-type (0001) plane AlGaN (Al = 6–55%) with high hole density. Phys. Status Solidi B 249, 459–463.

Khan, M.A., Skogman, R.A., Schulze, R.G., Gershenson, M., 1983a. Electrical properties and ion implantation of epitaxial GaN, grown by metalorganic chemical vapor deposition. Appl. Phys. Lett. 42, 430–432.
Khan, M.A., Skogman, R.A., Schulze, R.G., Gershenson, M., 1983b. Properties and ion implantation of $Al_xGa_{1-x}N$ epitaxial single-crystal films prepared by low-pressure metalorganic chemical vapor deposition. Appl. Phys. Lett. 43, 492–494.
Khan, M.R.H., Koide, Y., Itoh, H., Sawaki, N., Akasaki, I., 1986. Edge emission of $Al_xGa_{1-x}N$. Sol. State Commun. 60, 509–512.
Kim, J., Ji, M.-H., Detchprohm, T., Ryou, J.-H., Dupuis, R.D., Sood, A.K., Dhar, N.K., 2015. $Al_xGa_{1-x}N$ ultraviolet avalanche photodiodes with avalanche gain greater than 10^5. IEEE Photon. Technol. Lett. 27, 642–645.
Kinoshita, T., Obata, T., Yanagi, H., Inoue, S., 2013. High p-type conduction in high-Al content Mg-doped AlGaN. Appl. Phys. Lett. 102, 012105-1–012105-3.
Kneissl, M., Rass, J. (Eds.), 2016. III-Nitride Ultraviolet Emitters. Springer International Publishing, AG, Cham, Switzerland.
Koide, Y., Itoh, H., Sawaki, N., Akasaki, I., Hashimoto, M., 1986. Epitaxial growth and properties of $Al_xGa_{1-x}N$ by MOVPE. J. Electrochem. Soc. 133, 1956–1960.
Kolbe, T., Knauer, A., Chua, C., Yang, Z., Einfeldt, S., Vogt, P., Johnson, N.M., Weyers, M., Kneissl, M., 2010. Optical polarization characteristics of ultraviolet (In)(Al)GaN multiple quantum well light emitting diodes. Appl. Phys. Lett. 97, 171105-1–171105-3.
Krishnankutty, S., Kolbas, R.M., Khan, M.A., Kuznia, J.N., van Hove, J.M., Olsen, D.T., 1992. Photoluminescence characterization of AlGaN-GaN pseudomorphic quantum wells and calculation of strain induced bandgap shifts. J. Electron. Mater. 21, 609–612.
Lambert, D.J.H., Wong, M.M., Chowdhury, U., Collins, C., Li, T., Kwon, H.K., Shelton, B.S., Zhu, T.G., Campbell, J.C., Dupuis, R.D., 2000. Back illuminated AlGaN solar-blind photodetectors. Appl. Phys. Lett. 77, 1900–1902.
Li, X.H., Detchprohm, T., Kao, T.T., Satter, M.M., Shen, S.C., Yoder, P.D., Dupuis, R.D., Wang, S., Wei, Y.O., Xie, H., Fischer, A.M., 2014. Low-threshold stimulated emission at 249 nm and 256 nm from AlGaN-based multiple-quantum-well lasers grown on sapphire substrates. Appl. Phys. Lett. 105, 141106-1–141106-4.
Li, X.H., Kao, T.T., Satter, M.M., Wei, Y.O., Wang, S., Xie, H., Shen, S.C., Yoder, P.D., Fischer, A.M., Ponce, F.A., Detchprohm, T., Dupuis, R.D., 2015a. Demonstration of transverse-magnetic deep-ultraviolet stimulated emission from AlGaN multiple-quantum-well lasers grown on a sapphire substrate. Appl. Phys. Lett. 106, 041115-1–041115-4.
Li, X.H., Wang, S., Xie, H., Wei, Y.O., Kao, T.T., Satter, M., Shen, S.C., Douglas Yoder, P., Detchprohm, T., Dupuis, R.D., Fischer, A.M., 2015b. Growth of high-quality AlN layers on sapphire substrates at relatively low temperatures by metalorganic chemical vapor deposition. Phys. Status Solidi B 252, 1089–1095.
Li, X.H., Wei, Y.O., Wang, S., Xie, H., Kao, T.T., Satter, M.M., Shen, S.C., Yoder, P.D., Detchprohm, T., Dupuis, R.D., Fischer, A.M., 2015c. Temperature dependence of the crystalline quality of AlN layer grown on sapphire substrates by metalorganic chemical vapor deposition. J. Cryst. Growth 414, 76–80.
Limb, J., Yoo, D., Zhang, Y., Ryou, J.-H., Shen, S.-C., Dupuis, R.D., 2008. GaN ultraviolet avalanche photodiodes grown on 6H-SiC substrates with SiN passivation. Electron. Lett. 44, 313–314.
Liu, Y.S., Kao, T.T., Satter, M., Lochner, Z., Shen, S.C., Detchprohm, T., Yoder, P.D., Dupuis, R.D., Ryou, J.-H., Fischer, A.M., Wei, Y.O., Ponce, F.A., 2015. Inverse-tapered p-waveguide for vertical hole transport in high-[Al] AlGaN emitters. IEEE Photon. Technol. Lett. 27, 1768–1771.

Lochner, Z., Kao, T.-T., Liu, Y.-S., Li, X.-H., Satter, M.M., Shen, S.-C., Yoder, P.D., Ryou, J.-H., Dupuis, R.D., Wei, Y., Xie, H., Fischer, A., Ponce, F.A., 2013. Deep-ultraviolet lasing at 243 nm from photo-pumped AlGaN/AlN heterostructure on AlN substrate. Appl. Phys. Lett. 102, 101110-1–101110-3.

Manasevit, H.M., 1968. Single-crystal gallium arsenide on insulating substrates. Appl. Phys. Lett. 12, 156–159.

Manasevit, H.M., Simpson, W.I., 1964. Single-crystal silicon on a sapphire substrate. J. Appl. Phys. 35, 1349–1351.

Manasevit, H.M., Erdmann, F.M., Simpson, W.I., 1971. The use of metalorganics in the preparation of semiconductor materials. IV. The nitrides of aluminum and gallium. J. Electrochem. Soc. 118, 1864–1868.

Manning, I.C., Weng, X., Acord, J.D., Fanton, M.A., Snyder, D.W., Redwing, J.M., 2009. Tensile stress generation and dislocation reduction in Si-doped $Al_xGa_{1-x}N$ films. J. Appl. Phys. 106, 023506-1–023506-7.

Martens, M., Mehnke, F., Kuhn, C., Reich, C., Kueller, V., Knauer, A., Netzel, C., Hartmann, C., Wollweber, J., Rass, J., Wernicke, T., Bickermann, M., Weyers, M., Kneissl, M., 2014. Performance characteristics of UV-C AlGaN-based lasers grown on sapphire and bulk AlN substrates. IEEE Photon. Technol. Lett. 26, 342–345.

Mehnke, F., Wernicke, T., Pingel, H., Kuhn, C., Reich, C., Kueller, V., Knauer, A., Lapeyrade, M., Weyers, M., Kneissl, M., 2013. Highly conductive n-$Al_xGa_{1-x}N$ layers with aluminum mole fractions above 80%. Appl. Phys. Lett. 103, 212109-1–212109-3.

Miederer, W., Ziegler, G., Dötzer, R., 1965. Method of crucible-free production of gallium arsenide rods from alkyl gallium and arsenic compounds at low temperatures. Patent 3,226,270, Dec. 28, 1965.

Moe, C.G., Wu, Y., Piprek, J., Keller, S., Speck, J.S., DenBaars, S.P., Emerson, D., 2006. AlGaN/AlN distributed Bragg reflectors for deep ultraviolet wavelengths. Phys. Status Solidi A 203, 1915–1919.

Morita, M., Isogai, K., Tsubouchi, K., Mikoshiba, N., 1981a. Characteristics of the metal insulator semiconductor structure: AlN/Si. Appl. Phys. Lett. 38, 50–52.

Morita, M., Uesugi, N., Isogai, S., Tsubouchi, K., Mikoshiba, N., 1981b. Epitaxial growth of aluminum nitride on sapphire using metalorganic chemical vapor deposition. Jpn. J. Appl. Phys. 20, 17–23.

Nakarmi, M.L., Kim, K.H., Khizar, M., Fan, Z.Y., Lin, J.Y., Jiang, H.X., 2005. Electrical and optical properties of Mg-doped $Al_{0.7}Ga_{0.3}N$ alloys. Appl. Phys. Lett. 86, 092108-1–092108-3.

Nam, K.B., Li, J., Nakarmi, M.L., Lin, J.Y., Jiang, H.X., 2002. Achieving highly conductive AlGaN alloys with high Al contents. Appl. Phys. Lett. 81, 1038–1040.

Northrup, J.E., Chua, C.L., Yang, Z., Wunderer, T., Kneissl, M., Johnson, N.M., Kolbe, T., 2012. Effect of strain and barrier composition on the polarization of light emission from AlGaN/AlN quantum wells. Appl. Phys. Lett. 100, 021101-1–021101-4.

Onishi, T., Imafuji, O., Nagamatsu, K., Kawaguchi, M., Yamanaka, K., Takigawa, S., 2012. Continuous wave operation of GaN vertical cavity surface emitting lasers at room temperature. IEEE J. Quantum Electron. 48, 1107–1112.

Paduano, Q., Weyburne, D., 2005. Optimized coalescence method for the metalorganic chemical vapor deposition (MOCVD) growth of high quality Al-polarity AlN films on sapphire. Jpn. J. Appl. Phys. 44, L150–L152.

Parish, G., Keller, S., Kozodoy, P., Ibbetson, J.P., Marchand, H., Fini, P.T., Fleischer, S.B., DenBaars, S.P., Mishra, U.K., Tarsa, E.J., 1999. High-performance (Al, Ga)N-based solar-blind ultraviolet p–i–n detectors on laterally epitaxially overgrown GaN. Appl. Phys. Lett. 75, 247–249.

Pau, J.L., McClintock, R., Minder, K., Bayram, C., Kung, P., Razeghi, M., Silversmith, D., 2007. Geiger-mode operation of back-illuminated GaN avalanche photodiodes. Appl. Phys. Lett. 91, 041104-1–041104-3.

Pau, J.L., Bayram, C., McClintock, R., Razeghi, M., Silversmith, D., 2008. Back-illuminated separate absorption and multiplication GaN avalanche photodiodes. Appl. Phys. Lett. 92, 101120-1–101120-3.

Pearton, S., Abernathy, C., Ren, F., 2006. Gallium Nitride Processing for Electronics, Sensors and Spintronics. Springer-Verlag, London.

Pernot, C., Hirano, A., Iwaya, M., Detchprohm, T., Amano, H., Akasaki, I., 2000. Solar-blind UV photodetectors based on GaN/AlGaN p-i-n photodiodes. Jpn. J. Appl. Phys. 39, L387–L389.

Reine, M.B., Hairston, A., Lamarre, P., Wong, K.K., Tobin, S.P., Sood, A.K., Cooke, C., Pophristic, M., Guo, S., Perez, B., Singh, R., Eddy Jr., C.R., Chowdhury, U., Wong, M.M., Dupuis, R.D., Li, T., DenBaars, S.P., 2006. Solar-blind AlGaN 256 x 256 p-i-n detectors and focal plane arrays. SPIE Proc. 6119, 611901-1–611901-15.

Ren, Z., Sun, Q., Kwon, S.-Y., Han, J., Davitt, K., Song, Y.K., Nurmikko, A.V., Cho, H.-K., Liu, W., Smart, J.A., Schowalter, L.J., 2007. Heteroepitaxy of AlGaN on bulk AlN substrates for deep ultraviolet light emitting diodes. Appl. Phys. Lett. 91, 051116-1–051116-3.

Rodak, L.E., Sampath, A.V., Gallinat, C.S., Chen, Y., Zhou, Q., Campbell, J.C., Shen, H., Wraback, M., 2013. Solar-blind $Al_xGa_{1-x}N$/AlN/SiC photodiodes with a polarization-induced electron filter. Appl. Phys. Lett. 103, 071110-1–071110-4.

Roder, C., Einfeldt, S., Figge, S., Hommel, D., 2005. Temperature dependence of the thermal expansion of GaN. Phys. Rev. B 72, 085218-1–085218-6.

Romanov, A.E., Speck, J.S., 2003. Stress relaxation in mismatched layers due to threading dislocation inclination. Appl. Phys. Lett. 83, 2569–2571.

Rubio, A., Corkill, J.L., Cohen, M.L., Shirley, E.L., Louie, S.G., 1993. Quasiparticle band structure of AlN and GaN. Phys. Rev. B 58, 11810–11816.

Ruehrwein, R.A., 1965. Altering proportions in a vapor deposition process to form a mixed crystal graded energy gap. US Patent 3,224,913, Dec. 21, 1965.

Ruehrwein, R.A., 1967. Production of epitaxial films of semiconductor compound material, US Patent 3,312,570, Apr. 4, 1967.

Ryou, J.-H., Yoder, P.D., Liu, J., Lochner, Z., Kim, H.S., Choi, S., Kim, H.-J., Dupuis, R.D., 2009. Control of quantum-confined stark effect in InGaN-based quantum wells. IEEE J. Spec. Top. Quantum Electron. 15, 1080–1091.

Satter, M., Ryou, J.-H., Shen, S.-C., Dupuis, R.D., Yoder, P.D., 2012. Design and analysis of 250-nm AlInN laser diodes on AlN substrates using tapered electron blocking layers. IEEE J. Quantum Electron. 48, 703–711.

Satter, M.M., Lochner, Z., Kao, T.T., Liu, Y.S., Li, X.H., Shen, S.C., Dupuis, R.D., Yoder, P.D., 2014. AlGaN-based vertical injection laser diodes using inverse tapered p-waveguide for efficient hole transport. IEEE J. Quantum Electron. 50, 166–173.

Schweitz, K.O., Wang, P.K., Mohney, S.E., Gotthold, D., 2002. V/Al/Pt/Au Ohmic contact to n-AlGaN/GaN heterostructures. Appl. Phys. Lett. 80, 1954–1956.

Scott, T.R., King, G., Wilson, J.M., 1957. Improvements in or relating to the production of material for semi-conductors. British patent 778,383, July 3, 1957.

Shen, S.-C., Zhang, Y., Yoo, D., Limb, J., Ryou, J.-H., Yoder, P.D., Dupuis, R.D., 2007. Performance of deep ultraviolet GaN avalanche photodiodes grown by MOCVD. IEEE Photon. Technol. Lett. 19, 1744–1746.

Shimahara, Y., Miyake, H., Hiramatsu, K., Fukuyo, F., Okada, T., Takaoka, H., Yoshida, H., 2011. Growth of high-quality Si-doped AlGaN by low-pressure metalorganic vapor phase epitaxy. Jpn. J. Appl. Phys. 50, 095502-1–095502-4.

Sridharan, S., Yoder, P.D., 2008. Anisotropic transient and stationary electron velocity in bulk wurtzite GaN. IEEE Electron Device Lett. 29, 1190–1192.

Sridharan, S., Yoder, P.D., Shen, S.-C., Ryou, J.-H., Dupuis, R.D., 2009. Geiger mode simulation of GaN homojunction avalanche photodetectors. Phys. Status Solidi C 6, 662–665.

Sun, W.H., Zhang, J.P., Yang, J.W., Maruska, H.P., Khan, M.A., Liu, R., Ponce, F.A., 2005. Fine structure of AlN/AlGaN superlattice grown by pulsed atomic-layer epitaxy for dislocation filtering. Appl. Phys. Lett. 87, 211915-1–211915-3.

Takano, T., Narita, Y., Horiuchi, A., Kawanishi, H., 2004. Room-temperature deep-ultraviolet lasing at 241.5 nm of AlGaN multiple-quantum-well laser. Appl. Phys. Lett. 84, 3567. http://dx.doi.org/10.1063/1.1737061.

Taniyasu, Y., Kasu, M., Makimoto, T., 2006. Increased electron mobility in n-type Si-doped AlN by reducing dislocation density. Appl. Phys. Lett. 89, 182112-1–182112-3.

Tarsa, E.J., Kozodoy, P., Ibbetson, J., Keller, B.P., Parish, G., Mishra, U., 2000. Solar-blind AlGaN-based inverted heterostructure photodiodes. Appl. Phys. Lett. 77, 316–318.

Verghese, S., McIntosh, K.A., Molnar, R.J., Mahoney, L.J., Aggarwal, R.L., Geis, M.W., Molvar, K.M., Duerr, E.K., Melngailis, I., 2001. GaN avalanche photodiodes operating in linear-gain mode and Geiger mode. IEEE Trans. Electron Devices 48, 502–511.

Wagner, J., Obloh, H., Kunzer, M., Maier, M., Köhler, K., Johs, B., 2001. Dielectric function spectra of GaN, AlGaN, and GaN/AlGaN heterostructures. J. Appl. Phys. 89, 2779–2785.

Wang, H.-M., Zhang, J.-P., Chen, C.-Q., Fareed, Q., Yang, J.-W., Khan, M.A., 2002. AlN/AlGaN superlattices as dislocation filter for low-threading-dislocation thick AlGaN layers on sapphire. Appl. Phys. Lett. 81, 604–606.

Wang, W., Hu, W., Pan, M., Hou, L., Xie, W., Xu, J., Li, X., Chen, X., Lu, W., 2014. Study of gain and photoresponse characteristics for back-illuminated separate absorption and multiplication GaN avalanche photodiodes. J. Appl. Phys. 115, 013103-1–013103-8.

Wong, M.M., Chowdhury, U., Collins, C.J., Yang, B., Denyszyn, J.C., Kim, K.S., Campbell, J.C., Dupuis, R.D., 2001. High quantum efficiency AlGaN/GaN Solar-blind photodetectors grown by metalorganic chemical vapor deposition. Phys. Status Solidi A 188, 333–336.

Wunderer, T., Chua, C.L., Yang, Z., Northrup, J.E., Johnson, N.M., Garrett, G.A., Shen, H., Wraback, M., 2011. Pseudomorphically grown ultraviolet C photopumped lasers on bulk AlN substrates. Appl. Phys. Express 4, 092101-1–092101-3.

Xie, J., Mita, S., Bryan, Z., Guo, W., Hussey, L., Moody, B., Schlesser, R., Kirste, R., Gerhold, M., Collazo, R., Sitar, Z., 2013. Lasing and longitudinal cavity modes in photo-pumped deep ultraviolet AlGaN heterostructures. Appl. Phys. Lett. 102, 171102-1–171102-4.

Yoder, P.D., Fichtner, W., 1998. Effects of scaling and lattice heating on n-MOSFET performance via electrothermal Monte Carlo simulation. In: Proceedings of SISPAD 1998, pp. 165–168.

Yoder, P.D., Flynn, E.J., 2006. Linear theory of the quasi-unipolar photodiode. IEEE J. Lightwave Technol. 24, 1937–1945.

Yoshida, H., Yamashita, Y., Kuwabara, M., Kan, H., 2008. Demonstration of an ultraviolet 336 nm AlGaN multiple-quantum-well laser diode. Appl. Phys. Lett. 93, 241106-1–241106-3.

Yoshimoto, N., Matsuoka, T., Sasaki, T., Katsui, A., 1991. Photoluminescence of InGaN films grown at high temperatures by metalorganic vapor phase epitaxy. Appl. Phys. Lett. 59, 2251–2253.

Youtsey, C., Adesida, I., Bulman, G., 1997. Highly anisotropic photoenhanced wet etching on n-type GaN. Appl. Phys. Lett. 71, 2151–2153.

Yu, H., Strupinski, W., Butun, S., Ozbay, E., 2006. Mg-doped AlGaN grown on an AlN/sapphire template by metalorganic chemical vapour deposition. Phys. Status Solidi A 203, 868–873.

Zeimer, U., Kueller, V., Knauer, A., Mogilatenko, A., Weyers, M., Kneissl, M., 2013. High quality AlGaN grown on ELO AlN/sapphire templates. J. Cryst. Growth 377, 32–36.

Zhang, J.P., Wang, H.M., Gaevski, M.E., Chen, C.Q., Fareed, Q., Yang, J.W., Simin, G., Khan, M.A., 2002. Crack-free thick AlGaN grown on sapphire using AlN/AlGaN superlattices for strain management. Appl. Phys. Lett. 80, 3542–3544.

Zhang, J.P., Wang, H.M., Sun, W.H., Adivarahan, V., Wu, S., Chitnis, A., Chen, C.Q., Shatalov, M., Kuokstis, E., Yang, J.W., Khan, M.A., 2003. High-quality AlGaN layers over pulsed atomic-layer epitaxially grown AlN templates for deep ultraviolet light-emitting diodes. J. Electron. Mater. 32, 364–370.

Zhang, Y., Shen, S.-C., Kim, H.-J., Choi, S., Ryou, J.-H., Dupuis, R.D., Narayan, B., 2009. Low-noise GaN ultraviolet p-i-n photodiodes on GaN substrates. Appl. Phys. Lett. 94, 221109-1–221109-3.

Zheng, T.C., Lin, W., Liu, R., Cai, D.J., Li, J.C., Li, S.P., Kang, J.Y., 2016. Improved p-type conductivity in Al-rich AlGaN using multidimensional Mg-doped superlattices. Sci. Rep. 6, 21897-1–21897-10.

Zhuang, D., Edgar, J., 2005. Wet etching of GaN, AlN and SiC: a review. Mater. Sci. Eng. R48, 1–46.

CHAPTER FIVE

Al(Ga)N Nanowire Deep Ultraviolet Optoelectronics

S. Zhao*, Z. Mi†,[1]
*McGill University, Montreal, QC, Canada
†University of Michigan, Ann Arbor, MI, United States
[1]Corresponding author e-mail address: zetian.mi@mcgill.ca

Contents

1. INTRODUCTION

An efficient light source in the deep ultraviolet (UV) band is required for a wide range of applications, including water purification, sterilization, medical diagnostics, and bio/chemical sensing. The current technologies are limited to mercury vapor lamps, a few gas lasers, and lasers based on frequency conversion. These devices are inefficient, have large ecological footprints, and require the use of hazardous materials. In this regard, significant attention has been paid to compound semiconductor devices, which can offer an enormous number of advantages, including high electrical efficiency, high electron–photon conversion rate, small size, and greener device manufacture process. Al(Ga)N alloys, which cover the bandgap energies from ~3.4 to 6.2 eV (~200 to 364 nm), are the materials of choice for

Semiconductors and Semimetals, Volume 96
ISSN 0080-8784
http://dx.doi.org/10.1016/bs.semsem.2016.05.001
© 2017 Elsevier Inc.
All rights reserved.

solid-state deep UV optoelectronic devices. While tremendous progress has been made on GaN-based quantum well light-emitting diodes (LEDs) and lasers in the blue/green and near-UV wavelength range, it has remained challenging to achieve efficient deep UV light sources (Akasaki and Amano, 1997; Hirayama et al., 2010; Khan et al., 2008; Kim et al., 2004; Liao et al., 2011; Taniyasu et al., 2006; Yoshida et al., 2008b). Relatively high performance LEDs have been demonstrated in the UV-A (400–320 nm) and UV-B (320–280 nm) spectral range (Allerman et al., 2004; Zhang et al., 2004). However, the device performance degrades considerably in the UV-C band (below 280 nm) (Hirayama et al., 2014; Khan et al., 2008; Kneissl et al., 2011; Muramoto et al., 2014). For example, UV LEDs operating near 400 nm can reach an external quantum efficiency (EQE) around 40% (Khan et al., 2008; Kneissl et al., 2011). However, the EQE generally drops to well below 10% for LEDs operating in the UV-C band (Adivarahan et al., 2004a,b; Bhattacharyya et al., 2009; Hirayama et al., 2009, 2010; Taniyasu and Kasu, 2010; Taniyasu et al., 2006; Zhang et al., 2002; Zhou et al., 2006). The performance of electrically injected (Al, In)GaN quantum well lasers also degrades considerably with decreasing wavelength. To date, the shortest wavelength electrically injected lasers are limited to ~330 nm, with threshold current densities on the order of tens of kA/cm^2 (Kneissl et al., 2007; Yoshida et al., 2008a,b, 2011). Only optically pumped lasers have been demonstrated in the UV-B and UV-C bands, with threshold power densities larger than 61 kW/cm^2 (Li et al., 2014; Martens et al., 2014; Takano et al., 2004; Xie et al., 2013).

There are a number of reasons accounting for the device performance degradation in the UV-C band. Firstly, to get a shorter emission wavelength, more Al needs to be incorporated, which leads to large dislocation and defect densities in Al-rich AlGaN epilayers, due to the large lattice and thermal mismatches with the commonly available substrates. Different from InGaN, where strong carrier localization is present due to the existence of In-rich clusters, such strong carrier confinement has remained challenging to realize in Al-rich AlGaN, which not only limits the emission efficiency, but also makes it difficult to reach the transparency carrier density for lasing. Secondly, *p*-type doping is extremely inefficient in AlN and Al-rich AlGaN. The Mg activation energy in AlN is 0.5–0.6 eV (Nakarmi et al., 2009), which is significantly larger than that in GaN. As a consequence, the concentration of thermally activated valence band holes from Mg acceptors is extremely low at room temperature, preventing the realization of electrically injected deep UV LEDs and lasers with high electrical efficiency. Additionally, the unique transverse magnetic (TM) ($E \parallel c$) light polarization in AlN and Al-rich AlGaN prevents the light extraction from the top surface of conventional *c*-plane deep UV LEDs (Li et al., 2003).

Significant efforts have been devoted to the growth and characterization of Al(Ga)N nanowires as an alternative approach for deep UV photonics. Al(Ga)N nanowires can be grown/synthesized using various approaches, including chemical vapor deposition (CVD), metalorganic vapor phase epitaxy (MOVPE)/metalorganic CVD (MOCVD), and molecular beam epitaxy (MBE) (Bertness et al., 2006; Carnevale et al., 2012; Chen et al., 2014a,b, 2015; He et al., 2011; Himwas et al., 2014; Hong et al., 2006; Kent et al., 2014; Li et al., 2015; Park et al., 2006; Pierret et al., 2013b,c; Sivadasan et al., 2015; Su et al., 2005b; Wang et al., 2012, 2013; Ye et al., 2015; Zhao et al., 2015a,b,c,d,e). Fundamentally, due to the efficient strain relaxation associated with the large surface-to-volume ratio, nearly dislocation-free III-nitride nanowires can be grown directly on Si, SiO*x*, or other foreign substrates with high internal quantum efficiency (IQE) (Consonni et al., 2011; Sarwar et al., 2015; Zhao et al., 2013a). Moreover, the formation energy of Mg dopant can be significantly reduced in the near-surface region of nanowire structures (Zhao et al., 2013b), thereby leading to enhanced Mg-dopant incorporation and more efficient current injection. In addition, the large surface area and the light multiple scattering effect can drastically enhance the light extraction efficiency of TM polarized photons in Al-rich AlGaN nanowire LEDs (Zhao et al., 2015b).

In this chapter, we review the recent progress of deep UV optoelectronic devices based on Al(Ga)N nanowires, including LEDs and lasers. It is also noted that a lot of work has been done with GaN nanowires in the near-UV spectral range, e.g., LEDs (Brubaker et al., 2013a,b; Hersee et al., 2009; Kim et al., 2003; Lee et al., 2007; Motayed et al., 2007) and optically pumped lasers (Inose et al., 2013; Johnson et al., 2002; Li et al., 2012; Sakai et al., 2010; Wright et al., 2014; Xu et al., 2012; Zhang et al., 2014), which will not be covered in this chapter. In Section 2, we discuss the growth and structural characterization of various Al(Ga)N nanowire structures. In Section 3, Al(Ga)N nanowire deep UV LEDs are presented. In Section 4, we discuss the recently achieved electrically injected ternary AlGaN nanowire deep UV lasers. Other emerging devices based on Al(Ga)N nanowires are presented in Section 5, followed by conclusion in Section 6.

2. GROWTH AND CHARACTERIZATION OF Al(Ga)N NANOWIRES

In this section, we discuss various approaches to grow/synthesize Al(Ga)N nanowires and their structural, optical, and electrical properties, which are relevant to the performance of Al(Ga)N nanowire deep UV LEDs and lasers. The growth and characterization of ternary AlGaN nanowires are discussed first, followed by the discussion on AlN nanowires. In the end, the epitaxy and properties of quantum-confined nanostructures in Al(Ga)N nanowires are also presented.

2.1 Ternary AlGaN Nanowires

To date, AlGaN nanowires can be grown by CVD (Chen et al., 2014a,b, 2015; He et al., 2011; Hong et al., 2006; Sivadasan et al., 2015; Ye et al., 2015), and MOCVD (MOVPE) and MBE (Bertness et al., 2006; Carnevale et al., 2012; Himwas et al., 2014; Kent et al., 2014; Park et al., 2006; Pierret et al., 2013b,c; Su et al., 2005b; Wang et al., 2013; Zhao et al., 2015a,b,c,e). In this section, these growth techniques and the properties of ternary AlGaN nanowires are discussed.

2.1.1 Ternary AlGaN Nanowires by CVD

He et al. reported the growth of AlGaN nanocones on Si substrate by conventional CVD technique (He et al., 2011). Through detailed structural characterization, Al content from 0% to 100% was derived. In addition, cathodoluminescence experiments were performance at room temperature, and emission peaks from ~300 to 470 nm were measured. Chen et al. further demonstrated the morphology-controlled synthesis of AlGaN nanostructures including nanonails, nanoneedles, nanorods, and nanoflowers (Chen et al., 2014a, 2015). In this work, the hexagonal nano structures were grown by a halide CVD process on bare Si (100) substrate, and the growth was suggested through a catalyst-free vapor-solid (VS) mechanism combined with surface diffusion. As illustrated in Fig. 1A, AlGaN nanoneedles with sharp tips can be formed under high NH_3 flow rate, and a higher evaporation temperature of $GaCl_3$ can lead to the formation of AlGaN nanorods with hexagonal flat tops (shown in Fig. 1B). Al composition up to 100% was reported by these authors. It is noted, however, that the photoluminescence (PL) emission peaks were only measured in the visible spectral range, due to the presence of extensive defects (Chen et al.,

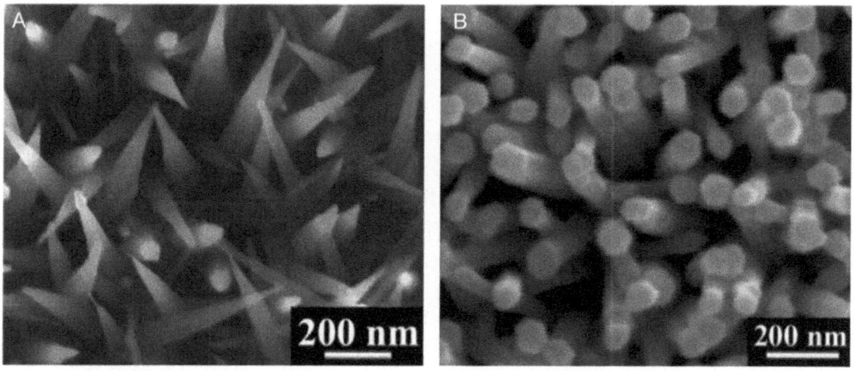

Fig. 1 Controlling the morphology of AlGaN nanostructures through a halide catalyst-free CVD process with (A) AlGaN nanoneedles and (B) AlGaN nanorods (Chen et al., 2015).

2014b). Ye et al. also reported the growth of AlGaN nanowires on sapphire (0001) substrate (Ye et al., 2015). Through detailed studies they suggested that Ga acted as the catalyst for the formation of AlGaN nanowires. In particular, the island growth of GaN in N-rich conditions assisted the formation of AlGaN nanowires. Additionally, Au-catalyzed growth of monodispersed AlGaN nanowires was reported (Sivadasan et al., 2015).

2.1.2 Ternary AlGaN Nanowires by MBE

On the other hand, research efforts have been devoted to the formation of AlGaN nanowires by epitaxial reactors. With MOVPE, Su et al. reported the flexible synthesis of AlGaN nanowires on alumina substrate through a vapor–liquid–solid (VLS) growth mechanism (Su et al., 2005a). They further reported GaN/AlGaN core-shell nanowires with an emission wavelength around 370 nm (Su et al., 2005b). To date, the majority of AlGaN nanowires are grown by MBE, which are described below.

AlGaN nanowires by MBE are generally grown through a diffusion-driven self-organized process under N-rich conditions. Ristic et al. reported the growth of free-standing, self-organized AlGaN nanocolumns by plasma-assisted MBE on Si (111) substrate (Ristic et al., 2002). The Al nominal concentration ranging from ~16% to 40% was realized by controlling III/V ratio, and the diameters of such AlGaN nanowires varied from 30 to 60 nm. Park et al. further reported the growth of hexagonal AlGaN nanorods on Si (001) substrate by plasma-assisted MBE (Park et al., 2006). In this work, Al mole fraction up to 15% was realized. In addition, Bertness et al. reported the growth of AlGaN nanowires on Si (111) substrate by gas-source MBE (Bertness et al., 2006). Al mole fraction up to 30% was realized. It was also found that the AlGaN nanowires coalesced with increasing Al content.

Up to here, it is worth noting that for these early attempts, AlGaN nanowires are grown *directly* on the underlying substrates. Recently, *with the usage of GaN nanowire template*, AlGaN nanowires, with PL emission across the UV-A, UV-B, and UV-C bands, have been demonstrated (Carnevale et al., 2012, 2013; Kent et al., 2014; Pierret et al., 2013b,c; Zhao et al., 2015a,e). A complete PL emission wavelength tuning of ternary AlGaN nanowires at low temperatures was reported by Pierret et al. (2013b). Fig. 2A shows a typical SEM image of such a self-organized AlGaN nanowire sample grown on GaN nanowire template by MBE under N-rich conditions on Si substrate. It is seen that, with the usage of GaN nanowire template and optimized growth conditions, highly uniform and dense AlGaN nanowires can be realized without any coalescence. In addition, by changing the Al/Ga ratio, the Al composition can be widely tuned from 29% to 80%, with the corresponding low-temperature PL emission wavelength varying in the range of 310–235 nm, as illustrated in Fig. 2B.

Similar wavelength tunability from ternary AlGaN nanowires at low temperatures was also reported by Zhao et al. (2015a). Moreover, they

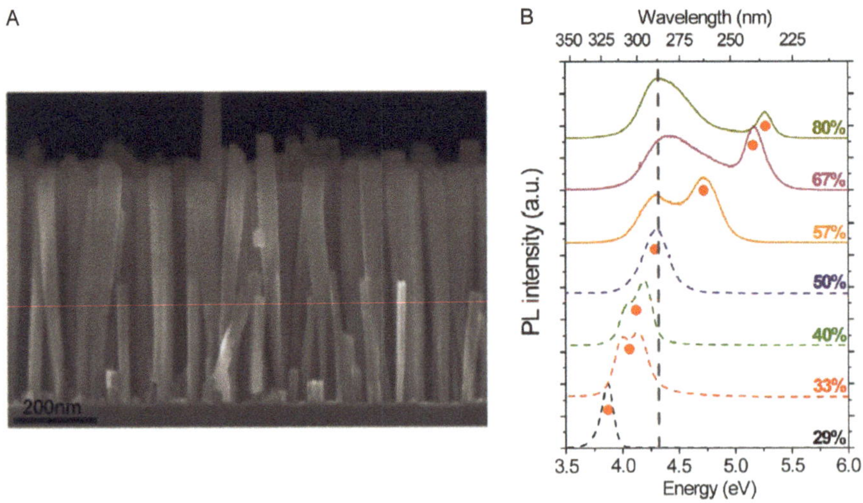

Fig. 2 (A) SEM image of ternary AlGaN nanowires grown by MBE under N-rich conditions on GaN nanowire template. (B) The tunable PL emission wavelength measured at 5 K by changing alloy composition. The numbers denote Al content (Pierret et al., 2013c).

further performed detailed structural and PL emission property characterizations of ternary AlGaN nanowires grown on GaN nanowire template on Si substrate by scanning transmission electron microscopy (STEM). The *top panel* of Fig. 3A shows the high-angle annular dark-field (HAADF) image of a single AlGaN nanowire, with the *bottom panel* showing the corresponding RGB pseudocolor image of Ga- and Al-signals. It is clearly seen that there exists a low Al concentration in the nanowire bulk region and a high Al concentration in the near-surface region. The presence of Al-rich AlGaN shell is attributed to the unique Al migration kinetics under N-rich conditions (Pierret et al., 2013b; Wang et al., 2012). Such an Al-rich AlGaN shell can suppress the surface nonradiative recombination, thereby leading to significantly enhanced carrier injection efficiency and emission efficiency of AlGaN nanowire LEDs (to be discussed in Section 3).

Fig. 3B shows a high-magnification STEM image of an AlGaN nanowire with an average Al content of 10% (estimated by the PL peak position). It is seen that the nanowire bulk region exhibits uniform compositional distribution. However, with further increasing Al composition, strong *atomic scale compositional modulation* was measured. As illustrated in Fig. 3C, the bright stripes correspond to Ga-rich AlGaN regions. Detailed studies further

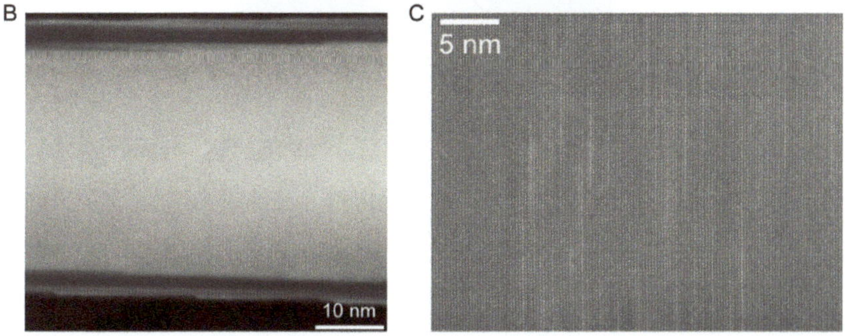

Fig. 3 (A) A STEM-HAADF image of a single AlGaN nanowire, with the corresponding RGB pseudocolor elemental mapping also shown. (B) and (C) High-resolution STEM-HAADF images of a low Al content nanowire and a high Al content nanowire, respectively (Zhao et al., 2015e).

suggested that Ga-rich AlGaN nanoclusters had dimensions of ~0.25 to 2 nm along the growth direction and of ~2 to 10 nm along the lateral direction (perpendicular to the growth direction) (Zhao et al., 2015e). Detailed studies further indicated that such atomic scale compositional fluctuation could possess quantum-dot/dash-like structural characteristics within the nanowires, and lead to discrete density of states and three-dimensional quantum confinement of charge carriers (Zhao et al., 2015e).

Illustrated in Fig. 4A are the PL spectra measured from a few AlGaN nanowire samples. It is seen that with increasing Al composition, the PL spectral linewidth becomes broader, which is consistent with the inhomogeneous broadening due to the presence of quantum-dot-like nanostructures. Also, due to the three-dimensional quantum confinement of charge carriers in nearly defect-free nanowires, IQE in the range of 70–80% has been achieved at room temperature over a wide range of excitation powers, as illustrated in Fig. 4B (Zhao et al., 2015e). As discussed in Section 4, these quantum-dot-like nanostructures and the resulting discrete density of states and three-dimensional quantum confinement of charge carriers can lead to electrically injected deep UV lasers, with low threshold.

Similar quantum-dot-like AlGaN nanostructures were also reported by Pierret et al. (2013a,b) and Belloeil et al. (2016). Fig. 5 shows the micro-PL spectra measured from Al-rich ternary AlGaN nanowires at low temperatures under an excitation of 244 nm with a spot size of 1 μm. Compared to the broad PL spectra measured from ensemble AlGaN nanowires, a spiky

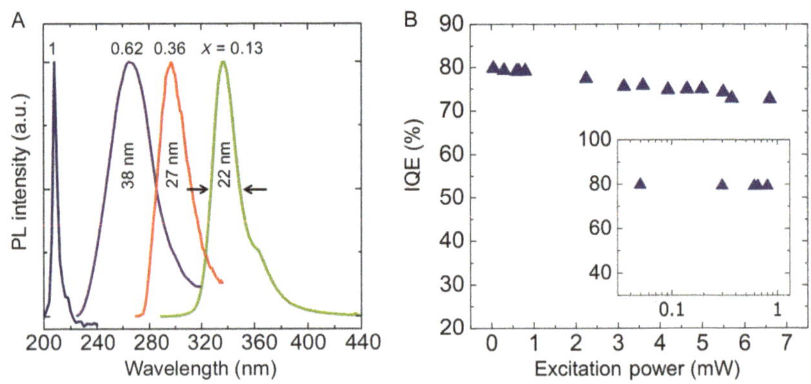

Fig. 4 (A) PL spectra of ternary AlGaN nanowires measured at room temperature. The PL spectrum of AlN nanowires is also shown for a comparison. *x* denotes the Al content. (B) Derived IQE of Al-rich ternary AlGaN nanowires (emitting around 260 nm) as a function of excitation power. The *inset* shows the semilogarithmic plot for low excitation powers (Zhao et al., 2015e).

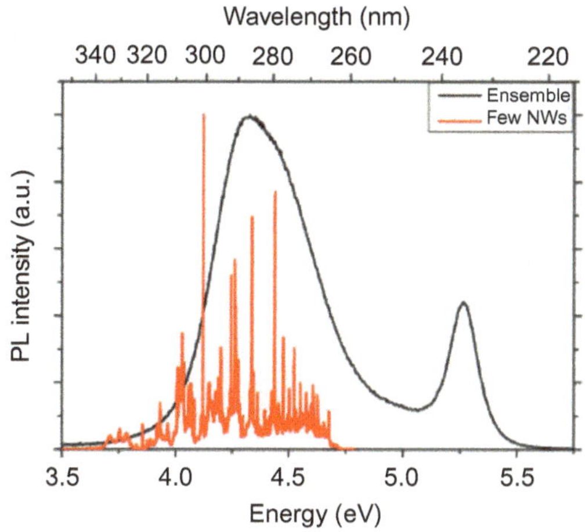

Fig. 5 A comparison of macro- and micro-PL characteristics of AlGaN quantum-dot-like nanostructures (Pierret et al., 2013c).

PL spectrum with many sharp peaks (linewidth around a few meV) can be seen from few AlGaN nanowires. Detailed studies suggested that these PL emission peaks originated from quantum-dot-like nanoclusters in the nanowires (Belloeil et al., 2016). It is noted that quantum-dot-like nanostructures have also been previously observed in AlGaN planar structures, and the formation mechanism is attributed to liquid phase epitaxy with the presence of excess Ga (Moustakas and Bhattacharyya, 2012).

2.1.3 Ternary AlGaN Nanowires on Patterned Substrate

Site-controlled AlGaN nanowires can also be formed on patterned substrate by both MOVPE and MBE (Nam et al., 1997; Yamano et al., 2015). Yamano et al. reported the growth of AlGaN nanocolumns on GaN nanopillars on sapphire substrate by plasma-assisted MBE (Yamano et al., 2015). In this work, the nanopillars were fabricated on a GaN template by e-beam lithography and inductively coupled-plasma (ICP) etching. These high-density GaN nanopillars in consequence provided shadow effect, such that AlGaN nanocolumns were selectively grown on the top of the underlying nanopillars. This process is schematically shown in Fig. 6.

The *top panel* of Fig. 7A shows the top-view and tilted-angle view of GaN nanopillars, which exhibit a flat top surface. The *bottom panel* of Fig. 7A shows the AlGaN nanocolumns grown on these GaN nanopillars, and faceted tops can be seen. In this work, by changing Al beam equivalent pressure (BEP), different PL emission wavelengths can be realized, as

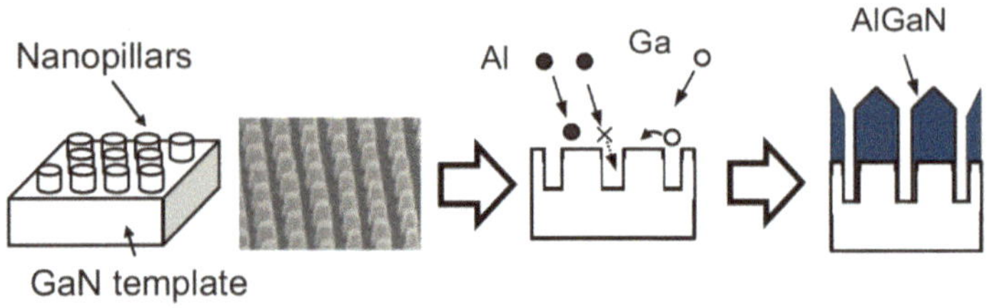

Fig. 6 The growth process of AlGaN nanocolumns on GaN nanopillars. *From* left *to* right, *it shows the schematic of GaN nanopillars on GaN template, SEM image of GaN nanopillars formed by ICP etching, shadow effect introduced by GaN nanopillars, and the formation of AlGaN nanocolumns on GaN nanopillars (Yamano et al., 2015).*

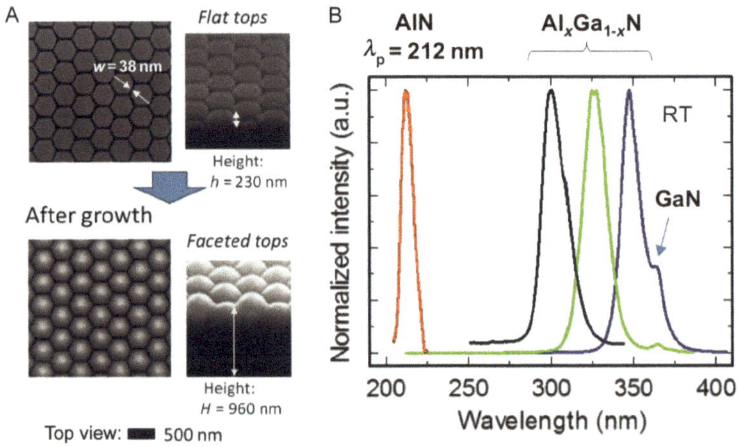

Fig. 7 (A) *Top panel*: SEM images of GaN nanopillars. *Bottom panel*: SEM images of AlGaN nanocolumns on GaN nanopillars. (B) PL spectra measured from Al(Ga)N nanocolumns (Yamano et al., 2015).

illustrated in Fig. 7B. The Al composition was further estimated by X-ray diffraction experiments, and Al compositions of ~0.13, 0.22, 0.43, and 1 were realized.

2.2 AlN Nanowires

To achieve high performance Al(Ga)N nanowire deep UV optoelectronic devices, it is essential to study the growth and properties of AlN nanowires. Additionally, due to its wide bandgap energy and excellent thermal and mechanical properties, AlN nanowires can be potentially used in many device applications such as power transistors, sensors, resonators, and piezoelectric nanogenerators, to name a few. In this regard, there are tremendous ongoing efforts in the synthesis and characterization of AlN nanowires. The synthesis approach has been varied widely from template-confined method

(Liu et al., 2001; Zhang et al., 2001), *dc* arc plasma method (Balasubramanian et al., 2004), to CVD techniques either with catalyst-assisted VLS mechanism (Tang et al., 2001) or catalyst-free VS mechanism (Zhao et al., 2005). The optical and electrical properties of these AlN nanowires have also been investigated. A comprehensive summary for these early research activities can be found in the review paper by Kenry et al. (2012). In these early studies, although defect-free regions can be found locally, the overall emission characteristics are dominated by defect energy levels in the wide bandgap AlN nanowires. In what follows, we show that AlN nanowires with much-improved optical and electrical properties can be realized by MBE.

The first band edge emission from AlN nanowires was reported by Landré et al. (2010a,b). In this work, the AlN nanowires were grown by MBE *directly* on SiO_2 dielectric layers under N-rich conditions. PL emission around 206 nm, with two phonon replica, was measured at 10 K. Zhao et al. further investigated the detailed growth and characterization of AlN nanowires (Connie et al., 2015; Wang et al., 2014; Zhao et al., 2015a,b). In these studies, catalyst-free AlN nanowires were grown on Si substrate using a plasma-assisted MBE system under N-rich conditions. *To promote the formation of AlN nanowires, GaN nanowire template was first grown on Si substrate.* The utilization of the GaN nanowire template can provide an additional dimension in controlling AlN nanowire size, density, and quality, and is fully compatible with processes for fabricating electrically injected deep UV devices on Si platform. The effect of the growth temperature of GaN nanowire template has been carefully investigated. From the SEM image shown in Fig. 8A, it is seen that only a film-like structure can be formed when the growth temperature of GaN nanowire template is around 830°C (sample A). This is ascribed to the formation of extremely low-density GaN nanowires at

Fig. 8 (A) SEM image of AlN film-like structure, due to the high growth temperature of the bottom GaN nanowire template. (B) SEM image of AlN nanowires with optimized growth conditions including the GaN nanowire template growth temperature. (C) SEM image of coalesced AlN nanowires on GaN nanowire template. The formation of coalescence is due to the high Al BEP as described in the text.

the relatively high growth temperature (Zhao et al., 2013a). As a consequence, the majority of the wafer surface area is directly exposed to AlN, leading to the formation of an AlN film-like structure, instead of nanowires (Bourret et al., 1998; Landré et al., 2010a,b). When the growth temperature of GaN nanowire template is reduced to around 780°C (sample B), high-density GaN nanowires can be formed, due to the reduced Ga adatom desorption rate, promoting the formation of relatively uniform AlN nanowires, as illustrated in Fig. 8B. The effect of Al BEP has also been investigated. As shown in Fig. 8C, AlN nanowires become highly coalesced when the Al BEP changes from $\sim 2 \times 10^{-8}$ Torr (in sample B) to $\sim 5 \times 10^{-8}$ Torr (sample C) while keeping other growth parameters identical to sample B.

The bottom curve in Fig. 9 shows the PL spectrum of undoped AlN nanowires measured *at room temperature*. A strong band edge emission ~207 nm can be seen, which is nearly identical to that of AlN bulk crystal, suggesting that such AlN nanowires grown on Si substrate is strain-free (Wang et al., 2014). Detailed studies further suggest that surface optical phonons contribute to phonon replica in such defect-free AlN nanowires (Wang et al., 2014). In addition, with optimized growth conditions, high relative IQE around 80% was realized at room temperature if taking the IQE at low temperature (20 K) to be unity (Zhao et al., 2015a). This high IQE is comparable to that of state-of-the-art GaN-based blue quantum well LEDs, and is largely attributed to the very efficient radiative recombination, due to the large exciton binding energy (~60 meV) and the formation of defect-free AlN nanowires.

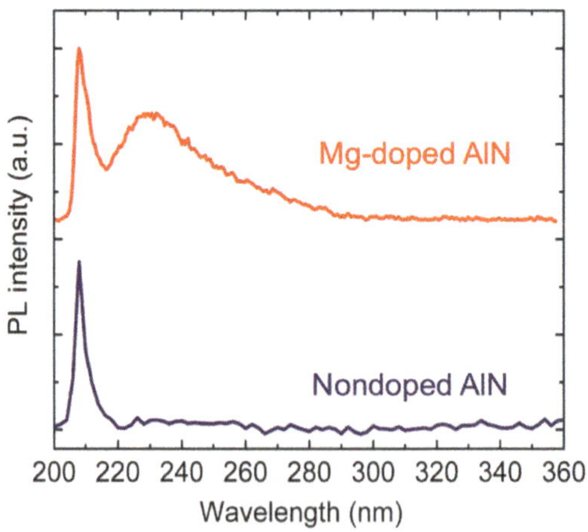

Fig. 9 PL spectra of undoped and Mg-doped AlN nanowires measured at room temperature.

Also shown in Fig. 9 is the PL spectrum of Mg-doped AlN nanowires (the top curve) *at room temperature*. To reduce Mg surface desorption, the substrate temperature was reduced compared to the growth temperature of AlN nanowires (Zhao et al., 2015a). Under optimized growth conditions, Mg-doped AlN nanowires exhibited relatively uniform size distribution. It is seen that besides the excitonic emission PL peak at ~207 nm, another low-energy PL peak appears, which is separated from the band edge emission by ~0.5 eV. This energy separation is consistent with the Mg activation energy in AlN (Nakarmi et al., 2009), suggesting that this low-energy PL peak is due to Mg acceptor energy level related radiative recombination.

To further understand Mg incorporation in AlN nanowires, first-principle calculation was performed (Zhao et al., 2015a). It was found that the Al-substitutional Mg formation energy was much lower in the near-surface region compared to the bulk region, due to the reduced strain associated with the large surface area of nanowires. As a consequence, Mg-dopant concentration could be significantly enhanced in AlN nanowires, compared to AlN planar structures (Taniyasu et al., 2006). This was further confirmed by studying the electrical properties of Mg-doped AlN nanowires (Connie et al., 2015). In this work, the average free hole concentration of Mg-doped AlN nanowires was estimated by large-area AlN nanowire devices on Si. The device structure is schematically shown in the *inset* of Fig. 10. In such a structure, the dominant resistance is from Mg-doped AlN segment. Free hole concentration on the

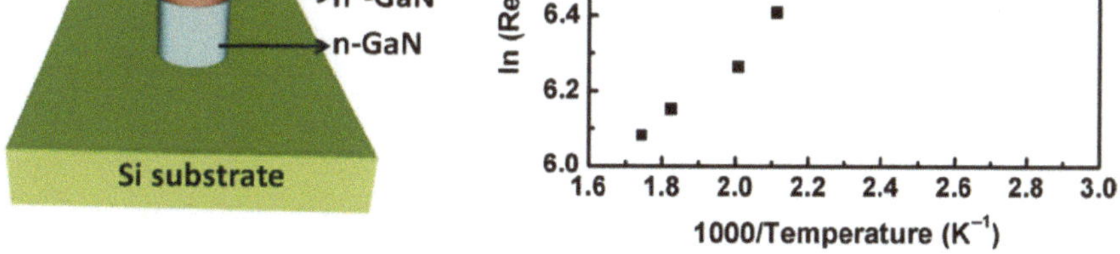

Fig. 10 (A) Layer-by-layer structure of the nanowire device used to characterize the electrical properties of *p*-AlN. In this structure, the device resistance is largely determined by the *p*-AlN segment. (B) Temperature-dependent resistivity of the device (Connie et al., 2015).

order of 10^{16} cm^{-3} was derived at room temperature. Additionally, detailed temperature-dependent studies have further indicated that the p-type conduction is associated with an activation energy of only ~20–30 meV (illustrated in Fig. 10), which is much smaller than the Mg activation energy in AlN (0.5–0.6 eV) and is consistent with impurity band hole conduction (Connie et al., 2015). This efficient hole conduction in Mg impurity band is a natural consequence of the significantly enhanced Mg doping in nanowire structures. Moreover, the formation of nitrogen vacancy-related defects (donors) can be greatly minimized in such AlN nanowires, due to N-rich conditions.

2.3 Quantum-Confined Nanostructures in Al(Ga)N Nanowires

Besides the efforts on the growth and characterization of ternary AlGaN nanowires and AlN nanowires, significant efforts have also been devoted to studying the epitaxy and properties of quantum-confined nanostructures in Al(Ga)N nanowires. GaN disk with AlN barrier structure is first described. Such GaN disk typical has a width of ~10 nm. By varying the height from ~1 to 4 nm, the PL emission wavelength can be tuned from ~290 to 460 nm (Renard et al., 2008, 2009; Songmuang et al., 2011). Moreover, the temperature-dependent PL studies have suggested that such GaN disk can behave as quantum dot (Renard et al., 2009). In addition, it has been found that due to the efficient strain relaxation from the nanowire large surface area, the piezoelectric effect, which is commonly measured in planar structures, can be largely eliminated in nanowires (Renard et al., 2009). Therefore, the overall polarization effect in such GaN quantum disk can be greatly reduced. The major contribution to the polarization field has been identified to be the spontaneous polarization-induced electric field, originated from the strong ionic bonding of III-nitrides.

Yoshizawa et al. further investigated GaN disk with AlGaN barrier structure (Yoshizawa et al., 1998). In this work, the nanostructure was grown by plasma-assisted MBE through a self-organized process on sapphire (0001) substrate. The height of such GaN disk was around 6 nm, while the lateral size was around 46 nm. The Al composition in AlGaN barrier was around 18%. PL emission ~360 nm was measured. AlGaN disk with AlN barrier structure was also reported by Himwas et al. (2013, 2014). By varying Al/N ratio and the height of AlGaN disk, tunable PL emission was realized from ~330 to 241 nm, with IQE in the range of 25–40%. In addition, multiple PL peaks were measured, which was attributed to alloy inhomogeneity (Himwas et al., 2013, 2014).

3. Al(Ga)N NANOWIRE LEDs

Al(Ga)N nanowire LEDs, with emission wavelengths covering the UV-A (400–320 nm), UV-B (320–280 nm), and UV-C (below 280 nm) bands, have been demonstrated (Carnevale et al., 2012, 2013; Kent et al., 2014; Qian et al., 2005; Sekiguchi et al., 2008a; Tang et al., 2013; Wang et al., 2013; Zhao et al., 2015a,b,c,e). Tang et al. reported single *n*-ZnO/*p*-ternary AlGaN nanowire LEDs (Tang et al., 2013). The nanowire LED structure was grown by CVD technique on Si (010) substrate. The device exhibited a turn on voltage of around 3 V, and emission around 394 nm was measured at room temperature. To the best of our knowledge, this is the only demonstration of functional LEDs fabricated from CVD grown AlGaN nanowires.

On the other hand, UV LEDs have also been realized with MBE grown Al(Ga)N nanowires. Wang et al. reported large-area ternary AlGaN nanowire core-shell LEDs (Wang et al., 2013). The nanowire structure was grown by MBE under N-rich conditions through a self-assemble process. The structure consisted of *n*- and *p*-GaN contact layers, and $n\text{-}Al_{0.36}Ga_{0.64}N/i\text{-}Al_{0.12}Ga_{0.88}N/p\text{-}Al_{0.36}Ga_{0.64}N$ double heterostructure as the active region. The presence of Al-rich AlGaN shell effectively passivated the nanowire surface, and an IQE of ~60% was realized. The device exhibited a turn on voltage of around 3 V. At a forward current density of 20 A/cm^2, the voltage was around 6.2 V. Strong electroluminescence (EL) emission around 340 nm was measured at room temperature. UV LEDs fabricated from AlGaN nanowires grown on patterned substrate by MBE were also reported (Sekiguchi et al., 2008a). In this work, GaN was inserted into AlGaN to form multiple quantum well. In addition, the top *p*-AlGaN layer was coalesced to improve current spreading and to simplify device fabrication. The device showed a turn on voltage of around 4 V, and at a forward current of 20 mA, the forward voltage was 4.5 V. Strong emission around 354 nm was measured at room temperature.

Recently, with further improvement in the MBE growth process of Al(Ga)N nanowires, LEDs emitting in the deep UV band (below 300 nm) have been achieved. In what follows, we will discuss these progresses, including the shortest wavelength LEDs (emitting around 207 nm at room temperature) fabricated from AlN nanowires.

3.1 Deep UV LEDs with Ternary AlGaN Nanowires

The first deep UV LEDs with ternary AlGaN nanowires were realized through a polarization doping technique to overcome the large Mg activation energy in Al-rich AlGaN (Carnevale et al., 2012). Later on, with further optimized growth conditions, improved electrical performance and emission wavelength as short as 250 nm were realized (Kent et al., 2014). For such polarization-doped ternary AlGaN nanowire LEDs, the nanowire structures were grown by MBE under N-rich conditions, and the electrical doping was realized by linearly grading the structure from GaN to AlN and then from AlN to GaN, forming a polarization-induced *p–n* junction, as illustrated in the *upper panel* of Fig. 11A. Both *n*- and

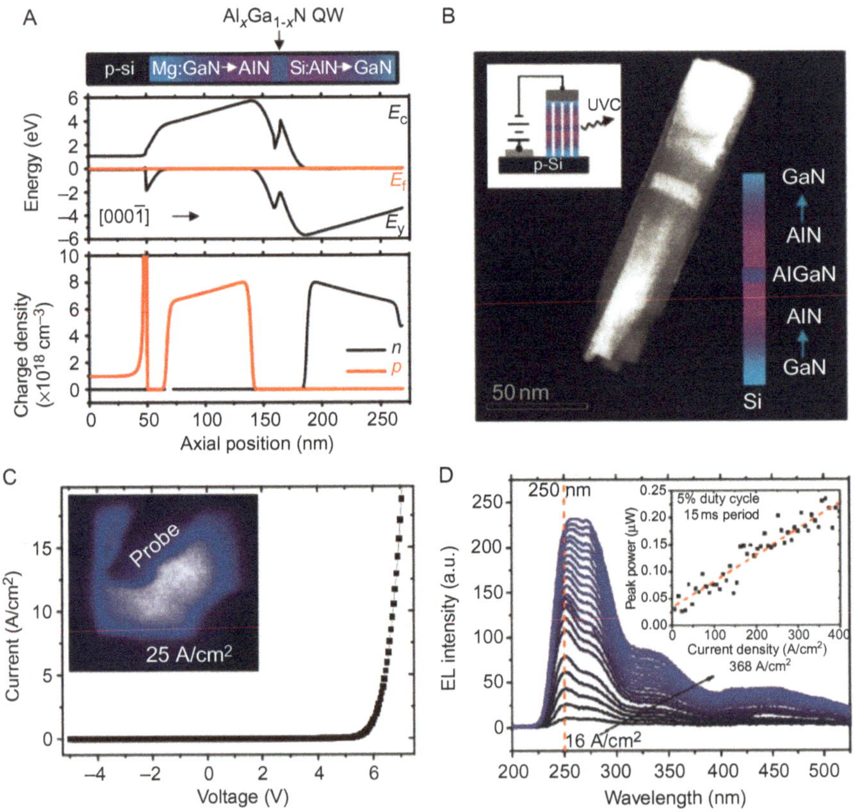

Fig. 11 (A) Schematic of the polarization doping in AlGaN nanowires. (B) STEM image of a single AlGaN nanowire, highlighting the nanowire LED structure. (C) *I–V* characteristics of polarization-doped AlGaN nanowire LEDs. The *inset* shows an optical image of the device. (D) The current-dependent EL spectra of polarization-doped AlGaN nanowire LEDs measured under a pulsed biasing, with the *inset* showing the light output power vs current density (Kent et al., 2014).

p-type dopant species on the order of 10^{19} cm^{-3} were realized, as shown in the *bottom panel* of Fig. 11A. A STEM image is shown in Fig. 11B, highlighting the nanowire LED structure. The *I–V* characteristics are shown in Fig. 11C. A turn on voltage of around 6 V can be seen; and at 20 A/cm^2, the voltage is around 7 V. An optical image of such AlGaN nanowire LEDs is shown in the *inset* of Fig. 11C. Fig. 11D shows the EL emission spectra under a pulsed operation. It is seen that at low injection levels, the EL peak around 250 nm dominates; while with further increasing injection currents, EL emission spectra become broad. The *inset* shows the light output power vs current density. It is seen that the light output power increases linearly; and at 400 A/cm^2, the output power is around 200 nW.

Recently, with the drastically improved MBE growth process and the enhanced *p*-type doping, ternary AlGaN nanowire LEDs with excellent electrical performance were reported (Zhao et al., 2015a). In this work, the device structure consisted of AlGaN *p–i–n* junction, and *n*- and *p*-GaN contact layers. The *I–V* characteristics are shown in Fig. 12. It is seen that the turn on voltage is around 5 V; and at a forward current density of 20 A/cm^2, the voltage is around 6 V. The *inset* of Fig. 12 shows the EL emission spectra under different injection currents at room temperature, and strong emission around 292 nm can be seen.

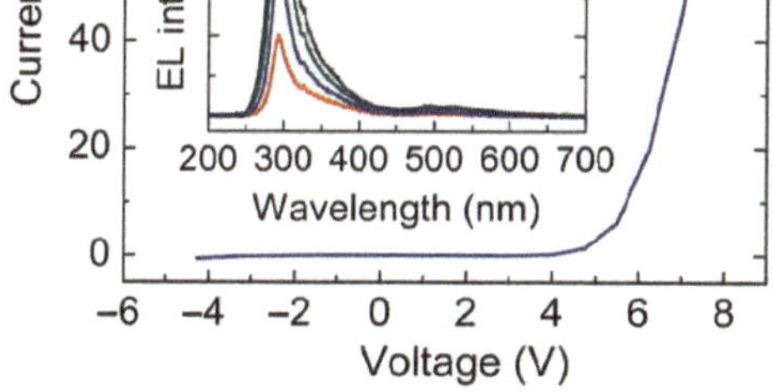

Fig. 12 Room-temperature *I–V* characteristics of AlGaN nanowire LEDs with enhanced Mg doping. The *inset* shows the room-temperature current-dependent EL spectra under a continuous-wave biasing (Zhao et al., 2015a).

3.2 AlN Nanowire LEDs

With the development of nearly defect-free AlN nanowires (discussed in Section 2), AlN nanowire LEDs were also achieved recently (Zhao et al., 2015a,b). The device structure is shown in Fig. 13A, which consists of *n*-GaN, *p–i–n* AlN, and *p*-AlGaN contact layers. A direct metal deposition at a tilting angle was employed to form *p*-contact on the nanowire top surface, and no polymer passivation was performed, in order to avoid any deep UV light absorption. Schematic of the fabricated device is also shown in Fig. 13A. A top-view SEM image after the top *p*-metal deposition is shown in Fig. 13B. It is seen that such metal contact connects the tops of nanowire very well. Fig. 13C shows the *I–V* characteristics of AlN nanowire LEDs measured at different temperatures. Excellent *I–V* characteristics can be seen. At room temperature the turn on voltage is around 5.5 V, which is consistent with the bandgap energy of AlN. Moreover, at 20 A/cm^2, the

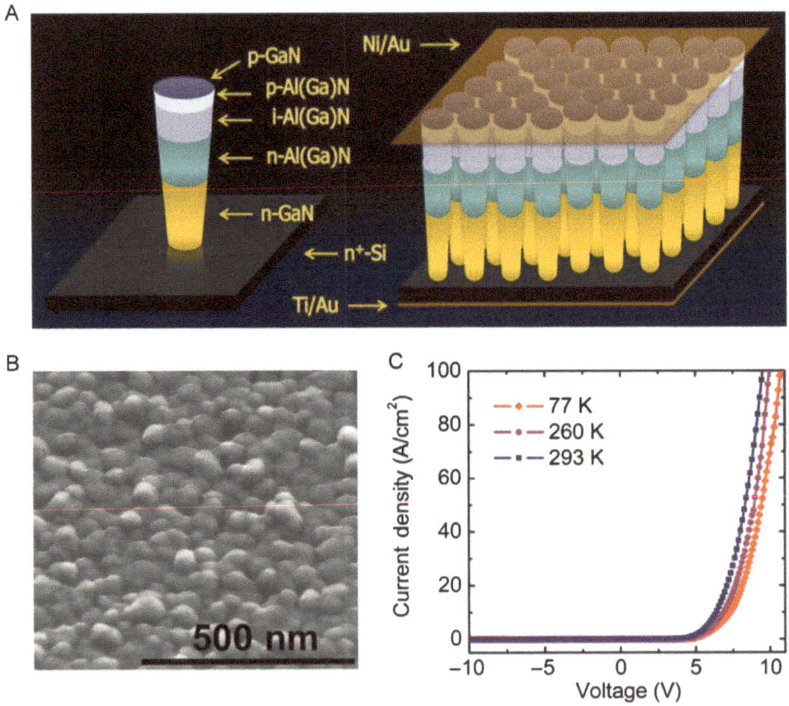

Fig. 13 (A) Schematic of AlN nanowire LEDs (*Left*: the structure; *Right*: the fabricated device). (B) SEM image of the device top surface after *p*-metal deposition. (C) *I–V* characteristics of AlN nanowire LEDs measured at different temperatures (Zhao et al., 2015b).

forward voltage is only around 7 V, corresponding to an electrical efficiency of around 85% considering the photon energy being around 6 eV. This electrical performance is drastically improved compared to the previously reported planar AlN LEDs and is largely ascribed to the significantly enhanced Mg-dopant incorporation in nanowire structures and the resulting efficient impurity band hole conduction (Connie et al., 2015; Zhao et al., 2015a,b). The impurity band hole conduction is further supported by the nearly temperature-independent conductance, as shown in Fig. 13C.

EL emission characteristics at room temperature were further investigated. The EL spectra were collected by an optical fiber. The measurement setup is schematically shown in Fig. 14A. By changing the fiber position EL

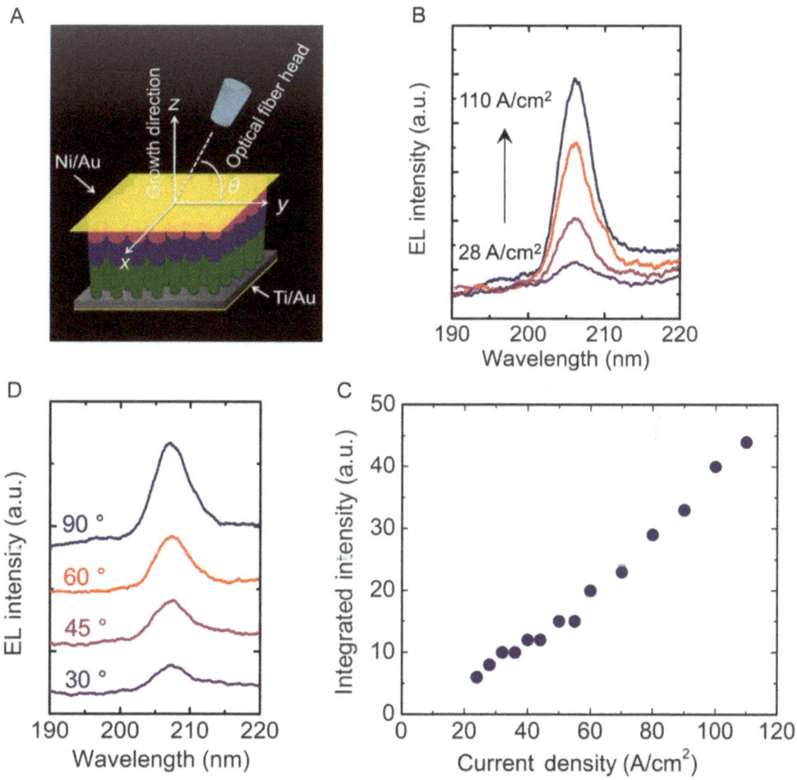

Fig. 14 (Clockwise) (A) Schematic of the EL measurement configuration for AlN nanowire LEDs. The detection angle is with respect to the *y*-axis. (B) The current-dependent EL spectra measured from the device top surface ($\theta = 90$ degree) under a pulsed biasing at room temperature. (C) The corresponding integrated EL intensity vs current density. (D) The angle-dependent EL spectra measured at 36 A/cm^2 under a continuous-wave biasing at room temperature (Zhao et al., 2015b).

emission from different angles can be measured. The 90 degree detection angle (with respective to the *y*-axis) indicates that the fiber is positioned directly on the top surface of the device. The EL spectra measured directly from the top surface under a pulsed bias are shown in Fig. 14B. A strong EL emission peak around 207 nm can be seen. Furthermore, there is no shift of the EL emission wavelength with increasing current, suggesting a robust exciton emission. The integrated EL intensity is shown in Fig. 14C. It is seen that the intensity increases almost linearly with injection current. The angle-dependent EL emission was also measured. As illustrated in Fig. 14D, the EL intensity decreases when the detection angle is reduced, indicating that the light is emitted primarily from the top surface. It is noted that light emission in AlN is intrinsically TM polarized (Li et al., 2003), with the electric field parallel along the nanowire *c*-axis ($E \parallel c$), which generally prevents light extraction from the top surface. In AlN nanowire LEDs, however, the light extraction from the nanowire top surface, i.e., along the *c*-axis, can be made possible due to the multiple light coupling and scattering process among nanowires (Zhao et al., 2015b).

The EL emission characteristics were also studied at 77 K. Illustrated in Fig. 15A are set of EL spectra measured from the device top surface under different injection currents. Strong EL peak, centered around 204 nm, can be seen. Compared to the room-temperature EL spectra, the peak wavelength is blue shifted, due to the increased bandgap energy (Connie et al., 2015). In addition, it is noted that EL emission can be measured below the voltage corresponding to the bandgap energy of AlN, suggesting

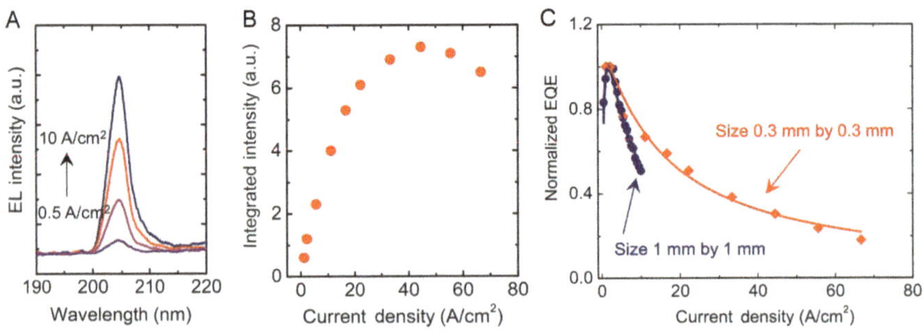

Fig. 15 (A) The current-dependent EL spectra of AlN nanowire LEDs measured at 77 K under a continuous-wave biasing. (B) and (C) The corresponding integrated EL intensity and the normalized relative EQE vs current density, respectively. The *solid lines* in (C) are fitting curves by the ABC model as described in the text for two different device sizes (Zhao et al., 2015b).

extremely efficient current injection into the device active region. The integrated EL intensity vs injection current density is shown in Fig. 15B. It is seen that with increasing current, the light output power first increases rapidly, followed by a saturation, and then drops. The derived relative EQE (defined by the light output power divided by current density) is shown in Fig. 15C. It is seen that the EQE exhibits a sharp increase with increasing current, and then followed by a rapid decrease, which is mainly attributed to the lack of carrier confinement in the active region and the resulting large electron overflow. This EQE trend can be well simulated by the ABC model:

$$\eta_i = \frac{BN^2}{AN + BN^2 + CN^3} \tag{1}$$

where η_i is the IQE, N is the carrier density in the device active region, and A, B, and C are the Shockley–Read–Hall nonradiative recombination coefficient, radiative recombination coefficient, and Auger recombination coefficient and/or electron overflow, respectively. A, B, and C parameters in the range of $5\times10^6\ \mathrm{s}^{-1}$, $4\times10^{-9}\ \mathrm{cm}^3\ \mathrm{s}^{-1}$, and $6\times10^{-26}\ \mathrm{cm}^6\ \mathrm{s}^{-1}$ were derived, respectively. Compared to InGaN/GaN nanowire LEDs, wherein A is in the range of 10^7–$10^8\ \mathrm{s}^{-1}$ (Guo et al., 2011; Nguyen et al., 2012), the A value in AlN nanowire LEDs is much smaller. This is ascribed to the smaller surface recombination velocity of AlN compared to InGaN. The C value in AlN nanowire LEDs, however, is much larger compared to InGaN/GaN nanowire LEDs, which is mainly attributed to the electron overflow in the homojunction structure.

It is worth noting that, although the electrical performance in such AlN nanowire LEDs is drastically improved compared to AlN planar LEDs (Taniyasu et al., 2006), the light extraction efficiency is still very low. The light extraction efficiency of TM polarized light from such AlN nanowire LEDs was simulated using the finite-difference time-domain method. A light extraction efficiency of ~7% was realized by only considering the light absorption by the underlying GaN nanowire template and Si substrate. This number, however, was further reduced to ~0.03% when the top p-metal electrode was considered. It is thus of critical importance to develop deep UV-transparent conductive electrode as well as eliminate the light absorption by the underlying GaN nanowire template and Si substrate, in order to improve the light extraction efficiency of such AlN nanowire LEDs.

4. ELECTRICALLY INJECTED LASERS WITH TERNARY AlGaN NANOWIRES

Compared to conventional AlGaN multiple quantum well lasers, AlGaN nanowire lasers are much less explored. Optically pumped lasers based on GaN nanowires, including random lasers, Fabry–Perot lasers, and plasmonic lasers have been demonstrated (Inose et al., 2013; Johnson et al., 2002; Li et al., 2012; Sakai et al., 2010; Wright et al., 2014; Xu et al., 2012; Zhang et al., 2014). Recently, by exploiting the Anderson localization of light in nearly defect-free AlGaN nanowires grown by MBE, electrically injected lasers in the UV-AII band were demonstrated at 6 K (Li et al., 2015). The threshold current densities were in the range of tens of A/cm^2, which were nearly two to three orders of magnitude lower compared to the previously reported AlGaN multiple quantum well lasers in the similar wavelength range (Yoshida et al., 2008a,b). More recently, by further increasing Al content electrically injected lasers with ternary AlGaN nanowire in the UV-B and UV-C bands were also demonstrated (Zhao et al., 2015c,e). In what follows, electrically injected AlGaN nanowire lasers in the UV-B and UV-C bands are described.

The AlGaN nanowire laser structure is schematically shown in Fig. 16A, which consists of GaN:Si (~250 nm), AlGaN:Si (~100 nm), AlGaN (~100 nm), AlGaN:Mg (~100 nm), and GaN:Mg (~10 nm) layers. In such vertically aligned, randomly distributed AlGaN nanowire arrays, light confinement can exist due to the Anderson localization of light (Inose et al., 2013; Li et al., 2015; Liu et al., 2012; Sakai et al., 2010; Zhao et al., 2015c,e); and the degree of such confinement depends on the nanowire size, spacing, and filling factor. By optimizing these factors, high Q-factor optical cavities can be obtained (Zhao et al., 2015e). Fig. 16B shows the simulated electrical field distribution ($\lambda \sim 290$ nm). It is seen that strong light confinement can be achieved, due to the recurrent multiple scattering process. Another important factor that needs to be considered is the optical confinement along the nanowire vertical direction, which can be made possible by the inversely tapered nanowire morphology, i.e., increasing nanowire diameter along the growth direction (Li et al., 2015; Zhao et al., 2015c). As illustrated in Fig. 16C, the effective refractive index reaches a minimum at GaN nanowire/Si interface, due to the inversely tapered morphology. Because of this effect, the light is mostly confined in the AlGaN segment, shown by the optical field profile along the nanowire vertical direction in Fig. 16C, and

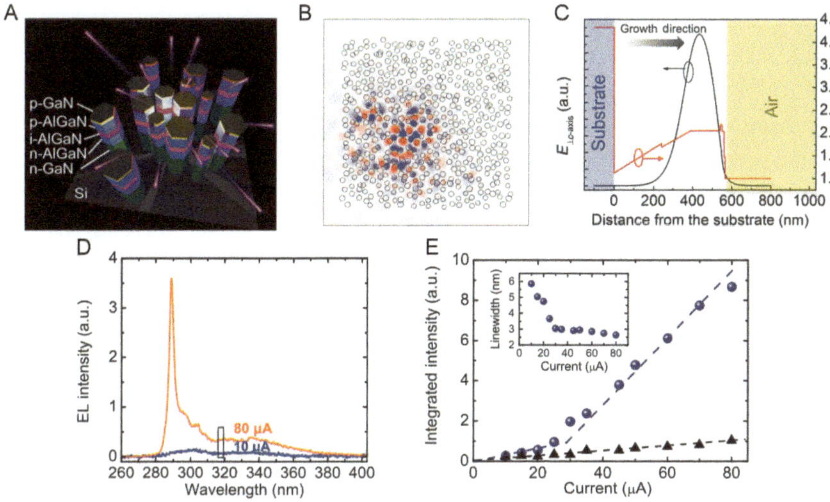

Fig. 16 (A) Schematic of the AlGaN nanowire lasers on Si substrate, with each layer labeled. (B) The simulated optical field distribution at a wavelength of 290 nm for a nanowire array cavity. (C) The electric field (*left* axis) and effective refractive index (*right* axis) as a function of the distance from substrate. (D) The room-temperature EL spectra measured at 10 and 80 μA under a continuous-wave biasing. (E) The integrated EL intensity vs injection current. The *blue circles* represent the lasing peak, and *black triangles* denote the background emission from the boxed region in (D) with a linewidth of 2 nm. The *inset* shows the EL spectral linewidth vs injection current. *Dashed lines* are guide-for-eye (Zhao et al., 2015e).

thus greatly minimizing the optical loss through the underlying GaN nanowire template and Si substrate.

The device fabrication process involved standard photolithography and contact metallization techniques (Li et al., 2015; Zhao et al., 2015c,e). The emitted light was collected from the nanowire top with a deep UV objective, and detected by a charge-coupled device that was attached to a high-resolution spectrometer. The room-temperature EL spectra under a continuous-wave biasing are shown in Fig. 16D. At 10 μA (the *bottom curve*), only a very broad spontaneous emission spectrum can be seen. As the injection current increases, a sharp EL peak around 289 nm appears (the *top curve* was measured at 80 μA). The integrated EL intensity vs injection current is shown in Fig. 16E, which exhibits a distinct threshold around 30 μA. As a comparison, the background emission was also investigated. The *black triangles* in Fig. 16E are the integrated EL intensity of the *boxed region* from Fig. 16D. It is seen that the background emission remains nearly constant, due to the clamping of carrier concentration above threshold. In addition, it is noted that a very small increase of the background emission exists with further increasing current, which has been commonly measured from quantum-dot lasers, mainly due to the inhomogeneous broadening and hot carrier effect (Tatebayashi et al., 2015). The *inset* of Fig. 16E shows the EL spectral linewidth vs injection current. A significant reduction in the EL spectral linewidth at threshold can be seen. For a lasing area of around 10 μm^2, a threshold current density was derived to be 300 A/cm^2. This number is significantly smaller compared to AlGaN quantum well lasers operating at 336 nm (Yoshida et al., 2011). This low lasing threshold is largely attributed to the discrete density of states associated with the quantum-dot-like nanostructures as discussed in Section 2, the nearly defect-free AlGaN nanowires, and the high Q-factor of the optical cavity.

By further changing alloy composition, electrically injected AlGaN nanowire lasers in the UV-C band have also been realized. Illustrated in Fig. 17A are the EL spectra taken at 77 K under a continuous-wave biasing. It is seen that with increasing current, a sharp EL peak centered around 262 nm appears. The light–current characteristics are shown in Fig. 17B, manifesting a clear threshold around 200 A/cm^2. Illustrated in the *inset* of Fig. 17B are the light–current characteristics in a logarithmic scale. The S-shaped curve, corresponding to linear spontaneous emission, superlinear amplified spontaneous emission, and linear lasing emission with increasing current, provides unambiguous evidence for the achievement of lasing.

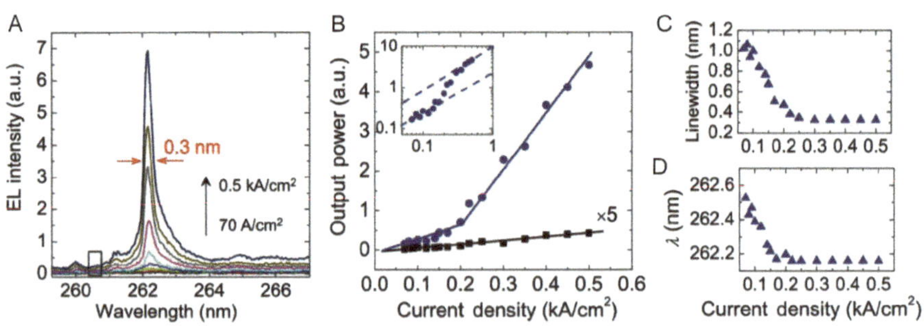

Fig. 17 (A) The current-dependent EL spectra of an AlGaN nanowire laser operating at 262 nm under a continuous-wave biasing at 77 K. (B) The integrated EL intensity vs injection current. The *blue circles* represent the lasing peak, and the *black squares* denote the background emission from the boxed region in (A) with a linewidth of 0.3 nm. The *inset* shows the plot in the logarithmic scale. Both *solid* and *dashed lines* are guide-for-eye. (C) and (D) Linewidth and peak wavelength vs current density, respectively (Zhao et al., 2015c).

Variations of the integrated EL intensity of the background emission (*boxed region* in Fig. 17A) vs current are also shown in Fig. 17B by *black squares*, which stay nearly constant above threshold. The EL spectral linewidth is reduced from ~1.2 to 0.3 nm near threshold, as shown in Fig. 17C. The lasing peak remains highly stable above threshold, as shown in Fig. 17D.

It is worthwhile mentioning that Al(Ga)N nanowires can provide a unique platform for deep UV photon management. For example, manifested by AlN nanowire LEDs in Section 3, strong TM polarized light emission was measured from the device top surface by exploiting the light multiple scattering and coupling effect among nanowires. On the other hand, deep UV photons could also be strongly confined within the nanowire arrays for certain nanowire size and spacing variations due to the Anderson localization of light, leading to electrically injected deep UV lasers, as shown earlier. Evidently, a precise control of the nanowire diameter and spacing is crucial for the design and performance of Al(Ga)N nanowire deep UV optoelectronic devices, which can be made possible by the selective area epitaxy on patterned substrates (Gacevic et al., 2015; Sekiguchi et al., 2008b; Yamano et al., 2015).

5. OTHER DEVICES AND APPLICATIONS WITH Al(Ga)N NANOWIRES

Photodetectors based on Ga(In)N nanowires in the near-UV and visible spectral range have been previously investigated (Bugallo Ade et al., 2010; Bugallo et al., 2011; Gonzalez-Posada et al., 2012). Efforts have also been spent on developing photodetectors with Al(Ga)N nanowire structures. Rigutti et al. reported GaN/AlN quantum disk photodetectors (Rigutti et al., 2010). The room-temperature responsivity of such single nanowire photodetectors was measured to be as high as $\sim 2 \times 10^3$ A/W at 300 nm. Moreover, due to the quantum-confined Stark effect in such GaN/AlN quantum disk structure, response to photons with energies below the bangap energy of GaN was also measured.

Single photon sources based on Al(Ga)N nanowire structures have also been developed. Holmes et al. reported single photon emission ~290 nm from a single GaN/$Al_{0.8}Ga_{0.2}N$ quantum dot under optical pumping (Holmes et al., 2014, 2016). In this work, the nanostructure was grown by selective area epitaxy using MOCVD. The GaN quantum dot had a height of ~1 nm and width of ~10 nm. Excellent single photon emission characteristics, including temperature-independent second-order correlation function

($g^2[0] \sim 0.13$), were achieved. In addition, highly polarized emission, with a polarization ratio of 97%, was measured (Holmes et al., 2016). Saniya et al. further demonstrated blue color single photon emission with ternary AlGaN nanowires in which a single InGaN quantum dot was inserted (Deshpande et al., 2013). The operation temperature was measured up to 200 K.

Due to the large band offset between GaN and AlN (~1.8 eV), it is also of practical interest to develop intraband devices with Al(Ga)N nanowires utilizing the transition among the quantum-confined states in the conduction or valence band. In addition, due to large phonon energies of GaN and AlN, they promise high temperature intraband optoelectronic devices. In this regard, Beeler et al. performed detailed studies on the intraband absorption of Ge-doped GaN/AlN nanodisk structure (Beeler et al., 2014). In this work, the nanostructure was grown by plasma-assisted MBE under N-rich conditions on Si (111) substrate. The intraband transition-induced TM polarized absorption line was found to be dependent on both Ge doping and GaN disk thickness, and absorption wavelengths in the range of ~1.58 to 1.95 μm were measured. This study provided important information for the intraband optoelectronic devices based on Al(Ga)N nanowires.

Additionally, due to the large bandgap energies of AlN and GaN, and excellent thermal and mechanical properties, great efforts have also been devoted to developing Al(Ga)N nanowire-based electronic devices, including GaN/AlN/AlGaN radial nanowire high electron mobility transistors (Li et al., 2006), AlGaN/GaN nanowire transistors (Chen and Wu, 2013), AlGaN/GaN nanowire fin field effect transistor by top-down approach (Im et al., 2015), and nanogenerators (Wang et al., 2010), to name a few. Additionally, the magnetic properties of Mg-doped AlN nanowires have also been investigated (Xu et al., 2013). The saturation magnetization and coercivity were measured to be 0.051 emu/g and 127 Oe, respectively. The Al vacancy and substitutional Mg were suggested to play important roles in this room-temperature ferromagnetism.

6. CONCLUSION

In summary, in this chapter we have provided an overview on the recent advances in the synthesis/growth of Al(Ga)N nanowires, their structural, optical, and electrical characteristics, and their applications in deep UV LEDs and lasers. With the improved MBE growth process, defect-free Al(Ga)N nanowires have been realized directly on Si substrate. More importantly, it has been found that Mg-dopant incorporation is significantly enhanced in Al(Ga)N nanowire structures compared to their planar counterparts, leading to efficient impurity band hole conduction. Al(Ga)N nanowire LEDs with excellent performance have been further demonstrated. In addition, detailed structural and optical characterizations have confirmed the presence of quantum-dot-like nanostructures in ternary AlGaN nanowires, leading to ultralow threshold electrically injected ternary AlGaN nanowire lasers operating in the UV-B and UV-C bands. These defect-free Al(Ga)N nanowires therefore can provide a viable path towards the realization of high efficiency deep UV optoelectronic devices.

REFERENCES

Adivarahan, V., Sun, W.H., Chitnis, A., Shatalov, M., Wu, S., Maruska, H.P., Khan, M.A., 2004a. 250 Nm AlGaN light-emitting diodes. Appl. Phys. Lett. 85, 2175.

Adivarahan, V., Wu, S., Zhang, J.P., Chitnis, A., Shatalov, M., Mandavilli, V., Gaska, R., Khan, M.A., 2004b. High-efficiency 269 nm emission deep ultraviolet light-emitting diodes. Appl. Phys. Lett. 84, 4762.

Akasaki, I., Amano, H., 1997. Crystal growth and conductivity control of group III nitride semiconductors and their application to short wavelength light emitters. Jpn. J. Appl. Phys. 36, 5393.

Allerman, A.A., Crawford, M.H., Fischer, A.J., Bogart, K.H.A., Lee, S.R., Follstaedt, D.M., Provencio, P.P., Koleske, D.D., 2004. Growth and design of deep-UV (240–290 nm) light emitting diodes using AlGaN alloys. J. Cryst. Growth 272, 227.

Balasubramanian, C., Godbole, V.P., Rohatgi, V.K., Das, A.K., Bhoraskar, S.V., 2004. Synthesis of nanowires and nanoparticles of cubic aluminium nitride. Nanotechnology 15, 370.

Beeler, M., Hille, P., Schormann, J., Teubert, J., De La Mata, M., Arbiol, J., Eickhoff, M., Monroy, E., 2014. Intraband absorption in self-assembled Ge-doped GaN/AlN nanowire heterostructures. Nano Lett. 14, 1665.

Belloeil, M., Gayral, B., Daudin, B., 2016. Quantum dot-like behavior of compositional fluctuations in AlGaN nanowires. Nano Lett. 16, 960.

Bertness, K.A., Roshko, A., Sanford, N.A., Barker, J.M., Davydov, A.V., 2006. Spontaneously grown GaN and AlGaN nanowires. J. Cryst. Growth 287, 522.

Bhattacharyya, A., Moustakas, T.D., Zhou, L., Smith, D.J., Hug, W., 2009. Deep ultraviolet emitting AlGaN quantum wells with high internal quantum efficiency. Appl. Phys. Lett. 94, 181907.

Bourret, A., Barski, A., Rouvière, J.L., Renaud, G., Barbier, A., 1998. Growth of aluminum nitride on (111) silicon: microstructure and interface structure. J. Appl. Phys. 83, 2003.

Brubaker, M.D., Blanchard, P.T., Schlager, J.B., Sanders, A.W., Herrero, A.M., Roshko, A., Duff, S.M., Harvey, T.E., Bright, V.M., Sanford, N.A., Bertness, K.A., 2013a. Toward discrete axial P–N junction nanowire light-emitting diodes grown by plasma-assisted molecular beam epitaxy. J. Electron. Mater. 42, 868.

Brubaker, M.D., Blanchard, P.T., Schlager, J.B., Sanders, A.W., Roshko, A., Duff, S.M., Gray, J.M., Bright, V.M., Sanford, N.A., Bertness, K.A., 2013b. On-chip optical interconnects made with gallium nitride nanowires. Nano Lett. 13, 374.

Bugallo Ade, L., Tchernycheva, M., Jacopin, G., Rigutti, L., Julien, F.H., Chou, S.T., Lin, Y.T., Tseng, P.H., Tu, L.W., 2010. Visible-blind photodetector based on P-I-N junction GaN nanowire ensembles. Nanotechnology 21, 315201.

Bugallo, A.D.L., Rigutti, L., Jacopin, G., Julien, F.H., Durand, C., Chen, X.J., Salomon, D., Eymery, J., Tchernycheva, M., 2011. Single-wire photodetectors based on InGaN/GaN radial quantum wells in GaN wires grown by catalyst-free metal-organic vapor phase epitaxy. Appl. Phys. Lett. 98, 233107.

Carnevale, S.D., Kent, T.F., Phillips, P.J., Mills, M.J., Rajan, S., Myers, R.C., 2012. Polarization-induced Pn diodes in wide-band-gap nanowires with ultraviolet electroluminescence. Nano Lett. 12, 915.

Carnevale, S.D., Kent, T.F., Phillips, P.J., Sarwar, A.T., Selcu, C., Klie, R.F., Myers, R.C., 2013. Mixed polarity in polarization-induced pn junction nanowire light emitting diodes. Nano Lett. 13, 3029.

Chen, C.-Y., Wu, Y.-R., 2013. Studying the short channel effect in the scaling of the AlGaN/GaN nanowire transistors. J. Appl. Phys. 113, 214501.

Chen, F., Ji, X., Lu, Z., Shen, Y., Zhang, Q., 2014a. Shape-controlled synthesis of one-dimensional $Al_xGa_{1-x}N$ nanotower arrays: structural characteristics and growth mechanism. J. Phys. D: Appl. Phys. 47, 065311.

Chen, F., Ji, X., Lu, Z., Shen, Y., Zhang, Q., 2014b. Structural and raman properties of compositionally tunable $Al_xGa_{1-x}N$ ($0.66 \leq X \leq 1$) nanowires. Mat. Sci. Eng. B 183, 24.

Chen, F., Ji, X., Zhang, Q., 2015. Morphology-controlled synthesis and structural characterization of ternary $Al_xGa_{1-x}N$ nanostructures by chemical vapor deposition. CrystEngComm 17, 1249.

Connie, A.T., Zhao, S., Sadaf, S.M., Shih, I., Mi, Z., Du, X., Lin, J., Jiang, H., 2015. Optical and electrical properties of Mg-doped AlN nanowires grown by molecular beam epitaxy. Appl. Phys. Lett. 106, 213105.

Consonni, V., Hanke, M., Knelangen, M., Geelhaar, L., Trampert, A., Riechert, H., 2011. Nucleation mechanisms of self-induced GaN nanowires grown on an amorphous interlayer. Phys. Rev. B 83, 035310.

Deshpande, S., Das, A., Bhattacharya, P., 2013. Blue single photon emission up to 200 k from an InGaN quantum dot in AlGaN nanowire. Appl. Phys. Lett. 102, 161114.

Gacevic, Z., Gomez Sanchez, D., Calleja, E., 2015. Formation mechanisms of GaN nanowires grown by selective area growth homoepitaxy. Nano Lett. 15, 1117.

Gonzalez-Posada, F., Songmuang, R., Den Hertog, M., Monroy, E., 2012. Room-temperature photodetection dynamics of single GaN nanowires. Nano Lett. 12, 172.

Guo, W., Zhang, M., Bhattacharya, P., Heo, J., 2011. Auger recombination in III-nitride nanowires and its effect on nanowire light-emitting diode characteristics. Nano Lett. 11, 1434.

He, C., Wu, Q., Wang, X., Zhang, Y., Yang, L., Liu, N., Zhao, Y., Lu, Y., Hu, Z., 2011. Growth and characterization of ternary AlGaN alloy nanocones across the entire composition range. ACS Nano 5, 1291.

Hersee, S.D., Fairchild, M., Rishinaramangalam, A.K., Ferdous, M.S., Zhang, L., Varangis, P.M., Swartzentruber, B.S., Talin, A.A., 2009. GaN nanowire light emitting diodes based on templated and scalable nanowire growth process. Electron. Lett. 45, 75.

Himwas, C., Hertog, M.D., Donatini, F., Dang, L.S., Rapenne, L., Sarigiannidou, E., Songmuang, R., Monroy, E., 2013. AlGaN/AlN quantum dots for UV light emitters. Phys. Status Solidi C 10, 285.

Himwas, C., Den Hertog, M., Dang, L.S., Monroy, E., Songmuang, R., 2014. Alloy inhomogeneity and carrier localization in AlGaN sections and AlGaN/AlN nanodisks in nanowires with 240–350 nm emission. Appl. Phys. Lett. 105, 241908.

Hirayama, H., Fujikawa, S., Noguchi, N., Norimatsu, J., Takano, T., Tsubaki, K., Kamata, N., 2009. 222–282 nm AlGaN and In AlGaN-based deep-UV leds fabricated on high-quality AlN on sapphire. Phys. Status Solidi A 206, 1176.

Hirayama, H., Noguchi, N., Kamata, N., 2010. 222 nm deep-ultraviolet AlGaN quantum well light-emitting diode with vertical emission properties. Appl. Phys. Express 3, 032102.

Hirayama, H., Maeda, N., Fujikawa, S., Toyoda, S., Kamata, N., 2014. Recent progress and future prospects of AlGaN-based high-efficiency deep-ultraviolet light-emitting diodes. Jpn. J. Appl. Phys. 53, 100209.

Holmes, M.J., Choi, K., Kako, S., Arita, M., Arakawa, Y., 2014. Room-temperature triggered single photon emission from A III-nitride site-controlled nanowire quantum dot. Nano Lett. 14, 982.

Holmes, M.J., Kako, S., Choi, K., Arita, M., Arakawa, Y., 2016. Linearly polarized single photons from small site-controlled GaN nanowire quantum dots. Proc. SPIE 9748, 97481e.

Hong, L., Liu, Z., Zhang, X.T., Hark, S.K., 2006. Self-catalytic growth of single-phase AlGaN alloy nanowires by chemical vapor deposition. Appl. Phys. Lett. 89, 193105.

Im, K.-S., Sindhuri, V., Jo, Y.-W., Son, D.-H., Lee, J.-H., Cristoloveanu, S., Lee, J.-H., 2015. Fabrication of AlGaN/GaN omega-shaped nanowire fin-shaped fets by a top-down approach. Appl. Phys. Express 8, 066501.

Inose, Y., Ema, K., Sakai, M., Kikuchi, A., Kishino, K., Ohtsuki, T., 2013. Anderson localization of light in two-dimensional random arrays of semiconductor nanocolumns. AIP Conf. Proc. 1566, 548.

Johnson, J.C., Choi, H.J., Knutsen, K.P., Schaller, R.D., Yang, P., Saykally, R.J., 2002. Single gallium nitride nanowire lasers. Nat. Mater. 1, 106.

Kenry, Youg, K.T., Yu, S.F., 2012. Aln nanowires: synthesis, physical properties, and nanoelectronics applications. J. Mater. Sci. 47, 5341.

Kent, T.F., Carnevale, S.D., Sarwar, A.T., Phillips, P.J., Klie, R.F., Myers, R.C., 2014. Deep ultraviolet emitting polarization induced nanowire light emitting diodes with Alxga(1)-Xn active regions. Nanotechnology 25, 455201.

Khan, A., Balakrishnan, K., Katona, T., 2008. Ultraviolet light-emitting diodes based on group three nitrides. Nat. Photonics 2, 77.

Kim, H.M., Kang, T.W., Chung, K.S., 2003. Nanoscale ultraviolet-light-emitting diodes using wide-bandgap gallium nitride nanorods. Adv. Mater. 15, 567.

Kim, K.H., Fan, Z.Y., Khizar, M., Nakarmi, M.L., Lin, J.Y., Jiang, H.X., 2004. AlGaN-based ultraviolet light-emitting diodes grown on AlN epilayers. Appl. Phys. Lett. 85, 4777.

Kneissl, M., Yang, Z., Teepe, M., Knollenberg, C., Schmidt, O., Kiesel, P., Johnson, N.M., Schujman, S., Schowalter, L.J., 2007. Ultraviolet semiconductor laser diodes on bulk AlN. J. Appl. Phys. 101, 123103.

Kneissl, M., Kolbe, T., Chua, C., Kueller, V., Lobo, N., Stellmach, J., Knauer, A., Rodriguez, H., Einfeldt, S., Yang, Z., Johnson, N.M., Weyers, M., 2011. Advances in group III-nitride-based deep UV light-emitting diode technology. Semicond. Sci. Technol. 26, 014036.

Landré, O., Fellmann, V., Jaffrennou, P., Bougerol, C., Renevier, H., Cros, A., Daudin, B., 2010a. Molecular beam epitaxy growth and optical properties of AlN nanowires. Appl. Phys. Lett. 96, 061912.

Landré, O., Fellmann, V., Jaffrennou, P., Bougerol, C., Renevier, H., Daudin, B., 2010b. Growth mechanism of catalyst-free [0001] GaN and AlN nanowires on Si by molecular beam epitaxy. Phys. Status Solidi C 7, 2246.

Lee, S.K., Kim, T.H., Lee, S.Y., Choi, K.C., Yang, P., 2007. High-brightness gallium nitride nanowire UV–blue light emitting diodes. Philos. Mag. 87, 2105.

Li, J., Nam, K.B., Nakarmi, M.L., Lin, J.Y., Jiang, H.X., Carrier, P., Wei, S.-H., 2003. Band structure and fundamental optical transitions in wurtzite AlN. Appl. Phys. Lett. 83, 5163.

Li, Y., Xiang, J., Qian, F., Gradecak, S., Wu, Y., Yan, H., Yan, H., Blom, D.A., Lieber, C.M., 2006. Dopant-free GaN/AlN/AlGaN radial nanowire heterostructures as high electron mobility transistors. Nano Lett. 6, 1468.

Li, Q., Wright, J.B., Chow, W.W., Luk, T.S., Brener, I., Lester, L.F., Wang, G.T., 2012. Single-mode GaN nanowire lasers. Opt. Express 20, 17873.

Li, X.-H., Detchprohm, T., Kao, T.-T., Satter, M.M., Shen, S.-C., Douglas Yoder, P., Dupuis, R.D., Wang, S., Wei, Y.O., Xie, H., Fischer, A.M., Ponce, F.A., Wernicke, T., Reich, C., Martens, M., Kneissl, M., 2014. Low-threshold stimulated emission at 249 nm and 256 nm from AlGaN-based multiple-quantum-well lasers grown on sapphire substrates. Appl. Phys. Lett. 105, 141106.

Li, K.H., Liu, X., Wang, Q., Zhao, S., Mi, Z., 2015. Ultralow-threshold electrically injected AlGaN nanowire ultraviolet lasers on Si operating at low temperature. Nat. Nanotechnol. 10, 140.

Liao, Y., Thomidis, C., Kao, C.-K., Moustakas, T.D., 2011. AlGaN based deep ultraviolet light emitting diodes with high internal quantum efficiency grown by molecular beam epitaxy. Appl. Phys. Lett. 98, 081110.

Liu, J., Zhang, X., Zhang, Y., He, R., Zhu, J., 2001. Novel synthesis of AlN nanowires with controlled diameters. J. Mater. Res. 16, 3133.

Liu, X.Y., Shan, C.X., Wang, S.P., Zhang, Z.Z., Shen, D.Z., 2012. Electrically pumped random lasers fabricated from Zno nanowire arrays. Nanoscale 4, 2843.

Martens, M., Mehnke, F., Kuhn, C., Reich, C., Kueller, V., Knauer, A., Netzel, C., Hartmann, C., Wollweber, J., Rass, J., Wernicke, T., Bickermann, M., Weyers, M., Kneissl, M., 2014. Performance characteristics of UV-C AlGaN-based lasers grown on sapphire and Bulk AlN substrates. IEEE Photonic. Tech. Lett. 26, 342.

Motayed, A., Davydov, A.V., He, M., Mohammad, S.N., Melngailis, J., 2007. 365 nm operation of N-nanowire/P-gallium nitride homojunction light emitting diodes. Appl. Phys. Lett. 90, 183120.

Moustakas, T.D., Bhattacharyya, A., 2012. The role of liquid phase epitaxy during growth of AlGaN by MBE. Phys. Status Solidi C 9, 580.

Muramoto, Y., Kimura, M., Nouda, S., 2014. Development and future of ultraviolet light-emitting diodes: UV-led will replace the UV Lamp. Semicond. Sci. Technol. 29, 084004.

Nakarmi, M.L., Nepal, N., Lin, J.Y., Jiang, H.X., 2009. Photoluminescence studies of impurity transitions in Mg-doped AlGaN alloys. Appl. Phys. Lett. 94, 091903.

Nam, O.H., Bremser, M.D., Ward, B.L., Nemanich, R.J., Davis, R.F., 1997. Growth of GaN and Al0.2ga0.8n on patterened substrates via organometallic vapor phase epitaxy. Jpn. J. Appl. Phys. 36, L532.

Nguyen, H.P., Djavid, M., Cui, K., Mi, Z., 2012. Temperature-dependent nonradiative recombination processes in GaN-based nanowire white-light-emitting diodes on silicon. Nanotechnology 23, 194012.

Park, Y.S., Hwang, B.R., Lee, J.C., Im, H., Cho, H.Y., Kang, T.W., Na, J.H., Park, C.M., 2006. Self-assembled Al(X)Ga(1-X)N nanorods grown on Si(001) substrates by using plasma-assisted molecular beam epitaxy. Nanotechnology 17, 4640.

Pierret, A., Bougerol, C., Gayral, B., Kociak, M., Daudin, B., 2013a. Probing alloy composition gradient and nanometer-scale carrier localization in single AlGaN nanowires by nanocathodoluminescence. Nanotechnology 24, 305703.

Pierret, A., Bougerol, C., Hertog, M.D., Gayral, B., Kociak, M., Renevier, H., Daudin, B., 2013b. Structural and optical properties of $Al_xGa_{1-x}N$ nanowires. Phys. Status Solidi Rapid Res. Lett. 7, 868.

Pierret, A., Bougerol, C., Murcia-Mascaros, S., Cros, A., Renevier, H., Gayral, B., Daudin, B., 2013c. Growth, structural and optical properties of AlGaN nanowires in the whole composition range. Nanotechnology 24, 115704.

Qian, F., Gradecak, S., Li, Y., Wen, C.-Y., Lieber, C.M., 2005. Core/multishell nanowire heterostructures as multicolor, high-efficiency light-emitting diodes. Nano Lett. 5, 2287.

Renard, J., Songmuang, R., Bougerol, C., Daudin, B., Gayral, B., 2008. Exciton and biexciton luminescence from single GaN/AlN quantum dots in nanowires. Nano Lett. 8, 2092.

Renard, J., Songmuang, R., Tourbot, G., Bougerol, C., Daudin, B., Gayral, B., 2009. Evidence for quantum-confined stark effect in GaN/AlN quantum dots in nanowires. Phys. Rev. B 80, R121305.

Rigutti, L., Tchernycheva, M., De Luna Bugallo, A., Jacopin, G., Julien, F.H., Zagonel, L.F., March, K., Stephan, O., Kociak, M., Songmuang, R., 2010. Ultraviolet photodetector based on GaN/AlN quantum disks in a single nanowire. Nano Lett. 10, 2939.

Ristic, J., Sanchez-Garcia, M.A., Calleja, E., Sanchez-Paramo, J., Calleja, J.M., Jahn, U., Ploog, K.H., 2002. AlGaN nanocolumns grown by molecular beam epitaxy: optical and structural characterization. Phys. Status Solidi A 192, 60.

Sakai, M., Inose, Y., Ema, K., Ohtsuki, T., Sekiguchi, H., Kikuchi, A., Kishino, K., 2010. Random laser action in GaN nanocolumns. Appl. Phys. Lett. 97, 151109.

Sarwar, A.T., Carnevale, S.D., Yang, F., Kent, T.F., Jamison, J.J., Mccomb, D.W., Myers, R.C., 2015. Semiconductor nanowire light-emitting diodes grown on metal: a direction toward large-scale fabrication of nanowire devices. Small 11, 5402.

Sekiguchi, H., Kato, K., Tanaka, J., Kikuchi, A., Kishino, K., 2008a. Ultraviolet GaN-based nanocolumn light-emitting diodes grown on N-(111) Si substrates by RF-plasma-assisted molecular beam epitaxy. Phys. Status Solidi A 205, 1067.

Sekiguchi, H., Kishino, K., Kikuchi, A., 2008b. Ti-mask selective-area growth of GaN by RF-plasma-assisted molecular-beam epitaxy for fabricating regularly arranged InGaN/GaN nanocolumns. Appl. Phys. Express 1, 124002.

Sivadasan, A.K., Patsha, A., Polaki, S., Amirthapandian, S., Dhara, S., Bhattacharya, A., Panigrahi, B.K., Tyagi, A.K., 2015. Optical properties of monodispersed AlGaN nanowires in the single-prong growth mechanism. Cryst. Growth Des. 15, 1311.

Songmuang, R., Kalita, D., Sinha, P., Den Hertog, M., André, R., Ben, T., González, D., Mariette, H., Monroy, E., 2011. Strong suppression of internal electric field in GaN/AlGaN multi-layer quantum dots in nanowires. Appl. Phys. Lett. 99, 141914.

Su, J., Cui, G., Gherasimova, M., Tsukamoto, H., Han, J., Ciuparu, D., Lim, S., Pfefferle, L., He, Y., Nurmikko, A.V., Broadbridge, C., Lehman, A., 2005a. Catalytic growth of group III-nitride nanowires and nanostructures by metalorganic chemical vapor deposition. Appl. Phys. Lett. 86, 013105.

Su, J., Gherasimova, M., Cui, G., Tsukamoto, H., Han, J., Onuma, T., Kurimoto, M., Chichibu, S.F., Broadbridge, C., He, Y., Nurmikko, A.V., 2005b. Growth of AlGaN nanowires by metalorganic chemical vapor deposition. Appl. Phys. Lett. 87, 183108.

Takano, T., Narita, Y., Horiuchi, A., Kawanishi, H., 2004. Room-temperature deep-ultraviolet lasing at 241.5 nm Of AlGaN multiple-quantum-well laser. Appl. Phys. Lett. 84, 3567.

Tang, C.C., Fan, S.S., De La Chapelle, M.L., Li, P., 2001. Silica-assisted catalytic growth of oxide and nitride nanowires. Chem. Phys. Lett. 333, 12.

Tang, X., Li, G., Zhou, S., 2013. Ultraviolet electroluminescence of light-emitting diodes based on single N-Zno/P-AlGaN heterojunction nanowires. Nano Lett. 13, 5046.

Taniyasu, Y., Kasu, M., 2010. Surface 210 nm light emission from an Aln P–N junction light-emitting diode enhanced by A-plane growth orientation. Appl. Phys. Lett. 96, 221110.

Taniyasu, Y., Kasu, M., Makimoto, T., 2006. An aluminium nitride light-emitting diode with a wavelength of 210 nanometres. Nature 441, 325.

Tatebayashi, J., Kako, S., Ho, J., Ota, Y., Iwamoto, S., Arakawa, Y., 2015. Room-temperature lasing in a single nanowire with quantum dots. Nat. Photonics 9, 501.

Wang, X., Song, J., Zhang, F., He, C., Hu, Z., Wang, Z., 2010. Electricity generation based on one-dimensional group-III nitride nanomaterials. Adv. Mater. 22, 2155.

Wang, Q., Nguyen, H.P.T., Cui, K., Mi, Z., 2012. High efficiency ultraviolet emission from $Al_xGa_{1-x}N$ core-shell nanowire heterostructures grown on si (111) by molecular beam epitaxy. Appl. Phys. Lett. 101, 043115.

Wang, Q., Connie, A.T., Nguyen, H.P., Kibria, M.G., Zhao, S., Sharif, S., Shih, I., Mi, Z., 2013. Highly efficient, spectrally pure 340 nm ultraviolet emission from AlxGa(1)-xN nanowire based light emitting diodes. Nanotechnology 24, 345201.

Wang, Q., Zhao, S., Connie, A.T., Shih, I., Mi, Z., Gonzalez, T., Andrews, M.P., Du, X.Z., Lin, J.Y., Jiang, H.X., 2014. Optical properties of strain-free AlN nanowires grown by molecular beam epitaxy on Si substrates. Appl. Phys. Lett. 104, 223107.

Wright, J.B., Campione, S., Liu, S., Martinez, J.A., Xu, H., Luk, T.S., Li, Q., Wang, G.T., Swartzentruber, B.S., Lester, L.F., Brener, I., 2014. Distributed feedback gallium nitride nanowire lasers. Appl. Phys. Lett. 104, 041107.

Xie, J., Mita, S., Bryan, Z., Guo, W., Hussey, L., Moody, B., Schlesser, R., Kirste, R., Gerhold, M., Collazo, R.N., Sitar, Z., 2013. Lasing and longitudinal cavity modes in photo-pumped deep ultraviolet AlGaN heterostructures. Appl. Phys. Lett. 102, 171102.

Xu, H., Wright, J.B., Luk, T.-S., Figiel, J.J., Cross, K., Lester, L.F., Balakrishnan, G., Wang, G.T., Brener, I., Li, Q., 2012. Single-mode lasing of GaN nanowire-pairs. Appl. Phys. Lett. 101, 113106.

Xu, Y., Yao, B., Liu, D., Lei, W., Zhu, P., Cui, Q., Zou, G., 2013. Room temperature ferromagnetism in new diluted magnetic semiconductor Aln:Mg nanowires. CrystEngComm 15, 3271.

Yamano, K., Kishino, K., Sekiguchi, H., Oto, T., Wakahara, A., Kawakami, Y., 2015. Novel selective area growth (SAG) method for regularly arranged AlGaN nanocolumns using nanotemplates. J. Cryst. Growth 425, 316.

Ye, F., Cai, X.-M., Zhong, X., Wang, H., Tian, X.-Q., Zhang, D.-P., Fan, P., Luo, J.-T., Zheng, Z.-H., Liang, G.-X., 2015. The role of Ga in the chemical vapor deposition growth of ternary $Al_xGa_{1-x}N$ nanowires. J. Alloys Compd. 620, 87.

Yoshida, H., Yamashita, Y., Kuwabara, M., Kan, H., 2008a. A 342-nm ultraviolet AlGaN multiple-quantum-well laser diode. Nat. Photonics 2, 551.

Yoshida, H., Yamashita, Y., Kuwabara, M., Kan, H., 2008b. Demonstration of an ultraviolet 336 nm AlGaN multiple-quantum-well laser diode. Appl. Phys. Lett. 93, 241106.

Yoshida, H., Kuwabara, M., Yamashita, Y., Uchiyama, K., Kan, H., 2011. The current status of ultraviolet laser diodes. Phys. Status Solidi A 208, 1586.

Yoshizawa, M., Kikuchi, A., Fujita, N., Kushi, K., Sasamoto, H., Kishino, K., 1998. Self-organization of $GaN/Al_{0.18}Ga_{0.82}N$ multi-layer nano-columns on (0001) Al_2O_3 by RF molecular beam epitaxy for fabricating GaN quantum disks. J. Cryst. Growth 189–190, 138–141.

Zhang, Y., Liu, J., He, R., Zhang, Q., Zhang, X., Zhu, J., 2001. Synthesis of aluminum nitride nanowires from carbon nanotubes. Chem. Mater. 13, 3899.

Zhang, J.P., Chitnis, A., Adivarahan, V., Wu, S., Mandavilli, V., Pachipulusu, R., Shatalov, M., Simin, G., Yang, J.W., Khan, M.A., 2002. Milliwatt power deep ultraviolet light-emitting diodes over sapphire with emission at 278 nm. Appl. Phys. Lett. 81, 4910.

Zhang, J.P., Hu, X., Bilenko, Y., Deng, J., Lunev, A., Shur, M.S., Gaska, R., Shatalov, M., Yang, J.W., Khan, M.A., 2004. AlGaN-based 280 nm light-emitting diodes with continuous-wave power exceeding 1 MW at 25 Ma. Appl. Phys. Lett. 85, 5532.

Zhang, Q., Li, G., Liu, X., Qian, F., Li, Y., Sum, T.C., Lieber, C.M., Xiong, Q., 2014. A room temperature low-threshold ultraviolet plasmonic nanolaser. Nat. Commun. 5, 4953.

Zhao, Q., Zhang, H., Xu, X., Wang, Z., Xu, J., Yu, D., Li, G., Su, F., 2005. Optical properties of highly ordered AlN nanowire arrays grown on sapphire substrate. Appl. Phys. Lett. 86, 193101.

Zhao, S., Kibria, M.G., Wang, Q., Nguyen, H.P.T., Mi, Z., 2013a. Growth of large-scale vertically aligned GaN nanowires and their heterostructures with high uniformity on SiOx by catalyst-free molecular beam epitaxy. Nanoscale 5, 5283.

Zhao, S., Le, B.H., Liu, D.P., Liu, X.D., Kibria, M.G., Szkopek, T., Guo, H., Mi, Z., 2013b. P-type InN nanowires. Nano Lett. 13, 5509.
Zhao, S., Connie, A.T., Dastjerdi, M.H., Kong, X.H., Wang, Q., Djavid, M., Sadaf, S., Liu, X.D., Shih, I., Guo, H., Mi, Z., 2015a. Aluminum nitride nanowire light emitting diodes: breaking the fundamental bottleneck of deep ultraviolet light sources. Sci. Rep. 5, 8332.
Zhao, S., Djavid, M., Mi, Z., 2015b. A surface emitting, high efficiency near-vacuum ultraviolet light source with aluminum nitride nanowires monolithically grown on silicon. Nano Lett. 15, 7006.
Zhao, S., Liu, X., Woo, S.Y., Kang, J., Botton, G.A., Mi, Z., 2015c. An electrically injected AlGaN nanowire laser operating in the ultraviolet-C band. Appl. Phys. Lett. 107, 043101.
Zhao, S., Nguyen, H.P.T., Kibria, M.G., Mi, Z., 2015d. III-nitride nanowire optoelectronics. Prog. Quant. Electron. 44, 14.
Zhao, S., Woo, S.Y., Bugnet, M., Liu, X., Kang, J., Botton, G.A., Mi, Z., 2015e. Three-dimensional quantum confinement of charge carriers in self-organized AlGaN nanowires: a viable route to electrically injected deep ultraviolet lasers. Nano Lett. 15, 7801.
Zhou, L., Epler, J.E., Krames, M.R., Goetz, W., Gherasimova, M., Ren, Z., Han, J., Kneissl, M., Johnson, N.M., 2006. Vertical injection thin-film AlGaN/AlGaN multiple-quantum-well deep ultraviolet light-emitting diodes. Appl. Phys. Lett. 89, 241113.

PART II

InGaN Nanostructures: Epitaxy, Properties, and Emerging Device Applications

CHAPTER SIX

Growth and Structural Characterization of Self-Nucleated III-Nitride Nanowires

T. Auzelle*,†,1, B. Daudin*,†,2
*Université Grenoble Alpes, INAC-PHELIQS, Grenoble, France
†CEA, INAC-PHELIQS, «Nanophysique et semiconducteurs group», Grenoble, France
[2]Corresponding author: e-mail address: bruno.daudin@cea.fr

Contents

1. INTRODUCTION

For almost two decades and the pioneering work of Kishino's and Calleja's groups (Calleja et al., 1999; Sanchez-Garcia et al., 1998; Yoshizawa et al., 1997, 1998), GaN nanowires (NWs) and related NW heterostructures are a subject of sustained interest related to their excellent optical and structural properties. Compared to other material families for which the vapor–liquid–solid method is commonly used to grow NWs, the case of III-nitride is unique: GaN NWs are most often grown by catalyst-free

[1] Present address: Paul-Drude-Institut für Festkörperelektronik, Hausvogteiplatz 5-7, 10117 Berlin, Germany.

Semiconductors and Semimetals, Volume 96
ISSN 0080-8784
http://dx.doi.org/10.1016/bs.semsem.2016.08.004

© 2017 Elsevier Inc.
All rights reserved.

plasma-assisted molecular beam epitaxy (PA-MBE), following a self-nucleated process on a variety of substrates ranging from Si (111) (Bertness et al., 2006, 2007, 2008; Meijers et al., 2006; Sanchez-Garcia et al., 1998; Yoshizawa et al., 1997) to Si (100) (Bertness et al., 2008; Guo et al., 2010), sapphire (Cherns et al., 2008; Chèze et al., 2010; Foxon et al., 2009; Yoshizawa et al., 1998), SiO_2 (Park et al., 2015; Stoica et al., 2008), diamond (Schuster et al., 2012), and even metals (Calabrese et al., 2016; Golam Sarwar et al., 2015; Wölz et al., 2015), making them a versatile alternative to conventional bidimensional (2D) layers grown on sapphire, Si, or SiC. More recently, the use of patterned substrates has been increasingly popular in order to overcome the intrinsic difficulties associated with spontaneous nucleation, namely density and morphology fluctuations of pure GaN NWs as well as composition variability of NW heterostructures. By contrast, selective area growth (SAG) of III-nitride NWs on patterned substrates has been shown to result in large collection of morphologically homogeneous, in-plane ordered, nanoobjects. Also in this case most of the literature refers to MBE grown material although it is worse stressing that the use of patterned substrates (including self-patterned ones, Chen et al., 2011) has alternatively made possible the use of metal-organic chemical vapor deposition as a growth technique for arrays of nitride nano/microwires (Hersee et al., 2006). In the case of PAMBE SAG of NW heterostructures, it has been demonstrated that chemical composition could be controlled by both mask opening size and pitch, paving the way to the realization, for instance, of light-emitting diodes (LEDs) with a spatially modulated wavelength emission (Sekiguchi et al., 2010). These pattern-related properties are tightly bound to the adatom diffusion on the base plane between the NWs and along their vertical walls toward the top as well as to shadow effects between adjacent NWs, a feature common to both SAG and self-nucleated NWs. The properties of selective area grown III-nitride NWs are reviewed in chapter "Selective area growth of InGaN/GaN nanocolumnar heterostructures by plasma assisted molecular beam epitaxy" by Albert et al. of this book. The present chapter will be limited to the case of self-nucleated NWs, with the goal of reviewing the details of their nucleation and of their structural properties.

2. NUCLEATION AND POLARITY

GaN NWs synthesis can be divided in two steps: nucleation and steady growth. Beyond its apparent triviality, this dichotomy is a consequence of the different growth conditions corresponding to each step. The full understanding of these specific growth conditions is the prerequisite for the optimization of NW nucleation and, to a certain extent, of their polarity. More generally, the full understanding of the physical parameters governing the different growth stages of NWs from the nucleation step to the steady growth step is a necessary condition, for instance, to increase the overall growth speed by deliberately switching from a set of growth parameters to another. The steady growth step is reached when the radial growth of the NW is close to zero, which usually occurs when the NW diameter exceeds 30 nm. In this regime, most of the GaN is incorporated on the top facet of the NW, hence perpetuating its 1D growth. The full picture to explain this anisotropic growth has not yet been fully provided but it is now accepted that it results from a combination of thermodynamical and kinetical mechanisms:

i. The thermodynamical approach relies on the fact that in the usual growth temperature range (800–850°C or higher) GaN NWs exhibit a wurtzite structure meaning that the primitive cell is anisotropic. It follows that the different facets of the NWs, namely the side facets (*m* plane—(1-100)) and the top facet (*c* or −*c* plane—(0001) or (000-1)) have different atomic structures, hence different energies resulting in different growth speeds under thermodynamic equilibrium. Facet energy can be usually provided by ab initio calculations if one can generate a volume only delimited by this type of facet and periodic boundaries. But this is not the case for *c* or −*c* planes. Nevertheless, Li et al. (2015) have partially overcome this issue by calculating the relative energies between several typical facets of GaN as a function of N and Ga chemical potential, eventually providing a thermodynamic Wulff plot. It clearly emphasizes that in nitrogen-rich conditions, the equilibrium shape of GaN as dictated by its relative surface energies, resembles to a rod-like structure. However, the aspect ratio of this rod (about two) is far below the aspect ratio of a typical NW (usually amounting to around 20), indicating that other mechanisms are accounting for their growth.

ii. A requisite for the growth of NWs is the use of high substrate temperatures (above 800°C) for which the desorption of Ga atoms as well as the decomposition of GaN are significant. This results in a strong dependence of the NW steady growth on kinetic mechanisms. However, purely geometrical features are also strongly influencing the NW morphology. The most striking experimental example has been brought by

Galopin et al. (2011) who have shown that increasing the angle between the nitrogen atomic beam and the NW axis leads to a drastic change in the NW structure: from a rod-like shape to an inverted cone-like shape. This dependence over the incident angle of atomic beams has been described by the geometrical model of Foxon et al. (2009) and later refined by Hestroffer and Daudin (2013) in the case of AlN grown on GaN NWs. It is based on the fact that most of the atomic beams have an incident angle relative to the substrate normal, θ, smaller than 45 degree, meaning that the effective flux of atoms impinging on the NW top facet ($\Phi_{top\ facet} = \Phi_0 \cos\theta$) is larger than the one impinging on the side facets ($\Phi_{top\ facet} = \Phi_0 \sin\theta$). In addition, in the case of NW assemblies of high density, a significant fraction of the NW side facets could be shadowed by neighboring NWs, hence locally reducing the effective flux of impinging atoms. The fact that more material is deposited on the NW top facet compared to the side facet eventually fosters the 1D growth of the NWs. As more specifically concerns the influence of kinetics, it appears that once Ga and N atoms have impinged on one of the NW facet, they are only physisorbed and still free to diffuse on the facet before to chemically bond to other atoms or desorb. Again, due to the different atomic structure of the NW facets, there are different diffusion barriers for N and Ga atoms which can kinetically affect the growth (Zywietz et al., 1998). As concern N atoms, this point is crucial as for lower diffusion barriers the recombination probability of two N neighboring atoms into N_2 increases, resulting in a loss of active N atoms because the N_2 molecule is unlikely to dissociate again and can easily desorb from the surface. An attempt to calculate from ab initio, this diffusion barrier for N atoms residing on the side facets of NWs has been carried on by Lymperakis and Neugebauer (2009). Their calculations have shown that N atoms are actually unstable on this facet and systematically recombine into N_2. It means that in spite of the overall N-rich growth conditions, the side facets of the GaN NWs are locally Ga-rich due to the loss of N atoms. As a result, the supernumerary Ga atoms residing on the side facets are diffusing around, eventually ending up on the NW top facet where they can be incorporated as GaN. This behavior could be nicely evidenced while growing InN on top of GaN NWs as the supernumerary In atoms could be ex situ observed by SEM as they reorganize into small droplets (see Fig. 1). Interestingly, Fig.1B tends to point out toward an accumulation of In along the side edges, consistent with the theoretical predictions of Lymperakis and

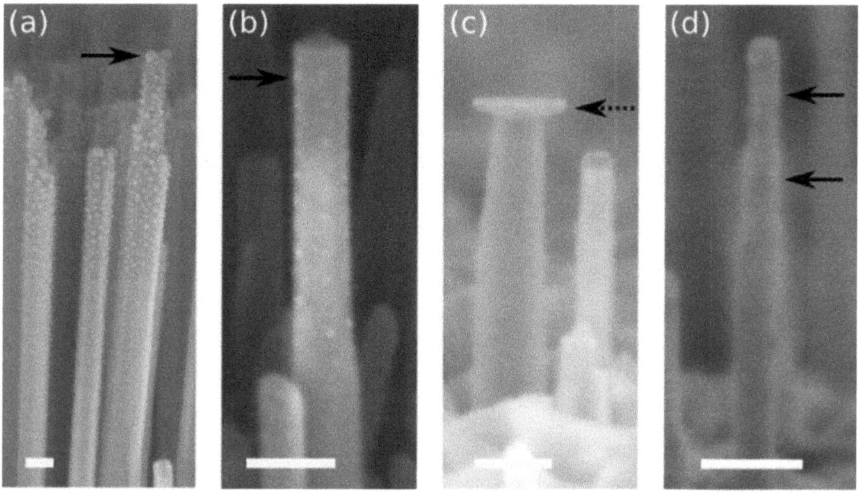

Fig. 1 Scanning electron microscopy (SEM) micrographs of different NW assemblies evidencing the ability of III atoms (In in this case) to diffuse along the GaN NW side walls. Scale bar: 50 nm. (A) Indium deposited at low temperature: In droplets are observed (one is pointed by the *arrow*) with a decreasing density toward the NW bottom due to a shadowing of the In flux by neighboring NWs; (B) indium deposited at higher temperature: In droplets are still visible but are now homogeneously spread along the NW because the thermal energy was sufficient to allow In to diffuse; (C) InN grown by exposing the In droplets to a N flux: the InN has grown exclusively on the NW top facet and is pointed by the *dotted arrow*; and (D) InN decomposed by thermal annealing: one can observe few In droplets (see *arrows*) homogeneously dispersed along the NW side walls.

Neugebauer (2009), who have shown that metal diffusion is preferentially occurring along (11-20) direction, with facet edge forming a diffusion barrier, leading as a whole to a "zigzag-like" progression of metal adatoms toward the NW top.

The important output of this review is the evidence that the formation of (1-100) side facets is a requirement to trigger the 1D growth of NWs. Therefore, in the following, the nucleation step of a GaN NW will be considered over once its initial nucleus has formed (1-100) side facets.

2.1 Growth of GaN NWs on Bare Silicon

Although GaN NWs can be grown on virtually any substrate including amorphous or metallic ones, Si (111) is the most often used. Because of the nonpolar character of Si, and because wurtzite structure of nitride materials is polar, the use of Si (111) then raises the issue of the GaN NW polarity, which will depend on the details of interface chemistry and could be a priori either N-polar or Ga-polar. Related to the variability of growth conditions and to the difficulty of determining the polarity of an ensemble of NWs, a large dispersion of results is found in the literature. As a matter of fact, both Ga polarity (Armitage and Tsubaki, 2010; Chèze et al., 2010; Furtmayr et al., 2008) and N-one (Hestroffer et al., 2011; Kong et al., 2011; Largeau et al., 2012) have been equally claimed. Ga polarity was also found by Brubaker et al. (2011). Interestingly, these authors have demonstrated that GaN NWs were actually growing on AlN columns exhibiting both N- and Al-polarity and the Ga polarity of GaN NWs growing on these columns was assigned in this case to a larger growth rate with respect to their N-polar counterparts. One key point of the papers of Brubaker et al. and Largeau et al. (2012) was to emphasize the role of AlN columns/pedestals as nucleation centers of GaN NWs, drawing attention to the Al/Si chemistry which will be discussed in details in Section 2.

2.2 Growth of GaN NWs on Silicon Using an AlN Buffer

The use of an AlN buffer to interface GaN NWs and Si probably came up first as a legacy from the field of thin film growth, where such AlN layer grown in between Si and GaN has been observed to drastically improve the crystallographic properties of the GaN. This comes from a 5/4 in-plane coincidence relationship between AlN and Si (111), which allows one to accommodate about 20% of the GaN/Si (111) in-plane lattice mismatch (Bourret et al., 1998). In the case of NWs, Largeau et al. (2008) and Songmuang et al. (2007) have shown that the presence of AlN noticeably increases the NWs relative orientation within an assembly, eventually limiting later coalescence events between NWs. In addition, thanks to the AlN, the incubation time prior to the nucleation of the NWs, i.e., the time associated with the probability of stabilizing a nucleus on the surface, is less sensitive to the substrate temperature compared to growth on bare Si. But, as shown in Fig. 2, the

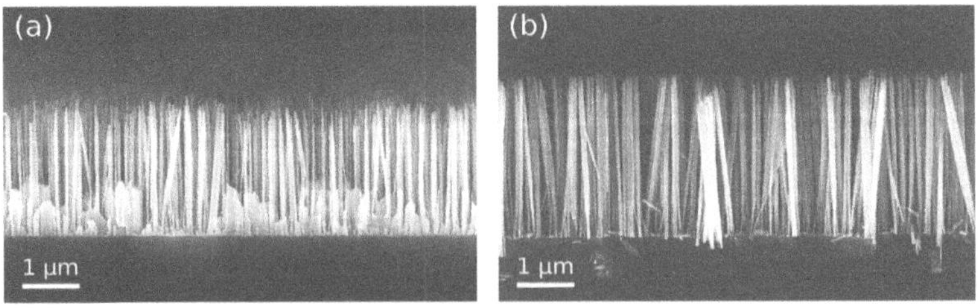

Fig. 2 SEM micrographs of GaN NWs assemblies grown on silicon with (A) an AlN buffer or (B) without. A parasitic layer at the bottom of the GaN NWs can be seen in (A).

main drawback and startling consequence of the use of an AlN buffer is, the appearance of a parasitic layer of GaN which is simultaneously growing along with the NWs. Indeed, it means that using similar growth conditions, two growth modes for GaN could be reached, either the NW one or the parasitic layer one.

This duality has not been fully accounted for in the first work dealing with the nucleation of NWs on AlN, carried on by Consonni et al. (2010). These authors have imaged by transmission electron microscopy (TEM) the morphology evolution of the GaN nucleus during the first minutes of the growth. They have found that GaN initially germs as a spherical capped nucleus, next matures into a truncated pyramid and later as a full pyramid in order to maximize the release of the epitaxial strain by elastic deformation. Next a significant shape transition is occurring, triggered by the plastic relaxation of the GaN nucleus on the AlN, eventually resulting in the formation of (1-100) side facets, which are assumed thermodynamically more stable than the previous {1-103} side facets of the pyramid. As a result, it is the plastic relaxation of the GaN which implies the formation of the necessary (1-100) facets for the NW growth. However, this morphological study is mostly supported by TEM images, which only provides a limited picture of the whole assembly of GaN nuclei, hence neglects the possible existence of different growth modes.

This model has been partly supported by the work of Landré et al. who have studied the first stages of GaN NW nucleation by in situ X-rays diffraction, using an MBE chamber coupled to a Synchrotron beamline (Landré et al., 2009). This study has put in evidence a sequence of three canonical steps, namely the incubation time, the nucleus ripening and, finally, the steady-state growth regime. In particular, it was demonstrated that the elastic strain relaxation of the nucleus was a necessary condition triggering the evolution of the nucleus into NWs, consistent with the results of Consonni et al. (2010).

Noticeably in these pioneering studies the AlN buffer has been assumed homogeneous, which, in the light of later works from Brubaker et al. (2011) and Largeau et al. (2012) should not be taken as granted. Indeed, the existence of so-called columnar protrusions or AlN pedestals having a density of about 10^9 cm^{-2} within the AlN buffer have been evidenced and even presumed as nucleation centers for NWs (which also have a density in the order of 10^9 cm^{-2}). Hence, in contrast to the nucleation model of Consonni et al. (2010), those two works suggest a strong implication of the AlN buffer structure in the NW nucleation process.

At last, the polarity of the AlN buffer has been suspected as well to play a role in the nucleation of GaN NWs. Hence, Fernández-Garrido et al. (2012) have tried to grow GaN NWs on N-polar and Al-polar AlN buffer obtained by using one or the other side of a SiC substrate. They have observed the growth of a regular GaN NW assembly on the N-polar AlN but the growth of only scarce NWs as well as a large parasitic layer on the Al-polar AlN. Hence, the authors have concluded that the N polarity of the GaN is a necessary condition for the growth of NWs, which is in contradiction with previous reports showing the existence of Ga-polar GaN NWs (Armitage and Tsubaki, 2010; Brubaker et al., 2011; Chèze et al., 2010; Furtmayr et al., 2008).

To sum up, those several works evidence the role of GaN relative facet energy as well as of the morphology and polarity of the AlN buffer in the self-organized nucleation of NWs, but without providing a complete picture able to explain all the experimental observations. In the following, a deep analysis of the AlN buffer growth is presented in order to clarify its subsequent role during NW nucleation.

2.2.1 AlN Buffer Growth on Si (111)

Exposition of a Si (111) surface to N quickly results in the formation of an amorphous SiN_x which is believed to hamper the epitaxial growth of material deposited on top. To overcome this issue, AlN growth on Si is usually initialized by first wetting the surface with a few monolayers of Al with the expectation to hamper impinging N atoms to bond with the buried Si atoms. This growth procedure is commonly used, although several reports have for long demonstrated that SiN_x is actually unstable in the presence of Al atoms (Hu et al., 2013; Le Louarn et al., 2009; Nikishin et al., 1999; Vézian et al., 2007). Indeed, upon exposure to pure Al, the SiN_x is reduced into an AlN having an excellent crystallographic quality. In addition, Nikishin et al. (1999) have also reported that the Al predeposition in not sufficient to fully hamper Si nitridation. Therefore, in the following, we will focus on two routes used for the synthesis of the AlN buffer: (1) one consists of predepositing first 12 ML_{AlN} of Al on the Si surface and will be called "Al-first" and (2) the other one consists of initially exposing the Si surface to N only in order to obtain SiN_x and it will be referred as "N-first."

2.2.1.1 Al-First AlN Buffer

Using a substrate temperature of 670°C, the deposited Al on Si is liquid and initially fully wets the surface, resulting in the formation of the γ-Al/Si phase (Michely et al., 1996; Saranin et al., 2002). However, once more than 0.68

ML_{Si} of Al have been deposited (1 ML_{Si} is the quantity of Al atoms necessary for completing the equivalent of one monolayer of Si on the Si (111) surface), the γ-Al/Si phase is fully formed and further deposited Al atoms are dewetting the surface, forming droplets (see Fig. 3). The consequences are twofold:

- The amount of Al atoms included in the γ-Al/Si phase is not sufficient to form a complete AlN ML but only corresponds to 0.55 ML_{AlN}. It means that the exposure of the γ-Al/Si phase to the N flux only will result both in the formation of AlN and of SiN_x.
- Si atoms from the substrate are diffusing into the formed Al droplets in order to satisfy the Al/Si phase diagram which features an eutectic at 577°C (Safarian et al., 2011). The concentration of Si atoms ranges from 12% at 577°C to up to 30% at 800°C. However, the maximum miscibility of Si into AlN is in the order of 12%, which means that once exposed to the N flux, a significant amount of Si will be released from the droplets and aggregates on the surface, forming SiN_x patches (dark area of Fig. 3D). In addition, as a result of the droplet geometry and/or of the surfactant behavior of Si, large AlN crystallites are grown at the previous location of the Al/Si droplet, which protrude above the average surface of the AlN film (see Fig. 3). Those objects have been referred as AlN pedestals and have been presumed as the nucleation centers for NWs (Largeau et al., 2012).

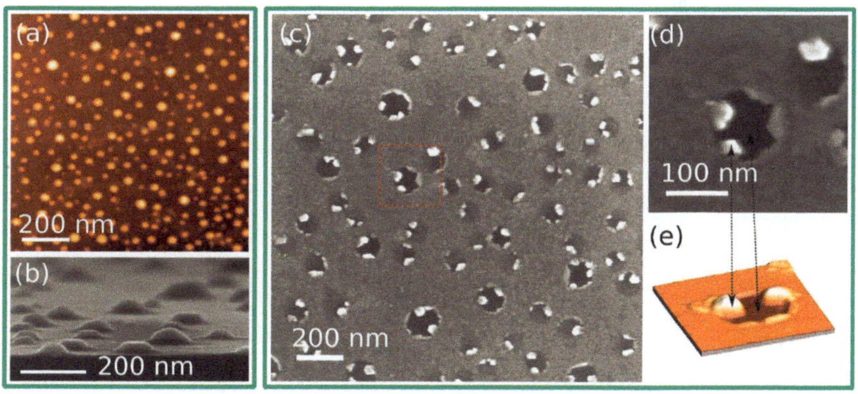

Fig. 3 (A and B) Atomic force microscopy (AFM) and SEM images of 12 ML_{AlN} of Al deposited on Si which have dewetted to form droplets; and (C) large-scale SEM image of an Al-first AlN buffer synthesized by first depositing 12 ML_{AlN} of Al. Singularities on the AlN surface are shown by SEM in (D) and AFM in (E). The protruding islands showing up with a white contrast in (C–E) are pedestals whereas the *dark areas* are Si released from the former Al/Si droplet.

As a conclusion, in spite of the predeposition of Al atoms on the Si (111) surface intended to minimize chemical reactions between Si and N atoms, it appears that the high reactivity between Al and Si leads to the formation of singularities in the Al-first AlN buffer. As GaN NWs are expecting to preferentially nucleate on singularities of the AlN buffer layer, the role of pedestals as nucleation centers cannot be disregarded at this stage.

2.2.1.2 N-First AlN Buffer

By exposing the Si surface to the N flux for a few minutes, a SiN_x amorphous layer is obtained with a self-regulated thickness of about 4 nm (Wierzbicka et al., 2013). Further deposition of Al on this layer has been found to result in the reduction of the SiN_x, i.e., in the formation of AlN out of the SiN_x (see the TEM image of the sample in Fig. 4E), and a release of Si atoms on the surface.

Indeed, an in situ reflected high energy electron diffraction (RHEED) monitoring of the surface (see Fig. 4A–C) reveals that the AlN has crystallized in one orientation (AlN (0001) [11-20] parallel to Si (111) [02-2]), whereas at the end of the SiN_x reduction, Si crystallites exhibiting two different orientations are observed. It suggests the existence of a $Si_{substrate}$/AlN/$Si_{crystallites}$ stacking sequence as cubic crystallites can occupy two different

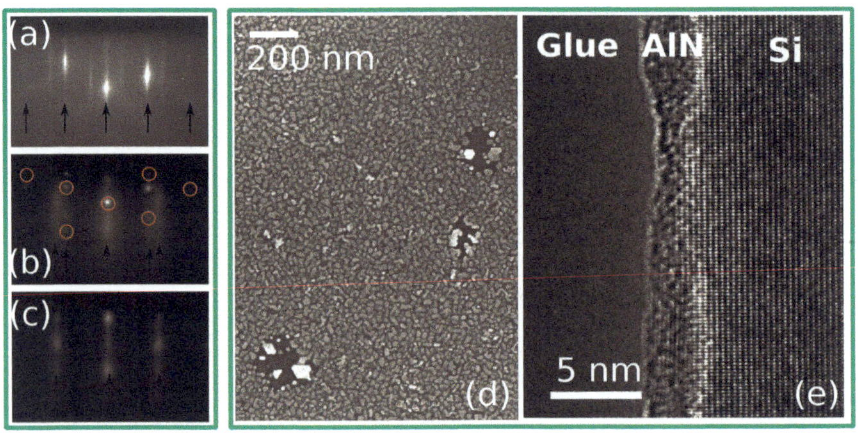

Fig. 4 (A–C) RHEED pattern of the surface during the N-first growth along the $[1\text{-}10]_{Si}=[11\text{-}20]_{AlN}$ zone axis: (A) initial SiN_x surface, (B) during the SiN_x reduction, and (C) final AlN. *Plain dark arrows* indicate the Si substrate diffraction rods, *dotted arrows* indicate AlN diffraction *rods* and *circles* are evidencing diffraction spots from one of the two Si crystallites orientation (see text); (D) SEM image of the N-first AlN buffer; and (E) transmission electron microscopy (TEM) image of the N-first AlN buffer.

orientations while sitting on an hexagonal lattice (i.e., the AlN) with $(111)_{\text{Si crystallites}} = (0001)_{\text{AlN}}$. Those Si crystallites could correspond to the dark features visible in the SEM top view of the AlN buffer (Fig. 4D), hence suggesting a density in the order of 10^9 cm^{-2}. It is this low density that makes unlikely their direct observation on a TEM cross-section as it is the case in Fig. 4E.

Therefore, similarly to the case of Al-first AlN buffers, Si is released on the surface during the buffer growth and can further affect its structure. For instance, during the exposition of the SiN_x to Al, part of the Al has dewetted the surface and organized into droplets where the released Si can eventually dissolve. Hence, the subsequent nitridation of the droplet will result in the formation of AlN pedestals, as visible in Fig. 4D.

Whether using the N-first or Al-first AlN buffer growth route, singularities related to the formation of Al/Si droplets are formed. Nevertheless, if using low growth temperature (670°C) for the Al-first AlN, one obtains 10^9 pedestals/cm^2, whereas if using high growth temperature (840°C) for the N-first AlN, one obtains only 10^8 pedestals/cm^2. Hence, the ability to use AlN buffers having different densities of pedestals opens a route to study their influence over the self-organized nucleation of NWs.

2.2.2 GaN NWs on AlN Buffer

N-rich conditions and high substrate temperature (>800°C) have been used to grow GaN NWs on the two different types of buffers previously introduced.

Benefiting from the large temperature gradients existing on the substrate (about 30°C from center to edge) (Mata et al., 2011), full GaN NW assemblies could be grown in the cooler areas of the substrate whereas no GaN could be nucleated in the hotter part of the substrate due to an incubation time exceeding the total growth duration (chosen short in this case, i.e., 30 min). This process allows one to ex situ picture the whole nucleation of GaN NWs, using only one substrate, by directing the focus from areas which were the hottest to the coolest during the growth, as shown in Figs. 5 and 6 for growth on the Al- and N-first AlN buffers, respectively. On the Al-first AlN (having 10^9 cm^{-2} pedestals), NWs were observed to nucleate exclusively on the pedestals (or their very close vicinity) but on the N-first AlN buffer (having 10^8 cm^{-2} pedestals) NWs were observe to nucleate both on the pedestals and in between. Hence, it appears that pedestals are not a requisite for NW nucleation. In addition, in the case of the Al-first AlN buffer, GaN crystallites nucleating in between the pedestals

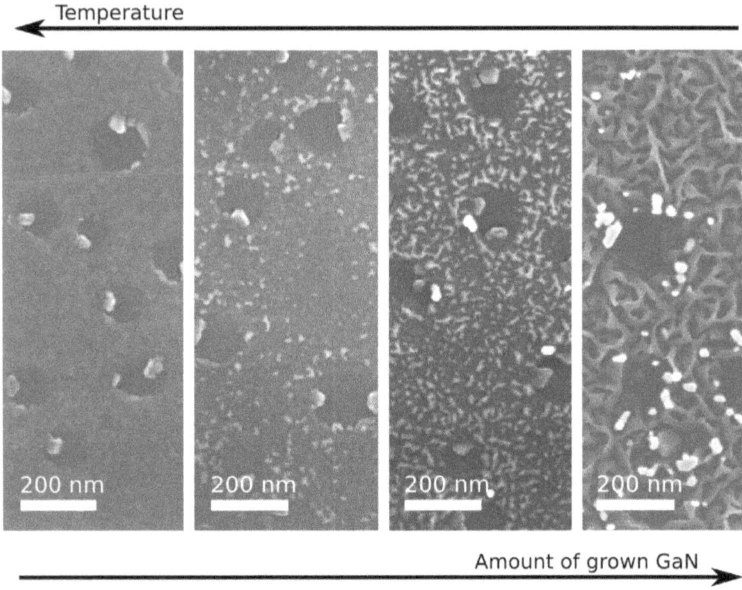

Fig. 5 SEM images and schematic of GaN NWs grown on an Al-first AlN buffer at different temperatures (i.e., at different stages through the nucleation process).

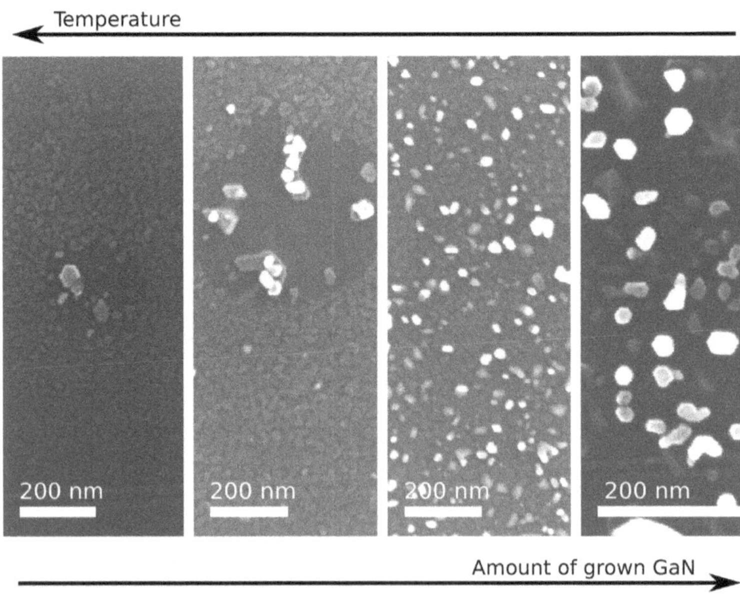

Fig. 6 SEM images and schematic of GaN NWs grown on an N-first AlN buffer at different temperature (i.e., at different stages through the nucleation process).

were all observed to feature a pyramidal morphology and were growing isotropically until coalescence with neighboring crystallites, eventually forming the typical parasitic layer observed in between NWs. If AlN pedestals themselves are not the driving force triggering the nucleation of NWs, one can wonder what differs between the GaN grown on the pedestals and in-between them in the case of the Al-first AlN buffer. To answer this question, high resolution scanning transmission electron microscopy (HR-STEM) images of GaN NW stumps (nucleated on a pedestal) and GaN pyramids (nucleated in between pedestals) have been acquired. Interestingly, they reveal that pyramids are always Ga-polar whereas NWs are always N-polar, possibly with a Ga-polar inversion domain isolated in their core (Auzelle et al., 2015a). This polarity asymmetry has been systematically observed and validated on larger scales noticeably by performing Kelvin probe force microscopy (KPFM) measurement on top of fully developed assemblies of NWs (Minj et al., 2015).

The requirement of the N polarity to trigger NW nucleation, which is also in agreement with the previous observations of Fernández-Garrido et al. (2012), can be understood as a result of the NW facet hierarchy. Indeed, as initially proposed by Consonni et al. (2010), once GaN crystallites have plastically relaxed on AlN, their shape is only determined by the facets energy. Hence, it could be deduced that Ga-polar crystallites are stable while featuring semipolar facets, eventually forming the observed pyramids, whereas N-polar crystallites are stable while featuring a combination of polar and nonpolar facets (i.e., (1-100) facets), the latter being the prerequisite for the unidimensional growth of NWs. This model, described in Fig. 7, is in agreement with the Wulff plot calculated by Li et al. (2015). At last, this nucleation model remains valid in the case of GaN crystallites hosting an

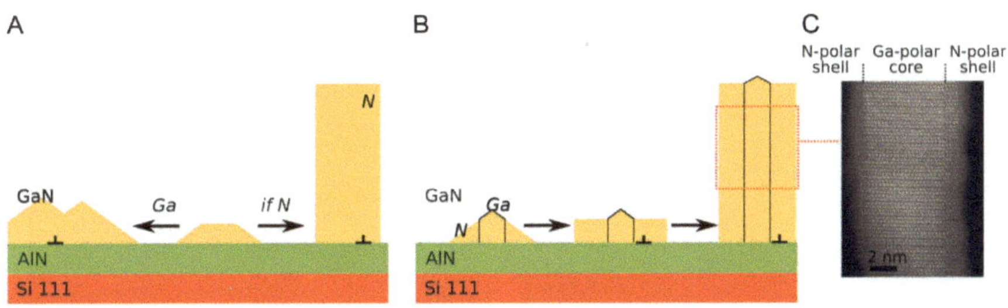

Fig. 7 (A and B) Schematic describing the nucleation of NWs, including NWs hosting an inversion domain boundary in their core—the half cross at the GaN/AlN interface represents the plastic relaxation of GaN crystallites and (C) high angular annular dark field STEM image of a NW hosting an inversion domain in its core.

inversion domain in their core as it only relies on the energy of the external facets.

N polarity as a necessary condition to trigger GaN NW nucleation means that AlN pedestals, which actually are nucleation centers for N-polar GaN, themselves exhibit a N polarity. More generally, such a conclusion obviously raises the question of the AlN buffer polarity. Several works tackling the polarity of AlN grown on Si have been published in the literature but all underline the importance of different criteria affecting the AlN polarity. For instance, in the case of Lebedev et al. (1999) the growth temperature is critical whereas in the case of Brubaker et al. (2016) the III/V ratio is of importance. Last but not least, Brubaker et al. have also reported the possible existence of basal inversion domains at the interface between the GaN and the AlN. Therefore, it seems legitimate to conclude that the polarity of the GaN grown on an AlN buffer is actually not controlled and most likely fluctuates from one AlN buffer to the other. This can explain the deviations between NW samples grown at different laboratories, especially as concern the ratio between NWs and the parasitic layer.

At last, polarity-selective etching of AlN thick layers by KOH has revealed the possible existence of very small Al-polar domains fully surrounded by a N-polar domain (see Fig. 8), which are then good candidates to host the type of NWs exhibiting an inversion domain in their core. If so, it means that the ratio of NWs hosting inversion domain is very dependent on the polarity pattern of the underneath AlN buffer and should then be a subject of large fluctuations between different samples. Based on this conclusion, one can assign the Ga-polar NWs reported by a few groups to actual N-polar NWs hosting a large Ga-polar inversion domain in their core.

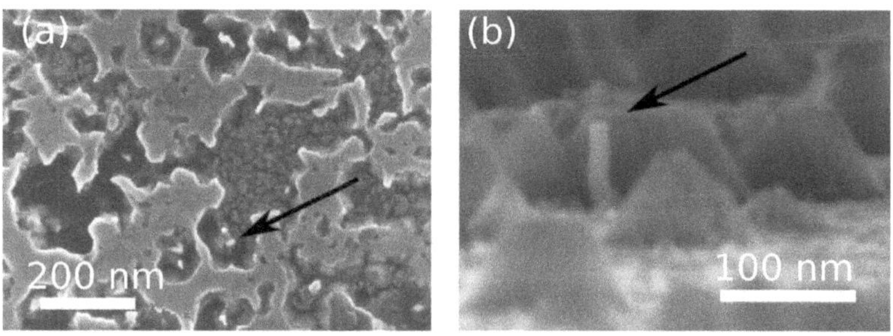

Fig. 8 (A and B), respectively, top view and side view of an AlN thick layer grown on Si whose N-polar domain have been selectively etched by KOH—the *two arrows* indicate Al-polar domains previously surrounded by N-polar AlN.

2.2.3 GaN NWs on Si (111) Without AlN Buffer

It has been previously reminded that the AlN buffer layer is convenient to reduce both tilt and twist of self-nucleated GaN NWs on Si (111). However, such a buffer layer is not strictly necessary, especially if targeting at applications for which the presence of an insulating AlN layer at the interface with Si would be detrimental. It has been shown that the growth of GaN NWs directly on Si (111) is associated with the formation of a thin amorphous Si_xN_y layer between GaN and Si and results in a large proportion of twisted and tilted NWs (Calleja et al., 1999; Furtmayr et al., 2008; Ristic et al., 2008; Stoica et al., 2008). More specifically, it has been further established that this amorphous Si_xN_y layer was a consequence of the exposure of bare Si (111) to N plasma, following the transitory formation of a crystalline β-Si_3N_4 layer (Hestroffer et al., 2012). In spite of the very different interface chemistry depending on the use or not of an AlN buffer, the GaN NWs directly grown on Si without AlN buffer layer were also found to be N-polar.

To conclude this section, it appears that GaN NW nucleation on AlN/Si (111) requires the N polarity, which then depends on the polarity of the AlN buffer itself. The study of the growth mechanism of the AlN buffers has revealed a strong interaction of the Al atoms with the substrate, which can lead to local release of Si atoms and likely disturb the polarity homogeneity of the layer. It is such phenomenon which is believed to be responsible for variability in the AlN buffer grown in the different laboratories, eventually explaining the different NW assembly structure and polarity reported in the literature. In absence of AlN buffer layers, the GaN NWs were also found to be N-polar, as a strong clue that N polarity is indeed a necessary condition to ensure the conversion of 3D precursors into NWs. Such a statement is supported by recent theoretical results on the relative surface energy of GaN bulk crystal facets, demonstrating that the high stability of {1-10-2} facets on the Ga-polar side is not favorable to the shape transition into NWs with vertical {1-100} side walls (Li et al., 2015).

3. FROM NUCLEATION TO STEADY-STATE GROWTH: THE ISSUE OF NUCLEI RIPENING

3.1 The Incubation Time

Three steps have been identified during the growth of self-nucleated NWs. Indeed, if measuring using an appropriate experimental technique the amount of GaN deposited after simultaneous opening of the Ga and N cells, an "incubation time" is observed, i.e., a delay before the formation of the first 3D GaN islands. Next, following the incubation time, the amount of GaN deposited as a function of time is found to obey a supralinear time dependence. Finally, this nucleation stage is followed by a steady-state regime for which the amount of GaN deposited is proportional to the deposition time. All three steps are schematically shown in Fig. 9.

The incubation stage and the transition to the nucleation stage have been studied in details by Consonni et al. (2011a) and Fernández-Garrido et al. (2015), who have shown that the length of the incubation time was strongly dependent on growth temperature and Ga flux. More precisely, as shown in Fig. 10, the incubation time dependence on Ga flux or on temperature is equivalent: for a given nominal Ga flux, the incubation time is longer and longer for higher and higher growth temperature. Similarly, for a given

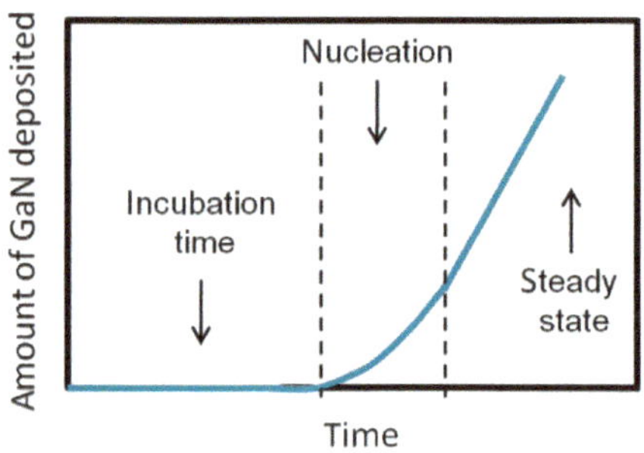

Fig. 9 Schematics of the three steps characterizing the growth of self-nucleated GaN nanowires.

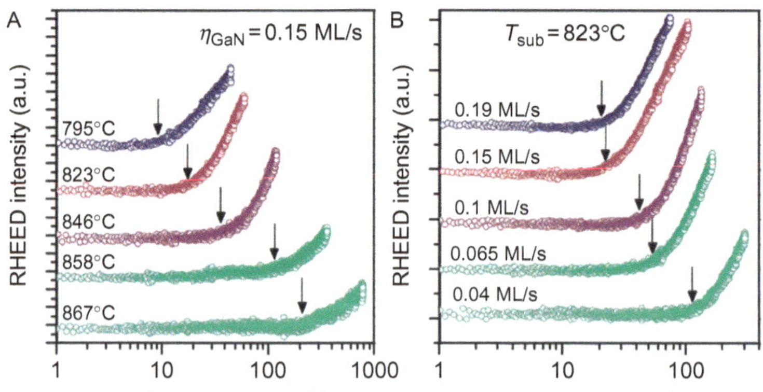

Fig. 10 Evolution in time of the GaN NW-related RHEED spots when providing Ga and N on an AlN buffer for different (A) substrate temperatures at constant Ga flux and (B) Ga flux at constant substrate temperature.

growth temperature, the incubation time is longer and longer for smaller and smaller Ga flux, establishing a direct relation between the duration of the incubation time and the Ga adatom density on the surface (Hestroffer, 2012). As shown by Consonni et al., such a dependence of the incubation time is consistent with standard island nucleation theory.

According to this theory, the incubation time is linked to the nucleation rate on the surface, i.e., to the probability to form stable nuclei above a critical size (Consonni et al., 2011a; Kashchiev et al., 1991). As a consequence, the incubation time is a very general feature and its duration is drastically dependent on the chemical nature of the surface on which GaN NWs are nucleating. For instance, when growing GaN on Si (111) without AlN buffer as described in Section 2.2.3, the formation of a SiN_x surface, because associated with a change in the Ga adatom diffusion length, is affecting the value of the incubation time, as shown in Fig. 11.

Interestingly, the density of self-nucleated NWs on Si is related to the incubation time through the variation of Ga diffusion length with growth temperature. The longer is the incubation time, i.e., the higher the growth temperature, the lower is the NW density (Mata et al., 2011). As a consequence, the density of self-nucleated GaN NWs can be controlled over several orders of magnitude by changing the growth temperature/incubation time, allowing one to easily tune it, depending on the targeted application (Zettler et al., 2015).

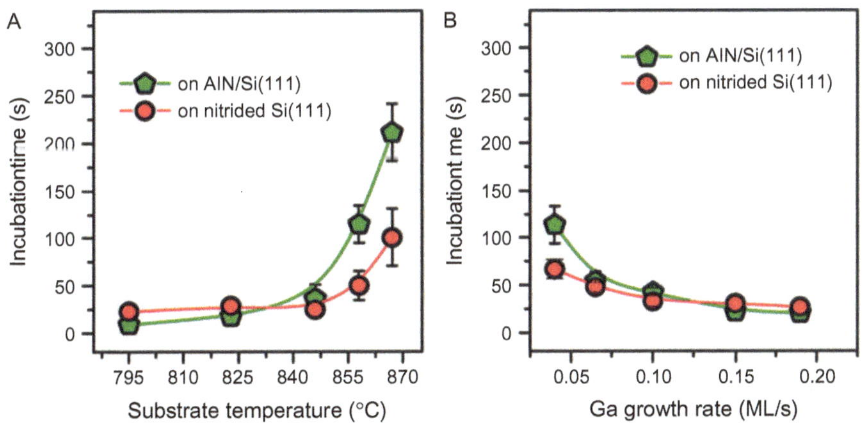

Fig. 11 (A) Incubation time duration as a function of substrate temperature for a Ga flux of 0.15 monolayer/s. (B) Incubation time duration as a function of Ga flux for a substrate temperature of 823°C. Note the difference between growth with and without AlN buffer layer.

3.2 The Nucleation Stage

Combined to growth temperature, the role played by chemistry at GaN/AlN/Si (111) and GaN/Si (111) interfaces has been extensively discussed in Sections 1 and 2 and shown to govern to a large extent the nucleation and the polarity of resulting NWs. However, independently of the details of interface chemistry, the next stage schematized in Fig. 9, namely the nucleation, has been found to exhibit common features. In particular, it has been demonstrated that the NW precursors starting to form at the end of the incubation time were elastically strained by the substrate below (Knelangen et al., 2010; Landré et al., 2009). During this stage, the supralinear dependence of the amount of GaN deposited has been assigned to a lateral growth of the nuclei, till reaching a critical size associated with plastic strain relaxation (Consonni et al., 2010, 2011b; Knelangen et al., 2010; Landré et al., 2009) and a concomitant shape transition into NWs with vertical walls, followed by a linear dependence of the amount of GaN deposited as a function of time.

In situ X-rays diffraction experiments in grazing incidence are reported in Fig. 12, showing the time evolution of the amount of GaN deposited (see Fig. 12A). It has to be noted that no significant evidence of GaN deposition is shown till about 18–20 min, corresponding to the incubation time in these experimental conditions (Fernández-Garrido et al., 2015; Landré et al., 2009). In addition, it is shown that the diffraction peak corresponding to GaN rapidly stabilizes around $h=3.62$, which is the value expected for relaxed GaN, pointing out toward a rapid strain relaxation of the precursors before the shape transition and further evolution into NWs with vertical walls. By contrast, as shown in Fig. 12B, the growth on GaN quantum dots (QDs) on AlN exhibits a drastically different sequence, with a GaN X-rays diffraction peak slowly evolving toward $h=3.62$, as an evidence of the compressive strain exerted by the AlN substrate (Coraux et al., 2006, 2007). Conversely, in the case of GaN QDs, the tensile strain exerted by GaN on AlN is put in evidence by the decrease in the intensity of the AlN peak, which turns to be asymmetric on the side of GaN. By contrast, in the case of NWs, neither asymmetry nor significant change in the intensity of the AlN peak are observed, putting in evidence the absence of mutual elastic interaction between AlN and GaN, assigned to the rapid plastic strain relaxation of GaN nuclei (Consonni et al., 2010, 2011b; Knelangen et al., 2010; Landré et al., 2009).

Finally, following the nucleation stage and the shape transition associated with strain relaxation of the precursors and formation of (1-101) facets (Consonni et al., 2010), a steady-state growth regime is observed, exhibiting a linear time dependence of the amount of GaN deposited.

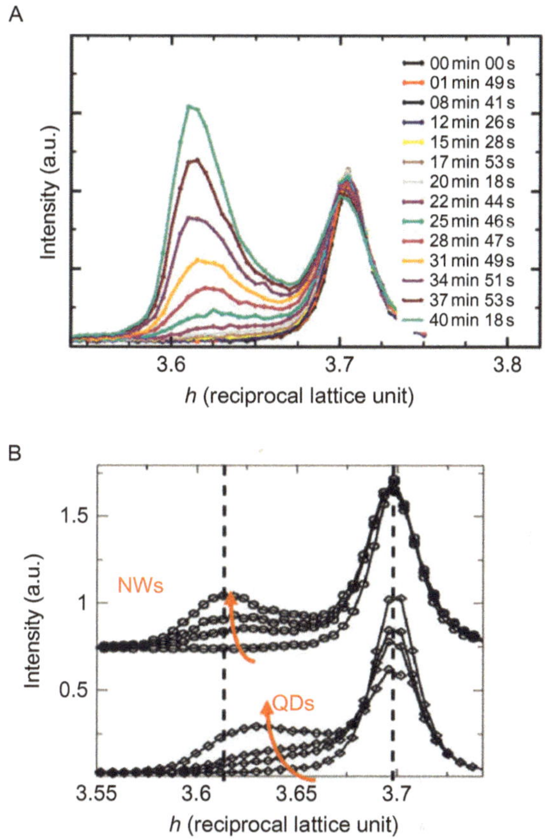

Fig. 12 (A) Evolution of the intensity of GaN diffraction peak at $h \sim 3.62$ as a function of time. Note the value of incubation time, between 18 and 20 min. The diffraction peak at $h = 3.7$ corresponds to the AlN buffer layer. (B) Comparison of the evolution of GaN diffraction peak position corresponding to the growth of GaN NWs on Si (111) and of GaN quantum dots (QDs) on AlN. Because of different growth conditions (Ga flux and temperature), successive scans for NWs correspond to deposition times of 0, 1500, 1850 and 2000 s. For QDs, successive scans correspond to deposition times of 0, 120, 160, and 200 s. In both cases, the total GaN deposited amounts to about 5–10 monolayers.

4. STRUCTURAL PROPERTIES OF GaN NWs

GaN NWs have been sometimes promoted as free of extended defects in order to emphasize their improved structural quality compared to thin films which usually include numerous dislocations. This difference would rely on the ability for NWs to release epitaxial strain through elastic deformation of free surfaces instead of undergoing a plastic relaxation, thanks to their high aspect ratio (Ertekin et al., 2005; Glas, 2006). Even in the case of NWs with a small aspect ratio, the dislocations formed at the interface between the NW and its substrate were observed to bend toward the closest side facet resulting in a dislocation free NW top part (Kishino and Ishizawa, 2015; Urban et al., 2014). Therefore, one can conclude that NWs are indeed free of extended defects such as dislocations in the general case. However, to fully account for the properties of GaN NWs, one cannot disregard the usual presence of two other types of extended defects: stacking faults and inversion domain boundaries (IDBs). At last, coalescence events between two neighboring NWs have also been incriminated to generate extended defects (Consonni et al., 2009).

4.1 Basal Stacking Faults

GaN NWs exhibit the usual wurtzite structure, except some small inclusions of zinc blende GaN that are usually forming during growth, although with a low density (down to 1 inclusion per micrometer). Because the latter can be described as an abnormal stacking order of the basal atomic planes compared to the regular wurtzite structure, they are referred as "basal stacking faults" (BSFs). The interested reader can refer to one of the only review covering the topic of BSFs done by Lähnemann et al. (2014). The most frequent BSFs encountered in NWs are of type I1 and consist of only one abnormal stacked plane. I1 BSFs are a radiative localization center for excitons (for temperatures below 50 K) resulting in the typical photoluminescence band centered at 3.42 eV. Their optical properties have been studied in a systematic manner by Corfdir et al. (2014), who have noticeably demonstrated that I1 BSFs behave as a perfect quantum well. Indeed, the intrinsic nature of the defect implies the absence of strain and homogeneous interfaces. In addition, the ability to probe single BSFs in single NWs has allowed to conduct a variety of original studies such as the determination of excitonic diffusion length in GaN NWs (Nogues et al., 2014) or the estimation of the spontaneous polarization of wurtzite GaN (Lähnemann et al., 2012).

4.2 Inversion Domain Boundaries

The discovery of IDBs in PA-MBE GaN NWs came as a side product from the systematic TEM investigation of GaN NW nucleation on AlN buffers (Auzelle et al., 2015a). The presence of a Ga-polar core in a few N-polar NWs implies the existence of IDBs which constitutes an extended defect.

Previous observation of such a defect in nitride films has fuelled various theoretical and experimental works, not always converging toward the same results. One can cite the work of Northrup et al. (1996), showing that two atomic configurations for IDBs can be envisaged, namely the holt IDB and the IDB*, the latter having a much lower formation energy although both types have been experimentally reported (Potin et al., 1998). However, the random localization of IDBs in GaN thin films complicates the preparation of TEM cross-section and hamper systematic characterizations. In the case of GaN NWs grown on AlN buffer, the characterization of IDBs can be eased as a simple mechanical dispersion (instead of a cross-section preparation) can be used to observe individual NWs by TEM, eventually allowing a statistical analysis. Hence, IDBs in NWs were reported to be of the IDB* type and their presence in a single NW have been correlated to a photoluminescence signal in the band centered at 3.45 eV (Auzelle et al., 2015b). This result is in slight contrast to previous calculations suggesting that IDBs were not optically active (Fiorentini, 2003).

Since the luminescence at 3.45 eV has been assigned to IDBs, it can be used to probe the presence of IDBs in NW assemblies grown on other types of substrates. Hence, GaN NWs grown on bare Si (111) should include inversion domains whereas GaN NWs grown on diamond (Schuster et al., 2012) or on Ti foils (Calabrese et al., 2016) substrates seem to be free of them.

4.3 NW Coalescence

As illustrated in Fig. 13, coalescence of thin NWs has been observed in the first stages of the growth of GaN NWs on bare Si (111), following the nucleation stage described in Fig. 9, eventually resulting in thicker NWs which are still exhibiting an hexagonal symmetry. This feature is not fully understood to date but likely results of the ability of closely spaced NWs to stick together and merge into a bigger one. Such a mechanism has been theoretically studied by Kaganer et al. (2016), who have shown that the evolution of NW density and morphology with time could be satisfactorily predicted if assuming that close NWs are continuously merging during the growth.

TEM results have shown that coalescence between neighboring NWs can result in the formation of a variety of defects. It includes BSFs, edge dislocations (Consonni et al., 2009) or even more complex defects such as "zipper array" of dislocations (Grossklaus et al., 2013). However, remarkably, this coalescence process is not affecting the crystalline quality and

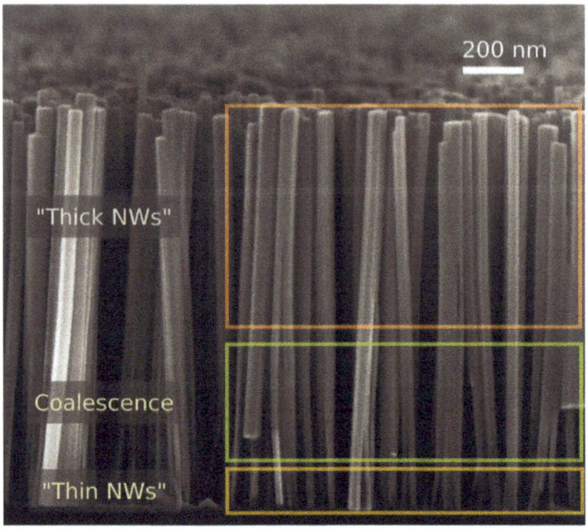

Fig. 13 SEM image of NWs showing the coalescence of thin NWs resulting on thicker NWs.

associated optical properties of the upper part of GaN NWs (Consonni et al., 2009). In addition, coalescence of enlarged NWs has been reported to generate a residual tensile strain amounting to 10^{-3} (Hugues et al., 2013).

5. CONCLUSION

As a matter of conclusion, it is worth pointing out that the ability of GaN NWs to self-nucleate in absence of any catalyst makes them peculiar if compared to NWs of other semiconductor families for which the use of a metallic catalyst is usually requested. This specificity of GaN NWs is related to the columnar growth mode identified for long as a general trend for GaN, with the consequent formation of a large number of threading edge dislocations for thin film growth. Along this line, NWs can be viewed as non-coalesced, vertically elongated grains free by nature of the high density of threading edge dislocations associated with grain coalescence. Furthermore, it has been shown above that the natural tendency to grow along the *c*-axis allows one to grow GaN NWs on a variety of crystalline and noncrystalline substrates. This feature, combined with the excellent structural and optical properties of GaN NWs opens the path to a wide range of applications. Although reviewing the issue of nitride NW heterostructures was out of the scope of this chapter, it is worth remarking that the easy elastic strain relaxation makes possible the realization of In-based or Al-based ternary alloys NW heterostructures in a wide range of chemical composition, potentially useful for light-emitting/detecting applications in the whole range of visible to ultraviolet. In particular, it has to be considered that n-type and p-type doping are potentially eased in NWs, because of the favored strain relaxation making the incorporation of doping impurities easier than in bulk layers. However, it has to be kept in mind that these potential advantages of NWs are partly counterbalanced by additional difficulties intrinsic to their high surface/volume ratio value. Indeed, the proximity of the surface, intrinsic to NWs, and the concomitant surface recombination of carriers have been found to affect both the electrical and optical properties. Surface passivation has to be considered to overcome these difficulties and optimize the physical properties of NWs. Despite this intrinsic complication if compared to bulk 2D heterostructures, NWs and NW heterostructures are expected to extend the range of nitride semiconductor applications, as discussed in details in other chapters of this book. Finally, the easy dispersion of NWs on auxiliary substrates makes them objects of particular interest for the basic study of single NW or NW heterostructures or for the practical realization of single devices such as nanoLEDs or single NW sensors, a field of interest in significant expansion, which is partly related to the improved control of NW nucleation and physical properties reviewed in the present chapter.

REFERENCES

Armitage, R., Tsubaki, K., 2010. Multicolour luminescence from InGaN quantum wells grown over GaN nanowire arrays by molecular-beam epitaxy. Nanotechnology 21, 195202.

Auzelle, T., Haas, B., Minj, A., Bougerol, C., Rouvière, J.-L., Cros, A., Colchero, J., Daudin, B., 2015a. The influence of AlN buffer over the polarity and the nucleation of self-organized GaN nanowires. J. Appl. Phys. 117 (24), 245303.

Auzelle, T., Haas, B., Den Hertog, M., Rouvière, J.-L., Daudin, B., Gayral, B., 2015b. Attribution of the 3.45 eV GaN nanowires luminescence to inversion domain boundaries. Appl. Phys. Lett. 107 (5), 51904.

Bertness, K.A., Roshko, A., Sanford, N.A., Barker, J.M., Davydov, A.V., 2006. Spontaneously grown GaN and AlGaN nanowires. J. Cryst. Growth 287, 522.

Bertness, K.A., Roshko, A., Mansfield, L.M., Harvey, T.E., Sanford, N.A., 2007. Nucleation conditions for catalyst-free GaN nanowires. J. Cryst. Growth 300, 94.

Bertness, K.A., Roshko, A., Mansfield, L.M., Harvey, T.E., Sanford, N.A., 2008. Mechanism for spontaneous growth of GaN nanowires with molecular beam epitaxy. J. Cryst. Growth 310, 3154.

Bourret, A., Barski, A., Rouvière, J.L., Renaud, G., Barbier, A., 1998. Growth of aluminum nitride on (111) silicon: microstructure and interface structure. J. Appl. Phys. 83 (4), 2003.

Brubaker, M.D., Levin, I., Davydov, A.V., Rourke, D.M., Sanford, N.A., Bright, V.M., Bertness, K.A., 2011. Effect of AlN buffer layer properties on the morphology and polarity of GaN nanowires grown by molecular beam epitaxy. J. Appl. Phys. 110, 053506.

Brubaker, M., Duff, S., Harvey, T., Blanchard, P., Roshko, A., Sanders, A., Sanford, N., Bertness, K., 2016. Polarity-controlled GaN/AlN nucleation layers for selective-area growth of GaN nanowire arrays on Si(111) substrates by molecular beam epitaxy. Cryst. Growth Des. 16 (2), 596.

Calabrese, G., Corfdir, P., Gao, G., Pfuller, C., Trampert, A., Brandt, O., Geelhaar, L., Fernandez-Garrido, S., 2016. Molecular beam epitaxy of single crystalline GaN nanowires on a flexible Ti foil. Appl. Phys. Lett. 108, 202101.

Calleja, E., Sánchez-García, M.A., Sánchez, F.J., Calle, F., Naranjo, F.B., Muñoz, E., Molina, S.I., Sánchez, A.M., Pacheco, F.J., García, R., 1999. Growth of III-nitrides on Si(111) by molecular beam epitaxy: doping, optical, and electrical properties. J. Cryst. Growth 201/202, 296.

Chen, X.J., Gayral, B., Sam-Giao, D., Bougerol, C., Durand, C., Eymery, J., 2011. Catalyst-free growth of high-optical quality GaN nanowires by metal-organic vapor phase epitaxy. Appl. Phys. Lett. 99 (25), 251910.

Cherns, D., Meshi, L., Griffiths, I., Khongphetsak, S., Novikov, S.V., Farley, N., Campion, R.P., Foxon, C.T., 2008. Defect reduction in GaN/(0001)sapphire films grown by molecular beam epitaxy using nanocolumn intermediate layers. Appl. Phys. Lett. 92, 121902.

Chèze, C., Geelhaar, L., Brandt, O., Weber, W.M., Riechert, H., Münch, S., Rothemund, R., Reitzenstein, S., Forchel, A., Kehagias, T., Komninou, P., Dimitrakopulos, G.P., Karakostas, T., 2010. Direct comparison of catalyst-free and catalyst-induced GaN nanowires. Nano Res. 3, 528.

Consonni, V., Knelangen, M., Jahn, U., Trampert, A., Geelhaar, L., Riechert, H., 2009. Effects of nanowire coalescence on their structural and optical properties on a local scale. Appl. Phys. Lett. 95, 241910.

Consonni, V., Knelangen, M., Geelhaar, L., Trampert, A., Riechert, H., 2010. Nucleation mechanisms of epitaxial GaN nanowires: origin of their self-induced formation and initial radius. Phys. Rev. B 81 (8), 085310.

Consonni, V., Trampert, A., Geelhaar, L., Riechert, H., 2011a. Physical origin of the incubation time of self-induced GaN nanowires. Appl. Phys. Lett. 99, 033102.

Consonni, V., Hanke, M., Knelangen, M., Geelhaar, L., Trampert, A., Riechert, H., 2011b. Nucleation mechanisms of self-induced GaN nanowires grown on an amorphous interlayer. Phys. Rev. B **83**, 035310.

Coraux, J., Amstatt, B., Budagoski, J.A., Bellet-Amalric, E., Rouvière, J.-L., Favre-Nicolin, V., Proietti, M.G., Renevier, H., Daudin, B., 2006. Mechanism of GaN quantum dots capped with AlN: an AFM, electron microscopy, and X-ray anomalous diffraction study. Phys. Rev. B 74, 195302.

Coraux, J., Favre-Nicolin, V., Proietti, M.G., Daudin, B., Renevier, H., 2007. Grazing incidence diffraction anomalous fine structure: application to the structural investigation of III-nitride quantum dots. Phys. Rev. B 75, 235312.

Corfdir, P., Hauswald, C., Zettler, J.K., Flissikowski, T., Lähnemann, J., Fernández-Garrido, S., Geelhaar, L., Grahn, H.T., Brandt, O., 2014. Stacking faults as quantum wells in nanowires: density of states, oscillator strength, and radiative efficiency. Phys. Rev. B 90 (19), 195309.

Ertekin, E., Greaney, P.A., Chrzan, D.C., Sands, T.D., 2005. Equilibrium limits of coherency in strained nanowire heterostructures. J. Appl. Phys. 97 (11), 114325.

Fernández-Garrido, S., Kong, X., Gotschke, T., Calarco, R., Geelhaar, L., Trampert, A., Brandt, O., 2012. Spontaneous nucleation and growth of GaN nanowires: the fundamental role of crystal polarity. Nano Lett. 12 (12), 6119–6125.

Fernández-Garrido, S., Zettler, J.K., Geelhaar, L., Brandt, O., 2015. Monitoring the formation of nanowires by line-of-sight quadrupole mass spectrometry: a comprehensive description of the temporal evolution of GaN nanowire ensembles. Nano Lett. 15 (3), 1930–1937.

Fiorentini, V., 2003. Origin of the efficient light emission from inversion domain boundaries in GaN. Appl. Phys. Lett. 82 (8), 1182.

Foxon, C.T., Novikov, S.V., Hall, J.L., Campion, R.P., Cherns, D., Griffiths, I., Khongphetsak, S., 2009. A complementary geometric model for the growth of GaN nanocolumns prepared by plasma-assisted molecular beam epitaxy. J. Cryst. Growth 311 (13), 3423–3427.

Furtmayr, F., Vielemeyer, M., Stutzmann, M., Arbiol, J., Estradé, S., Peirò, F., Morante, J.R., Eickhoff, M., 2008. Nucleation and growth of GaN nanorods on Si (111) surfaces by plasma-assisted molecular beam epitaxy—the influence of Si- and Mg-doping. J. Appl. Phys. 104, 034309.

Galopin, E., Largeau, L., Patriarche, G., Travers, L., Glas, F., Harmand, J.C., 2011. Morphology of self-catalyzed GaN nanowires and chronology of their formation by molecular beam epitaxy. Nanotechnology 22 (24), 245606.

Glas, F., 2006. Critical dimensions for the plastic relaxation of strained axial heterostructures in free-standing nanowires. Phys. Rev. B 74 (12). http://journals.aps.org/prb/abstract/10.1103/PhysRevB.74.121302.

Golam Sarwar, A.T.M., Carnevale, S.D., Yang, F., Kent, T.F., Jamison, J.J., McComb, D.W., Myers, R.C., 2015. Semiconductor nanowire light emitting diodes grown on metal: a direction towards large scale fabrication of nanowire devices. Small 11, 5402.

Grossklaus, K.A., Banerjee, A., Jahangir, S., Bhattacharya, P., Millunchick, J.M., 2013. Misorientation defects in coalesced self-catalyzed GaN nanowires. J. Cryst. Growth 371, 142–147.

Guo, W., Zhang, M., Banerjee, A., Bhattacharya, P., 2010. Catalyst-free InGaN/GaN nanowire light emitting diodes grown on (001) silicon by molecular beam epitaxy. Nano Lett. 10, 3355.

Hersee, S.D., Sun, X., Wang, X., 2006. The controlled growth of GaN nanowires. Nano Lett. 6, 1808.

Hestroffer, K., 2012. Growth and Characterization of GaN Nanowires and GaN/AlN Nanowire Heterostructures, PhD thesis, University of Grenoble.

Hestroffer, K., Daudin, B., 2013. A geometrical model for the description of the AlN shell morphology in GaN-AlN core-shell nanowires. J. Appl. Phys. 114 (24), 244305.

Hestroffer, K., Leclere, C., Bougerol, C., Renevier, H., Daudin, B., 2011. Polarity of GaN nanowires grown by plasma-assisted molecular beam epitaxy on Si(111). Phys. Rev. B 84, 245302.

Hestroffer, K., Leclere, C., Cantelli, V., Bougerol, C., Renevier, H., Daudin, B., 2012. *In situ* study of self-assembled GaN nanowires nucleation on Si(111) by plasma-assisted molecular beam epitaxy. Appl. Phys. Lett. 100, 212107.

Hu, J., Hao, Z., Niu, L., Yanxiong, E., Wang, L., Luo, Y., 2013. Atomically smooth and homogeneously N-polar AlN film grown on silicon by alumination of Si3N4. Appl. Phys. Lett. 102 (14), 141913.

Hugues, M., Shields, P.A., Sacconi, F., Mexis, M., Auf der Maur, M., Cooke, M., Dineen, M., Di Carlo, A., Allsopp, D.W.E., Zúñiga-Pérez, J., 2013. Strain evolution in GaN nanowires: from free-surface objects to coalesced templates. J. Appl. Phys. 114 (8), 84307.

Kaganer, V., Fernández-Garrido, S., Dogan, P., Sabelfeld, K., Brandt, O., 2016. Nucleation, growth, and bundling of GaN nanowires in molecular beam epitaxy: disentangling the origin of nanowire coalescence. Nano Lett. 16 (6), 3717–3725.

Kashchiev, D., Verdoes, D., Van Rosmalen, G.M., 1991. Induction time and metastability limit in new phase formation. J. Cryst. Growth 110, 373.

Kishino, K., Ishizawa, S., 2015. Selective-area growth of GaN nanocolumns on Si(111) substrates for application to nanocolumn emitters with systematic analysis of dislocation filtering effect of nanocolumns. Nanotechnology 26 (22), 225602.

Knelangen, M., Consonni, V., Trampert, A., Riechert, H., 2010. In situ analysis of strain relaxation during catalyst-free nucleation and growth of GaN nanowires. Nanotechnology 21 (24), 245705.

Kong, X., Ristić, J., Sanchez-Garcia, M.A., Calleja, E., Trampert, A., 2011. Polarity determination by electron energy-loss spectroscopy: application to ultra-small III-nitride semiconductor nanocolumns. Nanotechnology 22, 415701.

Lähnemann, J., Brandt, O., Jahn, U., Pfüller, C., Roder, C., Dogan, P., Grosse, F., Belabbes, A., Bechstedt, F., Trampert, A., Geelhaar, L., 2012. Direct experimental determination of the spontaneous polarization of GaN. Phys. Rev. B 86 (8), 081302(R).

Lähnemann, J., Jahn, U., Brandt, O., Flissikowski, T., Dogan, P., Grahn, H.T., 2014. Luminescence associated with stacking faults in GaN. J. Phys. D. Appl. Phys. 47 (42), 423001.

Landré, O., Bougerol, C., Renevier, H., Daudin, B., 2009. Nucleation mechanism of GaN nanowires grown on (111) Si by molecular beam epitaxy. Nanotechnology 20, 415602.

Largeau, L., Dheeraj, D.L., Tchernycheva, M., Cirlin, G.E., Harmand, J.C., 2008. Facet and in-plane crystallographic orientations of GaN nanowires grown on Si(111). Nanotechnology 19 (15), 155704.

Largeau, L., Galopin, E., Gogneau, N., Travers, L., Glas, F., Harmand, J.-C., 2012. N-polar GaN nanowires seeded by Al droplets on Si(111). Cryst. Growth Des. 12, 2724.

Le Louarn, A., Vézian, S., Semond, F., Massies, J., 2009. AlN buffer layer growth for GaN epitaxy on (111) Si: Al or N first? J. Cryst. Growth 311 (12), 3278–3284.

Lebedev, V., Schröter, B., Kipshidze, G., Richter, W., 1999. The polarity of AlN films grown on Si(111). J. Cryst. Growth 207 (4), 266–272.

Li, H., Geelhaar, L., Riechert, H., Draxl, C., Li, H., Geelhaar, L., Riechert, H., Draxl, C., 2015. Computing equilibrium shapes of wurtzite crystals: the example of GaN. Phys. Rev. Lett. 115 (8), 085503.

Lymperakis, L., Neugebauer, J., 2009. Large anisotropic adatom kinetics on nonpolar GaN surfaces: consequences for surface morphologies and nanowire growth. Phys. Rev. B 79 (24), 241308(R).

Mata, R., Hestroffer, K., Budagosky, J., Cros, A., Bougerol, C., Renevier, H., Daudin, B., 2011. Nucleation of GaN nanowires grown by plasma-assisted molecular beam epitaxy: the effect of temperature. J. Cryst. Growth 334 (1), 177–180.

Meijers, R., Richter, T., Calarco, R., Stoica, T., Bochem, H.P., Marso, M., Lüth, H., 2006. GaN-nanowhiskers: MBE-growth conditions and optical properties. J. Cryst. Growth 289, 381.

Michely, T., Reuter, M.C., Tromp, R.M., 1996. Al on Si (111): phase diagram and atomic mechanisms. Phys. Rev. B 53 (7), 4105.

Minj, A., Cros, A., Garro, N., Colchero, J., Auzelle, T., Daudin, B., 2015. Assessment of polarity in GaN self-assembled nanowires by electrical force microscopy. Nano Lett. 15 (10), 6770–6776.

Nikishin, S.A., Antipov, V.G., Francoeur, S., Faleev, N.N., Seryogin, G.A., Elyukhin, V.A., Temkin, H., Prokofyeva, T.I., Holtz, M., Konkar, A., Zollner, S., 1999. High-quality AlN grown on Si(111) by gas-source molecular-beam epitaxy with ammonia. Appl. Phys. Lett. 75 (4), 484.

Nogues, G., Auzelle, T., Den Hertog, M., Gayral, B., Daudin, B., 2014. Cathodoluminescence of stacking fault bound excitons for local probing of the exciton diffusion length in single GaN nanowires. Appl. Phys. Lett. 104 (10), 102102.

Northrup, J., Neugebauer, J., Romano, L., 1996. Inversion domain and stacking mismatch boundaries in GaN. Phys. Rev. Lett. 77 (1), 103.

Park, Y., Jahangir, S., Park, Y., Bhattacharya, P., Heo, J., 2015. InGaN/GaN nanowires grown on SiO_2 and light emitting diodes with low turn on voltages. Opt. Express 23, A650.

Potin, V., Nouet, G., Ruterana, P., 1998. Evidence for multiple atomic structure for the {10$\bar{1}$0} inversion domain boundaries in GaN layers. Appl. Phys. Lett. 74 (7), 947.
Ristic, J., Calleja, E., Fernancez-Garrido, S., Cerutti, L., Trampert, A., Jahn, U., Ploog, K.H., 2008. On the mechanisms of spontaneous growth of III-nitride nanocolumns by plasma-assisted molecular beam epitaxy. J. Cryst. Growth 310, 4035.
Safarian, J., Kolbeinsen, L., Tangstad, M., 2011. Liquidus of silicon binary systems. Metall. Mater. Trans. B 42 (4), 852–874.
Sanchez-Garcia, M.A., Calleja, E., Monroy, E., Sanchez, F.J., Calle, F., Muñoz, E., Beresford, S., 1998. The effect of the III/V ratio and substrate temperature on the morphology and properties of GaN- and AlN-layers grown by molecular beam epitaxy on Si(111). J. Cryst. Growth 183, 23.
Saranin, A., Kotlyar, V., Zotov, A., Kasyanova, T., Cherevik, M., Lifshits, V., 2002. Structure of domain walls in Al/Si gamma-phase. Surf. Sci. 517 (1–3), 151.
Schuster, F., Furtmayr, F., Zamani, R., Magén, C., Morante, J.R., Arbiol, J., Garrido, J.A., Stutzmann, M., 2012. Self-assembled GaN nanowires on diamond. Nano Lett. 12 (5), 2199–2204.
Sekiguchi, H., Kishino, K., Kikuchi, A., 2010. Emission color control from blue to red with nanocolumn diameter of InGaN/GaN nanocolumn arrays grown on same substrate. Appl. Phys. Lett. 96 (23), 231104.
Songmuang, R., Landré, O., Daudin, B., 2007. From nucleation to growth of catalyst-free GaN nanowires on thin AlN buffer layer. Appl. Phys. Lett. 91 (25), 251902.
Stoica, T., Sutter, E., Meijers, R.J., Debnath, R.K., Calarco, R., Luth, H., Grutzmacher, D., 2008. Interface and wetting layer effect on the catalyst-free nucleation and growth of GaN nanowires. Small 4, 751.
Urban, A., Malindretos, J., Klein-Wiele, J.-H., Simon, P., Rizzi, A., 2014. Corrigendum: Ga-polar GaN nanocolumn arrays with semipolar faceted tips (2013 *New J. Phys.* **15** 053045). New J. Phys. 16 (1), 19501.
Vézian, S., Le Louarn, A., Massies, J., 2007. Selective epitaxial growth of AlN and GaN nanostructures on Si(111) by using NH3 as nitrogen source. J. Cryst. Growth 303 (2), 419–426.
Wierzbicka, A., Zytkiewicz, Z.R., Kret, S., Borysiuk, J., Dluzewski, P., Sobanska, M., Klosek, K., Reszka, A., Tchutchulashvili, G., Cabaj, A., Lusakowska, E., 2013. Influence of substrate nitridation temperature on epitaxial alignment of GaN nanowires to Si(111) substrate. Nanotechnology 24 (3), 35703.
Wölz, M., Hauswald, C., Flissikowski, T., Gotschke, T., Fernandez-Garrido, S., Brandt, O., Grahn, H.T., Geelhaar, L., Riechert, H., 2015. Epitaxial growth of GaN nanowires with high structural perfection on a metallic TiN film. Nano Lett. 15, 3743–3747.
Yoshizawa, M., Kikuchi, A., Mori, M., Fujita, N., Kishino, K., 1997. Growth of self-organized GaN nanostructures on Al_2O_3(0001) by RF-radical source molecular beam epitaxy. Jpn. J. Appl. Phys. 36, L459.
Yoshizawa, M., Kikuchi, A., Fujita, N., Kushi, K., Sasamoto, H., Kishino, K., 1998. Self-organization of GaN/$Al_{0.18}Ga_{0.82}N$ multi-layer nano-columns on (0001) Al_2O_3 by RF molecular beam epitaxy for fabricating GaN quantum disks. J. Cryst. Growth 189/190, 138.
Zettler, J.K., Hauswald, C., Corfdir, P., Musolino, M., Geelhaar, L., Riechert, H., Brandt, O., Fernández-Garrido, S., 2015. High-temperature growth of GaN nanowires by molecular beam epitaxy: toward the material quality of bulk GaN. Cryst. Growth Des. 15, 4104.
Zywietz, T., Neugebauer, J., Scheffler, M., 1998. Adatom diffusion at GaN (0001) and (000-1) surfaces. Appl. Phys. Lett. 73 (4), 487.

CHAPTER SEVEN

Selective Area Growth of InGaN/GaN Nanocolumnar Heterostructures by Plasma-Assisted Molecular Beam Epitaxy

S. Albert[1], A.M. Bengoechea-Encabo, M.Á. Sánchez-García, E. Calleja
ETSI Telecomunicación, Universidad Politécnica de Madrid, Madrid, Spain
[1]Corresponding address: e-mail address: salbert@isom.upm.es

Contents

1. INTRODUCTION

It was shown several years ago that dislocation- and strain-free group-III nitrides, in particular GaN, can be grown on Si(100) and (111), as well as on Al_2O_3(0001) layers as quasi one-dimensional structures in a self-assembled (SA) fashion by molecular beam epitaxy (MBE) (Cerutti et al., 2006; Sanchez-Garcia et al., 1998; Yoshizawa et al., 1997). These structures are commonly referred to as nanowires, nanocolumns (NCs), or nanorods. Here the term NCs will be used.

SA GaN NCs grown by MBE are structures with a hexagonal cross-section, defined by *m*-plane side facets, which grow axially along the

Semiconductors and Semimetals, Volume 96
ISSN 0080-8784
http://dx.doi.org/10.1016/bs.semsem.2016.08.003
© 2017 Elsevier Inc.
All rights reserved.

c-direction. The first reports by Yoshizawa et al. (1997) date back to 1997, closely followed by Sanchez-Garcia et al. (1998). The term "self-assembled" (SA) refers to the formation of GaN NCs without the use of a foreign catalyst material which proceeds usually under N-rich conditions at growth temperatures beyond the onset of the GaN decomposition one. SA GaN NCs grown on Si(111) have excellent structural and optical properties compared to thin films (Calleja et al., 2000; Meijers et al., 2006; Park et al., 2004; Thillosen et al., 2006) due to the absence of extended defects in the NCs that are also intrinsically free of strain because of their small footprint and large surface-to-volume ratio. The small footprint allows NCs to "disconnect" from the substrate after a few monolayers by generating dislocations at the interface (Trampert et al., 2003) while the remaining crystal volume is free of defects.

In general, when growing nanocolumnar InGaN/GaN heterostructures a higher amount of strain, as compared to 2D layers can be accommodated, due to strain relief by lateral relaxation, before dislocations are formed (Ertekin et al., 2005). On the other hand, the use of SA InGaN/GaN NCs for optoelectronic devices such as light-emitting diodes (LEDs) has serious drawbacks related to unintentional polychromatic emission deriving from an inhomogenous axial and radial distribution of In%, geometry dispersion, and the inherent tendency of InGaN to develop composition fluctuations (Kishino et al., 2007; Lin et al., 2010). In addition, efficient and reliable LEDs based on SA NCs are hindered by limitations due to a strong electrical dispersion (Bavencove et al., 2011). In order to overcome these limitations, selective area growth (SAG) of GaN on polar GaN/sapphire substrates has been developed (Bengoechea-Encabo et al., 2011; Kishino et al., 2009; Sekiguchi et al., 2008) (Fig. 1). SAG of GaN NCs has to achieve simultaneously: (i) selectivity and (ii) the formation of a single NC that follows the mask dimension in each nanohole. In order to reach selectivity, the Ga atoms impinging on the mask must either desorb from it or diffuse to the nearest nanohole, that is, avoiding nucleation on the mask. Typical mask materials used are SiN_x (Bertness et al., 2010), SiO_2 (Calleja et al., 2007), Ti (Sekiguchi et al., 2008), or Mo (Urban et al., 2013).

SAG allows for the growth of nanostructures with well-controlled position and diameter, resulting in geometrical arrays with very little morphology dispersion, which are anticipated to be appealing building blocks for InGaN-based optoelectronic devices. In addition, SAG has been demonstrated on nonpolar (Albert et al., 2014a; Bengoechea-Encabo et al., 2012) and semipolar (Bengoechea-Encabo et, al., 2013) orientations of

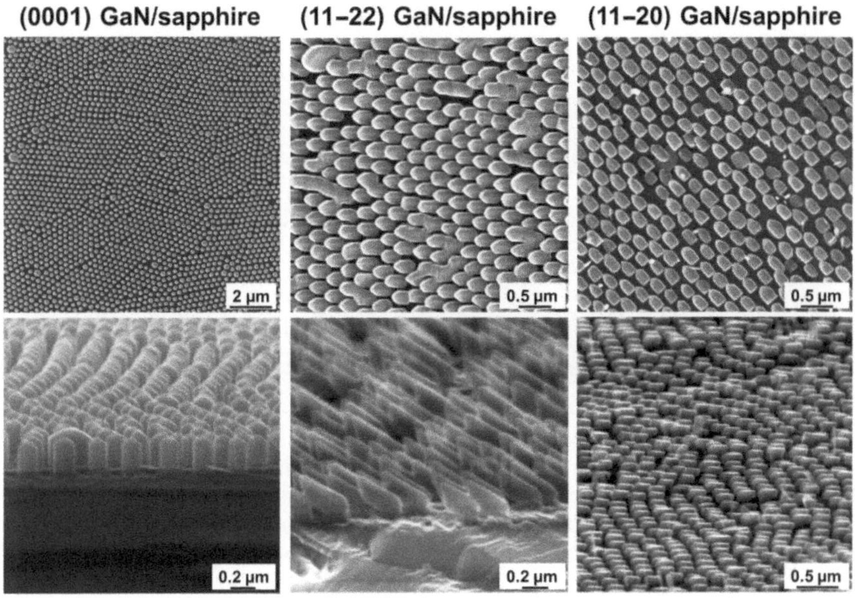

Fig. 1 Top view and cross-sectional SEM pictures of SAG GaN NCs grown on polar, semipolar, and nonpolar GaN/sapphire templates. Colloidal lithography was used as masking technique.

GaN/sapphire (Fig. 1) in order to reduce or even avoid internal polarization fields which affect detrimentally the internal quantum efficiency.

2. SAG OF InGaN/GaN NCs ON GaN/SAPPHIRE TEMPLATES

The most commonly used approach for SAG of InGaN-based devices such as LEDs is the use of InGaN quantum disks (Qdisks) (Albert et al., 2012b; Sekiguchi et al., 2010), which yield an increased oscillator strength due to a higher carrier confinement. In 2010, the emission color control from blue to red was demonstrated using SAG InGaN multiquantum disks (MQdisks) (Fig. 2A–F) by Sekiguchi et al. (2010).

Room temperature micro-PL spectra from these structures, as shown in Fig. 2G reveal a double peak emission where the low energy one monotonically red-shifts with increasing diameter from 137 to 270 nm, for a constant pitch of 400 nm. This emission wavelength dependence on diameter has been explained by different diffusion contributions of In- and Ga atoms to the growth depending on the NC diameter (Sekiguchi et al., 2010).

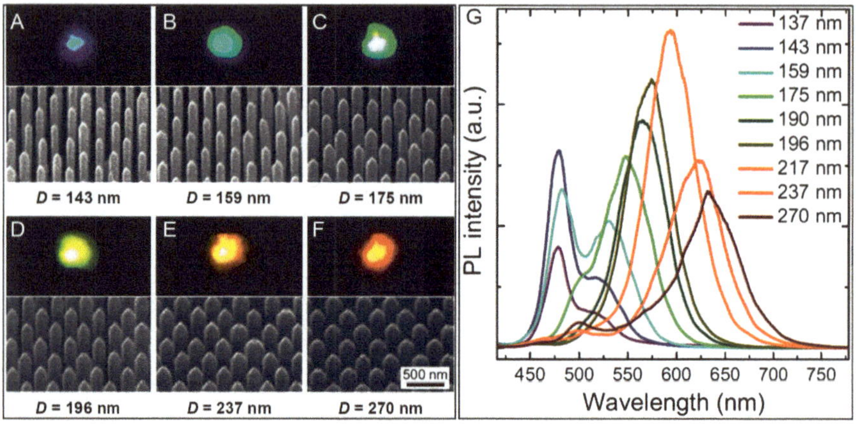

Fig. 2 (A)–(F) Bird's eye view SEM and emission images excited by He-Cd laser from InGaN/GaN NCs. (A) 143 nm, (B) 159 nm, (C) 175 nm, (D) 196 nm, (E) 237 nm, (F) 270 nm, and (G) PL spectra of InGaN/GaN NCs for patterns with different diameters from 137 to 270 nm. *Adapted with permission from Sekiguchi, H., Kishino, K., Kikuchi, A., 2010. Emission color control from blue to red with nanocolumn diameter of InGaN/GaN nanocolumn arrays grown on same substrate. Appl. Phys. Lett. 96, 231104.*

Later on the room-temperature operation of SAG InGaN-based LEDs emitting in the infrared (1.46 μm) under DC current injection was shown by Kishino et al. (2012). In this case, a triangular-lattice array Ti-mask was employed to grow Si-doped n-type GaN NCs on GaN/sapphire, including MQWs with p-type Ga-rich InGaN (300 nm thick) cladding and In-rich InGaN (80 nm thick) active layers. In 2013, the monolithic integration of InGaN-based three primary colors (RGB) LEDs was demonstrated by Kishino et al. (2013), where SAG was used for the growth of four distinct Qdisk-based NC LEDs on a MOCVD GaN/sapphire template.

A potential drawback of QWs and Qdisks-based LED structures is that they reach their maximum external quantum efficiency (EQE) at relatively low current injection, below 10 A/cm^2 (Karpov, 2011). At higher injection currents, an efficiency "*droop*" is observed even under pulsed current conditions, where self-heating effects are avoided. This "*droop*" has been attributed to several mechanisms such as carrier overflow and leakage as well as Auger recombination (Kioupakis et al., 2011; Piprek, 2010; Vampola et al., 2009). The impact of both carrier overflow and Auger recombination is thought to be drastically reduced by increasing the volume of the active region of the device (lower carrier density). The use of low defect density axial nanocolumnar morphologies with long InGaN sections (much higher volume than Qdisks) was then envisaged. This approach using SAG nanocolumnar "pin" structures with long (250–500 nm) InGaN sections will be presented next.

2.1 Growth and Characterization of InGaN/GaN Nanocolumnar Heterostructures for LEDs Emitting in the Blue, Green, and Red Spectral Range

This section addresses the SAG of axial InGaN/GaN NCs with long InGaN sections on Ti-masked GaN/sapphire templates. These heterostructures were processed into blue, green, and yellow emitting LEDs.

Following that the phosphor-free white light emission at RT from a monolithic integration of red, green, and blue emitting InGaN long sections in each NC will be discussed.

A first series of InGaN/GaN NCs (samples B1–B4) was grown under specific conditions (temperature, III/V ratio, In/Ga ratio) in order to obtain emission in the blue, green, and red spectral range. Since the InGaN growth temperature (650°C) is significantly lower than that of GaN (900°C), the SAG of InGaN requires the previous growth of long enough GaN NCs (typically 500 nm for a pitch of 270 nm) to prevent the nucleation of InGaN on the mask (taking advantage of *shadowing* effect).

Colloidal lithography was used as masking technique (Albert et al., 2012a; Bengoechea-Encabo et al., 2012) allowing a faster and clean patterning of larger areas in a short time compared to e-beam lithography. The main drawback of colloidal lithography is a higher dispersion in hole diameters as well as pitch (Fig. 1). Fig. 3A shows the cross-sectional SEM images of SAG InGaN/GaN NCs emitting in the UV, blue, green, and red spectral range (Albert et al., 2013b).

The PL spectrum of sample B1 (Fig. 3B) shows the characteristic strain-free GaN D^0X line at 3.472 eV, and the absence of any InGaN-related peak. PL spectra from samples B2 and B3 reveal InGaN-related peaks red-shifting as the growth temperature is decreased (B1: 750°C, B2: 700°C, B3: 650°C), corresponding to an increase of In%. Sample B4 (grown at 650°C) shows a further increase of the In% (PL peak at 2.04 eV) due to specific conditions with a higher In/Ga ratio and increased active N. It is worth to mention that the full width at half maximum values of the InGaN PL peaks in Fig. 3B show no trend with In-content (123 meV for B2; 132 meV for B3; and 99 meV for B4) being all of them significantly lower than those corresponding to SA InGaN/GaN NCs reported before (Albert et al., 2011), indicating a more homogeneous in In-incorporation among NCs.

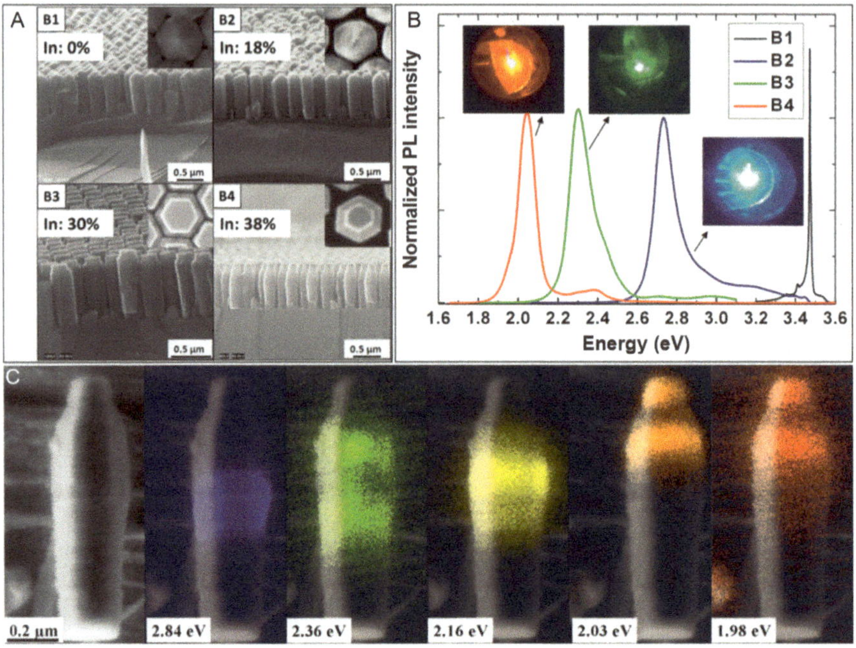

Fig. 3 (A) Cross-sectional SEM images of samples of series B with corresponding In-contents based on LT-PL measurements. *Insets* show top view SEM picture of the respective sample, (B) 12 K PL spectra of InGaN/GaN NCs of series B showing *blue*, *green*, and *red* emission (*insets*), and (C) spatially resolved CL measurements at 10 K of a single InGaN/GaN NC from sample B4. *Adapted with permission from Albert, S., Bengoechea-Encabo, A., Sanchez-Garcia, M.A., Calleja, E., Jahn, U., 2013b. Selective area growth and characterization of InGaN nanocolumns for phosphor-free white light emission. J. Appl. Phys. 113, 114306.*

Spatially resolved CL measurements performed on a single NC from sample B4 (Fig. 3C) show an In% increase toward the NC top (red-shift). A possible reason for this is the InGaN partial decomposition at the growth temperature (typically around 650°C) (Gallinat et al., 2007) with subsequent In-segregation toward the growth front yielding a higher In% at the NC top. In addition, a lower In-incorporation close to the InGaN/GaN interface may happen to minimize strain effects caused by the lattice mismatch (*lattice pulling*) (Kong et al., 2012, 2015). After a transition region with low In%, the onset of strain relaxation allows for a higher In-incorporation. The CL emission at 2.84 eV, seemingly originating from the InGaN/GaN interface region (Fig. 3C) can be attributed to the pulling effect.

Another explanation for the observed In% increase toward the top may relate to the change of the local III/V and In/Ga ratios due to morphology changes during the InGaN growth. In order to gain insight into this last mechanism, three new samples (series C) were grown on (0001) GaN/sapphire templates (Albert et al., 2015a), starting with a common 450 nm high GaN section with an average diameter of around 170 nm for all three samples. The InGaN sections were subsequently grown for 1 h on top of the GaN NCs using the same In- and Ga-flux at a growth temperature of 650°C. An In/Ga ratio of 1.96, with an In-flux of 6.3 nm/min (3.33×10^{14} atoms/(cm^2 s)) and a Ga-flux of 2.3 nm/min (1.7×10^{14} atoms/(cm^2 s)), was used for all samples. The only variable that was changed between the samples was the N-flux, i.e., C1: 14 nm/min (10.3×10^{14} atoms/(cm^2 s)), C2: 17 nm/min (12.5×10^{14} atoms/(cm^2 s)), and C3: 20 nm/min (14.7×10^{14} atoms/(cm^2 s)).

The GaN NCs grown first have a *c*-plane top facet, i.e., **no** pyramidal top in contrast to samples of series B. It has been shown in Albert et al. (2012a) that the NC top morphology (flat or pyramidal top) of SAG GaN NCs grown on GaN/sapphire can be controlled with the local (nanohole) Ga/N ratio at the NC nucleation stage. In all three samples, ordered InGaN/GaN NCs with a flat top facet, emitting in the green spectral range, were obtained (Fig. 4). Cross-sectional SEM pictures of the individual NCs show a diameter increase toward the NC top starting at (or slightly below) the nominal InGaN/GaN interface of about 40% and 20% for samples C1 and C2, respectively. Sample C3, on the other hand, shows a constant

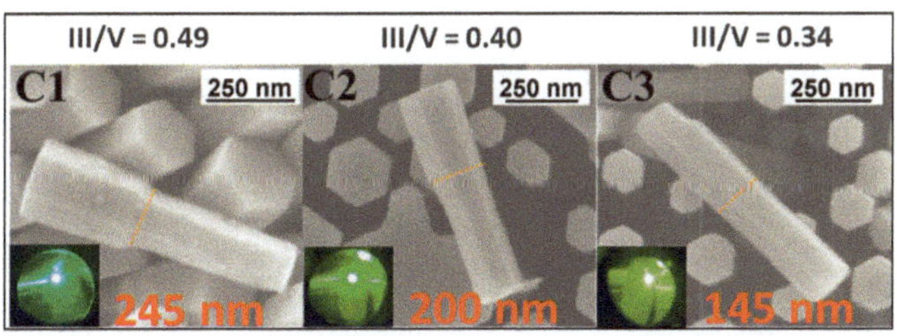

Fig. 4 Cross-sectional SEM pictures and corresponding III/V ratio during InGaN growth of individual NCs of series C with similar GaN section diameter (around 170 nm). The *numbers* in the respective pictures refer to the top diameter of the NCs. The *dotted lines* indicate the position of the nominal InGaN/GaN interface. *Insets* show PL emission. *Adapted with permission from Albert, S., Bengoechea-Encabo, A., Kong, X., Sanchez-Garcia, M.A., Trampert, A., Calleja, E., 2015a. Correlation among growth conditions, morphology, and optical properties of nanocolumnar InGaN/GaN heterostructures selectively grown by molecular beam epitaxy. Cryst. Growth Des. 15, 2661.*

diameter but for the topmost NC region where a sudden and asymmetric diameter reduction (20%) is observed. Notice that the comparison in Fig. 4 was done on NCs having the same starting GaN diameter of about 170 nm.

Fig. 5 shows the results of the structural characterization of samples C1 and C3 by TEM. The bright-field cross-sectional TEM image of sample C1 (Fig. 5A) reveals a high degree of uniformity among NCs all having a flat top. At about the InGaN/GaN interface an abrupt diameter increase is observed, concomitant with the presence of a stacking fault, detected by

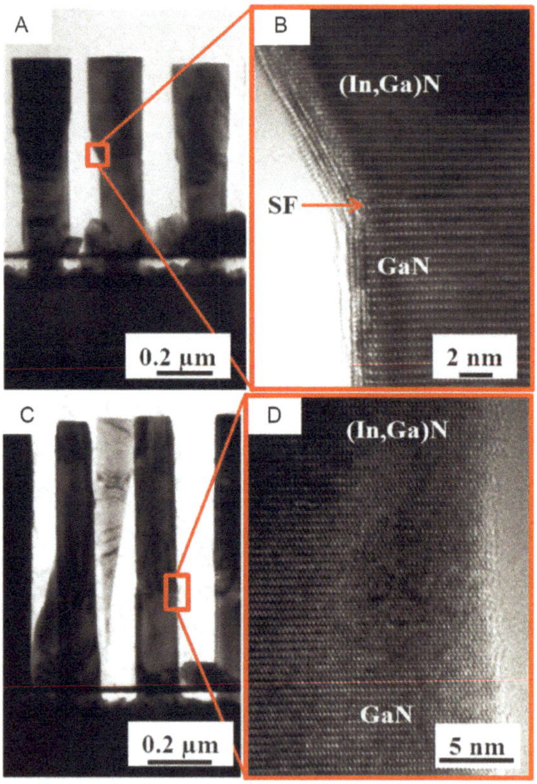

Fig. 5 Bright-field TEM picture of a NC array of (A) sample C1 and (C) sample C3. HRTEM pictures of the InGaN/GaN interface region in a single NC of (B) sample C1 and (D) sample C3. *Adapted with permission from Albert, S., Bengoechea-Encabo, A., Kong, X., Sanchez-Garcia, M.A., Trampert, A., Calleja, E., 2015a. Correlation among growth conditions, morphology, and optical properties of nanocolumnar InGaN/GaN heterostructures selectively grown by molecular beam epitaxy. Cryst. Growth Des. 15, 2661.*

high-resolution TEM (HRTEM) (Fig. 5B), that crosses the entire NC (no formation of partial dislocations) and which appears to be linked to the position of the surface step that leads to the diameter increase onset.

The bright-field cross-sectional TEM image of sample C3 (Fig. 5C) on the other hand reveals a uniform morphology across the InGaN/GaN interface region with no abrupt change in NC diameter. Only a slight decrease in NC diameter toward the top is observed.

The HRTEM images of samples C1 (Fig. 5B) and sample C3 (Fig. 5D) reveal a coherent interface between InGaN and GaN sections. In addition, selected area electron diffraction patterns (not shown) taken at the interface region show no in-plane diffraction spot splitting. Both results together with the absence of misfit dislocations indicate that the lattice mismatch at the InGaN/GaN interface is accommodated elastically within the strained region (transition) of the InGaN section.

Based on the observed dependence of the diameter change with the III/V ratio, the different behavior observed in samples C1 to C3 can be understood as a consequence of different "local" III/V ratios at the NC top. The "local" III/V ratio results from the addition of the directly impinging fluxes (top) and the diffusion fluxes along the NC sidewalls toward the top. It has been shown by Fernández-Garrido et al. (2013) that SA GaN NCs diameter is determined by the local Ga/N ratio at the growth front (NC top) where the system always moves toward stoichiometry during growth. A similar mechanism is found in the SAG InGaN/GaN NCs discussed here.

Since there is no significant active nitrogen diffusion flux along the NC sidewalls (Lymperakis and Neugebauer, 2009) due to the very fast desorption and N_2 formation, the N-flux to be considered is just the one that impinges directly on the NCs top. On the other hand, the total amount of metal reaching the NC top depends strongly on the diffusion contribution up the NC sidewalls and can be expressed as:

$$\Phi_{metal}^{total} = \Phi_{(In/Ga)}^{imp,\,top} + \Phi_{(In/Ga)}^{diff} + \Phi_{(In/Ga)}^{dec} - \Phi_{(In/Ga)}^{des} \tag{1}$$

where $\Phi_{(In/Ga)}^{dec}$ and $\Phi_{(In/Ga)}^{des}$ stand for metal fluxes arising from decomposition and desorption, respectively. Their relative importance depends very much on the growth conditions used. All relevant contributions are shown in Fig. 6.

While the direct top impinging metal fluxes ($\Phi_{In/Ga}^{imp,top}$) are independent of the NC diameter, the metal diffusion flux ($\Phi_{In/Ga}^{diff}$) arriving to the NC top

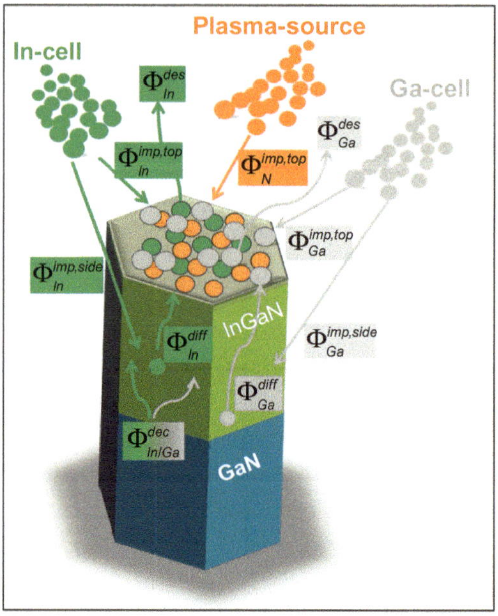

Fig. 6 Sketch summarizing the contributions from different processes to the overall metal flux toward the NC top.

does depend on the diameter as follows (from a simple geometrical argumentation):

$$\Phi_{(\mathrm{In/Ga})}^{diff} \propto \frac{\Phi_{(\mathrm{In/Ga})}^{imp,side} \times 2\pi r L_{(\mathrm{In/Ga})}^{diff}}{\pi r^2} = \Phi_{(\mathrm{In/Ga})}^{imp,side} \times \frac{2L_{(\mathrm{In/Ga})}^{diff}}{r} \tag{2}$$

with r being the NC radius and $L_{\mathrm{In/Ga}}^{diff}$ the diffusion lengths of Ga and In atoms along the *c*-direction on the NC sidewall. Based on Eq. (2), an increase (decrease) in NC diameter leads to a net decrease (increase) in the diffusion metal flux toward the top.

According to this, if for a given initial NC diameter there is no stoichiometry at the top growth front, say, by metal excess, the system reacts increasing the NC diameter to balance the metal/N ratio to reach unity (stoichiometry) at the top growth front. Once stoichiometry is reached, the NC diameter stabilizes at a fixed value.

At this point it is worth to distinguish between the two definitions of diffusion length in solids: (i) diffusion length for reevaporation and (ii) diffusion length for incorporation into the crystal. While the second one depends mostly on surface morphology and available active nitrogen, the first one depends exponentially on temperature and is smaller at temperatures that promote atoms reevaporation. When growing InGaN NCs at 650°C, In-desorption is efficient and the dominant In-diffusion length is the reevaporation one (shortest). Then, in Eq. (2) L_{Ga}^{diff} will be significantly higher than L_{In}^{diff}, meaning that the diffusion flux contribution to the overall metal amount at the NCs top is much higher for Ga (Sekiguchi et al., 2010). In other words, an "effective" *collection length* can be defined (from the NCs top as reference point) for the Ga and In atoms diffusing along the NCs sidewalls, being much shorter for In atoms at 650°C.

The diameter increase observed in the InGaN sections of samples C1 and C2 can now be understood as originating from the local III/V ratio at the NC top being too high to reach stoichiometry for the starting diameter defined by the GaN section underneath (although ratios of nominal impinging fluxes are <1), that is:

$$\frac{\Phi_{metal}^{total}}{\Phi_{\mathrm{N}}^{imp;top}} > 1 \tag{3}$$

at the beginning of the InGaN section growth. To compensate for the metal excess at the NC top, the NC diameter increases to decrease the net metal contribution by diffusion toward the top. The higher the metal excess, the higher the NCs diameter increase needed, as shown in Fig. 4.

On the other hand, a net nitrogen excess at the growth front should lead to a NC diameter decrease, following the same mechanism to reach stoichiometry. In that case the starting condition would be

$$\frac{\Phi_{metal}^{total}}{\Phi_{\mathrm{N}}^{imp;top}} < 1 \tag{4}$$

then causing a diameter decrease to enhance the diffusion metal flux contribution. Notice that in samples C1 to C3 the change in active nitrogen flux is responsible for the InGaN section diameter increase or decrease.

Following this model, the evolution of the InGaN section diameter should then depend on the fulfilled condition, defined either by Eqs. (3) or (4), until stoichiometry is reached. Fig. 7 shows this dependence for NCs of samples C1, C2, and C3 starting from different initial GaN diameters (because of diameter variations inherent to the colloidal lithography process). Results indicate that the diameter increase of the InGaN section in samples C1 and C2 is determined by the condition imposed by Eq. (3), i.e., local metal excess at the NC top. Furthermore, it can be seen that a

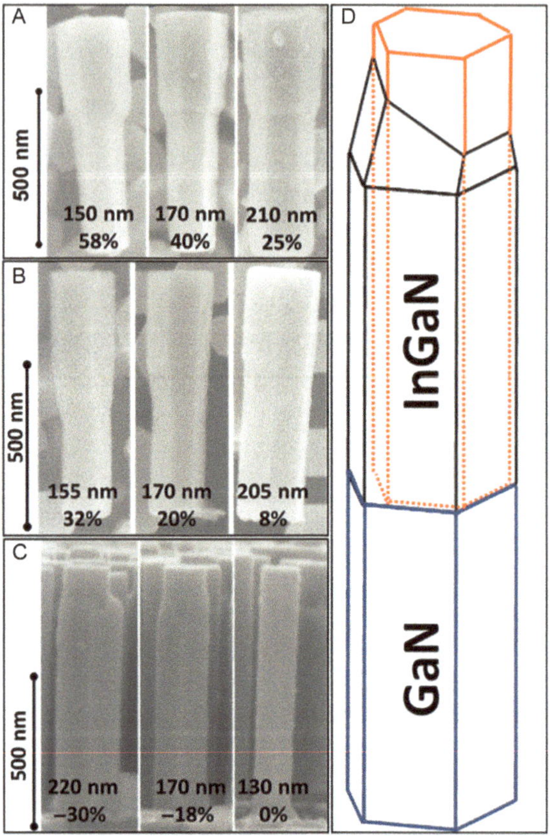

Fig. 7 Diameter variation of InGaN sections in NCs of samples (A) C1, (B) C2, and (C) C3 starting from different initial GaN diameters (as indicated). The InGaN sections diameter variations are indicated in *percent*; and (D) sketch showing the different contributions to the InGaN section growth in sample C3. *Adapted with permission from Albert, S., Bengoechea-Encabo, A., Kong, X., Sanchez-Garcia, M.A., Trampert, A., Calleja, E., 2015a. Correlation among growth conditions, morphology, and optical properties of nanocolumnar InGaN/GaN heterostructures selectively grown by molecular beam epitaxy. Cryst. Growth Des. 15, 2661.*

higher-metal excess (smaller GaN diameter) does lead to a greater relative InGaN diameter increase (58% to 40% to 25%).

The main difference in the morphology evolution of samples C1 and C2 is the GaN diameter at which the regime established by Eq. (3) starts but not the trend itself. In sample C1 (Fig. 7A), an initial diameter of around 210 nm still leads to a significant diameter change of 25% upon InGaN growth, whereas a similar diameter in sample C2 (Fig. 7B) is nearly enough to reach stoichiometry at the NC top (only small change of around 8%). This behavior difference is a consequence of the different nominal III/V ratios used, i.e., 0.49 in sample C1 and 0.4 in sample C2 (more active nitrogen).

InGaN/GaN NCs of sample C3 with a GaN starting diameter of around 170 nm and the highest amount of active nitrogen (III/V = 0.34) start growing under conditions given by Eq. (4) (nitrogen excess), thus coherent with the observed decrease of the InGaN section top diameter (18% smaller than the initial GaN one). The InGaN top diameter (around 140 nm) does not depend on the initial GaN diameter, whether 220, 170, or 130 nm because 140 nm is the diameter around which stoichiometry at the NCs top is established in all cases (sample C3). When the starting diameter is set to around 130 nm, stoichiometry is already fulfilled at the growth front, so no significant changes of the InGaN section diameter are expected.

Concerning the specific geometry of the InGaN NCs sections in sample C3, which have an irregular top but for the lowest initial diameter (130 nm), a possible explanation is illustrated in the sketch shown in Fig. 7D. For initial GaN diameters of 220 and 170 nm the InGaN section starts growing, directly at the InGaN/GaN interface, under the regime given by Eq. (4), thus forcing a diameter reduction to a value determined by the NC top stoichiometry condition. Then this smaller diameter will leave a portion of the initial GaN NC top surface exposed, where InGaN axial growth can also proceed at a different growth rate. The combination of the two processes would give rise to a complex structure in which two different diameters are observed, inner and outer. This model would also predict that, for a sufficiently small initial GaN diameter, the growth front at the NC top would be at stoichiometry, so that no diameter changes should be present. This is indeed observed for an initial GaN diameter of 130 nm.

Aside from the described effects on the NC morphology, changes in the local III/V and In/Ga ratios at the NC top should also affect the axial In distribution inside the NCs. The In distribution was checked on individual NCs from samples C1 and C3 by spatially resolved low-loss electron energy loss spectroscopy (EELS) scanning along the axial direction of the NCs with spot diameter of 10 nm (Fig. 8).

Bright-field TEM images of the selected NCs, including the positions of the EELS measurements marked by small dots, are shown in Fig. 8. A low In-content is observed close to the InGaN/GaN interface region in both samples, indicating the presence of a lattice pulled transition region. Beyond this region, the In-concentration tends to increase toward the top in the single NC of sample C1 (Fig. 8A), whereas it saturates and finally decreases for

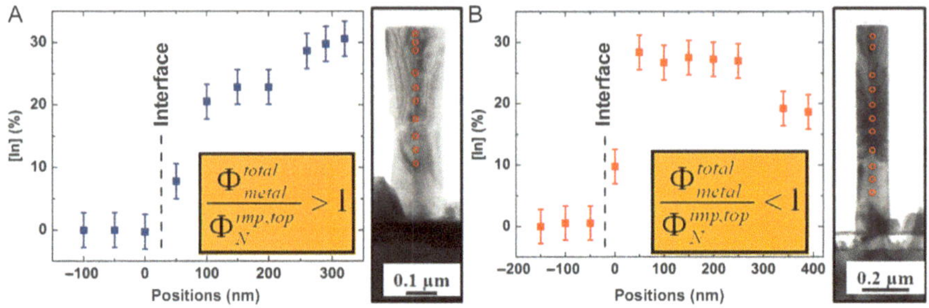

Fig. 8 Indium concentration profiles measured by low-loss EELS and bright-field TEM images of a single NC from: (A) sample C1 and (B) sample C3. *Circles* in bright-field TEM images indicate the points at which measurements were performed. *Adapted with permission from Albert, S., Bengoechea-Encabo, A., Kong, X., Sanchez-Garcia, M.A., Trampert, A., Calleja, E., 2015a. Correlation among growth conditions, morphology, and optical properties of nanocolumnar InGaN/GaN heterostructures selectively grown by molecular beam epitaxy. Cryst. Growth Des. 15, 2661.*

the NC of sample C3 (Fig. 8B). An increasing In-content toward the top was already shown by spatially resolved CL for red emitting NCs (Fig. 3C). The potential mechanisms invoked there, i.e., lattice pulling and thermal decomposition are present here as well, in addition to the effect of diameter changes.

However, the NC from sample C3 shows an opposite behavior beyond the transition region. A maximum In composition (28%) is reached rapidly and after saturation it diminishes close to the NC top. Notice that the NC selected from sample C3 has a kink (diameter decrease) close to the top and facing the front. Since all InGaN sections were grown at 650°C the same InGaN decomposition and In-segregation phenomena apply to all of them, so that, the observed variations in In-content must relate to differences in the local In/Ga ratio as a consequence of the morphology change.

For samples of series C, following Eq. (2), a NC diameter increase will reduce the Ga- and In-diffusion contribution along the sidewalls to the total metal amount at the top. Given the higher Ga diffusion flux contribution ($L_{In}^{diff} < L_{Ga}^{diff}$) to the overall metal amount at the NCs top, the overall effect is a net increase in the In/Ga ratio at the growth front as the NC diameter increases. If the NC diameter increase is gradual (see Fig. 8A) this will lead to a monotonic increase of In-content toward the NC top side.

When considering the NC from sample C3, a diameter decrease would lead to a decrease of the local In/Ga ratio on its top and consequently to a lower In-content. This can be seen in Fig. 8B where the measured

In composition at the two top most points refer to the NC portion with smaller diameter.

Nanocolumnar p–i–n (p-GaN/i-InGaN/n-GaN) LED structures, emitting in the blue (LED-1), green (LED-2), and yellow (LED-3) were grown using Si and Mg as dopants with undoped InGaN axial sections 250–500 nm long (Bengoechea-Encabo et, al., 2014). The devices have a circular shape (defined by mesa etching) of 100 μm diameter (photograph in Fig. 9). The n-type contact is a conventional Ti/Al/Ti/Au (20/200/40/55 nm) electrode deposited by electron beam evaporation on the n-type GaN template. A p-type Ni/Au (20/500 nm) electrode was deposited by thermal evaporation on top of the NCs. The use of 500 nm of Au ensures a proper current spreading.

Since light is collected through the sapphire substrate the extraction efficiency is only about 1% due to losses at the GaN/sapphire and sapphire/air interfaces (Vadivelu et al., 2013).

Normalized RT-EL spectra at the maximum injected current (150 mA nominal) are shown in Fig. 10A for LED-1, LED-2, and LED-3. At this maximum injected current, EL spectra peak at 441 nm (LED-1), 502 nm (LED-2), and 568 nm (LED-3). Fig. 10B shows the EL light intensity to current density (*L/J*) ratio as a function of the nominal current density for the

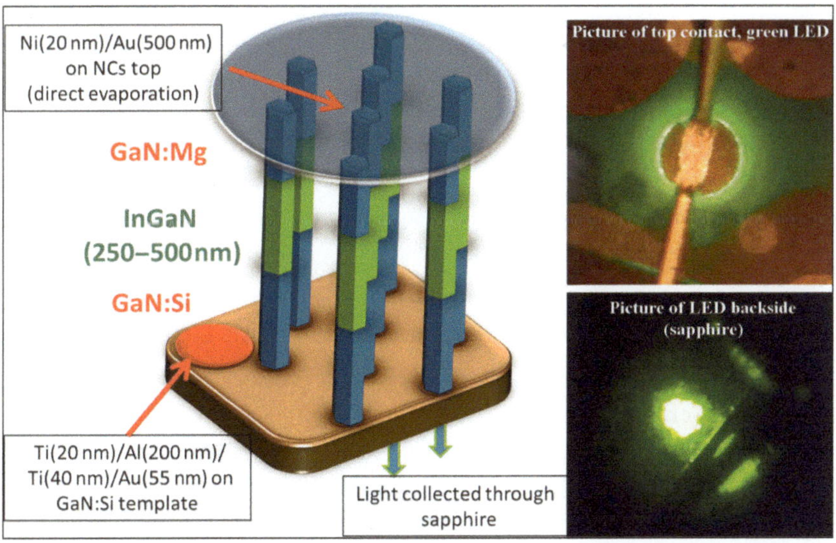

Fig. 9 Sketch of the GaN:Mg/InGaN/GaN:Si p–i–n structure fabricated together with photographs taken from the top and the backside of the working device.

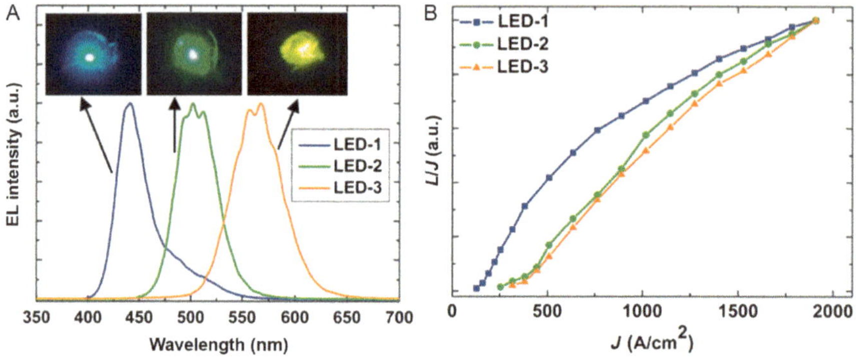

Fig. 10 (A) Normalized RT-EL spectra of devices LED-1, LED-2, LED-3, at an injected current of 150 mA (nominal). The *inset* shows photographs of the light emission during EL measurements and (B) integrated EL intensity (*L*)/current density (*J*) vs *J* dependence of LED-1, LED-2, and LED-3.

three devices, where a monotonous increase is observed. The first question arises about the very high current density needed and the fact that *L*/*J* values (roughly equivalent to EQE) do not peak. In standard planar LEDs EQEs rise very fast peaking at current densities around 20 A/cm^2 and then the *droop* effect is observed. Results on SA InGaN dot-in-a-wire LEDs also show no EQE peaking for current densities up to 200 A/cm^2 which remains stable up to 640 A/cm^2 (Nguyen et al., 2011). In the present case, the slow rise of *L*/*J* values with increasing current injection and the fact that NCs are not passivated, point to a very strong nonradiative surface recombination at the InGaN sidewalls. This would lead to a reduction of the injection efficiency and thus the quantum efficiency, as shown for SA dot-in-a-wire LEDs (Nguyen et al., 2013). Unfortunately, under these circumstances no conclusive statement regarding the absence of the droop effect in the p–i–n LEDs presented here can be made.

2.2 InGaN/GaN NCs for Phosphor Free White Light Emission

After demonstrating the controllability of emission wavelength of the InGaN NCs (emitting from red to blue) and single color NC-based LEDs, white light emission using SAG InGaN/GaN NCs will be discussed (Albert et al., 2013a,b). A stack of InGaN regions (similar to a MQW approach but using thicker InGaN regions) emitting in the basic colors red, green, and blue (R-G-B) were monolithically integrated into individual NCs to fabricate phosphor-free white light emitters. Two similar heterostructures were grown in which the growth times for each InGaN section was, either the same (20 min, sample L1) or modulated to weight the intensity of each color contribution (sample L2).

The PL spectra of the R-G-B sample L1 (Fig. 11B) show a blue-whitish emission at both RT and LT. In order to better tune R-G-B contributions toward real white light emission, different growth times for each color section were used in sample L2. The growth times (red: 41 min; green: 11 min; blue: 8 min) of the respective InGaN sections were modified based on the integrated room-temperature PL intensities of the single color emissions of samples R, G, and B (Fig. 11A). Fig. 11C shows LT and RT-PL spectra of the "balanced" R-G-B stacked sample L2 covering the energy range from 1.8 to 2.8 eV (almost the whole visible spectrum) yielding white light emission. Notice that in this approach each individual NC acts as a white light emitter, in contrast to white light emission as a result of In-content fluctuations among NCs in SA nanocolumnar white LEDs.

The bright-field STEM image in Fig. 12A shows the NCs pencil-like shape in cross-section revealing a high uniformity and the absence of threading dislocations or other extended defects. The NCs diameter increases slightly (as indicated by the black arrows) at the transition region from GaN to the InGaN sections, which is identified by the strain-related dark contrast. The high-angle annular dark-field (HAADF) STEM imaging is shown in Fig. 12B where the different InGaN regions are resolved as an increasing bright contrast, labeled as (I), (II), and (III) in a magnified image (Fig. 12C). However, well-defined boundaries between these three regions cannot be determined but for an *r*-plane at the InGaN/GaN interface. The reasons for these undefined interfaces (blurred) between InGaN sections can be both a lattice pulling effect that widens the interface region and the pencil-like morphology itself that leads to an umbrella-shaped heterostructure, causing an overlapping from regions with different In-contents in the TEM picture.

The local distribution of In within sample L2 was studied by spatially resolved low-loss EELS along the axial and radial directions using a beam diameter of about 10 nm. Fig. 13A displays a bright-field TEM image of a selected NC including the positions of the EELS measurements marked by small dots. Along the axial direction (Fig. 13B), a steep increase of the In-concentration is observed beyond point 3 reaching up to 45% at the NC top, which corresponds to region III. The region between points 1 and 3 refers to GaN and a low In% (5%) InGaN (due to pulling effect). The In% change between points 4 and 5 can be attributed to region II.

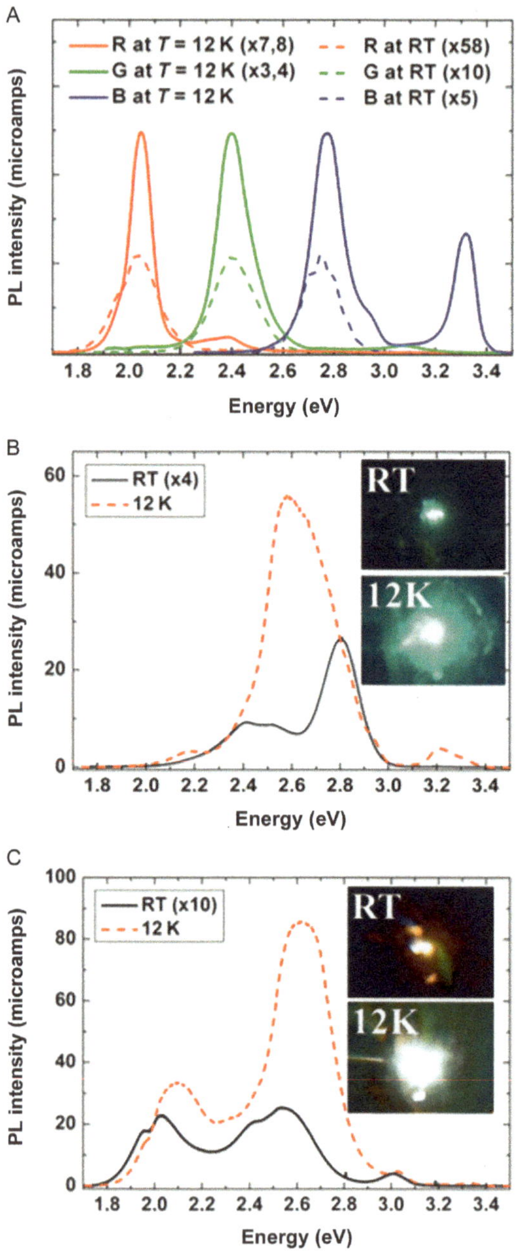

Fig. 11 Room- and low-temperature (12 K) PL spectra of (A) *red*, *green*, and *blue* emitting InGaN/GaN NCs structures labeled R, G, and B, respectively, (B) R-G-B stacked sample (L1), and (C) R-G-B stacked sample (L2) with modified growth times based on RT-PL intensity of individual InGaN segments (Fig. 13A). *Adapted with permission from Albert, S., Bengoechea-Encabo, A., Kong, X., Sanchez-Garcia, M.A., Calleja, E., Trampert, A., 2013a. Monolithic integration of InGaN segments emitting in the blue, green, and red spectral range in single ordered nanocolumns. Appl. Phys. Lett. 102, 181103.*

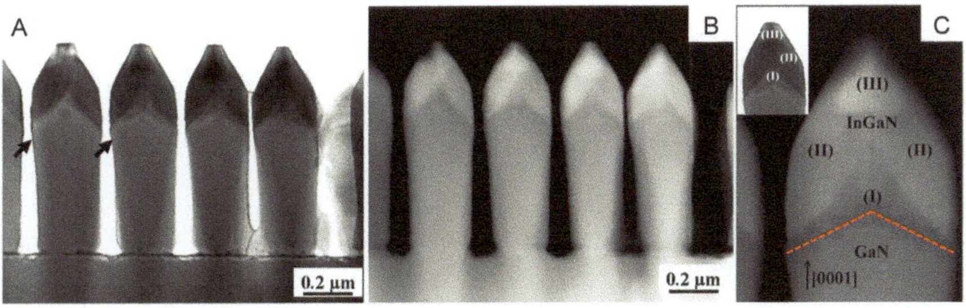

Fig. 12 (A) Bright-field STEM image of the RGB NCs array (sample L2) in cross-section, (B) corresponding HAADF Z-contrast image identifying the InGaN region in the upper part of the columns, and (C) magnified part of the top area with the different InGaN region marked I, II, and III (*inset*: appropriate bright-field STEM of the NC; *dashed line* corresponds to GaN/InGaN interface). *Adapted with permission from Albert, S., Bengoechea-Encabo, A., Kong, X., Sanchez-Garcia, M.A., Calleja, E., Trampert, A., 2013a. Monolithic integration of InGaN segments emitting in the blue, green, and red spectral range in single ordered nanocolumns. Appl. Phys. Lett. 102, 181103.*

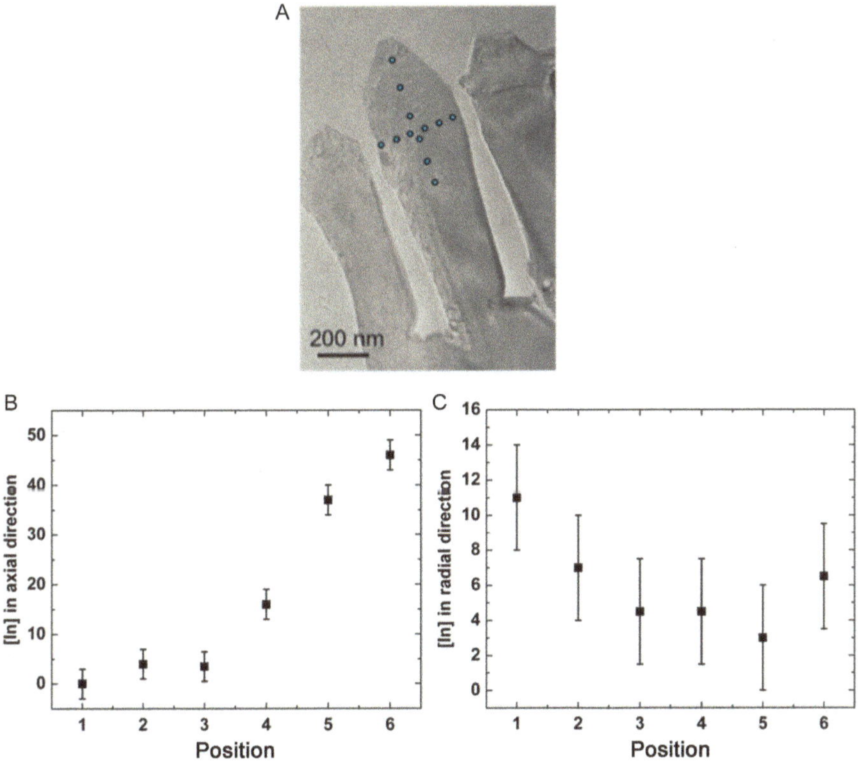

Fig. 13 (A) Bright-field TEM image along the [1–100] direction of a single RBG NC, (B) axial, and (C) radial In% scans measured by low-loss EELS at positions indicated by *small dots* in the bright-field image. *Adapted with permission from Albert, S., Bengoechea-Encabo, A., Kong, X., Sanchez-Garcia, M.A., Calleja, E., Trampert, A., 2013a. Monolithic integration of InGaN segments emitting in the blue, green, and red spectral range in single ordered nanocolumns. Appl. Phys. Lett. 102, 181103.*

The radial low-loss EELS scan (Fig. 13C) reveals an In-concentration variation along the NC diameter being maximum at the edges and minimum in the center. This result reflects the scan going through areas of different In% inside region I. The specific pencil-like morphology of the heterostructure as well as the ambiguities arising from the projection requirements of TEM in combination with the thin foil preparation make a more detailed EELS analysis difficult. Nevertheless, the results demonstrate the presence of regions with different In%, which is in agreement with PL measurements.

3. SAG OF InGaN/GaN CORE–SHELL MICROPILLARS

Axial InGaN/GaN nanostructures were studied thoroughly in Section 2. Some drawbacks of this approach are certain practical limitations due to the polar character of the axial direction (*c*-plane); the need to reduce surface recombination (strong because of the large exposed InGaN surface); and the potential increase of point defects and a rather small emission area (top part of the structure).

InGaN/GaN core–shell structures, in which the active layers are grown on top of nonpolar planes (*m*-planes), will benefit from the absence of internal electric fields and a much higher active area (perimeter) (Fig. 14).

In a core–shell structure the active LED area can be increased by a factor of around four times the aspect ratio in comparison to a planar LED (Li and Waag, 2012). In addition, if phosphor materials, whose grain size is in the micron range, are to be embedded between the columns for light conversion in order to achieve white light emission in LEDs, the diameter and pitch of the core–shell structures must be in the micron range, so that from now on

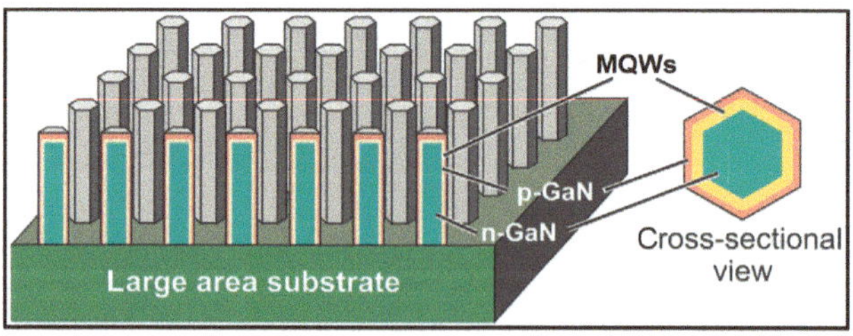

Fig. 14 Core–shell approach to fabricate GaN-based nanoLEDs. Sketches of a core–shell nanoLED ensemble and cross-sectional view of a core–shell nanoLED. The active LED area can be significantly increased with selected aspect ratio and NCs filling factor in comparison to planar LEDs. *Adapted with permission from Li, S., Waag, A., 2012. GaN based nanorods for solid state lighting. J. Appl. Phys. 111, 071101.*

these heterostructures are referred to as micropillars (MPs). This approach was followed by a number of authors using MOVPE (Dong et al., 2009; Jacopin et al., 2012; Koester et al., 2011; Qian et al., 2004, 2005; Wierer et al., 2012). However, no successful attempts were made by PAMBE until recently due to difficulties with the promotion of lateral InGaN growth at the MP sidewalls while keeping selectivity and avoiding merging.

The growth of InGaN/GaN core–shell structures by PAMBE would be appealing because of its potential for high In-content shell layers that are not easily obtained by MOVPE due to the much higher growth temperatures involved. Selectively grown GaN MPs of several micron in diameter are also achievable by PAMBE. However under typical SAG growth conditions, the growth rate is too small (50 nm/h) for achieving MPs several micron high in a reasonable time span (Albert et al., 2015b). Thus, different strategies were envisaged, like the conformal overgrowth on GaN MPs etched from a GaN template (4 μm thick) or MOVPE grown GaN MPs (Albert et al., 2014c, 2015c).

GaN MPs were first etched by inductively coupled plasma (ICP) out of a GaN/sapphire template 4 μm thick, as follows: (i) definition of an hexagonally ordered hole pattern with photolithography using AZ 5214E positive photoresist, (ii) deposition of Ni and removal of the resist with isopropyl alcohol and acetone resulting in an array of ordered Ni microdisks which were used as mask, (iii) ICP etching of the GaN template using Cl:Ar (10 sccm:25 sccm), a pressure of 5 mTorr, an RF power of 150 W and an ICP power of 300 W, resulting in an etch rate of around 400 nm/min, and (iv) removal of the Ni disk with HCl and HNO_3. Following that, the InGaN shell was grown with a Ga-flux of 3 nm/min (2.2×10^{14} atoms/(cm^2 s)), an In-flux of 9.6 nm/min (5.1×10^{14} atoms/(cm^2 s)), and a nitrogen flux of 12 nm/min (8.8×10^{14} atoms/(cm^2 s)) at 625°C.

Fig. 15A shows the sequence of steps from etched GaN MPs fabrication to the InGaN overgrowth on them. The etched GaN MPs have a cylindrical shape and rough sidewalls as a consequence of the etching process. However, after InGaN overgrowth, the MPs show a perfectly defined hexagonal geometry that develops due to surface energy minimization that favors *m*-planes (Gacevic et al., 2015; Urban et al., 2013).

The LT-PL spectrum taken from the InGaN/GaN MPs (Fig. 15B) reveals the GaN near band-edge emission and two InGaN-related peaks at around 3 and 2.4 eV. The inset of Fig. 15B also shows RT-CL measurements on individual InGaN/GaN MP revealing that the two InGaN-related emissions, observed by PL, originate from different regions, i.e., a high

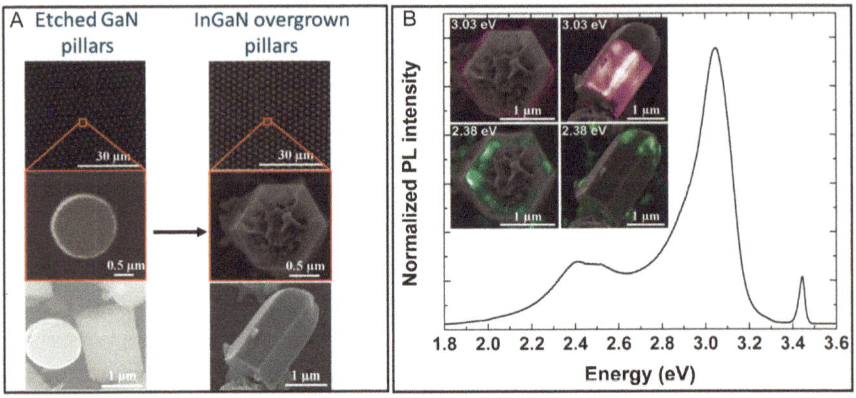

Fig. 15 (A) Top view and cross-sectional SEM pictures of etched GaN MPs and InGaN overgrown on them and (B) low-temperature (7 K) PL spectrum of InGaN/GaN MPs. The *inset* shows spatially resolved RT-CL measurements of a single InGaN/GaN MP with overlaid monochromatic CL intensity images and the secondary electron contrast.

energy one at 3.03 eV from the MP sidewalls (*m*-planes) and a lower energy one at 2.38 eV that stems from the MP top region. These results prove the successful fabrication of InGaN/GaN core–shell structures by PAMBE.

A drawback of InGaN/GaN core–shell structures grown from etched GaN MPs is the limit in aspect ratio, determined by the GaN template thickness and also the material quality before and after etching (etch damage, yellow band, dislocations, etc.). It is worth to mention that upon etching the GaN template, the MPs will fully relax from the template residual strain but they will "inherit" the existing dislocations and defects (Albert et al., 2014c).

Another approach is to start growing the InGaN shell from GaN MPs with high aspect ratio (>10) grown by MOVPE. Fig. 16 shows top and cross-sectional SEM images of the SAG MOVPE MPs and the InGaN overgrown MPs which from now on will be referred to as sample D1. Fig. 16B shows the MOVPE grown GaN MPs which have a hexagonal shape with smooth sidewalls. The subsequent growth of InGaN by PAMBE leads to a clear increase in diameter from the initial MOVPE GaN MPs (550 nm) to 640 nm at the InGaN/GaN MPs bottom and 770 nm at the MPs top for sample D1, while preserving well-defined hexagonal facets.

The LT-PL spectrum of sample D1 (Fig. 17A) reveals a main emission peak at around 3.1 eV, corresponding to an In-content of about 10%. In addition, a second InGaN-related broader emission centered at 2.3 eV can be observed. Similar results can be seen in Fig. 17B from RT-CL spectra of a single InGaN/GaN MP and an ensemble of ~300 MPs. Aside from the

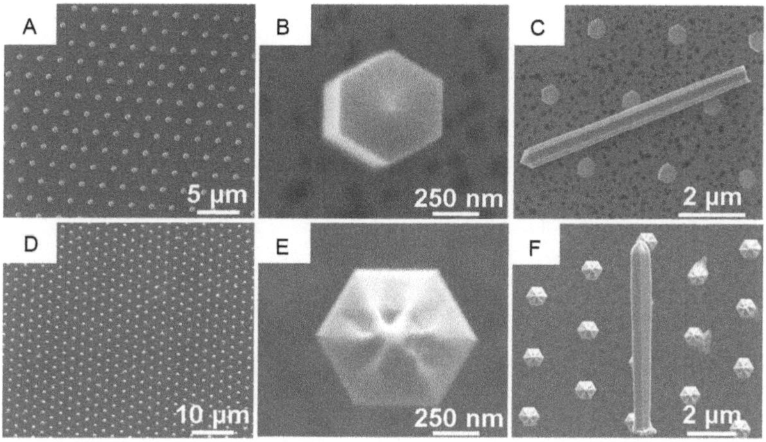

Fig. 16 SEM images of (A)–(C) MOVPE grown GaN MPs and (D)–(F) sample D1. A clear increase in MP diameter is observed upon InGaN/GaN overgrowth. *Adapted with permission from Albert, S., Bengoechea-Encabo, A., Ledig, J., Schimpke, T., Sanchez-Garcia, M.A., Strassburg, M., Waag, A., Calleja, E., 2015c. Demonstration of (In, Ga)N/GaN core-shell micro light-emitting diodes grown by molecular beam epitaxy on ordered MOVPE GaN pillars. Cryst. Growth Des. 15 (8), 3661.*

observation of the two InGaN-related peaks at 2.3 and 3.0 eV, the similarity of both CL spectra points to a very high MPs uniformity. Spatially resolved SEM-CL measurements (Fig. 17C) reveal that the emission at 3.0 eV originates at the MP side facets (*m*-plane), whereas the emission at 2.3 eV comes from the MPs topmost region. These two InGaN emission peaks are attributed to a different In-incorporation depending on the crystal plane, i.e., *m*-plane for the lateral shell and {10–11} planes for the top region (Mandl et al., 2013; Wernicke et al., 2012). In addition, a red-shift of the InGaN shell-related CL emission toward the MP top can be observed (Fig. 17C, upper row) pointing to a graded In composition of the shell toward the MP top.

In InGaN core–shell structures grown by PAMBE, the local III/V ratio at the growth front (and with that the In-content) is affected by a variety of factors. The first is the geometrical arrangement of the growth plane with respect to the impinging fluxes (effusion cells and plasma source) in combination with the sample rotation. All effusion cells and the plasma source are mounted with an angle of 45 degree with respect to the substrate top surface. Because of that, every molecular beam impinges at different MP facets as a function of time, leading to a growth process similar to migration-enhanced epitaxy. In addition, diffusion anisotropy of the metal adatoms on the

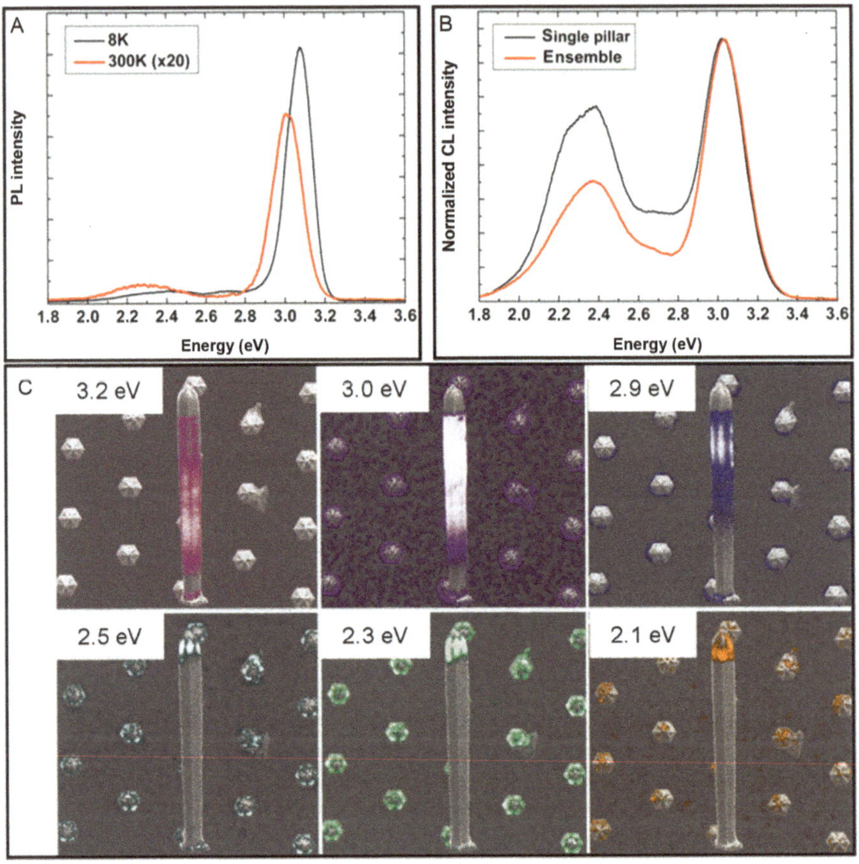

Fig. 17 (A) RT- and 8 K PL spectra of sample D1, (B) RT-CL spectra of a single InGaN/GaN MP and an ensemble of ~300 MPs of sample D1 measured in top view, and (C) spatially resolved CL measurements with monochromatic CL intensity images (*color*) overlaid on the secondary electron image (*grayscale*) of a single MP. *Adapted with permission from Albert, S., Bengoechea-Encabo, A., Ledig, J., Schimpke, T., Sanchez-Garcia, M.A., Strassburg, M., Waag, A., Calleja, E., 2015c. Demonstration of (In, Ga)N/GaN core-shell micro light-emitting diodes grown by molecular beam epitaxy on ordered MOVPE GaN pillars. Cryst. Growth Des. 15 (8), 3661.*

m-plane sidewalls as well as the shadowing effect by surrounding MPs play a role (Albert et al., 2015c).

The potential of PAMBE grown core–shell InGaN/GaN MPs for LED applications was confirmed by growing a p–i–n structure (sample D2), which is shown in SEM top and side views in Fig. 18A and B.

EL measurements were performed by current injection via a microprobe tip placed along the MP-LED sidewall up to the top, as shown in Fig. 18C–F.

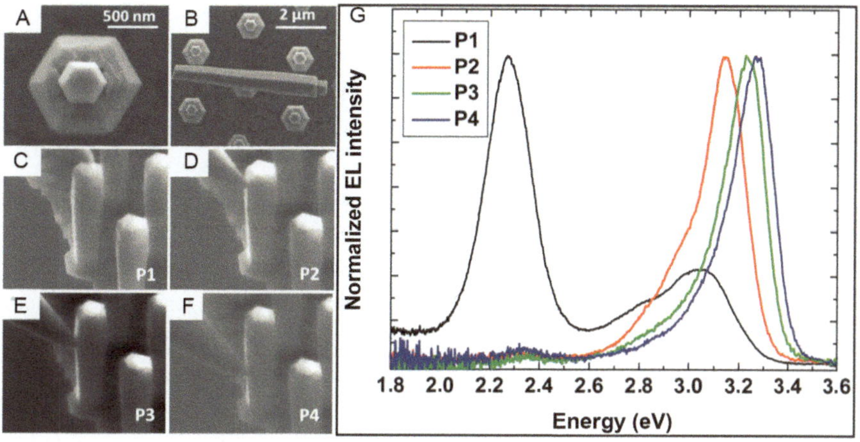

Fig. 18 (A) Top view and (B) cross-sectional SEM images of a single core–shell p–i–n structure (sample D2); (C)–(F) SEM images at 30 degree tilt showing the probe tip contact at four different positions on the top and sidewall of a single microLED to obtain EL; and (G) normalized EL spectra taken at the respective positions P1–P4 with an injection current of 50 μA. *Adapted with permission from Albert, S., Bengoechea-Encabo, A., Ledig, J., Schimpke, T., Sanchez-Garcia, M.A., Strassburg, M., Waag, A., Calleja, E., 2015c. Demonstration of (In, Ga)N/GaN core-shell micro light-emitting diodes grown by molecular beam epitaxy on ordered MOVPE GaN pillars. Cryst. Growth Des. 15 (8), 3661.*

Due to the absence of a current spreading layer, the current injection takes place at localized areas around the point contact resulting in a localized injection of carriers into the active region. EL spectra at 50 μA measured at different contact points on the MP-LED are shown in Fig. 18G. From points P4 to P2 (bottom to top) the dominant InGaN-related EL signal red-shifts from 3.27 to 3.13 eV, in agreement with the commented In% variation along the axial direction within the InGaN shell in sample D1. The EL signal from point P1 (MPs top) shows a dominant peak at 2.28 eV together with a second weaker and broader peak at 3.0 eV. In this case, the high energy peak at P1 is likely originating from the InGaN shell (*m*-plane) close to the MP top, which is weakly excited by the injected electrons from the top contact, while the emission at 2.28 eV relates to an axial diode grown on the MP top (*c*-plane) where a much higher In% incorporation takes place.

4. SAG OF InGaN/GaN NCs ON SILICON

The successful SAG of axial and core–shell InGaN/GaN columnar structures by PAMBE on GaN/sapphire templates was shown in the previous sections. Section 4 will focus on the results from similar experiments performed on Si(111) substrates, as a way to take advantage of cheaper substrates and the well-established Si-based technology.

A first attempt of position-controlled growth of GaN NCs by PAMBE directly on bare Si(111) substrates was performed using a SiO_2 mask of nanoholes (80 nm in diameter) defined by e-beam lithography and etched by ICP (Calleja et al., 2007). The result was partially successful since selectivity was preserved though several individual NCs grew within each nanohole. The nucleation of several NCs was a consequence of (i) an insufficient metal wetting due to the use of N-rich conditions typically employed in self-assembled growth and (ii) the lattice mismatch between GaN and Si that promoted 3D Volmer–Weber nucleation. In a similar approach, Ishizawa et al. (2008) performed SAG on nitridated Al nanodisks deposited on Si(111). Due to nitridation the Al nanodisks converted into AlN which promoted the nucleation of many GaN NCs within the AlN surface and periphery (SA growth). Again, no single NC per nanohole (or nanodisk) was achieved.

The successful SAG by PAMBE of ordered GaN NCs (one per nanohole) on Si(111) was reported by Bertness et al. (2010) using an AlN buffer (40 nm thick) covered by a SiN_x mask (70–220 nm thick) patterned with hole openings of 300 nm to 2 μm diameter. The GaN NCs were grown under highly nitrogen-rich conditions at temperatures close to 840°C. Typical growth times were extremely long (24–72 h). Fig. 19 shows results for the SAG of GaN NCs on AlN-buffered Si(111). The NCs have a well-defined hexagonal cross-sectional and pyramidal tops ((1–103) facets at 32 degree vs *c*-plane) for small diameters, preserving selectivity (Fig. 19A). However for wider diameters, hexagonal facets are not well defined and the MP surface looks rough, as if the MP was formed by the coalescence of several NCs. Indeed, larger MPs often display small voids at their base and faceting of sidewalls (Fig. 19C) which points to coalescence of GaN islands or NCs nucleating at multiple points within the microhole. The multiple nucleation within the microholes may derive from the highly N-rich regime (insufficient metal wetting within the microholes).

A different approach that may circumvent these problems is the use of GaN-buffered Si(111) substrates which would allow for homoepitaxial growth in a similar fashion as SAG on GaN/sapphire. In addition, the absence of AlN buffers would be advantageous for device performance since an AlN buffer could potentially hinder electrical conduction (vertical). The SAG of GaN NCs on GaN-buffered Si(111) was successfully demonstrated in Albert et al. (2013c).

Fig. 19 Arrays of SAG GaN NCs: (A) with different mask opening diameters from 500 to 1000 nm, (B) close-up of NCs for 500 nm opening, and (C) close-up of GaN micropillar for a 1000 nm opening. *Adapted with permission from Bertness, K.A., Sanders, A.W., Rourke, D.M., Harvey, T.E., Roshko, A., Schlager, J.B., Sanford, N.A., 2010. Controlled nucleation of GaN nanowires grown with molecular beam epitaxy. Adv. Funct. Mater. 20, 2911.*

For GaN NCs SAG Ti nanohole masks were deposited on GaN-buffered Si(111) substrates by colloidal lithography. Nanoholes were arranged in a compact hexagonal lattice, with an average pitch of 270 nm and diameters of around 170 nm. The NC growth was performed with a Ga-flux of 18.5 nm/min and N-flux of 5 nm/min, i.e., nominally highly metal-rich conditions, at 820°C for 3 h leading to the formation of SAG NCs of around 550 nm length.

One of the unique properties of NCs is their very high crystalline quality, even when grown on low quality substrates like GaN on Si(111), due to an efficient strain relaxation caused by the NCs large surface-to-volume ratio and by the bending of threading dislocations that may be present at the bottom of the NC toward the NC sidewalls (Harui et al., 2008; Li et al., 2010).

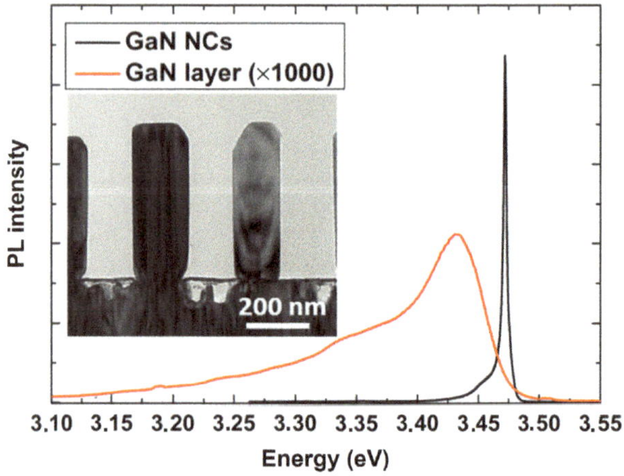

Fig. 20 Low-temperature (12 K) PL spectra of ordered GaN NCs and the GaN buffer, *inset*: bright-field TEM images of ordered GaN NCs.

This material improvement upon NC growth can be seen in Fig. 20, where LT-PL spectra of the GaN buffer layer and SAG GaN NCs grown on top are shown. The dominant PL emission from the GaN buffer is a broad peak centered at 3.42 eV with a low energy shoulder; the former being attributed to Si-donor-bound excitons under tensile strain. In contrast, SAG GaN NCs show sharp and intense (2000 times higher than the buffer) D^0X emission at 3.472 eV with a line width of 3 meV, typical of high quality, strain-free material. However, a line width value of 3 meV is still higher than the state of the art in SA GaN NCs (typically around 1 meV) which may be due to disorder by extended defects (arising from the buffer) within the GaN NCs volume (Albert et al., 2014b).

Indeed, the bright-field TEM image (inset in Fig. 20) reveals the presence of threading dislocations and inversion domains. Inversion domains were found to be caused by Ti incorporation at the homoepitaxial GaN interface (Kong et al., 2016).

The effects of the growth temperature, In/Ga ratio, and III/V ratio on the SAG of InGaN NCs was studied, aiming to achieve the whole range of In compositions (up to 100%). For this purpose, two series of samples were grown under the conditions described in Table 1 (series E and F). The InGaN NCs were grown for 1 h on top of 550 nm high GaN NCs grown by SAG on GaN-buffered Si(111).

The morphology of NCs emitting in the green spectral range was first optimized with samples of series E (E1–E6). Furthermore, the growth

Table 1 Growth Conditions of the InGaN Segments of Series E and F

Sample Number	Impinging Fluxes (nm/min) ($\times 10^{14}$ Atoms/(cm^2 s)) Ga	In	N	III/V Ratio	In/Ga Ratio	T_{sample} (°C)
E1	4.3 (3.17)	4.3 (2.27)	10 (7.37)	0.74	0.72	**700**
E2	4.3 (3.17)	4.3 (2.27)	10 (7.37)	0.74	0.72	**650**
E3	4.3 (3.17)	4.3 (2.27)	10 (7.37)	0.74	0.72	**600**
E4	4.3 (3.17)	4.3 (2.27)	10 (7.37)	0.74	0.72	**550**
E5	2.3 (1.7)	6.3 (3.33)	10 (7.37)	**0.68**	1.96	550
E6	2.3 (1.7)	6.3 (3.33)	14 (10.3)	**0.49**	1.96	550
F1	2.3 (1.7)	6.3 (3.33)	14 (10.3)	**0.49**	1.96	**480**
F2	2.3 (1.7)	6.3 (3.33)	10 (7.37)	**0.68**	1.96	**450**
F3	2.3 (1.7)	6.3 (3.33)	14 (10.3)	**0.49**	1.96	**450**
F4	–	4.3 (2.27)	10 (7.37)	**0.31**	–	**300**

Bold numbers indicate the growth conditions that were changed in the respective samples.

conditions were tuned with samples of series F (F1–F4) to reach 100% In-content while keeping a nanocolumnar morphology. More specific details about the results presented in this section can be found in Albert et al. (2013c).

SEM images from samples of series E are shown in Fig. 21, from where the height of the InGaN section can be estimated. From these values, a growth rate increase from 5.7 nm/min (sample E1) to 8.2 nm/min (sample E3) is determined (growth temperature decrease from 700 to 600°C), which can be explained by a decrease of In-desorption, as well as by a lower InN decomposition rate (Gallinat et al., 2007; Koblmüller et al., 2007).

In agreement with an increased growth rate with respect to sample E1, a higher In% can be found in samples E2 and E3, as indicated by a red-shift of the PL emission (no significant InGaN-related PL emission in sample E1) (Fig. 21).

When lowering the growth temperature from 600 to 550°C (sample E4), the PL peak red-shifted further (Fig. 21D). On the other hand, while keeping the temperature at 550°C, no significant PL shift is observed when changing either the In/Ga ratio (Fig. 21E) or the group-III/N ratio (Fig. 21F) in samples E5 and E6. The effect of an increased In/Ga ratio is an improvement in the NC morphology, that is, a more homogeneous

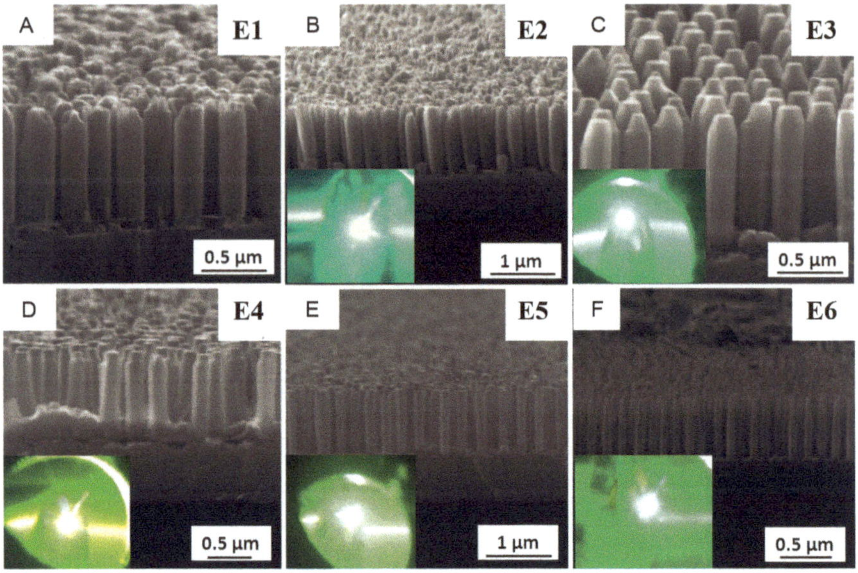

Fig. 21 Cross-sectional SEM pictures of InGaN/GaN samples: (A)–(F) E1–E6. *Insets* show the respective PL emission color.

NC diameter, as can be seen in Fig. 21E for sample E5. On the other hand, a reduction in group-III/N ratio led to a needle-like shape (sample E6 in Fig. 21F).

Finally the emission wavelength was pushed toward the infrared by increasing the In composition, up to 100% (samples of series F). Fig. 22 shows SEM images as well as PL spectra of these samples. Starting from the same growth conditions as for sample E6, but using a lower growth temperature of 480°C, sample F1 shows a broad PL emission at 2.1 eV. A further decrease of the growth temperature leads to a more pronounced red-shift of the PL emission down to 1.18 eV (sample F3 in Fig. 22).

Besides the growth temperature, the group-III/N ratio can be used as well for changing the In-content, as can be seen when comparing samples F2 and F3, grown at the same temperature. In order to achieve the SAG of InN NCs on top of GaN, sample F4 was grown at much lower temperature (300°C) to suppress InN decomposition. It can be seen in Fig. 22, that the nanocolumnar shape, as well as selectivity, is kept. The PL spectrum of F4 shows a peak at 0.78 eV, which is higher than the energy corresponding to fully relaxed high quality InN layers, which is known to be at 0.63 eV (Gallinat et al., 2006). Possible reasons for such a blue-shift in the PL emission from InN NCs are the presence of compressive strain (not commonly

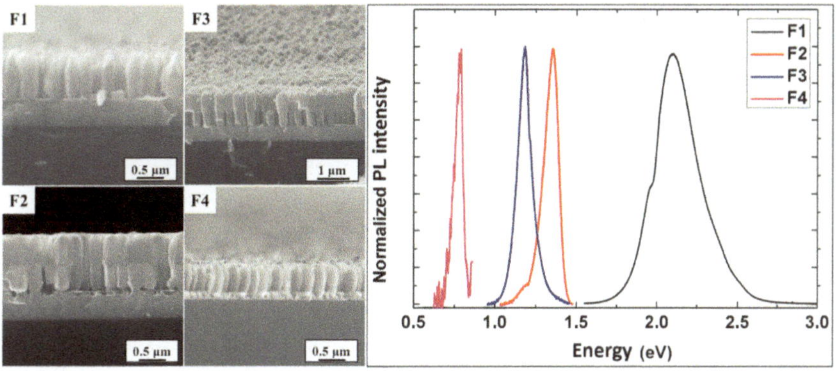

Fig. 22 Cross-sectional SEM images and LT-PL spectra of InGaN/GaN samples of series F.

observed in NCs) or a high electron concentration. A high electron concentration would account for both the measured PL line width of around 70 meV, and the observed emission energy of 0.78 eV (Song et al., 2008). It has to be noted that the growth conditions for InN NCs were extrapolated from samples with lower In-content, thus they may not have been the optimal ones.

5. SUMMARY AND CONCLUSIONS

In this chapter, we have given an overview of the SAG of axial, as well as core–shell InGaN/GaN nano- and microcolumns by plasma-assisted MBE, on GaN/sapphire and GaN-buffered Si(111).

We have shown that the In% of ordered InGaN NCs grown on both GaN/sapphire and GaN/Si(111) can be controlled by means of growth temperature, In/Ga ratio and III/V ratio allowing for the fabrication of emitters covering the whole spectral range from the infrared (InN) up to the ultraviolet (GaN). The evolution of the InGaN NC morphology as a function of III/V ratio and NC diameter has been explained by changes in the local III/V ratio at the column top $(Ga/N)_{top}$, i.e., an increasing column diameter for $(Ga/N)_{top} > 1$ and a decreasing column diameter for $(Ga/N)_{top} < 1$. In nominally composition-fixed InGaN NCs an In-gradient along the growth direction has been found, which was explained as a consequence of thermal InN decomposition, lattice pulling and changes in the local III/V and In/Ga ratios at the column top due to morphology changes during the growth. Furthermore, blue, green, and yellow emitting LEDs based on ordered NCs having bulk-like InGaN active regions (250–500 nm thickness) have been discussed. These nanoLEDs with embedded thick InGaN region are believed to have potential in lighting, since they combine advantages of ordered NCs (high homogeneity, quality, and light extraction efficiency) with the benefits of a thick InGaN active region (reduction of Auger recombination and current overflow). Following that, the potential of selectively grown InGaN NCs for phosphor-free white light emission was demonstrated by the monolithic integration of three InGaN portions emitting in the red, green, and blue in individual NCs. In these structures, the color temperature could be tuned by playing with the relative "weight" of the colors by means of thickness variation of the respective InGaN regions. Finally, in order to take advantage of an enhanced emission/absorption area and a reduced (or zero) internal electric field, recent results on the PAMBE growth of high aspect ratio InGaN/GaN core–shell structures and p–i–n microLED structures were discussed.

ACKNOWLEDGMENTS

We acknowledge partial financial support by EU FP7 Contract SMASH 228999-2, GECCO 280694-2 and by Spanish Projects CAM/P2009/ESP-1503 and MICINN MAT2011-26703.

REFERENCES

Albert, S., Bengoechea-Encabo, A., Lefebvre, P., Sanchez-Garcia, M.A., Calleja, E., Jahn, U., Trampert, A., 2011. Emission control of InGaN nanocolumns grown by molecular-beam epitaxy on Si(111) substrates. Appl. Phys. Lett. 99, 131108.

Albert, S., Bengoechea-Encabo, A., Sanchez-Garcia, M.A., Barbagini, F., Calleja, E., Luna, E., Trampert, A., Jahn, U., Lefebvre, P., Lopez, L.L., Estrade, S., Rebled, J.M., Peiro, F., Nataf, G., de Mierry, P., Zuñiga-Perez, J., 2012a. Ordered GaN/InGaN nanorods arrays grown by molecular beam epitaxy for phosphor-free white light emission. Int. J. High Speed Electron. Syst. 21, 1.

Albert, S., Bengoechea-Encabo, A., Lefebvre, P., Barbagini, F., Sanchez-Garcia, M.A., Calleja, E., Jahn, U., Trampert, A., 2012b. Selective area growth and characterization of InGaN nano-disks implemented in GaN nanocolumns with different top morphologies. Appl. Phys. Lett. 100, 231906.

Albert, S., Bengoechea-Encabo, A., Kong, X., Sanchez-Garcia, M.A., Calleja, E., Trampert, A., 2013a. Monolithic integration of InGaN segments emitting in the blue, green, and red spectral range in single ordered nanocolumns. Appl. Phys. Lett. 102, 181103.

Albert, S., Bengoechea-Encabo, A., Sanchez-Garcia, M.A., Calleja, E., Jahn, U., 2013b. Selective area growth and characterization of InGaN nanocolumns for phosphor-free white light emission. J. Appl. Phys. 113, 114306.

Albert, S., Bengoechea-Encabo, A., Sanchez-Garcia, M.A., Kong, X., Trampert, A., Calleja, E., 2013c. Selective area growth of In(Ga)N/GaN nanocolumns by molecular beam epitaxy on GaN-buffered Si(111): from ultraviolet to infrared emission. Nanotechnology 24, 175303.

Albert, S., Bengoechea-Encabo, A., Zuñiga-Perez, J., de Mierry, P., Val, P., Sanchez-Garcia, M.A., Calleja, E., 2014a. Selective area growth of GaN nanostructures: a key to produce high quality (11-20) a-plane pseudo substrates. Appl. Phys. Lett. 105, 091902.

Albert, S., Bengoechea-Encabo, A., Barbagini, F., Lopez-Romero, D., Sanchez-Garcia, M.A., Calleja, E., Lefebvre, P., Kong, X., Jahn, U., Trampert, A., Müller, M., Bertram, F., Schmidt, G., Veit, P., Petzold, S., Christen, J., 2014b. Advances in MBE selective area growth of III-nitride nanostructures: from nanoLEDs to pseudo substrates. Int. J. High Speed Electron. Syst. 23 (3–4), 1450020.

Albert, S., Bengoechea-Encabo, A., Sabido-Siller, M., Müller, M., Schmidt, G., Metzner, S., Veit, P., Bertram, F., Sanchez-Garcia, M.A., Christen, J., Calleja, E., 2014c. Growth of InGaN/GaN core-shell structures on selectively etched GaN rods by molecular beam epitaxy. J. Cryst. Growth 392, 5.

Albert, S., Bengoechea-Encabo, A., Kong, X., Sanchez-Garcia, M.A., Trampert, A., Calleja, E., 2015a. Correlation among growth conditions, morphology, and optical properties of nanocolumnar InGaN/GaN heterostructures selectively grown by molecular beam epitaxy. Cryst. Growth Des. 15, 2661.

Albert, S., Bengoechea-Encabo, A., Lopez-Romero, D., de Mierry, P., Zuniga-Perez, J., Kong, X., Trampert, A., Sanchez-Garcia, M.A., Calleja, E., 2015b. Selective area growth of III-nitride nanorods on polar, semi-polar, and non-polar orientations: device applications. In: Razeghi, M., Tournie, E., Brown, G.J. (Eds.), Proc. SPIE 9370, Quantum Sensing and Nanophotonic Devices XII, 937017.

Albert, S., Bengoechea-Encabo, A., Ledig, J., Schimpke, T., Sanchez-Garcia, M.A., Strassburg, M., Waag, A., Calleja, E., 2015c. Demonstration of (In, Ga)N/GaN core-shell micro light-emitting diodes grown by molecular beam epitaxy on ordered MOVPE GaN pillars. Cryst. Growth Des. 15 (8), 3661.

Bavencove, A.L., Tourbot, G., Garcia, J., Désières, Y., Gilet, P., Levy, F., André, B., Gayral, B., Daudin, B., Dang, L.S., 2011. Submicrometre resolved optical characterization of green nanowire-based light emitting diodes. Nanotechnology 22, 345705.

Bengoechea-Encabo, A., Barbagini, F., Fernandez-Garrido, S., Grandal, J., Ristic, J., Sanchez-Garcia, M.A., Calleja, E., Jahn, U., Luna, E., Trampert, A., 2011. Understanding the selective area growth of GaN nanocolumns by MBE using Ti nanomasks. J. Cryst. Growth 325, 89.

Bengoechea-Encabo, A., Albert, S., Sanchez-Garcia, M.A., Lopez, L.L., Estrade, S., Rebled, J.M., Peiro, F., Nataf, G., de Mierry, P., Zuniga-Perez, J., Calleja, E., 2012. Selective area growth of a- and c-plane GaN nanocolumns by molecular beam epitaxy using colloidal nanolithography. J. Cryst. Growth 353, 1.

Bengoechea-Encabo, A., Albert, S., Zuniga-Perez, J., de Mierry, P., Trampert, A., Barbagini, F., Sanchez-Garcia, M.A., Calleja, E., 2013. Selective area growth and characterization of GaN nanocolumns, with and without an InGaN insertion, on semi-polar (11-22) GaN templates. Appl. Phys. Lett. 103, 241905.

Bengoechea-Encabo, A., Albert, S., Lopez-Romero, D., Lefebvre, P., Barbagini, F., Torres-Pardo, A., Gonzalez-Calbet, J.M., Sanchez-Garcia, M.A., Calleja, E., 2014. Light-emitting-diodes based on ordered InGaN nanocolumns emitting in the blue, green and yellow spectral range. Nanotechnology 25, 435203.

Bertness, K.A., Sanders, A.W., Rourke, D.M., Harvey, T.E., Roshko, A., Schlager, J.B., Sanford, N.A., 2010. Controlled nucleation of GaN nanowires grown with molecular beam epitaxy. Adv. Funct. Mater. 20, 2911.

Calleja, E., Sánchez-García, M.A., Sánchez, F.J., Calle, F., Naranjo, F.B., Muñoz, E., Jahn, U., Ploog, K., 2000. Luminescence properties and defects in GaN nanocolumns grown by molecular beam epitaxy. Phys. Rev. B 62, 16826.

Calleja, E., Ristic, J., Fernández-Garrido, S., Cerutti, L., Sánchez-García, M.A., Grandal, J., Trampert, A., Jahn, U., Sánchez, G., Griol, A., Sánchez, B., 2007. Growth, morphology, and structural properties of group-III-nitride nanocolumns and nanodisks. Phys. Status Solidi B 244, 8.

Cerutti, L., Ristic, J., Fernández-Garrido, S., Calleja, E., Trampert, A., Ploog, K.H., Lazic, S., Calleja, J.M., 2006. Wurtzite GaN nanocolumns grown on Si(001) by molecular beam epitaxy. Appl. Phys. Lett. 88, 213114.

Dong, Y., Tian, B., Kempa, T.J., Lieber, C.M., 2009. Coaxial group III-nitride nanowire photovoltaics. Nano Lett. 9, 2183.

Ertekin, E., Greaney, P.A., Chrzan, D.C., Sands, T.D., 2005. Equilibrium limits of coherency in strained nanowire heterostructures. J. Appl. Phys. 97, 114325.

Fernández-Garrido, S., Kaganer, V.M., Sabelfeld, K.K., Gotschke, T., Grandal, J., Calleja, E., Geelhaar, L., Brandt, O., 2013. Self-regulated radius of spontaneously formed GaN nanowires in molecular beam epitaxy. Nano Lett. 13, 3274.

Gacevic, Z., Gómez Sánchez, D., Calleja, E., 2015. Formation mechanisms of GaN nanowires grown by selective area growth homoepitaxy. Nano Lett. 15, 1117.

Gallinat, C.S., Koblmüller, G., Brown, J.S., Bernardis, S., Speck, J.S., Chern, G.D., Readinger, E.D., Shen, H., Wraback, M., 2006. In-polar InN grown by plasma-assisted molecular beam epitaxy. Appl. Phys. Lett. 89, 032109.

Gallinat, C.S., Koblmüller, G., Brown, J.S., Speck, J.S., 2007. A growth diagram for plasma-assisted molecular beam epitaxy of In-face InN. J. Appl. Phys. 102, 064907.

Harui, S., Tamiya, H., Akagi, T., Miyake, H., Hiramatsu, K., Araki, T., Nanishi, Y., 2008. Transmission electron microscopy characterization of position-controlled InN nanocolumns. Jpn. J. Appl. Phys. 47, 5330.

Ishizawa, S., Kishino, K., Kikuchi, A., 2008. Selective-area growth of GaN nanocolumns on Si(111) substrates using nitrided Al nanopatterns by RF-plasma-assisted molecular-beam epitaxy. Appl. Phys. Express 1, 015006.

Jacopin, G., de Luna Bugallo, A., Lavenus, P., Rigutti, L., Julien, F.H., Zagonel, L.F., Kociak, M., Durand, C., Salomon, D., Jun Chen, X., Eymery, J., Tchernycheva, M., 2012. Single-wire light-emitting diodes based on GaN wires containing both polar and nonpolar InGaN/GaN quantum wells. Appl. Phys. Express 5, 014101.

Karpov, S.Y., 2011. Modeling of III-nitride light-emitting diodes: progress, problems, and perspectives. Gallium nitride materials and devices VI. In: Proc. SPIE, 7939, 79391C.

Kioupakis, E., Rinke, P., Delaney, K.T., Van de Walle, C.G., 2011. Indirect Auger recombination as a cause of efficiency droop in nitride light-emitting diodes. Appl. Phys. Lett. 98, 161107.

Kishino, K., Kikuchi, A., Sekiguchi, H., Ishizawa, S., 2007. InGaN/GaN nanocolumn LEDs emitting from blue to red. In: Proc. SPIE 6473, 64730T.

Kishino, K., Sekiguchi, H., Kikuchi, A., 2009. Improved Ti-mask selective-area growth (SAG) by rf-plasma-assisted molecular beam epitaxy demonstrating extremely uniform GaN nanocolumn arrays. J. Cryst. Growth 311, 2063.

Kishino, K., Kamimura, J., Kamiyama, K., 2012. Near-infrared InGaN nanocolumn light-emitting diodes operated at 1.46 μm. Appl. Phys. Express 5, 031001.

Kishino, K., Nagashima, K., Yamano, K., 2013. Monolithic integration of InGaN-based nanocolumn light-emitting diodes with different emission colors. Appl. Phys. Express 6, 012101.

Koblmüller, G., Gallinat, C.S., Speck, J.S., 2007. Surface kinetics and thermal instability of N-face InN grown by plasma-assisted molecular beam epitaxy. J. Appl. Phys. 101, 083516.

Koester, R., Hwang, J.S., Salomon, D., Chen, X., Bougerol, C., Barnes, J.P., Le Si Dang, D., Rigutti, L., de Luna Bugallo, A., Jacopin, G., Tchernycheva, M., Durand, C., Eymery, J., 2011. M-plane core-shell InGaN/GaN multiple-quantum-wells on GaN wires for electroluminescent devices. Nano Lett. 11, 4839.

Kong, X., Albert, S., Bengoechea-Encabo, A., Sanchez-Garcia, M.A., Calleja, E., Trampert, A., 2012. Plasmon excitation in electron energy-loss spectroscopy for determination of indium concentration in (In, Ga)N/GaN nanowires. Nanotechnology 23, 485701.

Kong, X., Albert, S., Bengoechea-Encabo, A., Sanchez-Garcia, M.A., Calleja, E., Trampert, A., 2015. Lattice pulling effect and strain relaxation in axial (In, Ga)N/GaN nanowire heterostrucutres grown on GaN-buffered Si(111) substrate. Phys. Status Solidi A 212, 736–739.

Kong, X., Li, H., Albert, S., Bengoechea-Encabo, A., Sanchez-Garcia, M.A., Calleja, E., Draxl, C., Trampert, A., 2016. Titanium induced polarity inversion in ordered (In, Ga)N/GaN nanocolumns. Nanotechnology 27, 065705.

Li, S., Waag, A., 2012. GaN based nanorods for solid state lighting. J. Appl. Phys. 111, 071101.

Li, S., Fündling, S., Sökmen, Ü., Merzsch, S., Neumann, R., Hinze, P., Weimann, T., Jahn, U., Trampert, A., Riechert, H., Peiner, E., Wehmann, H.-H., Waag, A., 2010. GaN and LED structures grown on pre-patterned silicon pillar arrays. Phys. Status Solidi C 7, 84.

Lin, H.-W., Lu, Y.-J., Chen, H.-Y., Lee, H.-M., Gwo, S., 2010. InGaN/GaN nanorod array white light-emitting diode. Appl. Phys. Lett. 97, 073101.

Lymperakis, L., Neugebauer, J., 2009. Large anisotropic adatom kinetics on nonpolar GaN surfaces: consequences for surface morphologies and nanowire growth. Phys. Rev. B 79, R241308.

Mandl, M., Wang, X., Schimpke, T., Kölper, C., Binder, M., Ledig, J., Waag, A., Kong, X., Trampert, A., Bertram, F., Christen, J., Barbagini, F., Calleja, E., Strassburg, M., 2013. Group III nitride core-shell nano- and microrods for optoelectronic applications. Phys. Status Solidi (RRL) 7, 800.

Meijers, R., Richter, T., Calarco, R., Stoica, T., Bochem, H.P., Marso, M., Lüth, H., 2006. GaN-nanowhiskers: MBE-growth conditions and optical properties. J. Cryst. Growth 289, 381.

Nguyen, H.P.T., Zhang, S., Cui, K., Han, X., Fathololoumi, S., Couillard, M., Botton, G.A., Mi, Z., 2011. p-type modulation doped InGaN/GaN dot-in-a-wire white-light-emitting diodes monolithically grown on Si(111). Nano Lett. 11, 1919.

Nguyen, H.P.T., Zhang, S., Connie, A.T., Kibria, M.G., Wang, Q., Shih, I., Mi, Z., 2013. Breaking the carrier injection bottleneck of phosphor-free nanowire white light-emitting diodes. Nano Lett. 13, 5437.

Park, Y.S., Park, C.M., Fu, D.J., Kang, T.W., Oh, J.E., 2004. Photoluminescence studies of GaN nanorods on Si(111) substrates grown by molecular-beam epitaxy. Appl. Phys. Lett. 85, 5718.

Piprek, J., 2010. Efficiency droop in nitride-based light-emitting diodes. Phys. Status Solidi A 207, 1–9.

Qian, F., Li, Y., Gradecak, S., Wang, D., Barrelet, C.J., Lieber, C.M., 2004. Gallium nitride-based nanowire radial heterostructures for nanophotonics. Nano Lett. 4, 1975.

Qian, F., Gradecak, S., Li, Y., Wen, C.Y., Lieber, C.M., 2005. Core/multishell nanowire heterostructures as multicolor, high-efficiency light emitting diodes. Nano Lett. 5, 2287.

Sanchez-Garcia, M.A., Calleja, E., Monroy, E., Sanchez, F.J., Calle, F., Muñoz, E., Beresford, R., 1998. The effect of the III/V ratio and substrate temperature on the morphology and properties of GaN- and AlN layers grown by molecular beam epitaxy on Si(111). J. Cryst. Growth 183, 23.

Sekiguchi, H., Kishino, K., Kikuchi, A., 2008. Ti-mask selective-area growth of GaN by RF-plasma-assisted molecular-beam epitaxy for fabricating regularly arranged InGaN/GaN nanocolumns. Appl. Phys. Express 1, 124002.

Sekiguchi, H., Kishino, K., Kikuchi, A., 2010. Emission color control from blue to red with nanocolumn diameter of InGaN/GaN nanocolumn arrays grown on same substrate. Appl. Phys. Lett. 96, 231104.

Song, D.Y., Holtz, M.E., Chandolu, A., Bernussi, A., Nikishin, S.A., Holtz, M.W., Gherasoiu, I., 2008. Effect of stress and free-carrier concentration on photoluminescence in InN. Appl. Phys. Lett. 92, 121913.

Thillosen, N., Sebald, K., Hardtdegen, H., Meijers, R., Calarco, R., Montanari, S., Kaluza, N., Gutowski, J., Luth, H., 2006. The state of strain in single GaN nanocolumns as derived from micro-photoluminescence measurements. Nano Lett. 6, 704.

Trampert, A., Ristic, J., Jahn, U., Calleja, E., Ploog, K.H., 2003. TEM study of (Ga, Al)N nanocolumns and embedded GaN nanodiscs. Inst. Phys. Conf. Ser. 180, 167.

Urban, A., Malindretos, J., Klein-Wiele, J.H., Simon, P., Rizzi, A., 2013. Ga-polar GaN nanocolumn arrays with semipolar faceted tips. New J. Phys. 15, 053045.

Vadivelu, R., Igawa, Y., Kishino, K., 2013. 633 nm red emissions from InGaN nanocolumn light-emitting diode by radio frequency plasma assisted molecular beam epitaxy. Jpn. J. Appl. Phys. 52, 8S.

Vampola, K.J., Iza, M., Keller, S., DenBaars, S.P., Nakamura, S., 2009. Measurement of electron overflow in 450 nm InGaN light-emitting diode structures. Appl. Phys. Lett. 94, 061116.

Wernicke, T., Schade, L., Netzel, C., Rass, J., Hoffmann, V., Ploch, S., Knauer, A., Weyers, M., Schwarz, U., Kneissl, M., 2012. Indium incorporation and emission wavelength of polar, nonpolar and semipolar InGaN quantum wells. Semicond. Sci. Technol. 27, 024014.

Wierer, J.J., Li, Q., Koleske, D.D., Lee, S.R., Wang, G.T., 2012. III-nitride core-shell nanowire arrayed solar cells. Nanotechnology 23, 194007.

Yoshizawa, M., Kikuchi, A., Mori, M., Fujita, N., Kishino, K., 1997. Growth of self-organized GaN nanostructures on Al_2O_3(0001) by RF-radical source molecular beam epitaxy. Jpn. J. Appl. Phys. 36, L459.

CHAPTER EIGHT

InN Nanowires: Epitaxial Growth, Characterization, and Device Applications

S. Zhao*, **Z. Mi**†,[1]
*McGill University, Montreal, QC, Canada
†University of Michigan, Ann Arbor, MI, United States
[1]Corresponding author: e-mail address: zetian.mi@mcgill.ca

Contents

Semiconductors and Semimetals, Volume 96
ISSN 0080-8784
http://dx.doi.org/10.1016/bs.semsem.2016.08.001

© 2017 Elsevier Inc.
All rights reserved.

1. INTRODUCTION

Compared to GaN and AlN, InN has a very narrow bandgap energy of ~0.65 eV, thus extending the optical bandgap of the III-nitride semiconductor family from the ultraviolet to the near-infrared (Bhuiyan et al., 2003; Davydov et al., 2002; Holtz et al., 2009; Huang et al., 2002; Matsuoka et al., 2002; Mi and Zhao, 2015; Taniyasu et al., 2004; Wu, 2009; Wu et al., 2002a,b; Zhao et al., 2015). InN also has very large electron mobility, with a theoretically predicted value up to ~14,000 $cm^2/V\ s$ at room temperature (Aydogu et al., 2012; Meyer and Bartoli, 1987; Polyakov et al., 2009; Wang et al., 2012b,c). A comparison of the basic physical parameters of InN, GaN and AlN is listed in Table 1. These unique properties render InN as a promising candidate for a wide range of electronic and photonic devices such as ultrahigh-speed transistors, multijunction solar cells, and near-infrared lasers and photodetectors. For these reasons, significant efforts have been devoted to the synthesis and characterization of InN in the past decade. Tremendous progress has been made in improving the

Table 1 Some Basic Physical Parameters of Wurtzite InN, AlN, and GaN (Wu, 2009)

Parameters	AlN	GaN	InN
Lattice constant a (300 K) (nm)	0.3112	0.3189	0.3533
Lattice constant c (300 K) (nm)	0.4982	0.5185	0.5693
Decomposition temperature (°C)	1040	850	630
Bandgap energy E_g (300 K) (eV)	6.14	3.43	0.64
Electron effective mass at band edge m_e^*/m_0	0.32	0.20	0.07
Electron mobility (300 K) ($cm^2/V\ s$)	141[a]	650[b]	8000–12,000[c]
			12,000–14,000[d]
Exciton binding energy (meV)	60	34	9
Exciton Bohr radius (nm)	1.4	2.4	8
Mg acceptor binding energy (eV)	0.51	0.17	0.06
E_2^H phonon (1/cm)	657	568	488

[a]Taniyasu et al. (2004).
[b]Measured from single nanowire transistor (Huang et al., 2002).
[c]Experimental value (Zhao et al., 2013d).
[d]Theoretically calculated low-field value (Polyakov et al., 2009).

quality of InN thin film/epilayers, including the drastically reduced background electron concentration and the improved electron mobility (Anderson et al., 2006; Arnaudov et al., 2004; Brown et al., 2008; Davydov et al., 2002; Fu et al., 2006b, 2007; Klochikhin et al., 2005; Look et al., 2002; Lu et al., 2001; Mi and Zhao, 2015; Wang et al., 2012c; Wu, 2009; Yamaguchi et al., 2011, 2013). With boundary-temperature-controlled epitaxy technique, the residual electron density is reduced to 1.8×10^{17} cm^{-3}, with an electron mobility of 3010 cm^2/V s (Wang et al., 2012c). This technique further produces background electron concentration of 1.4×10^{17} cm^{-3} and electron mobility of 3560 cm^2/V s from a 6.4-μm thick InN epilayer (Wang et al., 2012d). Background electron concentration around 1.5×10^{17} cm^{-3} and electron mobility around 3570 cm^2/V s have also been reported by Fehlberg et al. (2007). In addition, epitaxial lateral overgrowth technique has been developed for InN, leading to an estimated background electron concentration around 5.5×10^{16} cm^{-3} (Kametani et al., 2013). Major challenges, however, have remained for InN epilayers, which include the presence of electron accumulation on the grown surfaces and the lack of *p*-type conduction.

These challenges are directly related to the lack of suitable substrate and the very low conduction band edge of InN due to the extremely large electron affinity (~5.8 eV, the largest value for any known semiconductors) (King et al., 2008; Piper et al., 2008; Walle and Neugebauer, 2003). The commonly available substrates, including Si, GaN, SiC, and sapphire, have large lattice mismatches (in the range of ~7–27%) to InN. The resulting large dislocation and defect densities explain the extremely high electron concentration in *nominally* nondoped InN, as well as the surface electron accumulation and Fermi-level pinning (Calleja et al., 2007b; King et al., 2007; Mahboob et al., 2004a,b; Piper et al., 2004, 2006; Veal et al., 2004).[a] The high background electron concentration, together with the large dislocation and defect densities, also drastically reduces the electron mobility due to various scattering processes including electron–electron and electron–impurity scatterings. The large dislocation and defect densities in InN can, in principle, be eliminated in nanowire structures due to the efficient strain relaxation associated with the large surface area (Glas, 2006; Xiang et al., 2008). Furthermore, recent studies have suggested that dopant

[a] It is worth noting that although surface electron accumulation and Fermi-level pinning in the conduction band are thought to be intrinsic in early days, recent studies have suggested that these properties may *not* be intrinsic in InN, as will be discussed in Section 6.

incorporation can be more efficient in nanowire structures due to the much lower formation energy (Xie et al., 2009), providing great promise to address the *p*-type conduction challenge pertaining to InN. Additionally, the use of nanowire structures can drastically reduce the footprint of electronic and photonic devices.

In this regard, InN nanowires have been intensively studied (Calleja et al., 2007b; Dingman et al., 2000; Johnson et al., 2004; Kamimura et al., 2012; Richter et al., 2009; Sánchez-García et al., 2006; Segura-Ruiz et al., 2009, 2010; Shen et al., 2006; Stoica and Calarco, 2011; Stoica et al., 2006; Werner et al., 2009; Zhao et al., 2012a). The growth techniques vary broadly from chemical vapor deposition (CVD) to epitaxial growth on various substrates, such as Si and GaN, with and/or without the use of foreign catalysts. These *nominally* nondoped InN nanowires typically exhibit tapered morphology and strongly *n-type degenerate* characteristics. The background electron concentration is generally on the order of 10^{18} cm^{-3}, or higher (Calleja et al., 2007a; Richter et al., 2009; Shen et al., 2006; Werner et al., 2009). This *n*-type background doping strongly influences the optical and electrical properties of InN nanowires. For example, the commonly measured electron mobility is *only* in the range of 76–760 cm^2/V s (Cheng et al., 2005; Fehlberg et al., 2006; Richter et al., 2009; Werner et al., 2009), which is much smaller compared to that of high-quality InN epilayers (Fehlberg et al., 2007; Wang et al., 2012c) and theoretical calculations (Polyakov and Schwierz, 2006a,b, 2007; Polyakov et al., 2009). Recently, with the improved molecular beam epitaxial (MBE) growth process, the background electron concentration of *nontapered* InN nanowire structures is drastically reduced to $\sim 10^{13}$ cm^{-3} at room temperature, due to the control of the surface charge properties (Zhao et al., 2013d). This leads to excellent electrical and optical properties of InN nanowires, e.g., electron mobility up to $\sim$12,000 cm^2/V s has been measured at room temperature (Zhao et al., 2013d). More importantly, *p*-type InN nanowires and near-band-edge emitting InN nanowire light-emitting diodes (LEDs) have also been realized (Le et al., 2014).

In this chapter, we review the recent progress made in InN nanowires. Here, the term "nanowires" broadly refer to fibers, nanoneedles, nanorods, nanocolumns, and microcrystals named by different research groups. This chapter is organized as follows. In Section 2, we discuss the growth and synthesis of InN nanowires using various techniques. In Section 3, the electrical and optical properties of *n*-type degenerate InN nanowires are presented. Section 4 is focused on the electrical and optical properties of intrinsic InN nanowires. In Section 5, the achievement of *p*-type conduction in

InN nanowires is presented. In Section 6, we provide a general discussion on the surface charge properties of InN, including InN epilayers and nanowires. In Section 7, we discuss a broad range of devices and applications based on InN nanowires, followed by summary in Section 8.

2. GROWTH AND SYNTHESIS OF InN NANOWIRES

InN nanowires have been synthesized by various CVD processes and epitaxial techniques including MBE, metalorganic CVD (MOCVD, also called metalorganic vapor phase epitaxy, MOVPE), and chemical beam epitaxy (CBE). These techniques have also been used to realize more complex nanowire architectures such as InN core–shell nanowires and site-controlled InN nanowire arrays.

2.1 InN Nanowires Synthesized by CVD

A large body of research work for synthesizing InN nanowires utilizes various CVD techniques in a quartz tube furnace (Cai et al., 2008; He and Noor Mohammad, 2007; Hu et al., 2006; Kang et al., 2006; Lei et al., 2012; Liu and Cheng, 2011; Tang et al., 2004; Vaddiraju et al., 2005; Wang et al., 2008a; Yin et al., 2004b; Zhang et al., 2002, 2012). To the best of our knowledge, Dingman et al. reported the first synthesis of InN nanowires by CVD technique (Dingman et al., 2000). In this work, polycrystalline InN "fibers," with diameters around 20 nm and lengths varying from 100 to 1000 nm, were formed from azido-indium precursors through a solid–liquid–solid process. The synthesis temperature was only around 200°C, well below the decomposition temperature of InN, due to the reduced crystallization barriers with the utilization of metal indium catalyst. Zhang et al. reported single crystalline InN nanowires by CVD technique through a vapor–solid (VS) process (Zhang et al., 2002). In this work, hexagonal wurtzite InN nanowires were formed on Al_2O_3 membrane by mixing In metal and In_2O_3 powders in ammonia environment at around 700°C. Lei et al. further reported the growth of InN nanowires on 4H-SiC substrate. For this, high purity InN powders were used as source materials and the mixed vapor of N_2 and NH_3 was the growth atmosphere (Lei et al., 2012). Additionally, it was reported recently that by solely controlling the growth temperature, InN quantum dots could be evolved into nanorods in a catalyst free process (Madapu et al., 2015).

To achieve a better control of the InN nanowire lateral size, Au-catalyzed growth processes have also been developed on various substrates such as Si substrate coated with SiO*x*/SiN*x* dielectric layers (Cai et al., 2008; Hu et al., 2006; Lan, 2004; Liu and Cheng, 2011; Tang et al., 2004; Zhang et al., 2012), conductive and flexible carbon cloth (Yang et al., 2013), and GaN nanowire template (Lan et al., 2004). Such growth processes are considered through a vapor–liquid–solid (VLS) mechanism. High-resolution transmission electron microscopy (TEM) studies have clearly measured the presence of Au catalyst on the tip of InN nanowires (Cai et al., 2008; Liu and Cheng, 2011; Tang et al., 2004; Zhang et al., 2012). Moreover, a very high growth rate of ~30 μm/h has been reported (Cai et al., 2008).

2.2 InN Nanowires Grown by MBE

InN nanowires have also been grown by MOVPE (MOCVD) (Kim et al., 2012; Kryliouk et al., 2007; Song et al., 2009; Zhang et al., 2011), CBE (Chao et al., 2006), and MBE (Calarco, 2012; Calleja et al., 2007b; Fukunaga et al., 2009; Grandal et al., 2007, 2009; Richter et al., 2009; Shen et al., 2006; Stoica and Calarco, 2011; Stoica et al., 2006; Werner et al., 2009). Since the large body of epitaxial InN nanowires are grown by MBE, in what follows we focus on the formation of InN nanowires by MBE.[b]

Compared to CVD, MBE has several unique advantages including ultra-high purity of source materials, and precise control over substrate temperature, growth rate, and dopant incorporation. The formation of InN nanowires by MBE is generally conducted under a nitrogen-rich condition through a diffusion-driven process, due to the anisotropies of surface energy, sticking coefficient, and diffusion coefficient on different crystalline planes. In the past decade, significant efforts have been devoted toward achieving nearly dislocation-free InN nanowires by MBE (Calarco, 2012; Calleja et al., 2007b; Fukunaga et al., 2009; Grandal et al., 2007, 2009; Richter et al., 2009; Shen et al., 2006; Stoica and Calarco, 2011; Stoica et al., 2006; Werner et al., 2009). The SEM image of such MBE grown InN nanowires is shown in Fig. 1. It is seen that these InN nanowires exhibit tapered morphology, with the nanowire top considerably larger than the nanowire root. Detailed electrical and optical studies (described in Section 3) have further indicated that such tapered InN nanowires are *n*-type degenerate, with very large background electron concentrations and surface electron accumulation.

[b] As we will discuss later on, the MBE grown InN nanowires actually do *not* need the epitaxial relationship with the underlying substrate. Therefore, the term "epitaxy" here refers to epitaxial reactors (MBE), instead of epitaxial relationship.

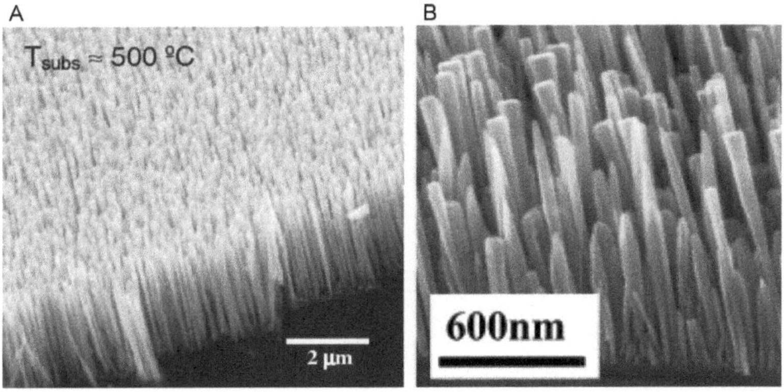

Fig. 1 SEM images of tapered InN nanowires grown by MBE under nitrogen-rich conditions by Calleja et al. (2007b) (A) and Stoica et al. (2006) (B). It is seen that the nanowires exhibit tapered morphology, without clear hexagonal facets.

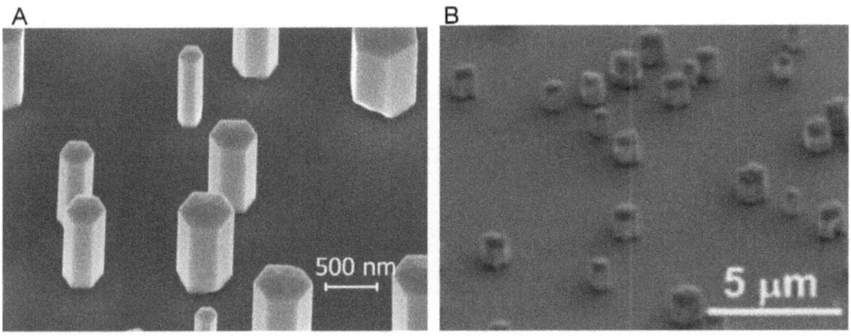

Fig. 2 SEM images of nontapered hexagonal InN nanowires grown by MBE with (A) on Si substrate (Zhao et al., 2012a) and (B) on GaN template (Wang et al., 2015).

Recently, InN nanowires with near-perfect hexagonal morphology have been realized on both Si (Chang et al., 2009; Golam Sarwar et al., 2015; Le et al., 2014; Zhao et al., 2012a, 2013b) and GaN template (Wang et al., 2015). Illustrated in Fig. 2 are SEM images of *nontapered* InN nanowires grown on Si substrate (Fig. 2A) and GaN template (Fig. 2B). Furthermore, it has been found that the nanowire morphology does *not* change significantly with the incorporation of Si and Mg dopants (Le et al., 2014; Zhao et al., 2012a, 2013b). As described in Section 4, compared to tapered InN nanowires, such nontapered InN nanowires can exhibit significantly improved optical and electrical properties (Zhao et al., 2012a,b, 2013d). Their charge properties, in both bulk and surface regions, can also be well controlled, which is central toward the realization of intrinsic and *p*-type InN (Zhao et al., 2012a,c, 2013b,d).

2.3 New Aspects of Synthesizing InN Nanowires

In this section, we discuss several special InN nanowire growth processes, which are important for device applications. For example, it is important to passivate InN nanowire surface to reduce the surface nonradiative recombination. For this purpose, the growth of InN core–shell nanowires has been developed. Moreover, site-controlled InN nanowires have been grown on patterned substrate, which can lead to a better control of the nanowire size and spacing. Additionally, the MBE grown InN nanowires on SiO_x are discussed.

2.3.1 InN Core–Shell Nanowires

Yin et al. reported the growth of InN/InP core–sheath nanowires in a tube furnace by CVD technique (Yin et al., 2004a). InN core with 30–40 nm and InP sheath with a thickness of 20–25 nm were identified by TEM studies. In addition, In nanoparticles were found on the tip of nanowires, suggesting a VLS growth mechanism. Nevertheless, it is noted that stacking faults and twins exist in such InN/InP core–sheath nanowires. Zhang et al. further reported the growth and structural characterization of InN/In_2O_3 core–shell nanowires by CVD technique through a VS mechanism (Zhang et al., 2006). The nanowires exhibited diameters around 20–80 nm, with lengths up to several tens of micrometers.

Besides CVD, InN core–shell nanowires have also been realized by MBE. Cui et al. reported the growth of InN/InGaN core–shell nanowires under nitrogen-rich conditions on Si substrate (Cui et al., 2012). The typical SEM image is shown in Fig. 3A. On comparing with nontapered hexagonal InN nanowires without InGaN shell, InN/InGaN core–shell nanowires exhibit pencil-like morphology due to the deposition of InGaN shell surrounding the InN nanowire core. Schematic of InN/InGaN core–shell nanowire is shown in Fig. 3B, highlighting the InN nanowire core and InGaN shell that is ~10 nm on the sidewall and ~30 nm on the nanowire top. Fig. 3C is the pseudocolor mapping of such InN/InGaN core–shell nanowires, which clearly shows that Ga mostly distributes on the periphery of the nanowire. In addition, the photoluminescence (PL) emission from the InGaN shell is measured at low temperatures. Illustrated in Fig. 3D, besides the InN near-band-edge emission around 1845 nm, another higher energy peak around 1685 nm can be seen, corresponding to the InGaN shell. Such core–shell nanowire structures possess relatively high internal quantum efficiency around 62% at room temperature.

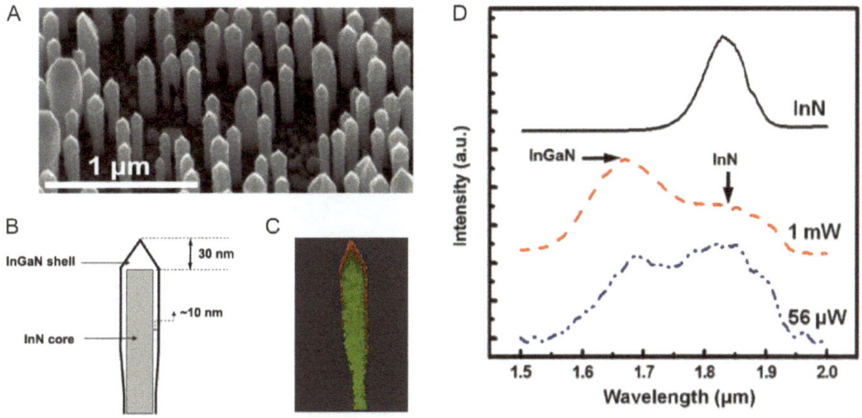

Fig. 3 (A) SEM image of InN/InGaN core–shell nanowires on Si by MBE reactor under nitrogen-rich conditions. (B) Schematic of the structure. (C) The corresponding pseudocolor mapping of In (*green*) and Ga (*red*). (D) The PL spectra measured at 5 K with two different powers. *Dashed lines*: InN/InGaN core–shell nanowires. *Solid line*: InN as a reference (Cui et al., 2012).

2.3.2 InN Nanowires by Selective Area Growth Process

To the best of our knowledge, the first site-controlled growth of InN nanowires was demonstrated by Liang et al. by CVD technique on Au-patterned Si (100) substrate (Liang et al., 2002). In this work, the location of InN nanowires was controlled by depositing Au in the desired areas. Au nanoparticles were clearly measured on the nanowire top from TEM, suggesting a VLS mechanism. Kamimura et al. further reported the growth of N-polar InN microcrystals on Mo-patterned sapphire (0001) substrate by MBE (Kamimura et al., 2010, 2012). In this work, hexagonal openings (area without Mo) were arranged in a triangular lattice with a period of 1 μm and a diameter of 433 nm, as illustrated in Fig. 4A and B. The SEM image of such InN microcrystals is shown in Fig. 4C. It is seen that InN microcrystals are formed in the opening area, with a clear hexagonal facet. TEM studies further confirmed that the laterally grown side areas were nearly free of dislocations, although large threading dislocation densities, on the order of 10^{9-10} cm^{-2}, were measured at the InN/sapphire interface and core region. More recently, vertically oriented InN nanorods on unintentionally patterned sapphire substrates under In-rich conditions were also been reported (Terziyska et al., 2015).

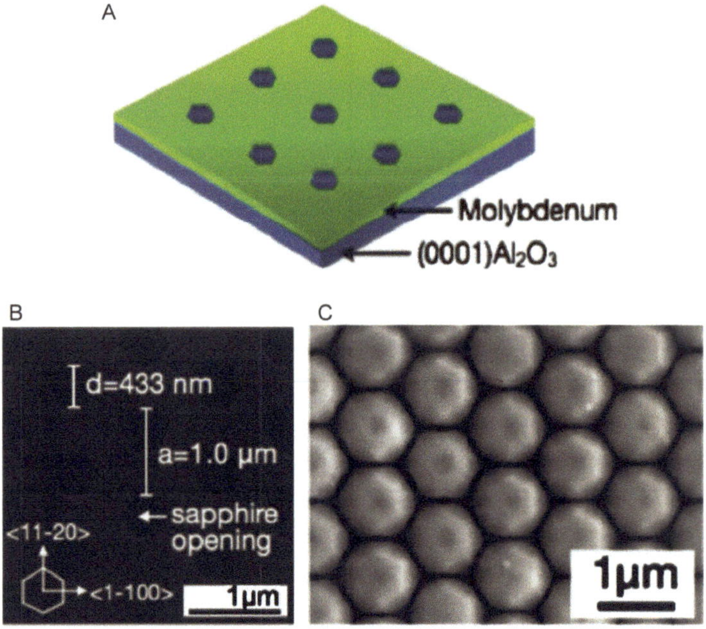

Fig. 4 (A) Schematic of Mo-patterned sapphire substrate. (B) The SEM image of patterned substrate. (C) Top view SEM image of InN microcrystals (Kamimura et al., 2010, 2012).

2.3.3 InN Nanowires Formed Directly on SiOx by MBE

During the MBE growth of InN epilayers (as well as many other semiconductor epilayers), an epitaxial relationship is generally maintained with the underlying substrate. However, for the growth of InN nanowires with MBE, a thin amorphous layer (2–3 nm) has been observed at the InN/Si interface (Chang et al., 2009), suggesting *no* epitaxial relationship with the substrate. It has been further demonstrated that high-quality InN nanowires can be grown directly on a SiO*x* template (Zhao, 2013). Their structural properties are similar to those grown directly on Si substrate. Moreover, strong near-band-edge PL emission has been measured, indicating excellent optical quality. These results, similar to what has been found in the growth studies of GaN nanowires on SiO*x* (Consonni et al., 2011; Ristić et al., 2008; Stoica et al., 2008; Zhao et al., 2013a), clearly indicate that *the epitaxial relationship with the substrate is not required for the MBE growth of InN nanowires*. This opens a new avenue for the MBE growth of InN nanowires.

3. ELECTRICAL AND OPTICAL PROPERTIES OF *n*-TYPE DEGENERATE InN NANOWIRES

The majority of research work on the electrical and optical characterization of InN nanowires has been performed on *n*-type degenerate InN nanowires, due to the difficulty in realizing intrinsic and *p*-type InN nanowires. In this section, the electrical and optical properties of *n*-type degenerate InN nanowires are discussed.

3.1 Electrical Properties

In the past decade, efforts have been devoted to characterizing the electron concentration and electron mobility of InN nanowires, which are predominantly carried out on single InN nanowires (Calleja et al., 2007a; Chang et al., 2005; Koley et al., 2011; Richter et al., 2009; Werner et al., 2009). The electrodes are typically defined by standard e-beam lithography process, and either back dielectric gate (Koley et al., 2011; Richter et al., 2009) or side polymer electrolyte gate (Khanal et al., 2009, 2011) is applied to modulate the electrical conduction in the nanowire channel. It has been commonly found that the electrical conduction is largely determined by the surface electron accumulation layer. Based on measurements performed on single nanowire transistors, the electron mobility is estimated to be in the range of 76–760 cm^2/V s, with the bulk electron concentration on the order of 10^{20} cm^{-3} (Richter et al., 2009; Werner et al., 2009) and the surface electron density in the range of 10^{11-14} cm^{-2} (Calleja et al., 2007a; Richter et al., 2009; Werner et al., 2009). Werner et al. have proposed that the surface conduction layer may come from the InN/In_2O_3 interface due to the existence of surface oxide layer, which is evidenced from X-ray photoelectron spectroscopy (XPS) (Werner et al., 2009). Additionally, as illustrated in Fig. 5, the electron mobility is reduced as the increase of electron concentration, due to the increased scattering events.

Besides the single InN nanowire transistor technique, Liu et al. further developed scanning current voltage microscopy (SIVM) and scanning gate microscopy techniques to characterize the electron concentration and mobility in InN nanowires (Liu et al., 2009). From SIVM mapping, they found strong correlation between surface barrier change and electrical conduction, and attributed it to the presence of a high density of electrons at the nanowire surface. An electron mobility of $\sim$150 cm^2/V s, with an electron concentration on the order of 10^{18} cm^{-3} and a surface electron density on

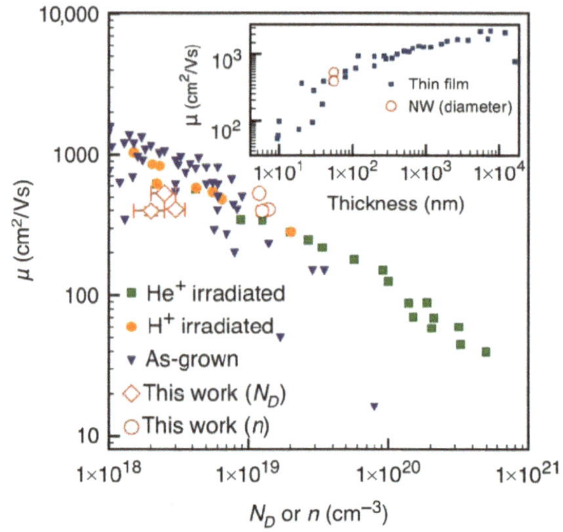

Fig. 5 Electron mobility as a function of donor/electron concentration measured from various InN epilayers and *n*-type degenerate nanowires (Khanal et al., 2009).

the order of 10^{13} cm^{-2}, was estimated. Moreover, some other fundamental electrical transport and/or magneto-transport phenomena in *n*-type degenerate InN nanowires have also been studied (Aravind et al., 2009; Blömers et al., 2008; Frielinghaus et al., 2009; Huang et al., 2011b; Lu et al., 2016; Petersen et al., 2009; Richter et al., 2008).

3.2 Optical Properties

Optical properties of InN nanowires depend strongly on the background electron concentration. For example, in some of the early InN nanowires, due to the extremely high background electron concentration, the Fermi edge is located deep in the conduction band, leading to a PL peak energy around 1.9 eV, which is much higher than the bandgap energy of InN (Liang et al., 2002; Tang et al., 2004). With improving material quality, near-band-edge emission is able to be observed (Johnson et al., 2004; Lan, 2004; Stoica et al., 2006). Nevertheless, the background electron concentration of those InN nanowires is still on the order of 10^{19} cm^{-3}, leading to nearly *temperature-independent* and *power-independent* PL spectra under typical optical pumping conditions. Moreover, PL spectra have remained broad, due to both inhomogeneous broadening and high background electron concentration. For example, Johnson et al. reported PL peak energy around 0.8 eV with a linewidth more than 100 meV (Johnson et al., 2004). With further improving material quality, the background electron concentration of InN nanowires is reduced to $\sim 10^{18}$ cm^{-3}, leading to the observation of *temperature-dependent* and *power-dependent* PL peak energies (Feneberg et al., 2008; Fukunaga et al., 2009; Kamimura et al., 2012; Segura-Ruiz et al., 2009, 2010). Detailed studies have further indicated that the PL emission in *n*-type degenerate InN nanowires involves the band-to-band recombination in the low-doped bulk region and Mahan exciton emission in the high-doped near-surface region (Zhao et al., 2012b). Time-resolved PL properties of *n*-type degenerate InN nanowires have also been investigated (Ahn et al., 2012; Chang and Gwo, 2009). It has been found that nanowires with small diameters (~30 nm) possess very fast initial carrier decay, in the range of only a few picoseconds, highlighting the surface-associated influence on carrier relaxation in semiconductor nanostructures.

The influence of high background electron concentration and surface electron accumulation on the optical properties has also been observed by Raman spectroscopy (Domènech-Amador et al., 2012; Jeganathan et al., 2010; Lazić et al., 2007; Lazic et al., 2008; Schafer-Nolte et al., 2010), e.g., the measurement of LO phonon–plasmon coupling modes (Lazić et al., 2007). Furthermore, studies on *single* InN nanowires have clearly suggested the *symmetry-forbidden* LO mode, which is ascribed to the presence of strong electric field in the surface electron accumulation layer (Schafer-Nolte et al., 2010). These measurements have provided additional evidences for the presence of surface electron accumulation on the lateral nonpolar surfaces of *n*-type degenerate InN nanowires. In addition, narrower Raman peaks have been observed with Mg doping, while an asymmetric broadening of the Raman peaks has been measured with Si doping; this is mainly due to the change of background electron concentration (Cuscó et al., 2010; Schafer-Nolte et al., 2010).

4. ELECTRICAL AND OPTICAL PROPERTIES OF INTRINSIC InN NANOWIRES

By improving the MBE growth process, intrinsic InN nanowires have been demonstrated (Zhao et al., 2012a,b,c, 2013d). Such intrinsic InN nanowires are free of surface electron accumulation and possess distinctly different electrical and optical properties compared to *n*-type degenerate InN nanowires.

4.1 Controlling the Surface Electron Accumulation

Angle-resolved XPS experiments have been performed to measure the surface charge properties on the nonpolar planes (nanowire sidewalls) of nondoped InN nanowires grown on Si substrate (Zhao et al., 2012a). The corresponding valence band spectrum is shown in Fig. 6A. It is seen that, the Fermi level is located ~0.5 eV above the valence band maximum (VBM). For a bandgap energy of ~0.65 eV, this indicates that the Fermi level is located in the band gap, i.e., *not* pinned in the conduction band. Evidently, there is *no* surface electron accumulation on the *grown nonpolar surface* of InN.

The near-surface Fermi level of InN nanowires grown under nonoptimized conditions has also been investigated as a comparison (Zhao et al., 2012a). The SEM image of such InN nanowires is shown in the inset of Fig. 6B. It is seen that the nanowires are highly tapered, in contrast to the nanowires grown under optimized conditions. The valence band spectrum is shown in Fig. 6B. It is seen that the Fermi level is pinned about 0.35 eV above the conduction band edge. These studies indicate that the surface electron accumulation and Fermi-level pinning in the conduction band are *not* fundamental properties of InN. The previously reported surface electron accumulation and large background electron concentration in InN nanowires are largely due to the presence of extensive surface defects and/or impurity incorporation (Zhao and Mi, 2014; Zhao et al., 2012a, 2013d).

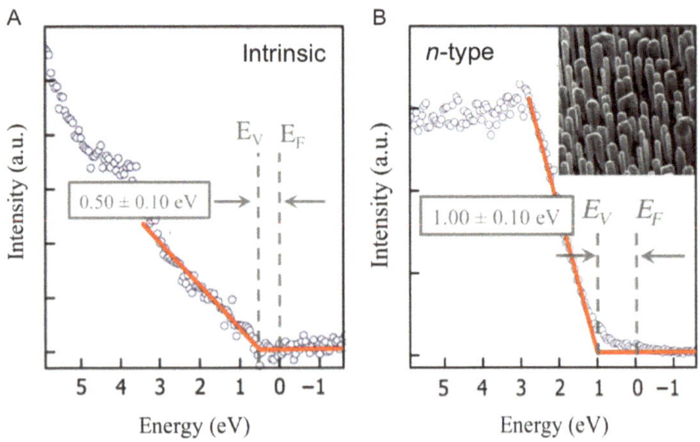

Fig. 6 Valence band spectrum of (A) intrinsic InN nanowires and (B) *n*-type degenerate InN nanowires, which are grown under nonoptimized conditions, with the *inset* showing the corresponding SEM image (Mi and Zhao, 2015).

4.2 Electrical Properties

The electrical transport properties of intrinsic InN nanowires have been studied by the nanoprobe technique, wherein a tungsten tip is utilized as the top contact and melted In droplets are used as the back contact (Zhao et al., 2013d). The top-left inset of Fig. 7 shows such a nanoprobing to a single InN nanowire. Fig. 7 shows the *I–V* characteristics, which evolve from Ohmic conduction at low bias to the space charge limited conduction with charge traps at high bias. With further increasing the bias, the conduction changes into charge-trap-free space charge limited conduction. The observation of space charge limited conduction strongly suggests the low background electron concentration in intrinsic InN nanowires, which is consistent with the absence of surface electron accumulation on the nanowire sidewalls measured by XPS. More importantly, the filling of charge traps as the increase of applied bias, as illustrated in the bottom-right inset of Fig. 7, indicates that the Fermi level is *neither* pinned in the conduction band *nor* in the band gap. The characteristic energy for charge traps has been further derived to be ~65 meV below the conduction band edge, and these charge traps are suggested from the ionic bonding nature of InN (Moustakas, 2013; Zhao et al., 2013d).

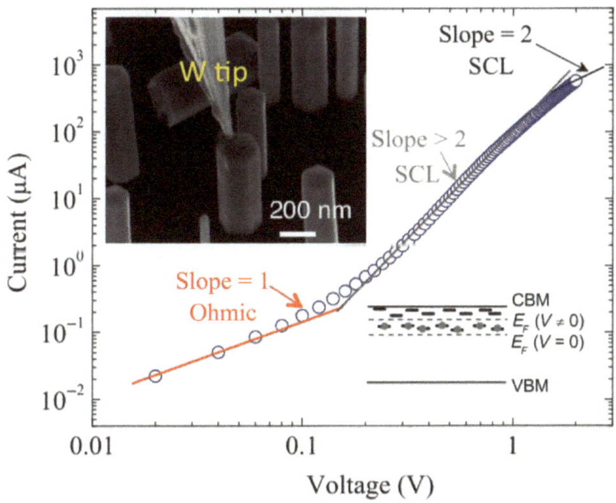

Fig. 7 The *I–V* characteristics of intrinsic InN nanowires measured by the nanoprobe technique. The *top-left inset* shows an SEM image of the measurement configuration. The *right-bottom inset* illustrates the progressive shift of the Fermi level due to the filling of charge traps with increasing the applied bias (Zhao and Mi, 2014).

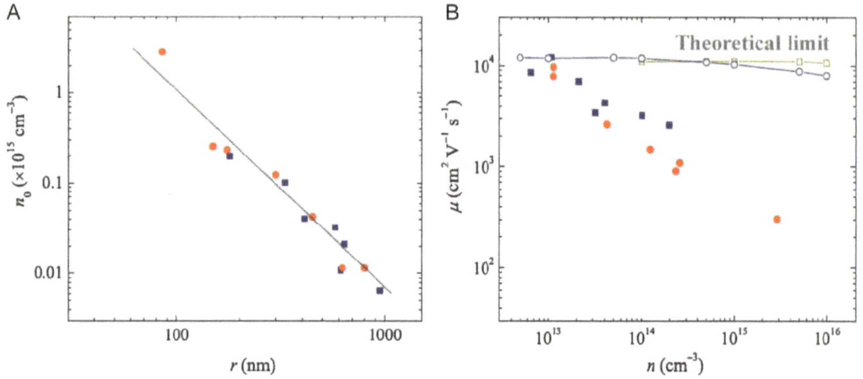

Fig. 8 (A) Electron concentration as a function of InN nanowire radius in a logarithmic scale. (B) Electron mobility as a function of electron concentration in InN nanowires. The *open symbols* are theoretical calculations (Mi and Zhao, 2015; Zhao et al., 2013d).

By further correlating Ohmic conduction with space charge limited conduction, the nanowire radius-dependent electron concentration can be derived (Zhao et al., 2013d). As illustrated in Fig. 8A, the background electron doping concentration is on the order of 10^{13} cm^{-3} for large diameter nanowires. Fig. 8B shows the electron mobility as a function of the background electron concentration. It is seen that the electron mobility reaches up to ~12,000 cm^2/V s, approaching the theoretically predicted maximum electron mobility at room temperature (Zhao et al., 2013d). Furthermore, it is also seen that the electron mobility decreases drastically as the increase of carrier concentration. This is because the relatively large electron concentration is from nanowires with smaller radii and, as a consequence, the electron mobility is limited by surface scatterings (Zhao et al., 2013d).

4.3 Optical Properties

Optical properties of intrinsic InN nanowires have also been found to be distinctly different compared to *n*-type degenerate InN nanowires. The Raman scattering spectrum measured from intrinsic InN nanowires is shown in Fig. 9, with the measurement configuration shown in the inset. The spectrum taken from *n*-type InN nanowires is also shown for a comparison. It is seen that compared to the spectrum of *n*-type InN nanowires, the intrinsic InN nanowires does *not* exhibit the coupled phonon mode around 445 cm^{-1} (Cuscó et al., 2009, 2010; Lazic et al., 2008), which is consistent with the absence of surface electron accumulation (Zhao et al., 2012c). Moreover, the main Raman shift peak $E_2{}^H$ of intrinsic InN nanowires, which appears

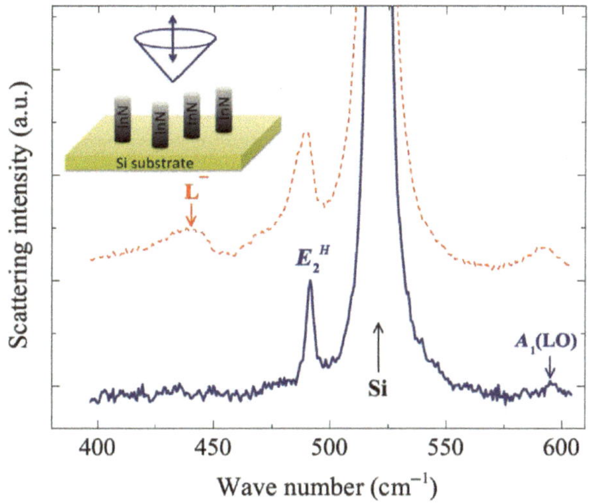

Fig. 9 Micro-Raman spectrum of intrinsic InN nanowires (*solid curve*). The spectrum of *n*-type InN nanowires is also shown for a comparison (*dashed curve*). The *inset* shows the experimental configuration, with the incident light in parallel to the nanowire *c*-axis (Zhao et al., 2012c).

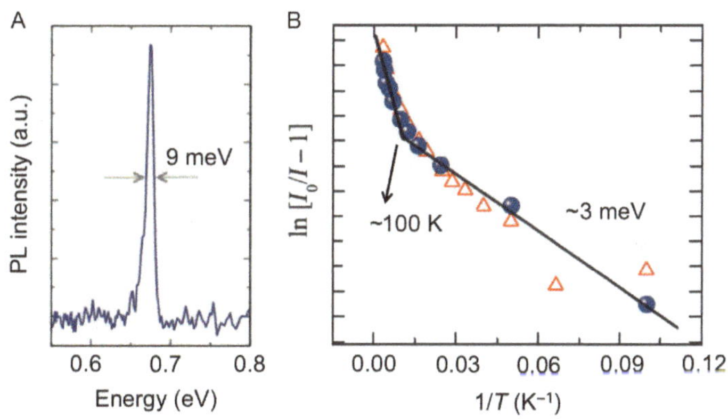

Fig. 10 (A) The micro-PL spectrum of intrinsic InN nanowires measured at 10 K under an excitation power of ~0.5 μW with a 632 nm laser. (B) The normalized integrated PL peak intensity vs the inverse temperature. *Different symbols* represent InN nanowires with different sizes (Zhao et al., 2012a,b).

around 491 cm^{-1}, is significantly narrower compared to that measured from *n*-type InN nanowires, consistent with low background electron concentration (Zhao et al., 2012c).

The PL properties of intrinsic InN nanowires have been investigated (Zhao et al., 2012a,b). As illustrated in Fig. 10A, the PL spectral linewidth (full width at half maximum) is only about 9 meV, which is the narrowest linewidth measured from any InN structures (Fu et al., 2006a; Segura-Ruiz et al., 2009; Shen et al., 2006; Stoica et al., 2006). By further analyzing the PL spectral linewidth, the background electron concentration on the order of 10^{15} cm^{-3} can be estimated, if assuming an inhomogeneous broadening of ~5 meV (Moret et al., 2009; Schlager et al., 2006; Stoica et al., 2006). This number is consistent with the estimation from the electrical transport properties measurements and is about 2–3 orders of magnitude smaller compared to *n*-type degenerate InN nanowires and epilayers (Fu et al., 2006a; Segura-Ruiz et al., 2009; Shen et al., 2006; Stoica et al., 2006).

Detailed analysis of the temperature- and power-dependent PL spectra further indicates free exciton emission. Fig. 10B shows $\ln[I_0/I-1]$ vs $1/T$, where I_0 is the maximum PL intensity at low temperature and I represents the integrated PL intensity measured at various temperature T. It is seen that there exist two distinct slopes, with a crossover temperature around 100 K (~8 meV). This slope change temperature indicates an upper limit of the exciton binding energy (Hoang et al., 2009; Mishra et al., 2007; Titova et al., 2006; Wu, 2009). Careful fittings in the low-temperature range (10–100 K) produce an exciton binding energy of around 3 meV. This small activation energy is consistent with the derived exciton binding energy in InN (Arnaudov et al., 2004; Fu et al., 2006b; Zhao et al., 2012b). Such a temperature-dependent PL emission thus suggests that the radiative recombination process changes from free exciton emission at low temperatures to electron–hole plasma emission at high temperatures in intrinsic InN nanowires. This exciton emission is also evidenced from the low-temperature power-dependent PL spectra, wherein the integrated PL intensity varies linearly with the excitation power (Zhao et al., 2012b).

Phonon replica or phonon sideband emission (PSB) has been observed (Zhao et al., 2012c). Fig. 11A shows the PL spectrum in a logarithmic scale, measured at 6 K under a 50 μW excitation. The inset shows a PL spectrum under an excitation of 500 μW. It is seen that besides the near-band-edge emission PL peak, another low-energy PL peak appears, with an energy separation of ~70 meV. This low-energy PL peak possesses an extremely narrow PL spectral linewidth (~9 meV), and the PL peak intensity is about *two orders of magnitude weaker* compared to the near-band-edge emission PL peak over a wide range of excitation power, as illustrated in Fig. 11B, suggesting that this low-energy PL peak is a phonon replica of the near-band-edge emission PL peak. The observation of phonon replica is consistent with the extremely low background electron concentration in intrinsic InN nanowires.

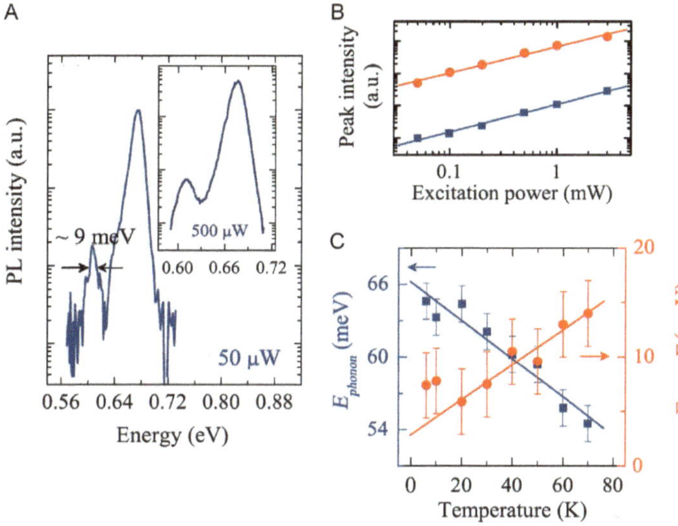

Fig. 11 (A) PL spectra of intrinsic InN nanowires in a logarithmic scale, measured at 6 K with an excitation power of 50 μW (the *inset* was under an excitation of 500 μW). (B) The derived power-dependent PL peak intensity. *Filled blue squares* and *filled red circles* represent PL peaks due to phonon sideband emission and near-band-edge emission, respectively. (C) The extracted temperature-dependent phonon energy (E_{phonon}) and the linewidth (in the form of $\Gamma_{PSB}-\Gamma$, where Γ is the spectral linewidth of the near-band-emission PL peak) measured under an excitation of 500 μW. *Solid lines* were linear fittings (Zhao et al., 2012c).

Temperature-dependent phonon replica characteristics in intrinsic InN nanowires have been further analyzed (Zhao et al., 2012c). Illustrated in Fig. 11C, the PL spectral linewidth of phonon sideband emission (Γ_{PSB}) becomes broader as the temperature increases while the phonon energy (E_{phonon}) becomes smaller as the temperature increases. LO phonon energy shift with temperature can be described by,

$$\omega_{PSBm}(T) = m\omega_0 - \left(\frac{5}{2} - m\right) k_B T - m\Delta(T) \tag{1}$$

where m is the PSB index (here, $m=1$[c]), ω_0 represents the PSB energy at $T=0$ K, and the third term is the anharmonic shift (Holtz et al., 2009; Song et al., 2006). The linear fit of the temperature-dependent phonon energy, as shown by the blue line in Fig. 11C, gives a slope of ~0.16 meV K^{-1},

[c] In this work, due to the cut-off wavelength of the detector, only the first PSB band was observed; hence $m=1$.

which is consistent with the calculated temperature coefficient if taking $m=1$ in Eq. (1). Similarly, the temperature-dependent PL spectral linewidth of phonon sideband emission can be described by:

$$\Gamma_{PSBm}(T)=2\sqrt{\left(\frac{5}{2}-m\right)k_{\mathrm{B}}T+\Gamma(T)+\Gamma_{phonon}(T)} \quad (2)$$

where $\Gamma(T)$ is the temperature-dependent interband PL emission spectral linewidth and $\Gamma_{phonon}(T)$ is the negligible temperature-dependent phonon broadening (Holtz et al., 2009). The plot of $(\Gamma_{PSB}(T)-\Gamma(T))$ vs T is shown in Fig. 11C. The red line is a linear fit, giving a slope of ~0.16 meV K^{-1}. This number is also consistent with the calculated temperature coefficient if taking $m=1$ in Eq. (2).

5. *p*-TYPE InN NANOWIRES

Mg, with activation energy of ~60 meV in InN, has been widely investigated for *p*-type doping in InN (Van de Walle et al., 2010; Yoshikawa et al., 2010). Mn and Zn have also been suggested as *p*-type dopants in InN (Chai et al., 2012; Wang et al., 2012a). In the past decade, significant progress has been made in the growth and characterization of Mg-doped InN epilayers. The presence of Mg acceptors has been clearly measured from a variety of experiments (Anderson et al., 2006; Khan et al., 2007; Mayer et al., 2011; Wang et al., 2007a). Furthermore, the buried *p*-type conduction underneath the surface electron accumulation layer has been detected by electrolyte-based capacitance voltage experiments (Ager et al., 2008; Anderson et al., 2006; Wang et al., 2007b, 2008b, 2011). Moreover, the existence of free holes has been suggested from thermoelectric measurements (Dmowski et al., 2009, 2012; Mayer et al., 2011; Wang et al., 2011). However, a direct measurement of *p*-type conduction and the demonstration of InN LEDs have *not* been realized using InN epilayers, due to the presence of surface electron accumulation. In this section, we discuss the progress made on Mg-doped InN nanowires, including the direction measurement of *p*-type conduction and the demonstration of InN nanowire LEDs.

5.1 PL Characteristics of Mg-Doped InN Nanowires

Previously, Mg-doped InN nanowires have been investigated by Raman spectroscopy. However, only reduced background electron concentration is suggested, and such Mg-doped InN nanowires have remained *n*-type degenerate (Cuscó et al., 2010; Schafer-Nolte et al., 2010). Recently, with the improved MBE growth process as discussed in Section 2, Mg acceptor-related radiative recombination has been clearly measured by PL spectroscopy (Zhao et al., 2013c). Illustrated in Fig. 12A are the power-dependent PL spectra. It is seen that at high excitation powers, besides the near-band-edge emission PL peak around 0.67 eV, another low-energy PL peak around 0.61 eV appears. This energy separation (~60 meV) is consistent with Mg activation energy in InN, suggesting that the low-energy PL peak is due to the Mg acceptor-related radiative recombination. In addition, from Fig. 12A it is seen that at the lowest excitation power only Mg acceptor-related radiative recombination PL peak exists, and with the increase of excitation power near-band-edge emission PL peak appears.

More features of these two PL peaks are revealed by the extracted excitation power-dependent integrated PL intensity (I_{PL}) and PL peak energy (E_{PL}). The superscripts L and H correspond to the low-energy PL peak and high-energy PL peak, respectively. As shown in Fig. 12B, both ${I_{PL}}^L$ and ${I_{PL}}^H$ increase as the increase of excitation power. However, ${I_{PL}}^L$ rises faster at low excitations, followed by a saturation trend at high excitations.

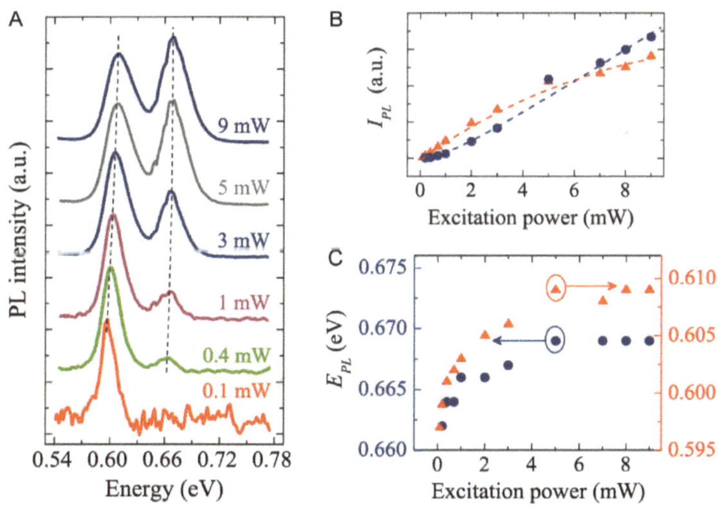

Fig. 12 (A) PL spectra of Mg-doped InN nanowires measured at 7 K. (B) and (C) The derived excitation power-dependent integrated PL intensities and PL peak energies, respectively. *Red triangles*: Mg acceptor-related emission. *Blue circles*: near-band-edge emission. *Dashed curves* are guide-for-eye (Zhao et al., 2013c).

This is in contrast to $I_{PL}{}^{H}$, which increases slower at low excitations but takes off faster at high excitations, with no saturation trend. This difference is attributed to the redistribution of hole population in the Mg acceptor energy levels and in the valence band (Klochikhin et al., 2005; Zhao et al., 2013c). Moreover, as illustrated in Fig. 12C, at low excitations both $E_{PL}{}^{L}$ and $E_{PL}{}^{H}$ exhibit blue-shift of ~12 and ~7 meV, respectively, and then stay nearly constant at high excitations. Detailed analysis has further suggested that the low-energy PL peak is due to neutral donor–acceptor pair recombination (D^0A^0) at low temperatures and free electron–acceptor transition (eA^0) at high temperatures (Zhao et al., 2013c).

In InN, the formation of defects such as nitrogen vacancies can be enhanced by Mg doping. These *n*-type donors have also been found to be stable in Mg-doped InN (Duan and Stampfl, 2009). However, the formation of nitrogen vacancies can be suppressed in InN nanowires grown under nitrogen-rich conditions by MBE.

5.2 *p*-Type InN Surface

The near-surface Fermi level of Mg-doped InN nanowires has been investigated by XPS experiments (Zhao et al., 2013b). Illustrated in Fig. 13, Mg-doped InN nanowires with a wide range of Mg doping concentration do *not* exhibit any surface electron accumulation, i.e., the near-surface Fermi level is located near or below the conduction band edge, which is in contrast to the presence of surface electron accumulation from Mg-doped InN epilayers (Ager et al., 2008; Anderson et al., 2006; Wang et al., 2007a, 2011). Another very important feature seen from Fig. 13 that the *p*-type surface is measured in Mg-doped InN nanowires with a few doping levels. For instance, InN nanowires with the highest Mg doping concentration investigated in this study (with a Mg cell temperature of 240°C) exhibit a near-surface Fermi level of ~0.15 eV above the VBM.

First-principle calculations have been further performed to understand the Mg doping into InN nanowires (Zhao et al., 2013b). It is found that the In-substitutional Mg dopant has a significantly lower surface formation energy compared to that in the bulk region, consequently leading to the preferential incorporation of Mg dopant into the near-surface region. However, the direct doping of Mg atoms into InN nanowires suffers considerably from the large surface desorption of Mg at elevated growth temperatures. This largely explains the nearly intrinsic surface measured from InN nanowires with relatively low Mg doping concentrations,

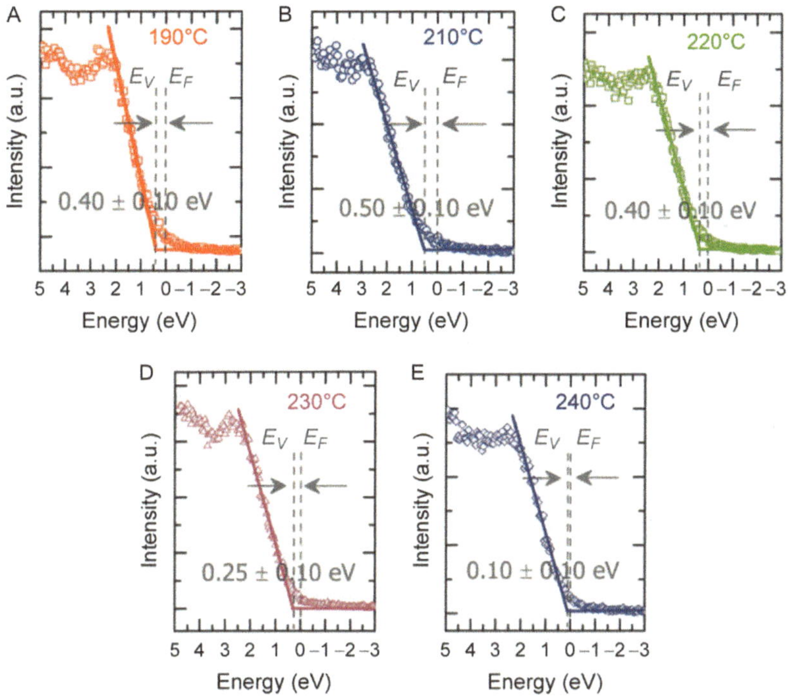

Fig. 13 XPS spectra of Mg-doped InN nanowires with different Mg doping concentrations. Panels (A)–(E) correspond to Mg cell temperatures of 190, 210, 220, 230, and 240°C, respectively (the higher Mg cell temperature, the higher Mg doping concentration) (Zhao et al., 2013b).

e.g., samples with Mg cell temperatures below 220°C shown in Fig. 13. As the increase of Mg doping concentration, Mg surface incorporation dominates over Mg surface desorption, leading to the achievement of *p*-type surface.

5.3 *p*-Type InN Nanowire Transistor

The direct evidence for *p*-type conduction has been measured from single Mg-doped InN nanowire transistors (Le et al., 2015; Zhao et al., 2013b). Schematic of the device is shown in Fig. 14A. The source–drain current (I_{SD}) vs the source–drain voltage (V_{SD}) under different back gate voltages (V_{GD}) is shown in Fig. 14B. It is seen that the channel conduction increases as more negative V_{GD} is applied. This increase in conductance (I_{SD}/V_{SD}) with increasingly negative V_{GD} provides unambiguous evidence for *p*-type conduction in Mg-doped InN nanowires. In addition, free holes in the

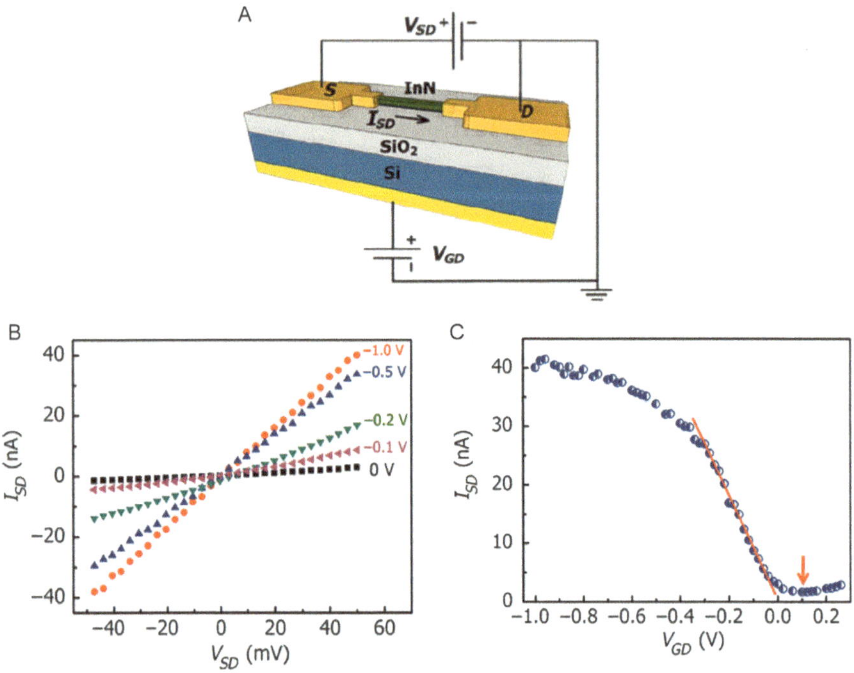

Fig. 14 (A) Schematic of Mg-doped InN nanowire field effect transistor. (B) $I_{SD}-V_{SD}$ characteristics under different backgate voltages. (C) $I_{SD}-V_{GD}$ dependence under $V_{SD}=0.05$ V, with the *arrow* denoting the minimum I_{SD} position. These measurements were performed at room temperature (Mi and Zhao, 2015; Zhao et al., 2013b).

conduction channel are depleted at a slightly positive V_{GD}, marked by the arrow in Fig. 14C, indicating that at zero gate voltage the nanowire has *p*-type conduction. By the $I_{SD}-V_{GD}$ dependence, the field effect hole mobility can be derived to be $\sim$100 cm^2/V s, which is comparable to theoretical calculations by the ensemble Monte Carlo method (Ma et al., 2011). With this hole mobility, the hole concentration (at $V_{GD}=0$ V) can be approximated to be 5×10^{15} cm^{-3}. Such a *direct* measurement of *p*-type conduction from Mg-doped InN nanowires is a consequence of the absence of surface electron accumulation on the nanowire nonpolar sidewalls.

5.4 *p–i–n* InN Nanowire LEDs

Another strong evidence for the achievement of *p*-type conduction in InN nanowires is the demonstration of *p–i–n* InN nanowire LEDs (Le et al., 2014). Schematic of the *single p–i–n* InN nanowire LED structure is shown in Fig. 15A. In this study, electrical injection was realized from two metal

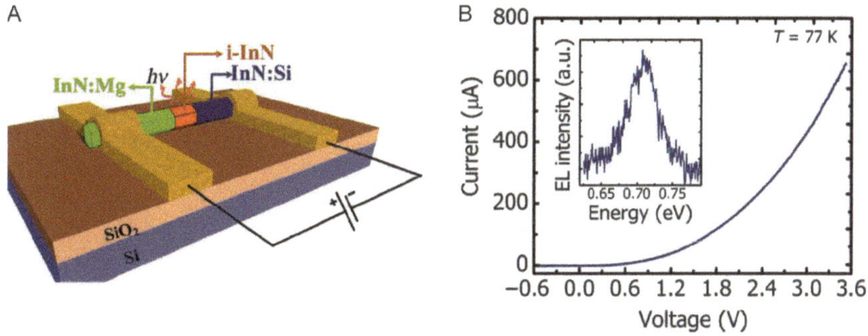

Fig. 15 (A) Schematic of *p–i–n* InN single nanowire LEDs. (B) The *I–V* characteristics of the device measured at 77 K, with the *inset* showing the EL spectrum at an injection current of 500 μA under a continuous-wave biasing (Le et al., 2014).

contacts, which were placed by standard e-beam lithography and metallization processes on the *n*- and *p*-InN segments, respectively. The Si and Mg doping concentrations were $\sim 5 \times 10^{17}$ cm^{-3} and $\sim 4 \times 10^{19}$ cm^{-3}, respectively. The *I–V* characteristics measured at 77 K are shown in Fig. 15B. It is seen that the device has a turn on voltage around 0.7 V. An electroluminescence (EL) spectrum measured under an injection current of 500 μA is shown in the inset of Fig. 15B, and a peak around 0.7 eV (corresponding to ~1.8 μm in wavelength) can be clearly seen. This near-band-edge EL emission from *p–i–n* InN nanowires provides unambiguous evidence for the *p*-type conduction in Mg-doped InN nanowires and further suggests the absence of surface electron accumulation on InN grown surfaces.

6. ON THE SURFACE CHARGE PROPERTIES OF InN

The surface electron accumulation and Fermi-level pinning in the conduction band have been previously considered to be intrinsic properties of InN (Calleja et al., 2007a; Mahboob et al., 2004a). Recent theoretical calculations, however, have suggested that the surface electron accumulation and Fermi-level pinning in the conduction band can be absent on the nonpolar planes of InN (Eisele and Ebert, 2012; Van de Walle and Segev, 2007). This is further confirmed by cross-sectional scanning photoelectron microscopy/spectroscopy (SPEM/S) (Wu et al., 2008) and tunneling *I–V* spectroscopy (Ebert et al., 2011; Eisele and Ebert, 2012). In these studies, the surface Fermi level was measured to locate 0.4–0.5 eV above the VBM, i.e., well below the conduction band edge and thus no surface electron accumulation. These measurements are consistent with the XPS experiments performed on the grown nonpolar surfaces of InN nanowires, as described in Section 4. Furthermore, as discussed in Sections 4 and 5, by controlling the dopant incorporation into InN nanowires, the surface charge properties can be widely tuned from *n*-type degenerate to nearly *p*-type degenerate (Zhao et al., 2012a, 2013b). These studies clearly indicate that the surface electron accumulation and Fermi-level pinning in the conduction band on InN nonpolar surfaces are *not* intrinsic properties of InN. Further studies by Zhao et al. have shed light on the potential causes, including the presence of extensive surface defects and impurities, for the commonly measured surface electron accumulation and Fermi-level pinning in the conduction band on InN grown nonpolar surfaces, as discussed in Section 4 (Zhao and Mi, 2014; Zhao et al., 2012a,b, 2013b,d).

Significant progress has also been made in understanding the surface charge properties of polar InN surfaces (Eisenhardt et al., 2013; Kuo et al., 2011; Linhart et al., 2012). For N-polar surfaces, it has been found that the near-surface Fermi level can be reduced from ~1.4 to ~0.8 eV above VBM by simply using HCl treatment (Kuo et al., 2011). For In-polar surfaces, it has been found that the near-surface Fermi level can be reduced significantly from ~1.3 to ~0.8 eV by varying the Mg dopant concentration (Linhart et al., 2012). These studies, therefore, indicate that the surface electron accumulation and Fermi-level pinning in the conduction band on both N- and In-polar planes may *not* be intrinsic properties of InN as well, if the epitaxy and surface quality of InN can be further improved.

7. InN NANOWIRE DEVICES AND APPLICATIONS

In the past decade, InN nanowire devices and applications have been intensively studied, including conventional optoelectronic devices such as light emitters, photodetectors, and solar cells, as well as emerging devices such as spintronic devices, nanogenerators (NGs), thermoelectric devices, and biosensors.

7.1 InN Nanowire Optoelectronic Devices

7.1.1 Light Emitters from Visible to Tera-Hertz

InN, due to its narrow bandgap energy, is an important material for near-infrared emitters. Furthermore, with suitable quantum confinement and by alloying with GaN, InN can also emit light efficiently in the visible spectral range (Bayerl and Kioupakis, 2014). Chen et al. have reported electrically excited infrared emission from InN nanowire transistors under a high electric field (Chen et al., 2007). As discussed in Section 5, *p–i–n* InN nanowire LEDs have also been demonstrated (Le et al., 2014). In addition, with the incorporation of a small amount of Ga, InGaN nanowire LEDs emitting around 1.46 μm have been reported (Kishino et al., 2012).

Additionally, due to the very large electron mobility and relatively small hole mobility, InN is one of the most promising candidates for tera-Hz emission (Ahn et al., 2007; Wilke et al., 2008). It has been found that compared to InN epilayers, InN nanowires can exhibit much stronger emission due to more efficient light absorption. Moreover, the performance of such InN nanowire tera-Hz emitters can be drastically improved by using nanowire arrays free of surface electron accumulation, since for these tera-Hz emitters the surface band bending induced electric field can significantly reduce the photo-Dember field that is responsible for the tera-Hertz emission (Ahn et al., 2007; Wilke et al., 2008).

7.1.2 Photodetectors

InN, due to its large absorption coefficient (10^5 cm^{-1} in the visible spectral range and 10^4 cm^{-1} in the near-infrared spectral range), and extremely large electron mobility (up to 14,000 $cm^2/V\ s$ at room temperature), holds tremendous promise for high-speed near-infrared photodetectors. Moreover, nearly defect-free InN and In-rich InGaN nanowires can be directly formed on Si substrate within the CMOS thermal budget, thus providing a viable alternative toward Si photonics. For comparison, the direct integration of conventional III–V semiconductors on Si is severely limited by the presence of large densities of dislocations.

The photoconduction properties of InN nanowires have been investigated. Chen et al. performed detailed studies on the photoconductivity of single InN nanowires under an excitation wavelength of 808 nm (Chen et al., 2009). A photoconductive gain around 8×10^7 was derived at room temperature. In addition, photocurrent generation devices were reported by Lee et al. (2008). Besides the studies on the photoconduction properties of InN nanowires, the light response of *p–i–n* InN nanowire photodiodes has also been investigated by Zhao et al. (2014). In this work, a light response up to 1.5 μm was measured, providing great promise for Si-integrated near-infrared photodetectors for chip-level optical communications. Additionally, Lai et al. demonstrated near-infrared photodetectors based on *n*-type InN nanowires and *p*-type ploy(3-hexylthiophene) (Lai et al., 2010). The hybrid device exhibited a diode behavior and a photoresponse in the wavelength range of 900–1260 nm under different reverse biases. An external quantum efficiency of 3.4% at 900 nm was measured at −10 V.

7.1.3 Solar Cells

Large-area InN nanowire solar cells have also been investigated by Nguyen et al. (2010). In this work, InN nanowire homojunctions, i.e., InN:Mg/i-InN/InN:Si, were used. The device structure was grown by MBE under nitrogen-rich conditions on n-Si (111) substrate. Short-circuit current of 14.4 mA/cm^2 and open-circuit voltage of 0.14 V were measured from such solar cells. A power conversion efficiency of ~0.68% was estimated under one-sun (AM 1.5G) illumination.

7.2 Emerging Devices with InN Nanowires

InN nanowire piezoelectric NGs have been investigated by a number of research groups. Huang et al. reported single InN nanowire NGs (Huang et al., 2010). The nanowire was grown by Au-catalyzed VLS process, and an output voltage of up to 1 V was realized. To further maximize the nanorod deformation, Ku et al. demonstrated large-area obliquely aligned InN nanowire NGs (Ku et al., 2012, 2013). In this work, the nanowires were grown on ZnO-coated Si (111) substrate, and an average output current of 205.6 nA/NW-scan under a tip deflection force of ~3 nN was obtained. Recently, Liu et al. further investigated large-area NGs made from *p*-type and intrinsic InN nanowires, and found that *p*-type InN nanowire NGs showed 160% more output current and 70% more output power compared to intrinsic InN nanowire NGs, indicating the importance of InN nanowire surface passivation for NG applications (Liu et al., 2016).

Spin degree of electrons has attracted great attention (Wolf et al., 2001). In this regard, Heedt et al. demonstrated the electrical injection of spin-polarized electrons into single InN nanowires (Heedt et al., 2012). In this work, the InN nanowires were grown by MBE under nitrogen-rich conditions, and lateral nanowire spin valves were realized. The spin relaxation length was derived to be around 200 nm. This work represents an important step toward nitride nanowire logic devices with electron spin.

The thermoelectric properties of InN nanowires have also been investigated theoretically by Huang et al. (2011a). These authors found that the thermoelectric properties of InN nanowires were strongly influenced by the background electron concentration, temperature, and nanowire size. At room temperature, the highest figure of merit factor (*ZT*) was calculated to be around 0.25 with a carrier density of around 10^{18} cm^{-3} and nanowire size of around 6 nm.

The use of InN nanowires for biosensing applications has been reported as well. Sofikiti et al. investigated InN nanowires as electrochemical sensing elements, and high pH sensitivity was found, which allows InN nanowires function as pH sensors under mild conditions (Sofikiti et al., 2009, 2010). These authors have further developed potentiometric urea biosensors through the entrapment of urease enzyme with InN nanowires.

8. SUMMARY

In this chapter, we have reviewed the recent advance on the growth, characterization, and device applications of InN nanowires. While early InN nanowires generally possess strong *n*-type characteristics and surface electron accumulation, due to the presence of extensive defects, recent InN nanowires by the improved MBE growth process exhibit extremely low background electron concentration (on the order of 10^{13} cm^{-3}) and high electron mobility (up to ~12,000 cm^2/V s). Moreover, detailed studies have further measured the absence of surface electron accumulation and Fermi-level pinning on the grown nonpolar surfaces of such InN nanowires. In addition, the grown surface of InN nanowires can be transformed into nearly *p*-type degenerate, with the surface Fermi level positioned near the valence band edge through controlled Mg dopant incorporation. More importantly, two most direct evidences for the achievement of *p*-type InN have been demonstrated by the *p*-type InN single nanowire transistors and near-band-edge emitting *p–i–n* InN single nanowire LEDs. With these remarkable progresses, it is believed that InN nanowires will provide great promise for a broad range of devices and applications, including near-infrared nanoscale LEDs, lasers, multijunction solar cells, ultrahigh-speed transistors, terahertz emitters, and biosensors, to name just a few.

REFERENCES

Ager, J.W., Miller, N., Jones, R.E., Yu, K.M., Wu, J., Schaff, W.J., Walukiewicz, W., 2008. Mg-doped InN and InGaN—photoluminescence, capacitance–voltage and thermopower measurements. Phys. Status Solidi B 245, 873.

Ahn, H., Ku, Y.P., Wang, Y.C., Chuang, C.H., Gwo, S., Pan, C.-L., 2007. Terahertz emission from vertically aligned InN nanorod arrays. Appl. Phys. Lett. 91, 132108.

Ahn, H., Yu, C.-C., Yu, P., Tang, J., Hong, Y.-L., Gwo, S., 2012. Carrier dynamics in InN nanorod arrays. Opt. Express 20, 769.

Anderson, P.A., Swartz, C.H., Carder, D., Reeves, R.J., Durbin, S.M., Chandril, S., Myers, T.H., 2006. Buried p-type layers in Mg-doped InN. Appl. Phys. Lett. 89, 184104.
Aravind, K., Su, Y.W., Ho, I.L., Wu, C.S., Chang-Liao, K.S., Su, W.F., Chen, K.H., Chen, L.C., Chen, C.D., 2009. Coulomb blockade behavior in an indium nitride nanowire with disordered surface states. Appl. Phys. Lett. 95, 092110.
Arnaudov, B., Paskova, T., Paskov, P., Magnusson, B., Valcheva, E., Monemar, B., Lu, H., Schaff, W., Schaff Amano, H., Akasaki, I., 2004. Energy position of near-band-edge emission spectra of InN epitaxial layers with different doping levels. Phys. Rev. B 69, 115216.
Aydogu, S., Akarsu, M., Ozbas, O., 2012. InN and ln1 − XGaX N: calculation of hall mobilities and effects of alloy disorder and dislocation scatterings. Mater. Sci. Semicond. Process. 15, 347.
Bayerl, D., Kioupakis, E., 2014. Visible-wavelength polarized-light emission with small-diameter InN nanowires. Nano Lett. 14, 3709.
Bhuiyan, A.G., Hashimoto, A., Yamamoto, A., 2003. Indium nitride (InN): a review on growth, characterization, and properties. J. Appl. Phys. 94, 2779.
Blömers, C., Schäpers, T., Richter, T., Calarco, R., Lüth, H., Marso, M., 2008. Phase-coherent transport in InN nanowires of various sizes. Phys. Rev. B 77, r201301.
Brown, G.F., Ager, J.W., Walukiewicz, W., Schaff, W.J., Wu, J., 2008. Probing and modulating surface electron accumulation in InN by the electrolyte gated hall effect. Appl. Phys. Lett. 93, 262105.
Cai, Z., Garzon, S., Chandrashekhar, M.V.S., Webb, R.A., Koley, G., 2008. Synthesis and properties of high-quality InN nanowires and nanonetworks. J. Electron. Mater. 37, 585.
Calarco, R., 2012. InN nanowires: growth and optoelectronic properties. Materials 5, 2137.
Calleja, E., Grandal, J., Sánchez-García, M.A., Niebelschütz, M., Cimalla, V., Ambacher, O., 2007a. Evidence of electron accumulation at nonpolar surfaces of InN nanocolumns. Appl. Phys. Lett. 90, 262110.
Calleja, E., Ristić, J., Fernández-Garrido, S., Cerutti, L., Sánchez-García, M.A., Grandal, J., Trampert, A., Jahn, U., Sánchez, G., Griol, A., Sánchez, B., 2007b. Growth, morphology, and structural properties of group-III-nitride nanocolumns and nanodisks. Phys. Status Solidi B 244, 2816.
Chai, J.H., Song, Y.-W., Reeves, R.J., Durbin, S.M., 2012. Optical and electrical characteristics of Mn-doped InN grown by plasma-assisted molecular beam epitaxy. Phys. Status Solidi A 209, 95.
Chang, Y.M., Gwo, S., 2009. Carrier and phonon dynamics of wurtzite InN nanorods. Appl. Phys. Lett. 94, 071911.
Chang, C.-Y., Chi, G.-C., Wang, W.-M., Chen, L.-C., Chen, K.-H., Ren, F., Pearton, S.J., 2005. Transport properties of InN nanowires. Appl. Phys. Lett. 87, 093112.
Chang, Y.L., Li, F., Fatehi, A., Mi, Z., 2009. Molecular beam epitaxial growth and characterization of non-tapered InN nanowires on Si(111). Nanotechnology 20, 345203.
Chao, C.K., Chyi, J.I., Hsiao, C.N., Kei, C.C., Kuo, S.Y., Chang, H.S., Hsu, T.M., 2006. Catalyst-free growth of indium nitride nanorods by chemical-beam epitaxy. Appl. Phys. Lett. 88, 233111.
Chen, J., Cheng, G., Stern, E., Reed, M.A., Avouris, P., 2007. Electrically excited infrared emission from InN nanowire transistors. Nano Lett. 7, 2276.
Chen, R.-S., Yang, T.-H., Chen, H.-Y., Chen, L.-C., Chen, K.-H., Yang, Y.-J., Su, C.-H., Lin, C.-R., 2009. High-gain photoconductivity in semiconducting InN nanowires. Appl. Phys. Lett. 95, 162112.
Cheng, G., Stern, E., Turner-Evans, D., Reed, M.A., 2005. Electronic properties of InN nanowires. Appl. Phys. Lett. 87, 253103.

Consonni, V., Hanke, M., Knelangen, M., Geelhaar, L., Trampert, A., Riechert, H., 2011. Nucleation mechanisms of self-induced GaN nanowires grown on an amorphous interlayer. Phys. Rev. B 83, 035310.

Cui, K., Fathololoumi, S., Golam Kibria, M., Botton, G.A., Mi, Z., 2012. Molecular beam epitaxial growth and characterization of catalyst-free InN/In(x)Ga(1-x)N core/shell nanowire heterostructures on Si(111) substrates. Nanotechnology 23, 085205.

Cuscó, R., Alarcon-Llado, E., Ibanez, J., Yamaguchi, T., Nanishi, Y., Artus, L., 2009. Raman scattering study of background electron density in InN: a hydrodynamical approach to the LO-phonon-plasmon coupled modes. J. Phys. Condens. Matter 21, 415801.

Cuscó, R., Domènech-Amador, N., Artús, L., Gotschke, T., Jeganathan, K., Stoica, T., Calarco, R., 2010. Probing the electron density in undoped, Si-doped, and Mg-doped InN nanowires by means of Raman scattering. Appl. Phys. Lett. 97, 221906.

Davydov, V.Yu., Klochikhin, A.A., Seisyan, R.P., Emtsev, V.V., Ivanov, S.V., Bechstedt, F., Furthmüller, J., Harima, H., Mudryi, A.V., Aderhold, J., Semchinova, O., Graul, A.J., 2002. Absorption and emission of hexagonal InN. Evidence of narrow fundamental band gap. Phys. Status Solidi B 229, r1.

Dingman, S.D., Rath, N.P., Markowitz, P.D., Gibbons, P.C., Buhro, W.E., 2000. Low-temperature, catalyzed growth of indium nitride fibers from azido-indium precursors. Angew. Chem. Int. Ed. 39, 1470.

Dmowski, L.H., Baj, M., Suski, T., Przybytek, J., Czernecki, R., Wang, X., Yoshikawa, A., Lu, H., Schaff, W.J., Muto, D., Nanishi, Y., 2009. Search for free holes in InN:Mg-interplay between surface layer and Mg-acceptor doped interior. J. Appl. Phys. 105, 123713.

Dmowski, L.H., Baj, M., Konczewicz, L., Suski, T., Maude, D.K., Grzanka, S., Wang, X.Q., Yoshikawa, A., 2012. Coexistence of free holes and electrons in InN: Mg with In- and N-growth polarities. J. Appl. Phys. 111, 093719.

Domènech-Amador, N., Cuscó, R., Calarco, R., Yamaguchi, T., Nanishi, Y., Artús, L., 2012. Double resonance Raman effects in InN nanowires. Phys. Status Solidi (RRL) Rapid Res. Lett. 6, 160.

Duan, X., Stampfl, C., 2009. Defect complexes and cluster doping of InN: first-principles investigations. Phys. Rev. B 79, 035207.

Ebert, P., Schaafhausen, S., Lenz, A., Sabitova, A., Ivanova, L., Dähne, M., Hong, Y.L., Gwo, S., Eisele, H., 2011. Direct measurement of the band gap and Fermi level position at InN(1120). Appl. Phys. Lett. 98, 062103.

Eisele, H., Ebert, P., 2012. Non-polar group-III nitride semiconductor surfaces. Phys. Status Solidi (RRL) Rapid Res. Lett. 6, 359.

Eisenhardt, A., Krischok, S., Himmerlich, M., 2013. Surface states and electronic structure of polar and nonpolar InN—an in situ photoelectron spectroscopy study. Appl. Phys. Lett 102, 231602.

Fehlberg, T.B., Umana-Membreno, G.A., Nener, B.D., Parish, G., Gallinat, C.S., Koblmuller, G., Rajan, S., Bernardis, S., Speck, J.S., 2006. Characterisation of multiple carrier transport in indium nitride grown by molecular beam epitaxy. Jpn. J. Appl. Phys. 45, L1090.

Fehlberg, T.B., Umana-Membreno, G.A., Gallinat, C.S., Koblmüller, G., Bernardis, S., Nener, B.D., Parish, G., Speck, J.S., 2007. Characterisation of multiple carrier transport in indium nitride grown by molecular beam epitaxy. Phys. Status Solidi C 4, 2423.

Feneberg, M., Däubler, J., Thonke, K., Sauer, R., Schley, P., Goldhahn, R., 2008. Mahan excitons in degenerate wurtzite InN: photoluminescence spectroscopy and reflectivity measurements. Phys. Rev. B 77, 245207.

Frielinghaus, R., Estévez Hernández, S., Calarco, R., Schäpers, T., 2009. Phase-coherence and symmetry in four-terminal magnetotransport measurements on InN nanowires. Appl. Phys. Lett. 94, 252107.

Fu, S.P., Chen, T.T., Chen, Y.F., 2006a. Photoluminescent properties of InN epifilms. Semicond. Sci. Technol. 21, 244.

Fu, S.P., Chen, Y.F., Tan, K., 2006b. Recombination mechanism of photoluminescence in InN epilayers. Solid State Commun. 137, 203.

Fu, S.P., Yu, C.J., Chen, T.T., Hsu, G.M., Chen, M.J., Chen, L.C., Chen, K.H., Chen, Y.F., 2007. Anomalous optical properties of InN nanobelts: evidence of surface band bending and photoelastic effects. Adv. Mater. 19, 4524.

Fukunaga, K., Kouyama, K., Hashimoto, M., Kamimura, J., Kunugita, H., Ema, K., Kikuchi, A., Kishino, K., 2009. Evaluation of bandgap energy and carrier density of InN nanocolumns. J. Lumin. 129, 1735.

Glas, F., 2006. Critical dimensions for the plastic relaxation of strained axial heterostructures in free-standing nanowires. Phys. Rev. B 74, r121302.

Golam Sarwar, A.T.M., Carnevale, S.D., Kent, T.F., Laskar, M.R., May, B.J., Myers, R.C., 2015. Molecular beam epitaxy of InN nanowires on Si. J. Cryst. Growth 428, 59.

Grandal, J., Sánchez-García, M.A., Calleja, E., Luna, E., Trampert, A., 2007. Accommodation mechanism of InN nanocolumns grown on Si(111) substrates by molecular beam epitaxy. Appl. Phys. Lett. 91, 021902.

Grandal, J., Sánchez-García, M.A., Calleja, E., Gallardo, E., Calleja, J.M., Luna, E., Trampert, A., Jahn, A., 2009. InN nanocolumns grown by plasma-assisted molecular beam epitaxy on A-plane GaN templates. Appl. Phys. Lett. 94, 221908.

He, M., Noor Mohammad, S., 2007. Novelty and versatility of self-catalytic nanowire growth: a case study with InN nanowires. J. Vac. Sci. Technol. B Microelectron. Nanometer Struct. 25, 940.

Heedt, S., Morgan, C., Weis, K., Burgler, D.E., Calarco, R., Hardtdegen, H., Grutzmacher, D., Schapers, T., 2012. Electrical spin injection into InN semiconductor nanowires. Nano Lett. 12, 4437.

Hoang, T.B., Moses, A.F., Zhou, H.L., Dheeraj, D.L., Fimland, B.O., Weman, H., 2009. Observation of free exciton photoluminescence emission from single wurtzite GaAs nanowires. Appl. Phys. Lett. 94, 133105.

Holtz, M.E., Gherasoiu, I., Kuryatkov, V., Nikishin, S.A., Bernussi, A.A., Holtz, M.W., 2009. Influence of phonons on the temperature dependence of photoluminescence in InN with low carrier concentration. J. Appl. Phys. 105, 063702.

Hu, M.S., Wang, W.M., Chen, T.T., Hong, L.S., Chen, C.W., Chen, C.C., Chen, Y.F., Chen, K.H., Chen, L.C., 2006. Sharp infrared emission from single-crystalline indium nitride nanobelts prepared using guided-stream thermal chemical vapor deposition. Adv. Funct. Mater. 16, 537.

Huang, Y., Duan, X., Cui, Y., Lieber, C.M., 2002. Gallium nitride nanowire nanodevices. Nano Lett. 2, 101.

Huang, C.T., Song, J., Tsai, C.M., Lee, W.F., Lien, D.H., Gao, Z., Hao, Y., Chen, L.J., Wang, Z.L., 2010. Single-InN-nanowire nanogenerator with upto 1 V output voltage. Adv. Mater. 22, 4008.

Huang, H.-H., Lu, I.L., Wu, Y.-R., 2011a. Study of thermoelectric properties of indium nitride nanowire. Phys. Status Solidi A 208, 1562.

Huang, L., Li, D., Chang, P., Chu, S., Bozler, H., Beloborodov, I., Lu, J., 2011b. Quantum transport in indium nitride nanowires. Phys. Rev. B 83, 245310.

Jeganathan, K., Purushothaman, V., Debnath, R.K., Calarco, R., Luth, H., 2010. Raman scattering on intrinsic surface electron accumulation of InN nanowires. Appl. Phys. Lett. 97, 093104.

Johnson, M.C., Lee, C.J., Bourret-Courchesne, E.D., Konsek, S.L., Aloni, S., Han, W.Q., Zettl, A., 2004. Growth and morphology of 0.80 eV photoemitting indium nitride nanowires. Appl. Phys. Lett. 85, 5670.

Kametani, T., Kamimura, J., Inose, Y., Kunugita, H., Kikuchi, A., Kishino, K., Ema, K., 2013. Fundamental optical properties of InN grown by epitaxial lateral overgrowth method. In: AIP Conference Proceedings, 1566, p. 107.

Kamimura, J., Kishino, K., Kikuchi, A., 2010. Dislocation reduction via selective-area growth of InN accompanied by lateral growth by rf-plasma-assisted molecular-beam epitaxy. Appl. Phys. Lett. 97, 141913.

Kamimura, J., Kishino, K., Kikuchi, A., 2012. Photoluminescence properties of selectively grown InN microcrystals. Phys. Status Solidi (RRL) Rapid Res. Lett. 6, 157.

Kang, T.-T., Liu, X., Zhang, R.Q., Hu, W.G., Cong, G., Zhao, F.-A., Zhu, Q., 2006. InN nanoflowers grown by metal organic chemical vapor deposition. Appl. Phys. Lett. 89, 071113.

Khan, N., Nepal, N., Sedhain, A., Lin, J.Y., Jiang, H.X., 2007. Mg acceptor level in InN epilayers probed by photoluminescence. Appl. Phys. Lett. 91, 012101.

Khanal, D.R., Walukiewicz, W., Grandal, J., Calleja, E., Wu, J., 2009. Determining surface Fermi level pinning position of InN nanowires using electrolyte gating. Appl. Phys. Lett. 95, 173114.

Khanal, D.R., Levander, A.X., Yu, K.M., Liliental-Weber, Z., Walukiewicz, W., Grandal, J., Sánchez-García, M.A., Calleja, E., Wu, J., 2011. Decoupling single nanowire mobilities limited by surface scattering and bulk impurity scattering. J. Appl. Phys. 110, 033705.

Kim, M.H., Moon, D.Y., Park, J., Nanishi, Y., Yi, G.-C., Yoon, E., 2012. Catalyst-free growth of InN nanorods by metal-organic chemical vapor deposition. Phys. Status Solidi A 209, 50.

King, P.D.C., Veal, T.D., Mcconville, C.F., Fuchs, F., Furthmüller, J., Bechstedt, F., Schley, P., Goldhahn, R., Schörmann, J., As, D.J., Lischka, D.J., Muto, D., Naoi, H., Nanishi, Y., Lu, H., Schaff, W.J., 2007. Universality of electron accumulation at wurtzite c- and a-plane and zinc-blende InN surfaces. Appl. Phys. Lett. 91, 092101.

King, P.D.C., Veal, T., Payne, D., Bourlange, A., Egdell, R., Mcconville, C., 2008. Surface electron accumulation and the charge neutrality level in In_2O_3. Phys. Rev. Lett. 101, 116808.

Kishino, K., Kamimura, J., Kamiyama, K., 2012. Near-infrared InGaN nanocolumn light-emitting diodes operated at 1.46 μm. Appl. Phys. Expr. 5, 031001.

Klochikhin, A., Davydov, V., Emtsev, V., Sakharov, A., Kapitonov, V., Andreev, B., Lu, H., Schaff, W., 2005. Acceptor states in the photoluminescence spectra of n-InN. Phys. Rev. B 71, 195207.

Koley, G., Cai, Z., Quddus, E.B., Liu, J., Qazi, M., Webb, R.A., 2011. Growth direction modulation and diameter-dependent mobility in InN nanowires. Nanotechnology 22, 295701.

Kryliouk, O., Park, H.J., Won, Y.S., Anderson, T., Davydov, A., Levin, I., Kim, J.H., Freitas, J.A., 2007. Controlled synthesis of single-crystalline InN nanorods. Nanotechnology 18, 135606.

Ku, N.J., Huang, J.H., Wang, C.H., Fang, H.C., Liu, C.P., 2012. Crystal face-dependent nanopiezotronics of an obliquely aligned InN nanorod array. Nano Lett. 12, 562.

Ku, N.J., Wang, C.H., Huang, J.H., Fang, H.C., Huang, P.C., Liu, C.P., 2013. Energy harvesting from the obliquely aligned InN nanowire array with a surface electron-accumulation layer. Adv. Mater. 25, 861.

Kuo, C.-T., Lin, S.-C., Chang, K.-K., Shiu, H.-W., Chang, L.-Y., Chen, C.-H., Tang, S.-J., Gwo, S., 2011. Is electron accumulation universal at InN polar surfaces? Appl. Phys. Lett. 98, 052101.

Lai, W.-J., Li, S.-S., Lin, C.-C., Kuo, C.-C., Chen, C.-W., Chen, K.-H., Chen, L.-C., 2010. Near infrared photodetector based on polymer and indium nitride nanorod organic/inorganic hybrids. Scr. Mater. 63, 653.

Lan, Z., 2004. Growth mechanism, structure and IR photoluminescence studies of indium nitride nanorods. J. Cryst. Growth 269, 87.

Lan, Z.H., Liang, C.H., Hsu, C.W., Wu, C.T., Lin, H.M., Dhara, S., Chen, K.H., Chen, L.C., Chen, C.C., 2004. Nanohomojunction (GaN) and nanoheterojunction (InN) nanorods on one-dimensional GaN nanowire substrates. Adv. Funct. Mater. 14, 233.

Lazić, S., Gallardo, E., Calleja, J., Agulló-Rueda, F., Grandal, J., Sánchez-Garcia, M., Calleja, E., Luna, E., Trampert, A., 2007. Phonon-plasmon coupling in electron surface accumulation layers in InN nanocolumns. Phys. Rev. B 76, 205319.

Lazic, S., Gallardo, E., Calleja, J., Agullorueda, F., Grandal, J., Sanchezgarcia, M., Calleja, E., 2008. Raman scattering by longitudinal optical phonons in InN nanocolumns grown on Si(111) and Si(001) substrates. Physica E Low Dimens. Syst. Nanostruct. 40, 2087.

Le, B.H., Zhao, S., Tran, N.H., Mi, Z., 2014. Electrically injected near-infrared light emission from single InN nanowire p-i-n diode. Appl. Phys. Lett. 105, 231124.

Le, B.H., Zhao, S., Ran, N.H., Szkopek, T., Mi, Z., 2015. On the Fermi-level pinning of InN grown surfaces. Appl. Phys. Expr. 8, 061001.

Lee, S., Lee, W., Seo, K., Kim, J., Han, S.H., Kim, B., 2008. Electronic transport mechanism and photocurrent generations of single-crystalline InN nanowires. Nanotechnology 19, 415202.

Lei, M., Huang, K., Zhang, R., Yang, H.J., Fu, X.L., Wang, Y.G., Tang, W.H., 2012. Catalyst-free chemical vapor deposition route to InN nanowires and their cathodoluminescence properties. J. Alloys Compd. 535, 50.

Liang, C.H., Chen, L.C., Hwang, J.S., Chen, K.H., Hung, Y.T., Chen, Y.F., 2002. Selective-area growth of indium nitride nanowires on gold-patterned Si(100) substrates. Appl. Phys. Lett. 81, 22.

Linhart, W.M., Chai, J., Morris, R.J.H., Dowsett, M.G., Mcconville, C.F., Durbin, S.M., Veal, T.D., 2012. Giant reduction of InN surface electron accumulation: compensation of surface donors by Mg dopants. Phys. Rev. Lett. 109, 247605.

Liu, H., Cheng, G., 2011. Indium predeposition-enabled vapor–liquid–solid growth of InN nanowires. Appl. Phys. Expr. 4, 105002.

Liu, J., Cai, Z., Koley, G., 2009. Charge transport and trapping in InN nanowires investigated by scanning probe microscopy. J. Appl. Phys. 106, 124907.

Liu, G., Zhao, S., Henderson, R.D., Leonenko, Z., Abdel-Rahman, E., Mi, Z., Ban, D., 2016. Nanogenerators based on vertically aligned InN nanowires. Nanoscale 8, 2097.

Look, D.C., Lu, H., Schaff, W.J., Jasinski, J., Liliental-Weber, Z., 2002. Donor and acceptor concentrations in degenerate InN. Appl. Phys. Lett. 80, 258.

Lu, H., Schaff, W.J., Hwang, J., Wu, H., Koley, G., Eastman, L.F., 2001. Effect of an AlN buffer layer on the epitaxial growth of InN by molecular-beam epitaxy. Appl. Phys. Lett. 79, 1489.

Lu, T.M., Wang, G.T., Pan, W., Zhao, S., Mi, Z., 2016. Electronic transport through Al/InN nanowire/Al junctions. Appl. Phys. Lett. 108, 063104.

Ma, N., Wang, X.Q., Liu, S.T., Chen, G., Pan, J.H., Feng, L., Xu, F.J., Tang, N., Shen, B., 2011. Hole mobility in wurtzite InN. Appl. Phys. Lett. 98, 192114.

Madapu, K.K., Dhara, S., Polaki, S., Amirthapandian, S., Tyagi, A.K., 2015. Growth of InN quantum dots to nanorods: a competition between nucleation and growth rates. CrystEngComm 17, 3139.

Mahboob, I., Veal, T., Mcconville, C., Lu, H., Schaff, W., 2004a. Intrinsic electron accumulation at clean InN surfaces. Phys. Rev. Lett. 92, 036804.

Mahboob, I., Veal, T., Piper, L., Mcconville, C., Lu, H., Schaff, W., Furthmüller, J., Bechstedt, F., 2004b. Origin of electron accumulation at wurtzite InN surfaces. Phys. Rev. B 69, r201307.

Matsuoka, T., Okamoto, H., Nakao, M., Harima, H., Kurimoto, E., 2002. Optical bandgap energy of wurtzite InN. Appl. Phys. Lett. 81, 1246.
Mayer, M.A., Choi, S., Bierwagen, O., Smith, H.M., Haller, E.E., Speck, J.S., Walukiewicz, W., 2011. Electrical and optical properties of p-type InN. J. Appl. Phys. 110, 123707.
Meyer, J.R., Bartoli, F.J., 1987. About InN mobility theoretical calculations. Phys. Rev. B 36, 5989.
Mi, Z., Zhao, S., 2015. Extending group-III nitrides to the infrared: recent advances in InN. Phys. Status Solidi B 252, 1050.
Mishra, A., Titova, L.V., Hoang, T.B., Jackson, H.E., Smith, L.M., Yarrison-Rice, J.M., Kim, Y., Joyce, H.J., Gao, Q., Tan, H.H., Jagadish, C., 2007. Polarization and temperature dependence of photoluminescence from zincblende and wurtzite InP nanowires. Appl. Phys. Lett. 91, 263104.
Moret, M., Ruffenach, S., Briot, O., Gil, B., 2009. The determination of the bulk residual doping in indium nitride films using photoluminescence. Appl. Phys. Lett. 95, 031910.
Moustakas, T.D., 2013. The role of extended defects on the performance of optoelectronic devices in nitride semiconductors. Phys. Status Solidi A 210, 169.
Nguyen, H.P.T., Chang, Y.-L., Shih, I., Mi, Z., 2010. InN p-i-n nanowire solar cells on Si. IEEE J. Sel. Top. Quantum electron. 1077, 2601.
Petersen, G., Hernandez, S.E., Calarco, R., Demarina, N., Schaepers, T., 2009. Spin-orbit coupling and phase-coherent transport in InN nanowires. Phys. Rev. B 80, 125321.
Piper, L., Veal, T., Mahboob, I., Mcconville, C., Lu, H., Schaff, W., 2004. Temperature invariance of InN electron accumulation. Phys. Rev. B 70, 115333.
Piper, L.F.J., Veal, T.D., Mcconville, C.F., Lu, H., Schaff, W.J., 2006. Origin of the n-type conductivity of InN: the role of positively charged dislocations. Appl. Phys. Lett. 88, 252109.
Piper, L., Colakerol, L., King, P., Schleife, A., Zúñiga-Pérez, J., Glans, P.-A., Learmonth, T., Federov, A., Veal, T., Fuchs, F., Muñoz-Sanjosé, V., Bechstedt, F., Mcconville, C., Smith, K., 2008. Observation of quantized subband states and evidence for surface electron accumulation in CdO from angle-resolved photoemission spectroscopy. Phys. Rev. B 78, 165127.
Polyakov, V.M., Schwierz, F., 2006a. Influence of interface roughness on two-dimensional electron gas streaming transport in GaN-based heterostructures. J. Appl. Phys. 100, 103704.
Polyakov, V.M., Schwierz, F., 2006b. Low-field electron mobility in wurtzite InN. Appl. Phys. Lett. 88, 032101.
Polyakov, V.M., Schwierz, F., 2007. Monte Carlo calculation of two-dimensional electron gas mobility in InN-based heterostructures. J. Appl. Phys. 101, 033703.
Polyakov, V.M., Schwierz, F., Fuchs, F., Furthmüller, J., Bechstedt, F., 2009. Low-field and high-field electron transport in zinc blende InN. Appl. Phys. Lett. 94, 022102.
Richter, T., Bloemers, C., Lueth, H., Calarco, R., Indlekofer, M., Marso, M., Schaepers, T., 2008. Flux quantization effects in InN nanowires. Nano Lett. 8, 2834.
Richter, T., Luth, H., Schapers, T., Meijers, R., Jeganathan, K., Estevez Hernandez, S., Calarco, R., Marso, M., 2009. Electrical transport properties of single undoped and n-type doped InN nanowires. Nanotechnology 20, 405206.
Ristić, J., Calleja, E., Fernández-Garrido, S., Cerutti, L., Trampert, A., Jahn, U., Ploog, K.H., 2008. On the mechanisms of spontaneous growth of III-nitride nanocolumns by plasma-assisted molecular beam epitaxy. J. Cryst. Growth 310, 4035.
Sánchez-García, M.A., Grandal, J., Calleja, E., Lazic, S., Calleja, J.M., Trampert, A., 2006. Epitaxial growth and characterization of InN nanorods and compact layers on silicon substrates. Phys. Status Solidi B 243, 1490.

Schafer-Nolte, E.O., Stoica, T., Gotschke, T., Limbach, F.A., Sutter, E., Sutter, P., Grutzmacher, D., Calarco, R., 2010. Enhanced light scattering of the forbidden longitudinal optical phonon mode studied by micro-Raman spectroscopy on single InN nanowires. Nanotechnology 21, 315702.
Schlager, J.B., Sanford, N.A., Bertness, K.A., Barker, J.M., Roshko, A., Blanchard, P.T., 2006. Polarization-resolved photoluminescence study of individual GaN nanowires grown by catalyst-free molecular beam epitaxy. Appl. Phys. Lett. 88, 213106.
Segura-Ruiz, J., Garro, N., Cantarero, A., Denker, C., Malindretos, J., Rizzi, A., 2009. Optical studies of MBE-grown InN nanocolumns: evidence of surface electron accumulation. Phys. Rev. B 79, 115305.
Segura-Ruiz, J., Molina-Sánchez, A., Garro, N., García-Cristóbal, A., Cantarero, A., Iikawa, F., Denker, C., Malindretos, J., Rizzi, A., 2010. Inhomogeneous free-electron distribution in InN nanowires: photoluminescence excitation experiments. Phys. Rev. B 82, 125319.
Shen, C.H., Chen, H.Y., Lin, H.W., Gwo, S., Klochikhin, A.A., Davydov, V.Y., 2006. Near-infrared photoluminescence from vertical InN nanorod arrays grown on silicon: effects of surface electron accumulation layer. Appl. Phys. Lett. 88, 253104.
Sofikiti, N., Chaniotakis, N., Grandal, J., Utrera, M., Sanchez-Garcia, M.A., Calleja, E., Iliopoulos, E., Georgakilas, A., 2009. Direct immobilization of enzymes in GaN and InN nanocolumns: the urease case study. Appl. Phys. Lett. 95, 113701.
Sofikiti, N., Chaniotakis, N., Grandal, J., Utrera, M., Sánchez-García, M.A., Calleja, E., 2010. GaN and InN nanocolumns as electrochemical sensing elements: potentiometric response to KCl, pH and urea. Mater. Lett. 64, 1332.
Song, D.Y., Basavaraj, M., Nikishin, S.A., Holtz, M., Soukhoveev, V., Usikov, A., Dmitriev, V., 2006. The influence of phonons on the optical properties of GaN. J. Appl. Phys. 100, 113504.
Song, H., Yang, A., Zhang, R., Guo, Y., Wei, H., Zheng, G., Yang, S., Liu, X., Zhu, Q., Wang, Z., 2009. Well-aligned Zn-doped InN nanorods grown by metal-organic chemical vapor deposition and the dopant distribution. Cryst. Growth Des. 9, 3292.
Stoica, T., Calarco, R., 2011. Doping of III-nitride nanowires grown by molecular beam epitaxy. IEEE J. Sel. Top. Quantum electron. 17, 859.
Stoica, T., Meijers, R.J., Calarco, R., Richter, T., Sutter, E., Lueth, H., 2006. Photoluminescence and intrinsic properties of MBE-grown InN nanowires. Nano Lett. 6, 1541.
Stoica, T., Sutter, E., Meijers, R.J., Debnath, R.K., Calarco, R., Luth, H., Grutzmacher, D., 2008. Interface and wetting layer effect on the catalyst-free nucleation and growth of GaN nanowires. Small 4, 751.
Tang, T., Han, S., Jin, W., Liu, X., Li, C., Zhang, D., Zhou, C., Chen, B., Han, J., Meyyapan, M., 2004. Synthesis and characterization of single-crystal indium nitride nanowires. J. Mater. Res. 19, 423.
Taniyasu, Y., Kasu, M., Makimoto, T., 2004. Electrical conduction properties of n-type Si-doped AlN with high electron mobility (>100 $cm^2V^{-1}\,s^{-1}$). Appl. Phys. Lett. 85, 4672.
Terziyska, P.T., Butcher, K.S.A., Rafailov, P., Alexandrov, D., 2015. Growth of vertically oriented InN nanorods from In-rich conditions on unintentionally patterned sapphire substrates. Appl. Surf. Sci. 353, 103.
Titova, L.V., Hoang, T.B., Jackson, H.E., Smith, L.M., Yarrison-Rice, J.M., Kim, Y., Joyce, H.J., Tan, H.H., Jagadish, C., 2006. Temperature dependence of photoluminescence from single core-shell GaAs–AlGaAs nanowires. Appl. Phys. Lett. 89, 173126.
Vaddiraju, S., Mohite, A., Chin, A., Meyyappan, M., Sumanasekera, G., Alphenaar, B.W., Sunkara, M.K., 2005. Mechanisms of 1D crystal growth in reactive vapor transport: indium nitride nanowires. Nano Lett. 5, 1625.

Van De Walle, C.G., Segev, D., 2007. Microscopic origins of surface states on nitride surfaces. J. Appl. Phys. 101, 081704.
Van De Walle, C.G., Lyons, J.L., Janotti, A., 2010. Controlling the conductivity of InN. Phys. Status Solidi A 207, 1024.
Veal, T.D., Mahboob, I., Piper, L.F.J., Mcconville, C.F., Lu, H., Schaff, W.J., 2004. Indium nitride: evidence of electron accumulation. J. Vac. Sci. Technol. B Microelectron. Nanometer Struct. 22, 2175.
Walle, C.G.V.D., Neugebauer, J., 2003. Universal alignment of hydrogen levels in semiconductors, insulators and solutions. Nature 423, 626.
Wang, X., Che, S.-B., Ishitani, Y., Yoshikawa, A., 2007a. Growth and properties of Mg-doped In-polar InN films. Appl. Phys. Lett. 90, 201913.
Wang, X., Che, S.-B., Ishitani, Y., Yoshikawa, A., 2007b. Systematic study on p-type doping control of InN with different Mg concentrations in both In and N polarities. Appl. Phys. Lett. 91, 242111.
Wang, N., Cai, Y., Zhang, R.Q., 2008a. Growth of nanowires. Mater. Sci. Eng. R Rep. 60, 1.
Wang, X., Che, S.-B., Ishitani, Y., Yoshikawa, A., 2008b. Hole mobility in Mg-doped p-type InN films. Appl. Phys. Lett. 92, 132108.
Wang, K., Miller, N., Iwamoto, R., Yamaguchi, T., Mayer, M.A., Araki, T., Nanishi, Y., Yu, K.M., Haller, E.E., Walukiewicz, W., Ager, J.W., 2011. Mg doped InN and confirmation of free holes in InN. Appl. Phys. Lett. 98, 042104.
Wang, J., Tang, G., Wu, X.S., Gu, M., 2012a. InN doped with Zn: bulk and surface investigation from first principles. Solid State Commun. 152, 1168.
Wang, S., Liu, H., Gao, B., Cai, H., 2012b. Monte Carlo calculation of electron diffusion coefficient in wurtzite indium nitride. Appl. Phys. Lett. 100, 142105.
Wang, X., Liu, S., Ma, N., Feng, L., Chen, G., Xu, F., Tang, N., Huang, S., Chen, K.J., Zhou, S., Shen, B., 2012c. High-electron-mobility InN layers grown by boundary-temperature-controlled epitaxy. Appl. Phys. Expr. 5, 015502.
Wang, X., Liu, S., Ma, N., Shen, B., Yoshikawa, A., 2012d. Growth of high-electron mobility InN and demonstration of p-type doping. IWN2012.
Wang, K., Araki, T., Yamaguchi, T., Chen, Y.T., Yoon, E., Nanishi, Y., 2015. InN nanocolumns grown by molecular beam epitaxy and their luminescence properties. J. Cryst. Growth 430, 93.
Werner, F., Limbach, F., Carsten, M., Denker, C., Malindretos, J., Rizzi, A., 2009. Electrical conductivity of InN nanowires and the influence of the native indium oxide formed at their surface. Nano Lett. 9, 1567.
Wilke, I., Ascazubi, R., Lu, H., Schaff, W.J., 2008. Terahertz emission from silicon and magnesium doped indium nitride. Appl. Phys. Lett. 93, 221113.
Wolf, S.A., Awschalom, D.D., Buhrman, R.A., Daughton, J.M., Molna, S.V., Roukes, M.L., Chtchelkanova, A.Y., Treger, D.M., 2001. Spintronics: a spin-based electronics vision for the future. Science 294, 1488.
Wu, J., 2009. When group-III nitrides go infrared: new properties and perspectives. J. Appl. Phys. 106, 011101.
Wu, J., Walukiewicz, W., Shan, W., Yu, K., Ager III, J., Haller, E., Lu, H., Schaff, W.J., 2002a. Effects of the narrow band gap on the properties of InN. Phys. Rev. B 66, 201403.
Wu, J., Walukiewicz, W., Yu, K.M., Ager, J.W., Haller, E.E., Lu, H., Schaff, W.J., Saito, Y., Nanishi, Y., 2002b. Unusual properties of the fundamental band gap of InN. Appl. Phys. Lett. 80, 3967.
Wu, C.-L., Lee, H.-M., Kuo, C.-T., Chen, C.-H., Gwo, S., 2008. Absence of Fermi-level pinning at cleaved nonpolar InN surfaces. Phys. Rev. Lett. 101, 106803.
Xiang, H., Wei, S.-H., Da Silva, J., Li, J., 2008. Strain relaxation and band-gap tunability in ternary $In_xGa_{1-x}N$ nanowires. Phys. Rev. B 78, 193301.
Xie, P., Hu, Y., Fang, Y., Huang, J., Lieber, C.M., 2009. Diameter-dependent dopant location in silicon and germanium nanowires. Proc. Natl. Acad. Sci. 106, 15254.

Yamaguchi, T., Araki, T., Nanishi, Y., 2011. Growth and fabrication of InN-based III-nitride device structure using droplet elimination process by radical beam irradiation. Proc. SPIE 7939, 793904.

Yamaguchi, T., Wang, K., Araki, T., Honda, T., Yoon, E., Nanishi, Y., 2013. Application of DERI method to InN/InGaN MQW, thick InGaN and InGaN/InGaN MQW structure growth. Proc. SPIE 8625, 862502.

Yang, Y., Ling, Y., Wang, G., Lu, X., Tong, Y., Li, Y., 2013. Growth of gallium nitride and indium nitride nanowires on conductive and flexible carbon cloth substrates. Nanoscale 5, 1820.

Yin, L.-W., Bando, Y., Zhu, Y., Golberg, D., Li, M., 2004a. Synthesis of InN/InP core/sheath nanowires. Appl. Phys. Lett. 84, 1546.

Yin, L.-W., Bando, Y., Golberg, D., Li, M.-S., 2004b. Growth of single-crystal indium nitride nanotubes and nanowires by a controlled-carbonitridation reaction route. Adv. Mater. 16, 1833.

Yoshikawa, A., Wang, X., Ishitani, Y., Uedono, A., 2010. Recent advances and challenges for successful p-type control of InN films with Mg acceptor doping by molecular beam epitaxy. Phys. Status Solidi A 207, 1011.

Zhang, J., Zhang, L., Peng, X., Wang, X., 2002. Vapor–solid growth route to single-crystalline indium nitride nanowires. J. Mater. Chem. 12, 802.

Zhang, J., Jiang, F., Yang, Y., Xu, B., Li, J., Wang, X., Wang, S., 2006. Growth and structural characterization of InN/In_2O_3 coaxial nanocables. Mater. Lett. 60, 2153.

Zhang, B., Song, H., Xu, X., Liu, J., Wang, J., Liu, X., Yang, S., Zhu, Q., Wang, Z., 2011. Well-aligned Zn-doped tilted InN nanorods grown on r-plane sapphire by MOCVD. Nanotechnology 22, 235603.

Zhang, J., Liu, H., Huang, R., Kong, T., Cheng, G., 2012. Controllable growth of InN nanostructures. J. Nanoeng. Nanomanuf. 2, 112.

Zhao, S., 2013. Molecular beam epitaxial growth, characterization, and nanophotonic device applications of InN nanowires on Si platform. PhD dissertation.

Zhao, S., Mi, Z., 2014. Is the Fermi-level pinned on InN grown surfaces? Phys. Status Solidi C 11, 412.

Zhao, S., Fathololoumi, S., Bevan, K.H., Liu, D.P., Kibria, M.G., Li, Q., Wang, G.T., Guo, H., Mi, Z., 2012a. Tuning the surface charge properties of epitaxial InN nanowires. Nano Lett. 12, 2877.

Zhao, S., Mi, Z., Kibria, M.G., Li, Q., Wang, G.T., 2012b. Understanding the role of Si doping on surface charge and optical properties: photoluminescence study of intrinsic and Si-doped InN nanowires. Phys. Rev. B 85, 245313.

Zhao, S., Wang, Q., Mi, Z., Fathololoumi, S., Gonzalez, T., Andrews, M.P., 2012c. Observation of phonon sideband emission in intrinsic InN nanowires: a photoluminescence and micro-Raman scattering study. Nanotechnology 23, 415706.

Zhao, S., Kibria, M.G., Wang, Q., Nguyen, H.P.T., Mi, Z., 2013a. Growth of large-scale vertically aligned GaN nanowires and their heterostructures with high uniformity on SiOx by catalyst-free molecular beam epitaxy. Nanoscale 5, 5283.

Zhao, S., Le, B.H., Liu, D.P., Liu, X.D., Kibria, M.G., Szkopek, T., Guo, H., Mi, Z., 2013b. p-Type InN nanowires. Nano Lett. 13, 5509.

Zhao, S., Liu, X., Mi, Z., 2013c. Photoluminescence properties of Mg-doped InN nanowires. Appl. Phys. Lett. 103, 203113.

Zhao, S., Salehzadeh, O., Alagha, S., Kavanagh, K.L., Watkins, S.P., Mi, Z., 2013d. Probing the electrical transport properties of intrinsic InN nanowires. Appl. Phys. Lett. 102, 073102.

Zhao, S., Nguyen, H.P.T., Mi, Z., 2014. Near-infrared InN nanowire optoelectronic devices on Si. IEEE Photonics Society Summer Topical Meeting Series, pp. 208–209.

Zhao, S., Nguyen, H.P.T., Kibria, M.G., Mi, Z., 2015. III-nitride nanowire optoelectronics. Prog. Quant. Electron. 44, 14.

CHAPTER NINE

Dynamic Atomic Layer Epitaxy of InN on/in GaN and Its Application for Fabricating Ordered Alloys in Whole III-N System

K. Kusakabe*, A. Yoshikawa*,†,1
*Center for SMART Green Innovation Research, Chiba University, Chiba, Japan
†Graduate School of Engineering, Kogakuin University, Tokyo, Japan
[1]Corresponding author: e-mail address: yoshi@faculty.chiba-u.jp

Contents

1. INTRODUCTION

1.1 Brief Introduction of Dynamic Atomic Layer Epitaxy in Monolayer-InN/GaN System

Over the last few decades, III-nitride semiconductors, AlN, GaN, InN, and their ternary/quaternary alloys, have received significant research attention and hence progressed intensively, since they have been

Semiconductors and Semimetals, Volume 96
ISSN 0080-8784
http://dx.doi.org/10.1016/bs.semsem.2016.10.001
© 2017 Elsevier Inc.
All rights reserved.

adapted/commercialized widely in optoelectronic and electronic devices including light-emitting diodes (LEDs), laser diodes (LDs), high temperature and high frequency field-effect transistors (FETs), heterojunction bipolar transistors (HBTs), and so on (Gil, 1998; Morkoc, 1999; Nakamura and Fasol, 1997; Pankove and Moustakas, 1998; Pearton, 1997). Thanks to these achievements, especially in blue-LEDs, it has been well known that outstanding pioneers, Professors Isamu Akasaki, Meijo University, Hiroshi Amano, Nagoya University, and Shuji Nakamura, University of California, Santa Barbara, won Nobel Prize in Physics in 2014 (http://www.kva.se/en/pressroom/Press-releases-2014/the-nobel-prize-in-physics-2014/).

During such the development history of III-nitrides, an important discovery on scientific aspect was made around 2002 that bandgap energy of InN was corrected from a previous/wrong value around 1.9 eV to new/reliable one around 0.65 eV (Davydov et al., 2002; Inushima et al., 2001; Ishitani et al., 2005; Matsuoka et al., 2002; Wu et al., 2002a). This fact immediately reminded us that the III-nitrides were not only wide bandgap semiconductors but also *wide-range bandgap* semiconductors covering wavelengths from λ ~200 nm to ~2 μm. Since this wavelength region almost overlaps a solar spectrum and particularly includes a fiber-communication wavelength band $\lambda = 1.3$–1.55 μm, researchers were inspired and attempted to investigate novel InN-based optical devices such as full-spectrum/high-efficiency solar cells and optical communication lasers. The authors believe this opportunity was a turning point to come the research on InN-based materials alive.

As one of the trailblazers, the authors intensively studied InN epitaxy both by RF plasma-assisted molecular beam epitaxy (RF-MBE) and low-pressure metalorganic vapor phase epitaxy (LP-MOVPE), and also fabrication/characterization of InN-based multiple quantum wells (MQWs) including InN/InGaN and InN/AlInN MQWs (Che et al., 2005a,b; Terashima et al., 2006). Although the authors succeeded in developing high-quality InN/InGaN MQWs of those days confirmed by X-ray diffraction (XRD), transmission electron microscopy (TEM) observations, and phtotoluminescence (PL) emissions (Che et al., 2007), whole-quality of the MQWs was still insufficient for device application. This is basically attributed to generation of high density defects at lattice-mismatched InN/(In)GaN heterointerfaces. In addition, another mismatch in optimum growth temperatures between InN and GaN limits low temperature epitaxy of InN-based ternary alloys that also makes it difficult to obtain device-quality MQWs.

It is a general practice in epitaxy to avoid defect generation that at least one of the layer thicknesses in the lattice-mismatched system is set to be less than the critical thickness. Besides, if the heterostructures consist of binary alloys, the growth conditions are free from compositional control that makes it easy/flexible to obtain high-quality layers. Upon reminding these, the authors proposed coherent monolayer-InN/GaN binary/binary nanostructures in which the critical thickness of InN is around two monolayers (MLs) for the lattice-mismatch as high as 11% (Yoshikawa et al., 2007). It is worth pointing out that optical and electrical properties of 1ML-InN/GaN MQWs are principally determined by those of GaN layers since the thickness of InN layer is ultimately thin. Furthermore, the whole-quality of 1ML-InN/GaN MQWs strongly depends on that of GaN layers. In other words, improvement of the whole-quality of 1ML-InN/GaN MQWs significantly relies upon increase in growth temperature of GaN layers. That is why, the phrase "*Increase in Growth Temperature*" has been posted as our guiding principle on epitaxy of III-N ordered alloys.

According to our proposal, the authors demonstrated successful achievement both of 1ML-InN and 2ML-InN layers coherently embedded in GaN-matrices grown by RF-MBE (Che et al., 2009; Hwang et al., 2008, 2009; Yoshikawa et al., 2007, 2008, 2009; Yuki et al., 2009). Fig. 1A and B shows group-III element images for 1ML-InN/GaN-matrix and

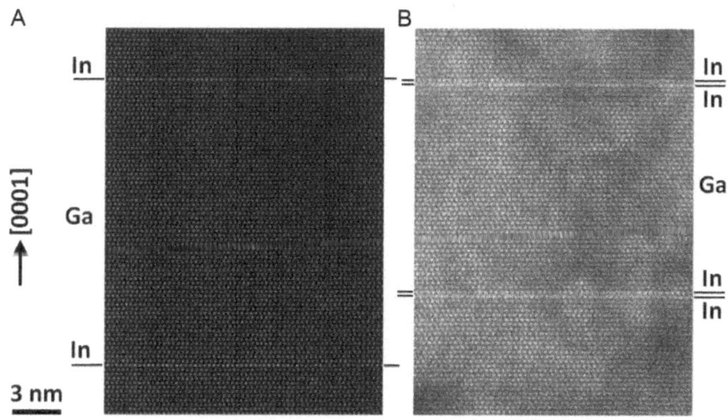

Fig. 1 Group-III element images taken by cross-sectional HAADF-STEM observation for (A) 1ML-InN/GaN-matrix grown at 650°C and (B) 2ML-InN/GaN-matrix grown at 620°C by RF-MBE. These growth temperatures are much higher than upper limit one (~500°C) of conventional +c-polar InN films. The *brighter* (*darker*) spots correspond to In-atoms (Ga-atoms). The InN layers were coherently embedded in GaN-matrices with atomically sharp interfaces both which were achieved under self-limiting and self-ordering processes by the D-ALEp.

2ML-InN/GaN-matrix structures taken by cross-sectional high-angle annular dark-field (DF) scanning transmission electron microscopy (HAADF-STEM), respectively. It was amazing that the 1ML-InN and 2ML-InN layers on/in GaN-matrices were grown at 620 and 650°C by RF-MBE, respectively, which were much higher temperatures than upper limit one of conventional InN films, ~500°C for +c-polar orientation (Che et al., 2009; Hwang et al., 2008, 2009; Yuki et al., 2009; Yoshikawa et al., 2007, 2008, 2009). The atomically flat 1–2ML-InN/GaN interfaces were observed in Fig. 1A and B even though the 1–2ML-InN/GaN-matrices were not grown at low temperature to suppress In/Ga intermixing.

Then, growth kinetics of the 1–2ML-InN/GaN-matrix structures even at the high temperatures is explained on the basis of understanding for fundamental behavior of growth fronts. Fig. 2 shows adsorption/desorption behavior of In-adlayers on a GaN surface at 630°C in MBE indicated by an imaginary part of pseudodielectric function $\langle \varepsilon_2 \rangle$ together with corresponding adlayer models. The pseudodielectric function $\langle \varepsilon_i \rangle$ is usually calculated by a spectroscopic ellipsometry (SE) system (J.A. Woollam, M2000 equipped with MBE in our case) and gives us information about

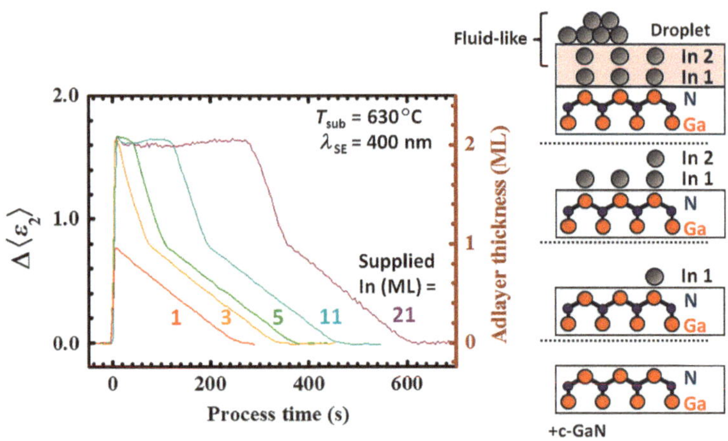

Fig. 2 Dynamic behavior of In-adlayer adsorption/desorption on the +c-polar GaN surface at 630°C in MBE indicated by the imaginary part of pseudodielectric function $\langle \varepsilon_2 \rangle$ together with corresponding adlayer models. The $\langle \varepsilon_2 \rangle$ was normalized by the adlayer thickness as $\Delta\langle \varepsilon_2 \rangle$. The SE probe wavelength was $\lambda_{SE}=400$ nm which was opaque for GaN at the temperature to neglect the effect of underlayer. The 1–21 ML In-metals were supplied with the rate $\kappa=0.5$ ML/s on the GaN surface. Each of adlayer models indicates coverage change of the monolayer (In 1) and the bilayer (In 2). The bilayer (In 2) behaves like fluid.

the growth front (In-adatoms) and the underlayer (GaN). The authors have introduced confident manners into in situ surface monitoring with SE analysis that: (1) a SE probe wavelength λ_{SE} is short enough to be opaque for the underlayer at the substrate temperature T_{sub} so that an optical interference effect is negligible in the $\langle \varepsilon_i \rangle$; and (2) the $\langle \varepsilon_2 \rangle$ is normalized by the adlayer thickness as $\Delta\langle \varepsilon_2 \rangle$ for the sake of easy understanding of the growth front phenomena without conducting complicated spectral analysis. The In-adatoms of 1–21 MLs were supplied with a rate $\kappa = 0.5$ ML/s on the GaN surface. A probe light of $\lambda_{SE} = 400$ nm was absorbed by the In-adatoms so that the $\Delta\langle \varepsilon_2 \rangle$ was proportional to coverage of the In-adatoms, detected with an atomic scale sensitivity and much thinner than a penetration depth of the probe light (Choi et al., 2006, 2008; Wu et al., 2007). Using this feature, the growth front phenomena changing moment by moment could be precisely traced and thus clearly understood. The observed $\Delta\langle \varepsilon_2 \rangle$ in Fig. 2 were described with four steps of surface processes each of which was attributed to (1) rapid increase in the $\Delta\langle \varepsilon_2 \rangle$ indicating deposition of the In-adatoms on GaN or increment of the In coverage, (2) the $\Delta\langle \varepsilon_2 \rangle$ plateaux corresponding to the In-adatoms saturated at 2ML-coverage during/after supplying In, (3) first quick decay of the $\Delta\langle \varepsilon_2 \rangle$ indicating desorption of the In-adatoms whose layer thickness decreased from 2 MLs to 1 ML, and (4) second slow decay of the $\Delta\langle \varepsilon_2 \rangle$ indicating desorption of the In-adatoms leading to the bare GaN surface. Therefore, the 1ML-In-supplied $\Delta\langle \varepsilon_2 \rangle$ indicated no plateau due to lack of the In-adatoms with 2 MLs. On the contrary, the other $\Delta\langle \varepsilon_2 \rangle$ showed only 2ML-coverage of In-adatoms even though excess In was supplied more than 2 MLs. These facts correspond to self-limiting and self-ordering processes to establish 2ML-In (In 1 + In 2) on the surface while all excess In-adatoms contribute to form small In-droplets on the bilayer (In 2). This situation coincides with the theoretically predicted model, the laterally contracted Ga/In bilayer on the GaN surface, where Ga/In-adlayers (In 1 + In 2), are likely ordered to the underlying crystalline structure. In this case, it is pointed out that the bilayer (In 2) easily migrates due to fluid-like behavior which arises beyond a fluid temperature (Neugebauer et al., 2003; Northrup et al., 2000; Northrup and Van de Walls, 2004).

As described in step (2), the self-limiting and self-ordering processes of In-adatoms appeared like atomic layer epitaxy (ALEp) so that the 1ML-InN and 2ML-InN layers could be embedded in GaN-matrices after appropriate crystallization into InN and the following capping by GaN even at the high temperatures. The authors would like to notice here that the growth kinetics is actually quite different from the typical ALEp. The self-limiting and self-ordering processes are available only at higher temperatures than the upper limit one of +c-InN epitaxy around 500°C. That is to say, it is impossible to realize continuous/successive deposition of InN layers thicker than 2 MLs, and also to crystallize In-adatoms on the InN surface, at the high temperatures. On the other hand, they become possible by a "*GaN matrix effect*" where the InN layers are grown on and capped by GaN both together. In addition, since the growth front changes quite dynamically, the GaN capping process for 1–2ML-InN layers must be immediately carried out during/before the desorption of InN. Reflecting these unique dynamic features, we named the above process "Dynamic Atomic Layer Epitaxy (D-ALEp)."

1.2 Proposal of III-N Ordered Alloys Developed by Dynamic-ALEp

The authors were absorbed in InN heteroepitaxy at the beginning, and then have proposed and demonstrated coherent 1–2ML-InN embedded in GaN-matrix structures. Through these investigations, the authors have deeply understood the growth kinetics and the behavior of growth front from in situ SE surface observation. Following that, the authors currently attempt to make the D-ALEp progress in fabricating ordered alloys, or digital alloys, in a whole III-nitride system.

Important remarks of the D-ALEp are summarized as (1) high temperature deposition/epitaxy that allows the growth front to coexist the fluid-like/mobile bilayer and the solid-like/ordered monolayer and (2) the "*matrix effect*" being available in embedding heterostructures. These are quite suitable for fabricating high-quality binary/binary MQWs and their extension to short-period superlattices (SPSs) in principle. In such a case, we can control the effective material properties by stacking structures. For example, if the 1ML-InN/GaN MQWs evolve into $(\mathrm{InN})_m/(\mathrm{GaN})_n$ SPSs where integers (m, n) denote each layer thickness in ML, the effective In-composition (X_{In}) is represented as $X_{\mathrm{In}} = m/(m+n)$ in an InN/GaN ordered alloy system. Of course, the features of D-ALEp and the extension to the ordered alloys are not restricted only in the InN/GaN system but applicable to the whole III-N system as quasiternary InN/AlN and GaN/AlN, and quasiquaternary InN/GaN/AlN ordered alloys. To realize the ordered alloys in the whole III-N system, it is necessary to universally understand the growth front phenomena of III-nitride semiconductors.

Fig. 3A shows equilibrium phase diagrams for group-III element binary systems, Al-In, Ga-Al, and In-Ga. Let us discuss the growth front of III-N

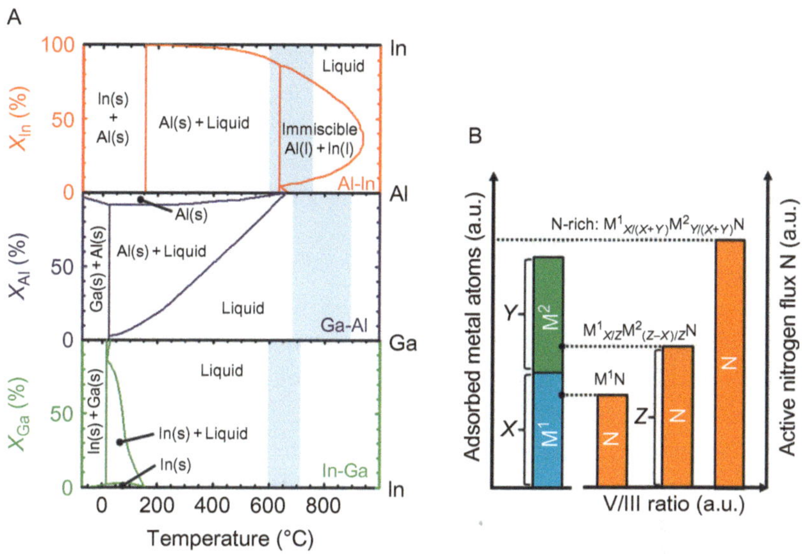

Fig. 3 (A) Equilibrium phase diagrams for the group-III element binaries of Al-In (upper), Ga-Al (middle), and In-Ga (lower). The *s* and *l* denote solid and liquid phases. The solid phases of Al, Ga, and In indicate face-centered cubic, orthorhombic, and tetragonal structures, respectively. The hatched regions depict the growth temperatures of their nitride ordered alloys which have been demonstrated and under planning by the D-ALEp in RF-MBE. The Al-In system includes the unique phase composed of immiscible Al-liquid and In-liquid above 636.5°C. (B) Schematic diagram of the composition ratio of ternary III-nitrides frozen/crystallized from "M^1+M^2+N" bilayer/monolayers under various V/III ratios where M^1 and M^2 correspond to any of group-III elements, Al, Ga, and In, and the equilibrium vapor pressure (sticking efficiency) of M^1 is lower (higher) than that of M^2. In other words, (M^1, M^2) = (Al, In), (Al, Ga), and (Ga, In), respectively.

ordered alloys in reference to Fig. 3A. The growth temperature regions are also indicated in Fig. 3A for their nitride ordered alloys, AlN/InN, GaN/AlN, and InN/GaN grown by the D-ALEp in RF-MBE. Since each of the element binary systems indicates a liquid phase for almost of the growth temperature regions, III-N ordered alloys seem to be grown under a liquid-phase epitaxy mode. We, however, examine the growth front phenomena so that the consideration should be slightly modified from bulk properties to surface adlayers. According to the laterally contracted Ga bilayer model (Neugebauer et al., 2003; Northrup et al., 2000), the surface fluid temperature of Ga on +c-GaN surface [T_{fluid}(Ga, +c-GaN)] is ~350°C which is much higher than the melting point of Ga [$T_{\text{m.p.}}$(Ga)]. Likewise, it is considered from our study that T_{fluid}(In, +c-GaN) is probably ~600°C (Yoshikawa et al., 2016a). Reviewing our previous reports (Cao et al., 2003; Xu et al., 2001), it is suggested that T_{fluid}(Al, −c-GaN) is ~820°C. Consequently, the T_{fluid}(In, +c-GaN) and T_{fluid}(Al, −c-GaN) are also much higher than the $T_{\text{m.p.}}$(In) and $T_{\text{m.p.}}$(Al), respectively. It is favorable to grow the III-N ordered alloys assisted with fluid-like surfaces due to their advantage of migration, resulting in flat surfaces and high-quality layers. In view of consideration of T_{fluid}, InN/GaN ordered alloys are grown at a hatched temperature region above T_{fluid}(In, +c-GaN) by RF-MBE. On the other hand, AlN/InN and GaN/AlN systems might be grown at below T_{fluid}(Al, −c-GaN). In this point, it should be carefully considered that fluid-like Al-adatoms appears within the hatched temperature regions whether or not. One possible solution is to utilize a unique phase composed of immiscible Al-liquid and In-liquid in the Al-In system (Clarke et al., 2013). When the monolayer consists of In (see In 1 in Fig. 2) and the bilayer consists of Al (see In 2 in Fig. 2), the Al-adatoms would mobile assisted by the In-monolayer without intermixing of Al/In even during desorption of the In-adatoms (Wang et al., 2016a).

Then, it is discussed how the composition ratio of ordered alloys is linked with bilayer/monolayers by the D-ALEp. Fig. 3B shows schematic diagram of the composition ratio of ternary III-nitrides frozen/crystallized from "M^1+M^2+N" bilayer/monolayers under various V/III ratios by RF-MBE. As an example, (M^1, M^2) are assumed here to be (Ga, In). GaN (M^1N) layers are grown under $X>Z$; $Ga_{X/Z}In_{(Z-X)/Z}N$ [$M^1_{X/Z}M^2_{(Z-X)/Z}N$] layers are grown under $X<Z<X+Y$; and $Ga_{X/(X+Y)}In_{Y/(X+Y)}N$ [$M^1_{X/(X+Y)}M^2_{Y/(X+Y)}N$] layers are grown under $Z>X+Y$. These relations are quite useful for achieving selective growth of InN (M^2N) and GaN (M^1N) SPSs from the Ga + In (M^1+M^2) mixed bilayer/monolayer, and suppressing inclusion of InGaN (M^2M^1N) ternary alloy at the interface. When growing InN/GaN (M^2N/M^1N) SPSs, the following points are necessary: (1) The capping of InN (M^2N) bilayer/monolayers with GaN (M^1N) should immediately be done after the deposition of InN (M^2N) under $X>Z$ since the rapid capping makes the effective InN (M^2N) coverage increase (Yoshikawa et al., 2016a). (2) Selective In (M^2) desorption avoids the deposition of unexpected InGaN (M^2M^1N) ternary alloys. (3) In order to establish binary InN (M^2N) in SPSs, residual Ga-adatoms (M^1) should be eliminated before the deposition of InN (M^2N) so that enough active N-atoms are supplied to dry-up.

Thanks to above tips, our achievement of InN/GaN ordered alloys is described in Section 3.1.

2. DEVELOPMENT OF DYNAMIC-ALEp IN HIGHLY MISMATCHED InN/GaN SYSTEM

2.1 Impact of Increase in Growth Temperature on InN/GaN Heteroepitaxy

In this section, development of the D-ALEp in the 1ML-InN/GaN system is discussed with keeping our policy "*Increase in Growth Temperature*" in mind.

It is generally known that an equilibrium vapor pressure of nitrogen over InN is much higher than those over other III-nitrides (Wang and Yoshikawa, 2004). This fact consequently requires much lower epitaxy temperatures of InN than those of other nitrides, resulting in a mismatch in epitaxy temperatures among them for fabricating InN-based heterostructures. Besides, the InN-based heterostructures inevitably belong to a lattice-mismatched system, for example, the lattice-mismatch between InN and GaN is $\Delta a_{InN/GaN} \sim 11\%$ in c-plane epitaxy. The mismatch in epitaxy is an undesirable factor in general that should be as small as possible to obtain high-quality layers.

As the first step, the authors carefully investigated how growth rate of continuous InN layers changes against temperatures and lattice polarities, and determined what temperature is the upper critical limit of thick InN layers. Fig. 4 shows growth rate of InN as a function of growth temperature

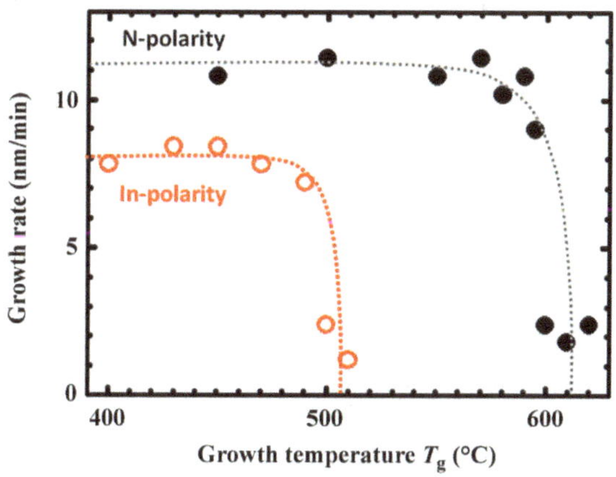

Fig. 4 Growth rate of InN as a function of growth temperature T_g for In-polarity (*open circles*) and N-polarity (*solid circles*). All InN layers were grown under very slightly N-rich stoichiometry at the constant N-flux condition. The critical temperatures of InN epitaxy were around 500 and 600°C for In-polarity and N-polarity regimes, respectively. Above these temperatures, no epitaxy of continuum InN layers was carried out while In-droplets were accumulated on the surfaces for both polarities.

T_g for both polar orientations grown by RF-MBE. All InN layers were grown under very slightly N-rich stoichiometry at a constant N-flux condition. The In-fluxes were adjusted as high as possible but not to appear In-droplets on the surface. It was clearly indicated that the critical temperatures of InN epitaxy were around 500 and 600°C for In-polarity and N-polarity, respectively. Above these temperatures, the growth rates dropped down and consequently no epitaxy of thick InN layers was carried out while In-droplets were accumulated on the surfaces for both polarities. In addition, it was found that the growth rates for both polarities are different from each other below the critical temperatures. These phenomena can be interpreted by paying attention to a sticking efficiency of N-atoms on the InN surface under an N-incorporation rate limiting epitaxy. The N-atoms are incorporated with three downward bonds for N-polarity but with only one bond for In-polarity with reference to hexagonal crystallography. Therefore, the sticking efficiency of N-atoms of N-polar InN is higher than that of In-polar InN, resulting in the higher critical temperature and growth rate of N-polar InN epitaxy. Upon taking a look at this interpretation at a glance, a certain researcher may be interested in the −c-polar InN-based heterostructures since the higher critical temperature of N-polar InN ~600°C seems suitable for mitigating the mismatch in epitaxy temperatures among them. As set forth in previous section, however, the whole-quality of those was fairly insufficient for device application. Consequently, the authors have decided to exclude the InN-based multilayers grown at below ~600°C by RF-MBE from research targets.

According to this decision and our policy, we have daringly set a battlefield higher than $T_g = 600$°C for fabricating device-grade InN-based heteronanostructures even though it is beyond the upper limit temperature of the thick InN epitaxy by RF-MBE. Then, how do we exceed the critical temperature of the thick InN layers? The answer can be found out by focusing on the growth front of matrices which peculiarly adsorbs source materials with thicknesses of a few MLs. Since the sticking efficiency depends both on the bond strength of matrices and the equilibrium vapor pressure of adsorbed materials, the net sticking efficiency of adsorbed materials becomes strong when the matrix is thermally robust or has strong bonds. Based on this "*Matrix effect*," the authors proposed the 1–2ML-InN/GaN matrices as the InN-based heteronanostructures (Yoshikawa et al., 2007). It is preferred that the 1–2ML-InN/GaN matrices be grown under +c-polar regime. As is discussed about the sticking efficiency of N-atoms on the InN surfaces above, since the In-adatoms have three back bonds in +c-polar regime but only one bond in −c-polar epitaxy, the In-adatoms are more tightly adsorbed onto the +c-polar GaN surface than the −c-polar case that makes the epitaxy temperatures higher in the former than the latter. It is noted from another point of view that epitaxy of the 1–2ML-InN layers on/in GaN matrices prefers +c-polarity to −c-polarity since the surface and interface are atomically flat and sharp due to general tendency of III-nitrides.

2.2 Coherent Monolayer-InN in GaN Matrix

Subsequently, the +c-polar epitaxy of 1–2ML-InN/GaN matrices grown at above 600°C by the D-ALEp is discussed.

The authors attempted first to fabricate InN/GaN MQWs with 40–48 periods at different growth temperatures from 500 to 650°C with a 50°C step by RF-MBE, leading certainly to 1–2ML-InN/GaN matrix structures due to self-limiting and self-ordering surface behavior which appeared beyond the upper limit temperature ~500°C of +c-polar thick InN epitaxy. Other growth conditions are found in Yoshikawa et al. (2007, 2008, 2009), Hwang et al. (2008, 2009), Yuki et al. (2009), and Che et al. (2009). The InN/GaN MQWs were characterized in view of structural and optical properties by XRD, TEM, and PL.

Fig. 5 shows XRD ω-2θ spectra around GaN 0002 of +c-polar InN/GaN MQWs grown at different temperatures T_g from 500 to 650°C by RF-MBE. The InN sources corresponding to 3MLs were supplied with

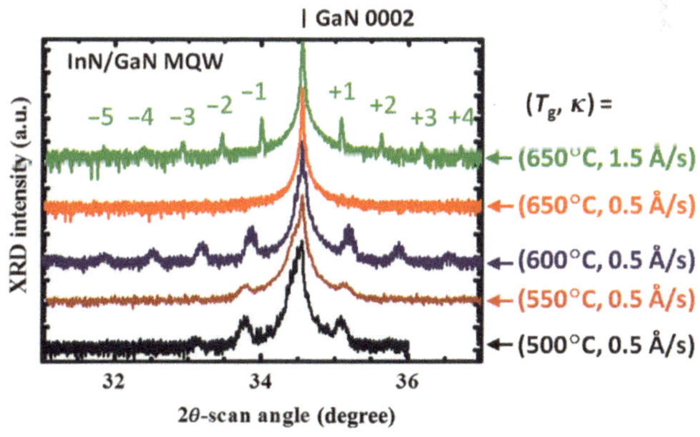

Fig. 5 XRD ω-2θ spectra around GaN 0002 of +c-polar InN/GaN MQWs with 40–48 periods grown at different temperatures T_g from 500 to 650°C by RF-MBE, which were higher than the upper limit one (~500°C) of +c-polar thick InN epitaxy. The 3ML-InN source was supplied with deposition rates $\kappa = 0.5$ and 1.5 Å/s.

deposition rates $\kappa = 0.5$ and 1.5 Å/s. It was found with observing satellite peaks that the MQW periodic structures under the InN deposition rate $\kappa = 0.5$ Å/s were fabricated at temperatures up to 600°C, but no periodic structure was observed at 650°C, reflecting that any InN layers did not freeze on/in GaN matrices during source supply and following GaN capping due to rapid decomposition/reevaporation of InN at the high temperature. Then, in order to insert InN layers into the GaN matrix at 650°C, the InN deposition rate was increased to $\kappa = 1.5$ Å/s. Consequently, it was demonstrated that the MQW was grown even at 650°C having significantly sharp satellite peaks up to sixth order.

Fig. 6A–D shows DF-TEM images taken under $\boldsymbol{g} = [0002]$ of +c-polar InN/GaN MQWs which coincident with those shown in Fig. 5. It is clearly indicated that the InN/GaN interface flatness and sharpness are remarkably improved with increase in the growth temperatures. These results basically originate from improvement of the quality of GaN matrices by enhanced migration of Ga-adatoms during epitaxy. Particularly, atomically flat interfaces appear at the temperatures of 600–650°C, which were formed by the self-limiting and self-ordering processes and further assisted by immiscible nature between InN and GaN. The impact of the growth temperature on dislocation densities was examined as well. The densities of screw dislocation (N_{screw}) and edge dislocation (N_{edge}) included in the MQW regions were evaluated from the DF-TEM images under $\boldsymbol{g} = [0002]$ and $\boldsymbol{g} = [11\text{–}20]$(not shown here), respectively. With increasing the growth temperature, the InN/GaN MQWs decreased in the dislocation densities as from $N_{screw} = 1\text{–}2 \times 10^{10}$ cm^{-2} at 500°C to $N_{screw} = 8\text{–}9 \times 10^{8}$ cm^{-2} at above 600°C, and from $N_{edge} = 5 \times 10^{10}$ cm^{-2} at 550°C to $N_{edge} = 7\text{–}9 \times 10^{9}$ cm^{-2} at above 600°C, while the N_{edge} at 500°C was too high to evaluate. Furthermore, the MQWs grown at above 600°C show coherently stacked structures and no serious generation of misfit dislocations at the interfaces at least in Fig. 6C and D. The integrated PL intensity (not shown here) was also remarkably increased with increase in the growth temperature. It is considered from these facts that the whole-quality of MQWs was drastically improved according to our policy "*Increase in Growth Temperature.*" The authors would like to notice that the InN/GaN MQW grown at 650°C shows the finest structure among those, while InN layers look discontinuous or fractional due to the growth front competition between the "*GaN matrix effect*" and rapid dissociation of InN during epitaxy at the high temperature. Such a critical condition

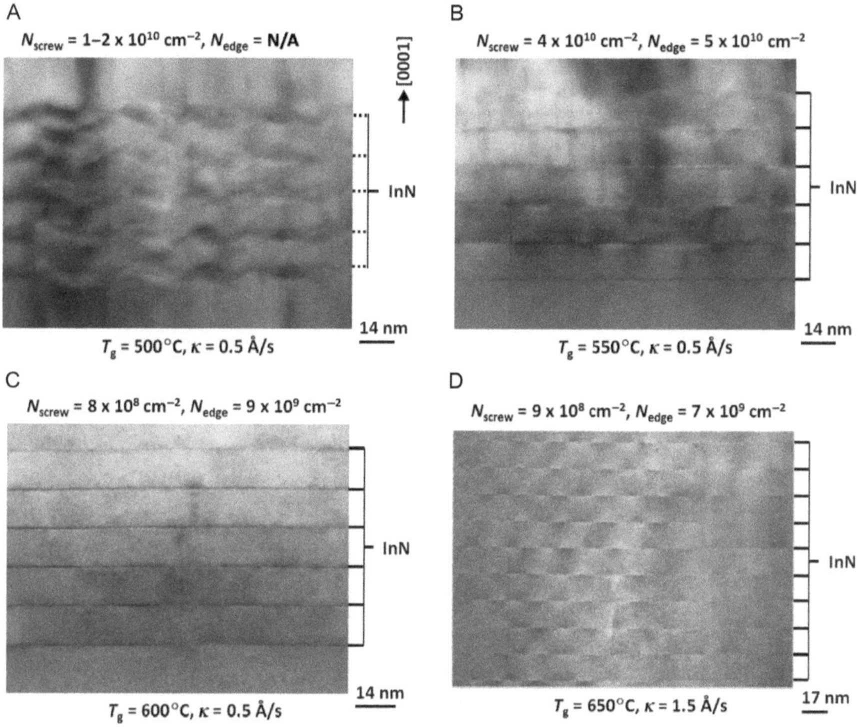

Fig. 6 Cross-sectional DF-TEM images under the diffraction condition $\boldsymbol{g}=[0002]$ for +c-polar InN/GaN MQWs corresponding to Fig. 5. The growth temperature T_g and the deposition rate κ of 3ML-InN sources were (A) 500°C and 0.5 Å/s, (B) 550°C and 0.5 Å/s, (C) 600°C and 0.5 Å/s, and (D) 650°C and 1.5 Å/s, respectively. The density of screw dislocation N_{screw} (edge dislocation N_{edge}) was evaluated from the DF-TEM images under $\boldsymbol{g}=[0002]$ ($\boldsymbol{g}=[11\text{–}20]$, not shown here). The contrast variation along with the growth direction is due to local distortion arising from high density of edge dislocations. The InN/GaN interfaces of (A) are unclear and undulate due to formation of facets.

results in fabrication of fractional ML (FML) InN, or 1ML-InN-disk, on/in the GaN matrix that can be controlled by the material supply rate, adlayer thickness, capping timing, and so on, by carefully observing the growth front with in situ SE observation.

In the above discussions, there has remained an ambiguous issue concerning with the precise layer thickness of InN embedded in the GaN matrices. As clearly indicated in Fig. 1A and B, as well as a part of Fig. 6, the authors have concluded that the layer thicknesses of InN are actually 2ML, 1ML, and FML grown at above 600°C by the D-ALEp. This fact

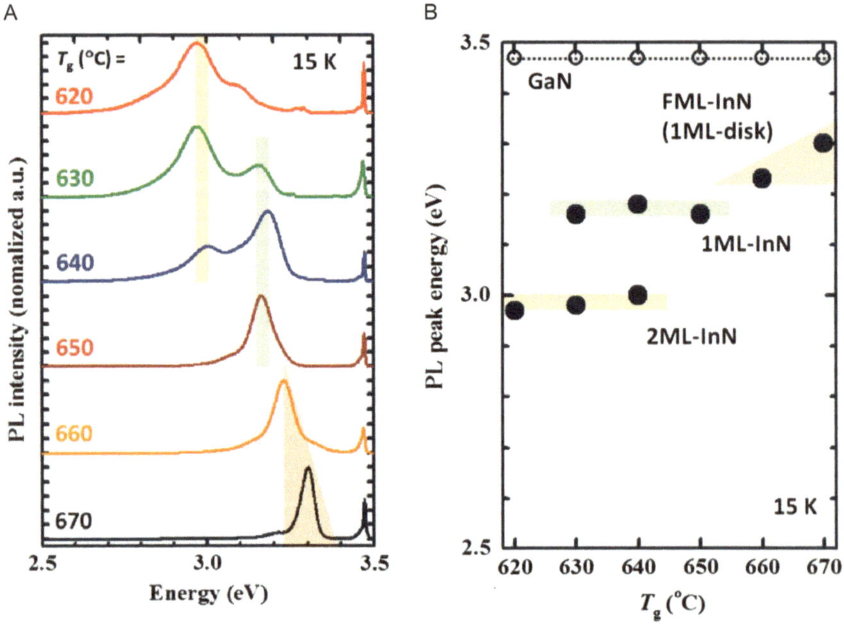

Fig. 7 (A) PL spectra at 15 K of InN/GaN MQWs including 2ML-InN, 1ML-InN, and/or FML-InN wells. The MQWs were grown at different temperatures T_g from 620 to 670° C by the D-ALEp in RF-MBE. (B) Dominant PL peak energies at 15 K as a function of growth temperature T_g, which are categorized in three energy regions, 2.96–3.00 eV, 3.14–3.18 eV, and the higher than those, together with that of GaN at 3.47 eV. They are attributed to transition energies of 2ML-InN/GaN, 1ML-InN/GaN, and FML-InN/GaN MQWs, respectively.

is also supported by PL peak energies as a cross check manner. Fig. 7A shows PL spectra at 15 K of InN/GaN MQWs which consist of 2ML-InN, 1ML-InN, and/or FML-InN wells. The MQWs were grown at different temperatures T_g from 620 to 670°C by RF-MBE in order to discretely change the layer thickness of InN via the self-limiting and self-ordering D-ALEp processes. If the MQW consists of ternary InGaN wells, the PL peak energy should gradually blue-shift with increase in the growth temperature from 620 to 650°C with the 10°C step by typical RF-MBE. On the other hand, the PL peak energies in Fig. 7A show hopping-blue-shift for the temperatures from 620 to 650°C. Fig. 7B shows dominant PL peak positions as a function of T_g. The dominant PL peak positions are categorized into three regions, 2.97–3.00 eV, 3.16–3.20 eV, and the higher than those, except that of GaN at 3.47 eV, which are attributed to transition energies of 2ML-InN/GaN, 1ML-InN/GaN, and FML-InN/GaN MQWs, respectively.

The authors believe this behavior supporting that: (1) InN wells grown by D-ALEp are composed of pure binary InN without intermixing of Ga/In and (2) the hopping-blue-shift is attributed to discrete change oflayer thickness of InN from 2MLs to 1ML. Further increasing growth temperature beyond 650°C, the PL peak energies gradually blue-shift that is attributed to an emission from FML-InN wells or 1ML-InN disks with decrease in the coverage or lateral size, as shown in Fig. 7A and B. Anyway, it is proved that 2ML-InN, 1ML-InN, and FML-InN layers can be grown in the GaN matrices depending on the growth temperature beyond the upper limit one ~500°C of +c-polar thick InN epitaxy by the D-ALEp.

For assistance of reader's easy understanding, a correlation map indicating how the InN layer thicknesses depend on the growth conditions is present. Fig. 8 shows the correlation diagram of the layer thickness of +c-polar InN on/in the GaN matrix for the growth temperature and the supplied InN thickness grown by RF-MBE. It is shown that thick InN layers are grown at below ~500°C while FML/1ML/2ML-InN layers can be grown at beyond ~600°C by the D-ALEp as discussed earlier. The boundary of correlation map is strongly relies on the InN deposition rate, GaN capping speed, and the other growth conditions. This diagram is one of the important key parts when we expand the D-ALEp developed in the InN/GaN system to fabricate ordered alloys in the whole III-nitride system. After writing down similar correlation diagrams in InN/AlN and GaN/AlN systems,

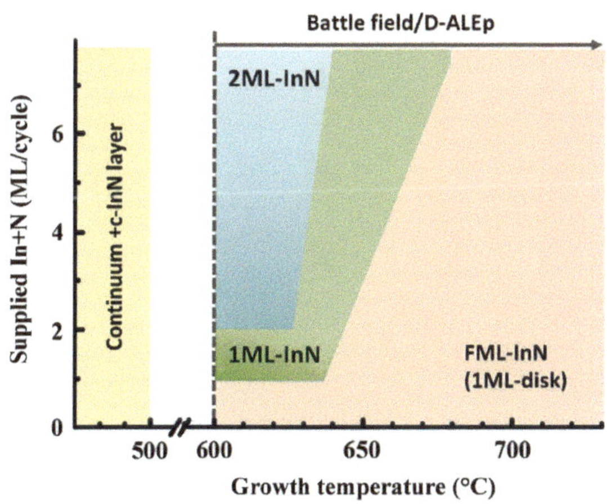

Fig. 8 Correlation diagram of the layer thickness of +c-polar InN in the GaN matrix between the growth temperature and the supplied In + N thickness grown by RF-MBE. The D-ALEp of InN/GaN has been developed at beyond ~600°C in this work.

we can arbitrary design and control material structures and properties of III-nitride ordered alloys.

Concerning with FML-InN, unusual emission behavior of FML-InN/GaN MQW found in cathodoluminescence (CL) observation is discussed. Fig. 9 shows panchromatic-CL images of the FML-InN/GaN MQW grown at 670°C. All CL images taken at 85 K depict the same scanned area but with the different penetration depth corresponding to the FML-InN/GaN MQW region for $V_{acc}=3$ and 6 kV and underlying GaN region for $V_{acc}=14$ and 18 kV, respectively. In the CL observation, the screw threading dislocations are found as dark spots since they play a role of nonradiative centers in III-nitrides (Abell and Moustakas, 2008). Likewise, the screw dislocations tend to strongly stick the FML-InN since the screw dislocation provides a relative high sticking efficiency on the growth surface.

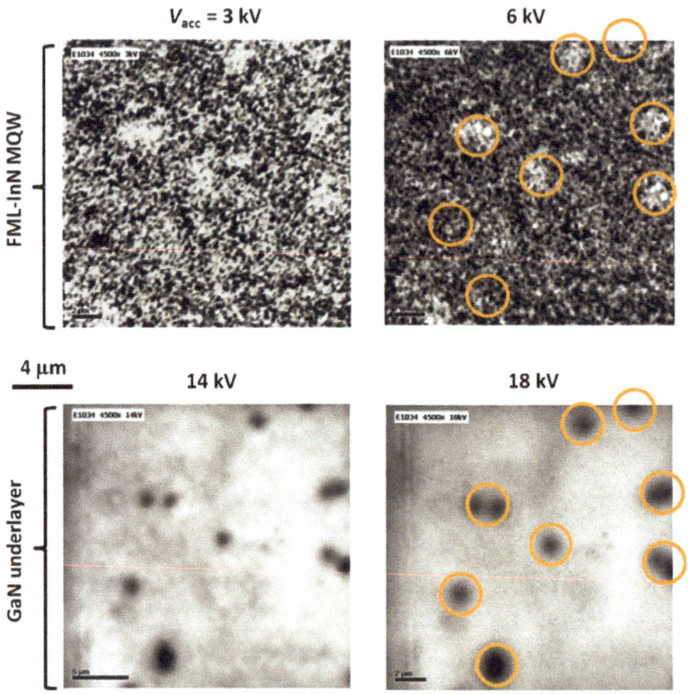

Fig. 9 Panchromatic–CL plane images at 85 K for the FML-InN/GaN MQW grown at 670°C. All images depict the same area but with the different depth position controlled by the e-beam accelerating voltage V_{acc}. The e-beam penetrates into the FML-InN/GaN MQW region for $V_{acc}=3$ and 6 kV, while it probes underlying GaN region for $V_{acc}=14$ and 18 kV. For $V_{acc}=6$ and 18 kV, the markers are added to identify the same locations where the FML-InN and the screw dislocation coexist.

Upon reminding these facts, it is unusually judged that no dark spots are observed in the FML-InN/GaN MQW region, and further, the FML-InN wells emit CL even though the screw dislocations actually exist and pass through the FML-InN wells. The detail mechanism why the screw dislocations loose nonradiative nature only in the FML-InN/GaN MQW has been still unclear. But, the FML-InN may potentially possess certain robustness against nonradiative and current leakage properties induced by dislocations.

2.3 Growth Front Analogy of Monolayer-In(N) on GaN by MOVPE

As has been discussed so far, the D-ALEp has been developed in the InN/GaN system grown by RF-MBE. Another technical approach should be taken into account to make universal understanding for the D-ALEp much deeper. In this section, the growth front behavior of In-adlayers on the GaN surface is discussed for the case of MOVPE. In general, the growth circumstance of MOVPE is quite different from that of MBE in view of experimental parameters including a reactor ambient/pressure, an optimum V/III ratio, a growth temperature region, and so on. On the other hand, the authors believe that the growth front of MOVPE is essentially in common with that of MBE with reference to in situ SE surface observations.

The adsorption/desorption dynamics of In-adatoms on GaN surfaces for both polar orientations were investigated in a horizontal LP-MOVPE reactor which equipped the SE system (Jobin Yvon, UVISEL). Prior to surface analysis, specular GaN surfaces were obtained by LP-MOVPE wherein: (1) +c-polar GaN layers with a typical step-flow surface morphology were grown on nitridated sapphire (0001) substrates by three-step polarity inversion epitaxy, in which 2ML-Al converted a −c-polar growth front into a +c-polar one (Cao et al., 2003; Lim et al., 2002; Xu et al., 2001) and (2) −c-polar GaN layers without pyramidal hillocks were grown on nitridated sapphire (0001) substrates having off-angels of 2–4 degree. During in situ surface observation, the In-adatoms were supplied by feeding trimethylindium (TMI) under N_2 ambient at 300 kPa, without NH_3.

Fig. 10 shows adsorption/desorption behavior of In-adlayers on the +c-polar GaN surface at different substrate temperatures $T_{sub}=770$–$900°C$ indicated by the imaginary part of pseudodielectric function $\Delta\langle\varepsilon_2\rangle$ normalized by the deposited layer thickness. The SE probe wavelength was $\lambda=385$ nm which was opaque for GaN at the temperatures to neglect the effect of underlayer. The In-adatoms of 20 MLs were supplied with the rate $\kappa=0.5$ ML/s on the +c-polar GaN surface. These conditions were

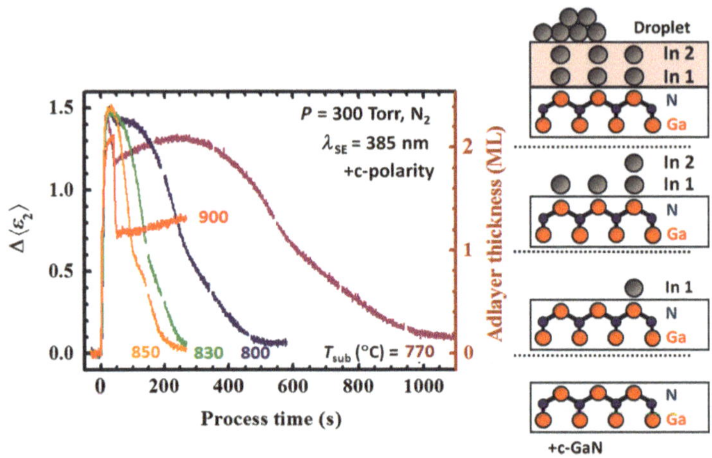

Fig. 10 Dynamic behavior of In-adlayer adsorption/desorption on the +c-polar GaN surface in LP-MOVPE indicated by $\Delta\langle\varepsilon_2\rangle$ together with adlayer deposition models. The processes were monitored for different substrate temperatures T_{sub} from 770 to 900°C under N_2 ambient at 300 kPa. The SE probe wavelength was $\lambda_{SE}=385$ nm which was opaque for GaN at the temperatures to neglect the effect of underlayer. The 20 ML In-metals were supplied at the rate $\kappa=0.5$ ML/s on the +c-polar GaN surface.

set by following MBE experiments as shown in Fig. 2. The observed $\Delta\langle\varepsilon_2\rangle$ in Fig. 10 were described with four steps of surface processes each of which was attributed to (1) rapid increase in the $\Delta\langle\varepsilon_2\rangle$ indicating deposition of the In-adatoms on GaN during In-supply, (2) the $\Delta\langle\varepsilon_2\rangle$ plateaux corresponding to the In-adatoms saturated at 2ML-coverage during/after In-supply, (3) first quick decay of the $\Delta\langle\varepsilon_2\rangle$ indicating desorption of the In-adatoms whose layer thickness decreased from 2 MLs to 1 ML, and (4) second slow decay of the $\Delta\langle\varepsilon_2\rangle$ indicating desorption of the In-adatoms leading to the bare GaN surface. These facts correspond to the self-limiting and self-ordering processes to form 2ML-In while all excess In-adatoms contribute to form small In-droplets on the topmost surface. This sequence was almost same as that of MBE shown in Fig. 2, and further the situation coincided with theoretically predicted models, the laterally contracted Ga (In) bilayer on the GaN surface, where Ga (In) adlayers were likely ordered to the underlying crystalline structure (Neugebauer et al., 2003; Northrup et al., 2000; Northrup and Van de Walls, 2004). However, a few irregular surface processes appeared in Fig. 10. One of the irregular processes was that the $\Delta\langle\varepsilon_2\rangle$ indicated overshooting behavior during the steps (1) and (2), and then converged on the plateau in step (2) at $T_{sub}=770°C$. This phenomenon was attributed to TMI cracking via a surface reaction followed by migration of In-droplets to form the uniform 2ML-adlayers which took relative long period ~300 s due to the low temperature. Another one was that the $\Delta\langle\varepsilon_2\rangle$ dropped down just after In-supply stopped at $T_{sub}=900°C$. This was attributed to decomposition of the GaN surface due to lack of NH_3 at the high temperature.

Then, how do those changes in the case of −c-polar growth front? Fig. 11 shows adsorption/desorption behavior of In-adlayers on the −c-polar GaN surface at different substrate temperatures $T_{sub}=770$–$900°$ C indicated by the $\Delta\langle\varepsilon_2\rangle$. Almost conditions were same as those of +c-polar regime for comparison. The observed $\Delta\langle\varepsilon_2\rangle$ in Fig. 11 were described with three steps of surface processes each of which was attributed to (1) rapid increase in the $\Delta\langle\varepsilon_2\rangle$ indicating deposition of the In-adatoms on GaN during In-supply, (2) the $\Delta\langle\varepsilon_2\rangle$ plateaux corresponding to the In-adatoms saturated at 1ML-coverage during/after In-supply, and (3) decay of the $\Delta\langle\varepsilon_2\rangle$ indicating desorption of the In-adatoms from 1 ML to 0 ML. The decomposition of −c-polar GaN surface was not detected at $T_{sub}=900°C$ though it was observed in the +c-polar GaN. This thermal stability in −c-polar GaN in LP-MOVPE was quite similar as the case of −c-polar InN grown by RF-MBE, as discussed and shown in Fig. 4. Thus, it can be explained by the sticking efficiency of N-atoms on the GaN surface. The N-atoms are incorporated with three downward bonds for −c-polarity but with only one bond for +c-polarity. Consequently, the thermally stable

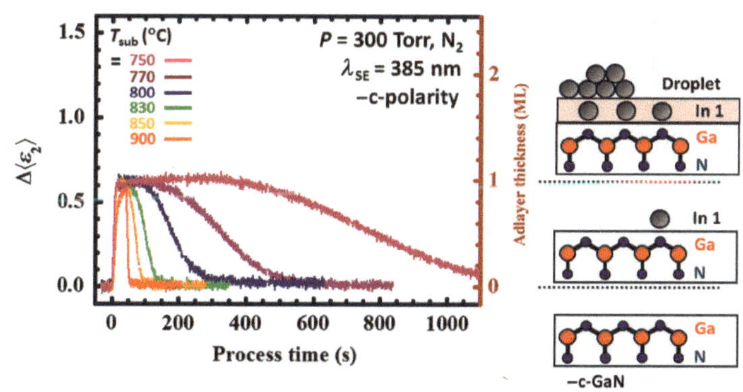

Fig. 11 Dynamic behavior of In-adlayer adsorption/desorption on the −c-polar GaN surface in LP-MOVPE indicated by $\Delta\langle\varepsilon_2\rangle$ together with adlayer deposition models. The processes were monitored for different substrate temperatures T_{sub} from 750 to 900°C under N_2 ambient at 300 kPa. The SE probe wavelength was $\lambda_{SE}=385$ nm which was opaque for GaN at the temperatures to neglect the effect of underlayer. The 20 ML In-metals were supplied at the rate $\kappa=0.5$ ML/s on the −c-polar GaN surface.

surface was obtained in −c-polar GaN rather than +c-polar GaN. However, the situation for sticking In-adatoms on the GaN surface was completely opposite. The thickness of In-adlayers were 2 MLs and 1 ML on the +c-polar and −c-polar GaN surfaces through the self-limiting and self-ordering processes, respectively, as shown in Figs. 10 and 11. This is attributed that the In-adatoms are more strongly stuck onto the +c-polar GaN surface than the −c-polar one (Feenstra et al., 2000).

As a summary of this section, the authors are convinced that the growth fronts of MBE and MOVPE are essentially equivalent to each other. In addition, the +c-polar growth front is suitable for fabricating 1ML-InN/GaN MQWs or SPSs to establish ordered alloys in III-nitrides.

3. III-N ORDERED ALLOYS GROWN BY DYNAMIC-ALEp

3.1 MBE Achievement of $(InN)_1/(GaN)_n$ Ordered Alloys

The authors proposed and successfully developed the D-ALEp and the resultant coherent 1–2ML-InN/GaN matrices structures grown by RF-MBE equipped with the in situ SE surface monitoring system (Che et al., 2009; Hwang et al., 2008, 2009; Yoshikawa et al., 2007, 2008, 2009; Yuki et al., 2009). Therefore, the problems in lattice-mismatch of 11% and epitaxy-temperature-mismatch > 200°C between InN and GaN could be overcome in principle. Subsequently, the authors have been extended the research target to fabricate coherent structure ordered InN/GaN alloys by decrease in the GaN layer thickness so that wavefunctions of neighboring electrons and holes interact with each other enough to establish continuum electron and hole states. In this section, the fabrication and structural control of $(InN)_1/(GaN)_n$ SPSs grown by RF-MBE are discussed where integers $n = 1$–20 denote GaN layer thickness in ML.

Fig. 12A shows the sample structure of $(InN)_1/(GaN)_n$ SPSs with 50–100 periods grown on MOVPE-grown GaN templates by RF-MBE. Fig. 12B shows the D-ALEp shutter sequence for growing $(InN)_1/(GaN)_n$ SPSs, respectively. Related discussions of growth kinetics are found in Yoshikawa et al. (2016b, submitted) and Kusakabe et al. (2016). All layers were grown under the slightly metal-rich condition, keeping at least 1ML-(In or Ga) adlayer to enhance surface migration since the bilayer on 1ML-(In or Ga) was fluid-like and mobile which was explained by the laterally contracted Ga (In) model (Neugebauer et al., 2003; Northrup et al., 2000; Northrup and Van de Walls, 2004). For growing SPSs, excess 3ML-In and 2ML-N were supplied at first to establish the 1ML-InN through the

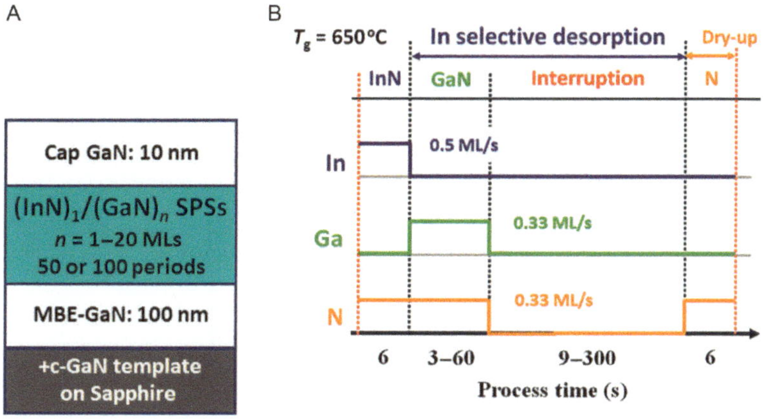

Fig. 12 (A) The sample structure of $(InN)_1/(GaN)_n$ SPSs grown on MOVPE-grown GaN template by RF-MBE and (B) the D-ALEp shutter sequence for growing $(InN)_1/(GaN)_n$ SPSs. All layers were grown under the slightly metal-rich condition, keeping at least 1ML-(In or Ga) adlayer. For growing SPSs, excess 3ML-In and 2ML-N were supplied and crystallized into the 1ML-InN by self-limiting and self-ordering processes after GaN capping at 650°C. The excess In-adatoms were selectively desorbed and the Ga-adatoms were remained during the GaN capping and interruption. Finally, all the Ga-adatoms were crystallized by dry-up process with active N-irradiation.

self-limiting and self-ordering processes by GaN capping at 650°C. Then, the growth interruption at 650°C allows us to selectively desorb excess In-adatoms from the growth front while the Ga-adatoms were remained on it, since the equilibrium vapor pressure of In is about two orders of magnitude higher than that of Ga. Finally, the Ga-adatoms were completely crystallized/frozen by a dry-up process with enough N-supply. The layer thickness of GaN n was varied by the deposition periods from 3 to 60 s, resulting in the thicknesses from $n=1$ to 20 MLs.

Such the growth front behavior was controlled as designed due to in situ monitoring in which the SE system (J.A. Woolam, M-2000) and a reflection high energy electron diffraction (RHEED) analyzer system (*k*SA, 400) were concurrently utilized (Choi et al., 2006; Kusakabe et al., 2016; Wu et al., 2007; Yoshikawa et al., 2016b, submitted). Fig. 13 shows typical evolution of SE and RHEED signals during D-ALEp of the $(InN)_1/(GaN)_4$ SPSs. They precisely traced each of processes so that it was possible to feed back an in situ correction of growth parameters. Especially, a careful attention was required for the selective desorption of In-adatoms and the dry-up of GaN surface. The former is necessary to avoid unintentional insertion of ternary InGaN since the InGaN easily introduces lattice relaxation in SPSs. The

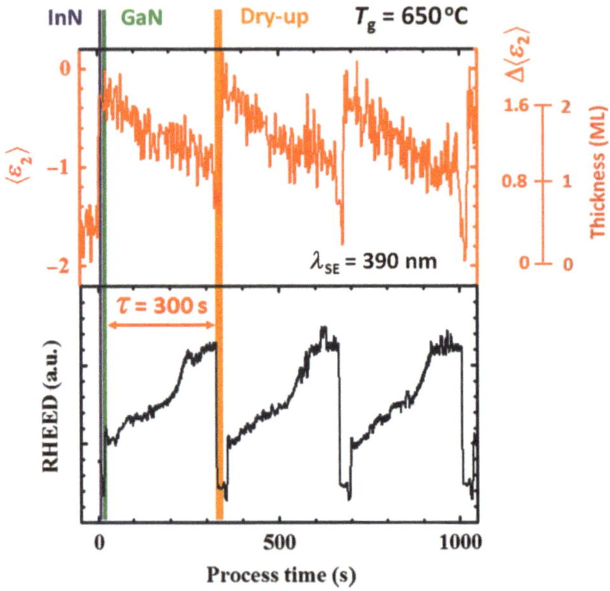

Fig. 13 Evolution of SE and RHEED signals during $(InN)_1/(GaN)_4$ SPSs growth under interruption period $\tau=300$ s at 650°C, where the SE signal is the imaginary part of pseudodielectric function $\langle\varepsilon_2\rangle$ and the RHEED one is specular spot intensity. The SE probe wavelength was $\lambda_{SE}=390$ nm. Briefly explaining, the $\langle\varepsilon_2\rangle$ is composed of varying and base components, indicating the adlayers and underlayers, respectively. In this case, the periodic increments of $\langle\varepsilon_2\rangle$ indicate deposition of InN and GaN on the GaN surface while the reduction and recovery of $\langle\varepsilon_2\rangle$ indicate desorption of the excess In-adlayers or crystallization of the Ga-adlayers. The RHEED intensity conversely changes against the $\langle\varepsilon_2\rangle$, but their interpretations are well coincident with each other.

latter is also necessary to ensure the "*GaN matrix effect*" for freezing or insertion of 1ML-InN into the GaN layers.

The selective desorption of In-adatoms was accomplished by taking long enough period at the growth interruption. Fig. 14 shows XRD for reciprocal space maps (RSMs) around GaN 11–24 of $(InN)_1/(GaN)_4$ SPSs grown under different growth interruption periods $\tau=90$–300 s. The coherent $(InN)_1/(GaN)_4$ SPSs were achieved for the growth interruption periods τ longer than 150 s. The further longer period $\tau=300$ s made the profile of zeroth satellite peak much sharper, indicating that the structural quality of $(InN)_1/(GaN)_4$ SPSs was improved by preventing the insertion of ternary InGaN at the interfaces.

Fig. 15A and B shows XRD ω-2θ spectra around GaN 0002 of $(InN)_1/(GaN)_n$ SPSs for the GaN layer thickness $n=1$–20 MLs. It was first demonstrated that coherent structure $(InN)_1/(GaN)_n$ SPSs were easily fabricated for

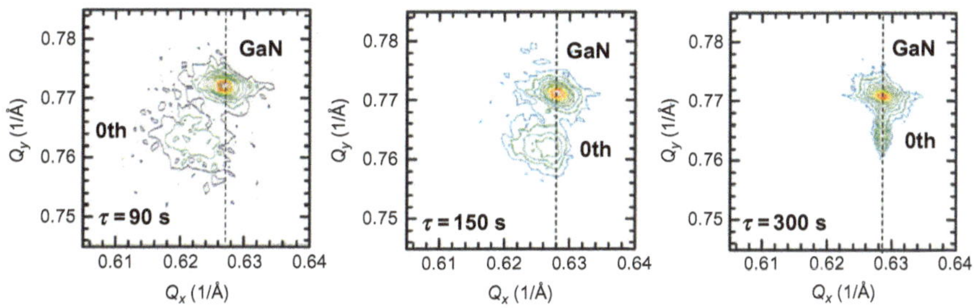

Fig. 14 XRD-RSMs around GaN 11–24 of $(InN)_1/(GaN)_4$ SPSs grown under different growth interruption periods $\tau = 90$, 150, and 300 s at 650°C. The coherent SPS structures were achieved for the growth interruption periods longer than $\tau = 150$ s. The further longer period of $\tau = 300$ s made the profile of zeroth satellite peak sharper, indicating that the structural quality of SPSs was improved by avoiding unintentional InGaN insertion.

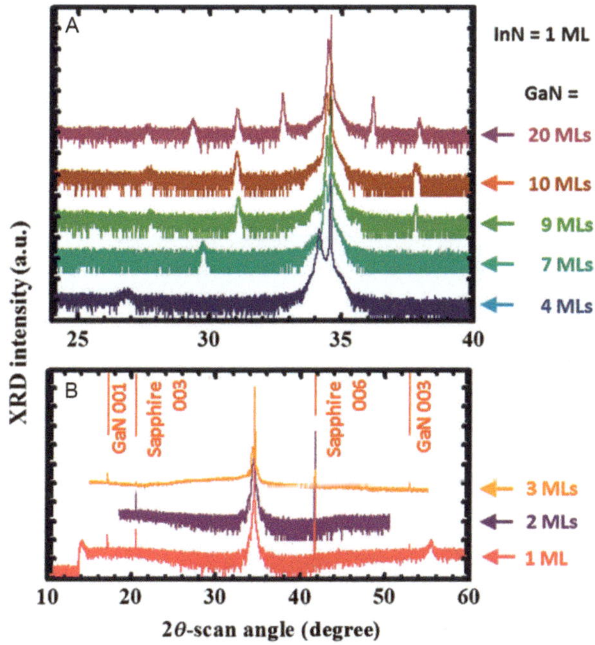

Fig. 15 XRD ω-2θ spectra around GaN 0002 of $(InN)_1/(GaN)_n$ SPSs grown at 650°C for the GaN layer thicknesses (A) $n = 4$–20 MLs and (B) $n = 1$–3 MLs, where 2θ scan ranges are (A) 24–40 degree and (B) wider than 19–50 degree, respectively. The satellite peaks were observed in (A), indicating that the SPSs had the fine periodicity as designed while no satellite peaks were observed in (B) probably due to the large strain preventing insertion of 1ML-InN. The spikes in (B) are attributed to the GaN and Sapphire diffractions including the multiple reflection effect at forbidden indices.

$n \geq 7$ MLs for which fine and clear satellite peaks were observed. Furthermore, although it was slightly difficult to grow coherent structure $(InN)_1/(GaN)_4$ SPSs, it was successfully achieved to fabricate those by paying careful attention for the growth interruption as shown in Fig. 15A. On the other hand, it was hardly to grow even periodical structures for $n \leq 3$ MLs. This was probably attributed to a large strain introduced by the lattice-mismatch between $(InN)_1/(GaN)_n$ SPSs and GaN templates. In Fig. 15B, there were spikes attributed to the GaN and Sapphire diffractions including a multiple reflection effect at forbidden indices, GaN 0001, 0003, and Sapphire 0003. In addition, although the $(InN)_1/(GaN)_1$ indicated subpeaks at $2\theta \sim 14$ and ~ 55 degree which looked like satellite peaks arising from the periodic structure, the effective In-composition was almost zero. Consequently, it is considered that the subpeaks are attributed to uncertain artificial effect at present.

Achievement of the coherent structure $(InN)_1/(GaN)_4$ SPSs are also supported by TEM observation as a cross check manner. Fig. 16 shows a cross-sectional HAADF–STEM image for the $(InN)_1/(GaN)_4$ SPS of which the growth interruption was $\tau = 150$ s. The successful insertion of 1ML-InN layers and fabrication of $(InN)_1/(GaN)_4$ SPS fine structure were confirmed in Fig. 16. It is noted that the 1ML-InN was fabricated through self-limiting

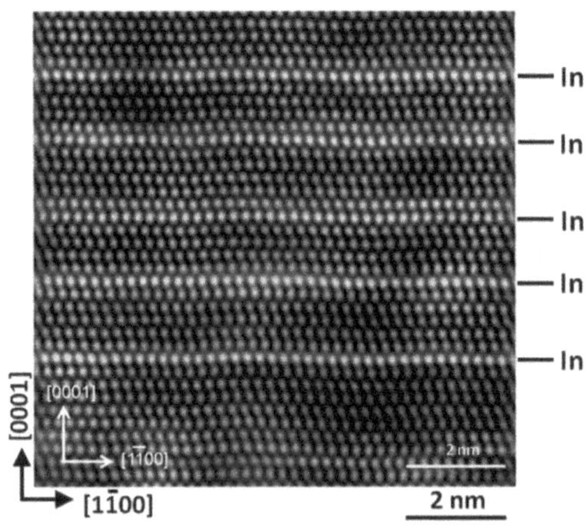

Fig. 16 Group-III element image taken by cross-sectional HAADF-STEM observation for the $(InN)_1/(GaN)_4$ SPS grown under the interruption period $\tau = 150$ s at 650°C. The *brighter* (*darker*) spots correspond to In-atoms (Ga-atoms). The effective thickness of GaN might be 5MLs in part. When the 4ML-GaN was designed and grown, slightly excess Ga-atoms were supplied to ensure the uniform 4ML thick coverage, leading to the partial formation of 5ML-GaN.

and self-ordering processes while the 4ML-GaN was grown by conventional MBE mode. Therefore, there may exist a little bit of ambiguity in controllability of the layer thickness of GaN. When the 4ML-GaN was designed and grown, slightly excess Ga-atoms were supplied to ensure the uniform 4-ML-thick coverage. Thus, the partial formation of 5ML-GaN appears in Fig. 16.

The authors have discussed about growth kinetics and structural analysis on the coherent structure $(InN)_1/(GaN)_n$ SPSs. Then, optical properties of the SPSs are discussed. Fig. 17A shows PL and PL excitation (PLE) spectra at 12 K of the $(InN)_1/(GaN)_n$ SPSs for different GaN layer thicknesses $n=4$, 7, and 9 MLs, together with the $(InN)_1/(GaN)_{40}$ MQW as a reference. The effective bandgap energies (E_g) of $(InN)_1/(GaN)_n$ were indicated by arrows that were evaluated from conventional PLE spectral fitting with the sigmoidal function (Martin et al., 1999; White et al., 2002). The PLE peaks at around 3.5 eV are attributed to photoabsorption by excitons in the GaN layers. For the $(InN)_1/(GaN)_n$ SPSs with $n=7$ and 9 MLs, and $(InN)_1/(GaN)_{40}$ MQW, the PL peak energies (E_{PL}) were almost same with one

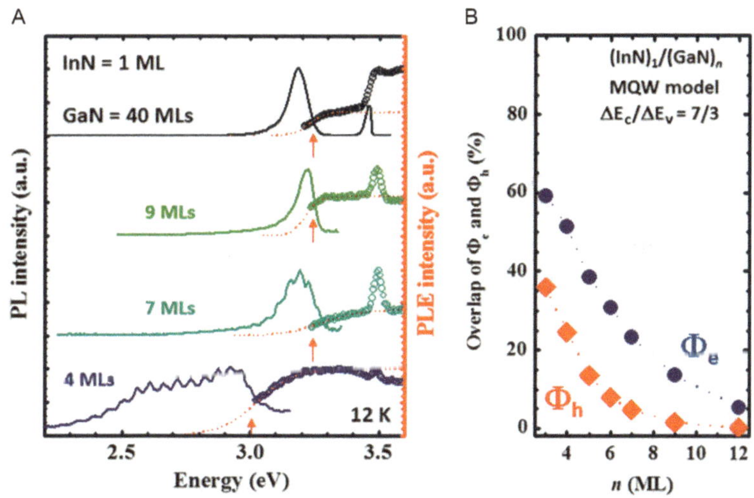

Fig. 17 (A) PL and PLE spectra at 12 K of the $(InN)_1/(GaN)_n$ SPSs for different GaN layer thicknesses $n=4$, 7, and 9 MLs, together with the $(InN)_1/(GaN)_{40}$ MQW as a reference. *Solid lines* and *open circles* indicate the PL and PLE spectra, respectively. The excitation source of PL was monochromatic light at 3.76 eV. The effective bandgap energies of $(InN)_1/(GaN)_n$ were indicated by the arrows which were evaluated from the PLE via conventional spectral fitting with the sigmoidal function (*dashed lines*). (B) Overlap of the electron and hole wavefunctions Φ_e (*circles*) and Φ_e (*diamonds*) in the SPSs as a function of the GaN layer thickness *n*.

another, and the effective bandgap energies E_g were as well. These facts suggest that the $(InN)_1/(GaN)_n$ SPSs with $n=7$ and 9 MLs can be considered as conventional MQWs having discrete energy states. On the other hand, the E_g and E_{PL} for $n=4$ MLs red-shifted to below those of $n=7$, 9, and 40 MLs. Besides, the PLE spectrum for $n=4$ MLs indicates broadening of photo-absorption region to below E_g for $n=7$, 9, and 40 MLs. These findings are attributed to formation of continuum electronic states in the $(InN)_1/(GaN)_4$ SPSs, and also consist with calculated results. Fig. 17B shows overlap of wavefunctions of neighboring electrons and holes in the SPSs as a function of the GaN layer thickness n, which was calculated by an opened software (http://www3.nd.edu/~gsnider/) using material parameters reported in Wang et al. (2008), Wu et al. (2002b), Vurgaftman and Meyer (2003), and King et al. (2008). The overlaps of wavefunctions of electrons and holes were around 20% and 5% at the GaN layer thickness $n=7$ MLs, respectively, and then drastically increased with decrease in the GaN layer thickness below $n<7$ MLs. Consequently, those of electrons and holes reached around 50% and 25% at the GaN layer thickness $n=4$ MLs, respectively. This calculation results probably reflected that the interaction of wavefunctions becomes strong enough to form the continuum electronic states in the $(InN)_1/(GaN)_4$ SPSs. According to both the experiments and the simulations, the authors concluded that $(InN)_1/(GaN)_4$ SPSs were equivalent with ordered InGaN ternary alloys. As was discussed, it was found interestingly that the GaN layer thickness $n=4$ MLs was the criterion in views of the structural control and the continuum-band formation.

3.2 MOVPE Trial for Growing $(InN)_1/(GaN)_n$ SPSs and Solar Cells

In this section, MOVPE trial for growing $(InN)_1/(GaN)_n$ SPSs is discussed. As one of the preferable device applications, the $(InN)_1/(GaN)_4$-designed layers (explained later) were adapted to InGaN-based solar cells grown on GaN substrates. As stated previously, the authors believe that the growth front of MOVPE is essentially in common with that of MBE. On the other hand, there are a lot of growth parameters in MOVPE more than in MBE, which are actually correlated with one another. Therefore, optimization of growth conditions in MOVPE requires lots of experiments in general. In order to move up those studies effectively, the D-ALEp processes have been developed with simultaneously fabricating InGaN-based solar cells in the horizontal LP-MOVPE reactor which equipped the SE system (Jobin Yvon, UVISEL).

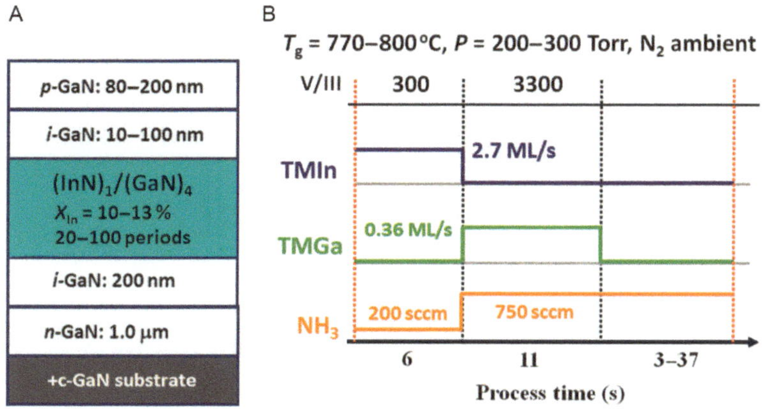

Fig. 18 (A) The sample structure of $(InN)_1/(GaN)_4$ solar cell grown on GaN substrate by LP-MOVPE and (B) the D-ALEp sequence for growing the $(InN)_1/(GaN)_4$ design structures under N_2 ambient. Excess InN were supplied and then immediately capped by 4ML-GaN with changing NH_3 flow rate from 200 to 750 sccm. After those, the growth interruption was carried out under NH_3 feeding to expect selective desorption of In-adatoms.

Fig. 18A shows a typical sample structure of InGaN solar cells grown on GaN substrates by LP-MOVPE, in which an InGaN photoabsorption region is designed as the $(InN)_1/(GaN)_4$ SPSs. Fig. 18B shows an example of D-ALEp sequence for growing the $(InN)_1/(GaN)_4$-designed layers under N_2 ambient at 770–800°C. During one cycle of $(InN)_1/(GaN)_4$ deposition, excess ~16ML-InN was supplied and then immediately capped by 4ML-GaN with changing V/III ratios from 300 to 3300. This is due to the fact that the low V/III ratio or the low NH_3 flow rate (ultimately no NH_3 supply) ensures self-limiting and self-ordering processes of In(N), as discussed in Figs. 10 and 11. After those, the growth interruption was carried out under NH_3 flow rate of 750 sccm to expect selective desorption of In-adatoms, of which the period was determined by in situ SE monitoring.

Fig. 19A–D shows evolutions of SE signals during D-ALEp of the $(InN)_1/(GaN)_4$-designed layers, where the SE signals are the real and imaginary parts of pseudodielectric function $\langle \varepsilon_1 \rangle$ and $\langle \varepsilon_2 \rangle$. In this case, periodic increments of $\langle \varepsilon_1 \rangle$ and $\langle \varepsilon_2 \rangle$ as spikes, more clearly in $\langle \varepsilon_2 \rangle$ than in $\langle \varepsilon_1 \rangle$, indicate the deposition of In(N) by supplying TMI and NH_3 while the reduction and recovery of $\langle \varepsilon_1 \rangle$ and $\langle \varepsilon_2 \rangle$ in the spikes indicate spending the residual In-atoms on the formation of InGaN layer together with arrived Ga-atoms during the GaN capping. The sinusoidal evolution in base components of the $\langle \varepsilon_1 \rangle$ and $\langle \varepsilon_2 \rangle$ supports the deposition of ternary InGaN layer since it is attributed to an optical interference within the InGaN layer whose growth

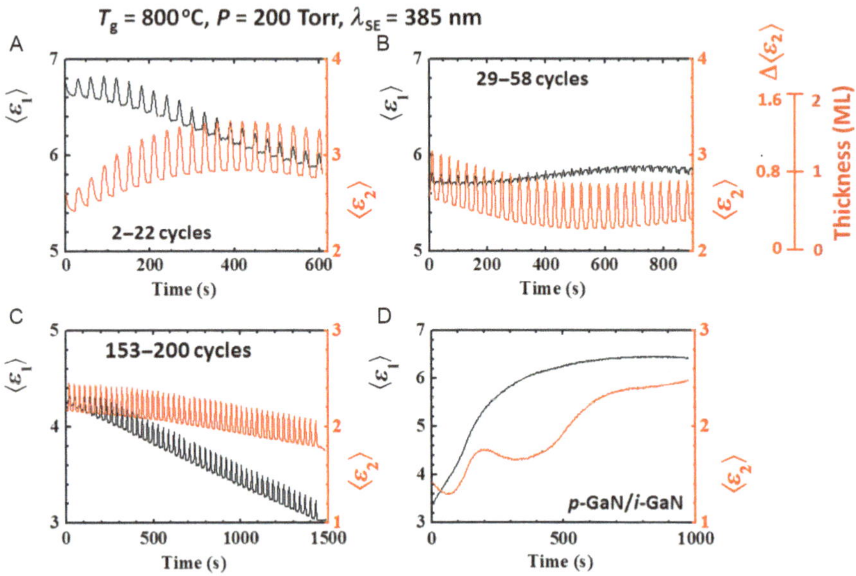

Fig. 19 Evolutions of SE signals during D-ALEp of the $(InN)_1/(GaN)_4$ design structures for (A) 2–22 cycles, (B) 29–58 cycles, (C) 153–200 cycles, and (D) the deposition of *p*-GaN/*i*-GaN after the 200 cycles. The SE signals were the real and imaginary parts of pseudodielectric function $\langle \varepsilon_1 \rangle$ and $\langle \varepsilon_2 \rangle$ at the probe wavelength $\lambda_{SE} = 385$ nm. Briefly explaining, the $\langle \varepsilon_1 \rangle$ and $\langle \varepsilon_2 \rangle$ are composed of varying and base components, indicating the adlayers and underlayers, respectively. In this case, spike-like increments of the $\langle \varepsilon_2 \rangle$ indicate the deposition of In(N) on the GaN surface while the reduction and recovery of $\langle \varepsilon_2 \rangle$ indicate the desorption of excess In-adlayers or the crystallization of In and Ga-atoms.

rate matches well with the period of sinusoidal evolution. As a current result, there have been a few issues to be overcome in the D-ALEp in LP-MOVPE: (1) the In(N)-adlayers partly disappears self-limiting and self-ordering behavior when TMI is supplied with NH_3, depending on the NH_3 flow rate, which is detected as insufficient increment of the $\langle \varepsilon_2 \rangle$, even though excess InN is supplied and (2) binary 1ML-InN layers cannot be inserted below the GaN layer, instead the In-atoms remain on the growth front forming small droplets. These are probably attributed to a certain thermal etching effect enhanced by N-atoms rather than H-atoms arising from an overpressure of NH_3. Since the arrived N-atoms penetrate and are incorporated below the In-adatoms in the D-ALEp process at the growth temperature beyond upper limit one of the InN epitaxy, the N-atoms make the sticking efficiency or bond strength of In-adatoms on the GaN surface much weaker (Yoshikawa et al., 2016a). As a result, the In-adatoms are free from atomic bonds and become the small droplets. If the lateral size of In-droplets is enough smaller than the SE probe wavelength λ_{SE}, it is difficult to detect them by the $\langle\varepsilon_2\rangle$, corresponding to the above issue (1) (Yoshikawa et al., 2016b, submitted). Finally, the In-droplets are incorporated into the ternary InGaN layer with In-compositions around 10–13%, as "*In-droplet-mediated epitaxy.*" Thus, the InGaN surface disappears the self-limiting and self-ordering processes observed on the GaN surface. This growth front behavior consequently induced in-plane compositional fluctuation, roughening the InGaN surface which was detected by decrease in the base component of $\langle\varepsilon_1\rangle$, in Fig. 19C. But the rough growth front was rapidly recovered during *p*-GaN/*i*-GaN epitaxy just after deposition of 200-period-$(InN)_1/(GaN)_4$, in Fig. 19D. Due to the above-mentioned issues, the $(InN)_1/(GaN)_4$-designed layers have no SPS periodicity measured by XRD so far. The authors believe that the $(InN)_1/(GaN)_n$ SPSs can be grown by MOVPE as well, after decreasing the growth temperature even though it is an opposite approach to our policy "*Increase in Growth Temperature.*"

Hereinafter, discussion is devoted to the InGaN-based solar cells. After finding that a monolithic InGaN material system covers almost entire of a solar spectrum (Davydov et al., 2002; Inushima et al., 2001; Ishitani et al., 2005; Matsuoka et al., 2002; Wu et al., 2002a), full-spectrum/high-efficiency InGaN solar cells have received significant interests. As an alternative, hybrid integration of III-nitride/non-III-nitride photovoltaic devices has been proposed to efficiently convert a short-wavelength region of sunlight suppressing a voltage loss, where InGaN top cells with absorption edge around 2.4 eV are mounted on conventional InGaP/(In)GaAs/Ge multijunction cells (Toledo et al., 2012). Likewise, the authors have proposed *SMART* tandem solar cells each subcell of which is composed of the $(InN)_m/(GaN)_n$ SPSs (Kusakabe and Yoshikawa, 2014). This is one of the reasons to have investigated the $(InN)_1/(GaN)_n$ SPSs and developed the D-ALEp both by RF-MBE and by LP-MOVPE in our group.

The $(InN)_1/(GaN)_4$-designed layer/*pn*-GaN solar cells were grown on GaN substrates by LP-MOVPE, in which the $(InN)_1/(GaN)_4$-designed layers were equivalent with ternary InGaN layers with In-compositions around 10–13% at present. After the epitaxy of solar cell structures, as shown in Fig. 18A, Au/Ni *p*-contacts interdigit-shaped were formed on the *p*-GaN surfaces by a typical manner, while *n*-contacts were conducted via an In/Ga solder on the backside of GaN substrates. In order to let the $(InN)_1/(GaN)_4$-designed layer locate within a depletion region as a photoabsorption region, *i*-GaN undoped layers were inserted below/above it. The total epitaxy process and device performance have been optimized in view of improving a junction property of solar cells, in which an open circuit voltage (V_{oc}) is set as an evaluation criteria. Through developing $(InN)_1/(GaN)_4$-designed layer/*pn*-GaN solar cells according to this criteria, several issues were revealed. The worst serious one was that almost as-given GaN substrates had damaged surfaces due to chemical mechanical polishing (CMP). This fact considerably degraded the quality of epitaxial films and also the performance of InGaN-based solar cells, since lots of macrodefects were generated and induced leakage currents as large as they hid the presence of *pn*-junctions. The residual damage of GaN substrate surfaces was effectively removed and recovered by thermal etching prior to growth, named dynamic atomic layer etching (D-ALEt). The D-ALEt is a reverse process of the D-ALEp, in which the evolution of etching and solid-phase diffusion was traced by the $\langle \varepsilon_1 \rangle$ in in situ SE monitoring. After this effort as well as so many others, the performance of $(InN)_1/(GaN)_4$-designed layer/*pn*-GaN solar cells has fairly been improved as is comparable with the world records. The device performances were measured by a semiconductor device analyzer (Keysight formerly Agilent, B1500A) under AM-1.5G irradiation (Asahi Spectra, HAL-320W).

Fig. 20 shows a relation of current–voltage (I–V) of the $(InN)_1/(GaN)_4$-designed layer/*pn*-GaN solar cell grown on GaN substrate under dark at

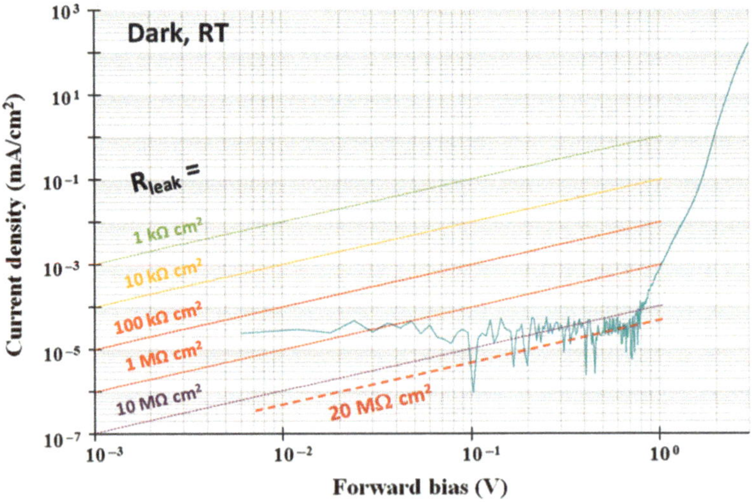

Fig. 20 Forward-biased I–V curve of the $(InN)_1/(GaN)_4$-designed layer/*pn*-GaN solar cell grown on GaN substrate under dark at RT. *Dashed lines* are given for eye guide indicating levels of the junction leakage resistance R_{leak}.

room temperature (RT). In order to evaluate a junction leakage resistance (R_{leak}), the I–V curve was measured in a forward-bias region. It was demonstrated that the R_{leak} was around 20 MΩ cm^2 and the current was below the noise level of our measurement system with decrease in the forward bias, indicating that the net R_{leak} was much higher than 20 MΩ cm^2 and the world record, to the best of our knowledge. Currently, the authors have investigated further increment of the R_{leak} even when the effective In-composition is increased more than 30%. The novel attempt has been proposed to block/passivate the defects within solar cells based on the D-ALEp in Wang et al. (2016a,b).

Fig. 21 shows I–V curves of the $(InN)_1/(GaN)_4$-designed layer/*pn*-GaN solar cell grown on GaN substrate under AM-1.5G irradiations. It was demonstrated that the conversion efficiency $\eta = 2.36\%$ was achieved under 1-sun irradiation, that was comparable with the world records (Valdueza-Felip et al., 2014; Young et al., 2013).

As a summary of Sections 3.1 and 3.2, the authors successfully demonstrated the $(InN)_1/(GaN)_4$ ordered alloys grown by RF-MBE and high-efficiency $(InN)_1/(GaN)_4$ solar cells grown by LP-MOVPE. Therefore, the authors have been convinced that the III-nitride order alloys are effective to fabricated high-performance devices.

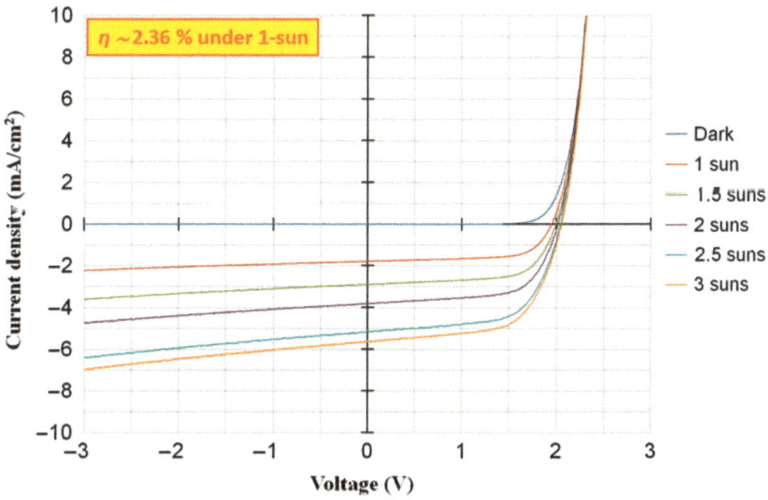

Fig. 21 I–V curves of the $(InN)_1/(GaN)_4$-designed layer/*pn*-GaN solar cell grown on GaN substrate under AM-1.5G irradiation with different concentrations from 1-sun to 3-suns. The conversion efficiency $\eta = 2.36\%$ is obtained under 1-sun irradiation.

3.3 AlGaN and AlInN Ordered Alloys Toward Electronic Device Application

In this chapter, the authors have described and discussed our recent achievements concerning epitaxy of the 1ML-InN on/in GaN matrix and the $(InN)_1/(GaN)_4$ SPSs by the D-ALEp developed with in situ SE monitoring, so far. Then, as a future prospect, our strategy is currently extended to establish ordered alloys, or digital alloys, in the whole III-nitride system such as ordered InN/AlN or GaN/AlN quasiternary alloys, and ordered InN/GaN/AlN quasiquaternary alloys. One of the advantages of extension to AlN-based ordered alloys is potential application to AlGaN and AlInN-based devices, where the growth conditions become more flexible than those of conventional alloys to obtain high-quality device structures. For example, AlGaN MQW UV-emitters, AlGaN/GaN MQW intersubband transition devices, and AlGaN/GaN or AlInN/GaN high electron mobility transistors (HEMTs) are potential candidates.

The authors would like to introduce the HEMT among them as our target device, whose electron channel is an atomically flat interface between the AlN/GaN or AlN/InN ordered alloys and the GaN underlayer. Fig. 22 shows a conceptual structure and remarks of proposed HEMT composed of AlN/GaN or AlN/InN ordered alloys grown by the D-ALEp. For achieving high-performance III-N HEMTs with physical limit

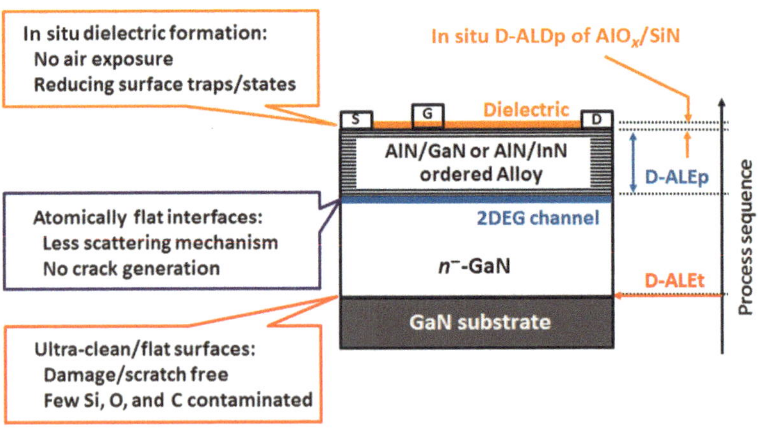

Fig. 22 Schematic diagram of proposed HEMT structure including AlN/GaN or AlN/InN ordered alloy grown on GaN substrate wherein: the surface of GaN substrate is cleaned/flattened by the D-ALEt; the ordered alloys are grown by the D-ALEp; and the surface dielectrics are deposited by the D-ALDp.

performance, there remain issues to be immediately overcome, for example, (1) current leakage through parallel conduction at the interface between epilayer and GaN substrate, (2) suppression of scattering factor in the two-dimensional electron gas (2DEG) channel as low as possible, and (3) reduction of current collapse arising from surface states, and so on. It is discussed one by one that how they overcome with reference to Fig. 22. For the issue (1), the surface of GaN substrate includes the CMP damage and is contaminated with Si, O, and C (Koblmüller et al., 2010). The CMP damage was successfully removed and recovered by the D-ALEt as described earlier. Likewise, it is considered that the D-ALEt can effectively remove the residual contaminations from the surface of GaN substrate, resulting in drastic reduction of leakage currents. For the issue (2), ordered alloys allow us to fabricate atomically flat surfaces/interfaces which are adapted to the 2DEG channels with extremely suppressed scattering factors. Crack generation could also be avoided using SPS structures (Zhang et al., 2002). Particularly, the authors have highly expected the quantum transport regime by less scattering factors which might pave the way for quantum entanglements in III-nitrides (Falson et al., 2015). Subsequently, for the issue (3), surface passivation with AlO_x/SiN dielectric is deposited by dynamic atomic layer deposition (D-ALDp) in the III-N growth chamber/reactor, where both N-atoms and O-atoms can be supplied (Wang et al., 2016b), in order to reduce surface traps/states by air exposure and/or unintentional contaminations. Therefore, the authors believe that our-proposed HEMT is the plausible candidate to break through underdeveloping issues in the high-power application area.

4. SUMMARY

In this chapter, the D-ALEp has been discussed in reference with our recent achievements, including coherent 1ML-InN on/in GaN-matrices, InN/GaN SPSs or ordered alloys, and InN/GaN layer/*pn*-GaN solar cells. All investigations have been made based on the in situ SE monitoring. As concluding remarks, the authors have surprised how the in situ SE monitoring is powerful tool to deeply understand the growth kinetics again and again through those works. The authors believe that the fundamental features of D-ALEp are universal and applicable to any other material systems so that it is strongly recommended to utilize the SE for any of epitaxy/deposition/etching/decoration fields at development stages.

ACKNOWLEDGMENTS

The authors would like to thank Drs. Song-Bek Che, Naoki Hashimoto, Eun-Sook Hwang, Daichi Imai, Ke Wang, Prof. Takaomi Itoi, and Dr. Takashi Mukai for their kind supports to achieve this work, technical assistances, and fruitful discussions. This work was partially supported by KAKENHI (A) 23246056, KAKENHI on Priority Areas 18069002, G-COE program at Chiba University (G-03, MEXT), JST-CREST, and JST-ALCA.

REFERENCES

Abell, J., Moustakas, T.D., 2008. The role of dislocations as nonradiative recombination centers in InGaN quantum wells. Appl. Phys. Lett. 92, 091901.

Cao, B., Xu, K., Seo, B.W., Arita, S., Nishida, S., Ishitani, Y., Yoshikawa, A., 2003. Dependences of GaN polarity on the growth temperatures of migration-enhanced-epitaxy-grown AlN in MOVPE. Phys. Status Solidi C, 2553–2556. http://dx.doi.org/10.1002/pssc.200303470.

Che, S.B., Terashima, W., Ishitani, Y., Matsuda, T., Ishii, H., Yoshida, S., Yoshikawa, A., 2005a. Fine-structure N-polarity InN/InGaN multiple quantum wells grown on GaN underlayer by molecular-beam epitaxy. Appl. Phys. Lett. 86, 261903.

Che, S.B., Shinada, T., Mizuno, T., Ishitani, Y., Yoshikawa, A., 2005b. Comparison between N- and in-polarity InN/InGaN MQWs structures grown by RF-MBE. Mat. Res. Soc. Symp. Proc. 892. FF06-03.

Che, S.B., Mizuno, T., Wang, X., Ishitani, Y., Yoshikawa, A., 2007. Fabrication and properties of coherent-structure In-polarity InN/$In_{0.7}Ga_{0.3}N$ multiquantum wells emitting at around 1.55 μm. J. Appl. Phys. 102, 083539.

Che, S.B., Yuki, A., Watanabe, H., Ishitani, Y., Yoshikawa, A., 2009. Fabrication of asymmetric GaN/InN/InGaN/GaN quantum-well light emitting diodes for reducing the quantum-confined stark effect in the blue-green region. Appl. Phys. Express 2, 021001.

Choi, S.J., Kim, T.H., Brown, A., Everitt, H.O., Losurdo, M., Bruno, G., Moto, A., 2006. Kinetics of gallium adsorption and desorption on (0001) gallium nitride surfaces. Appl. Phys. Lett. 89, 181915.

Choi, S.J., Kim, T.H., Wolter, S., Brown, A., Everitt, H.O., Losurdo, M., Bruno, G., 2008. Indium adlayer kinetics on the gallium nitride (0001) surface: monitoring indium segregation and precursor-mediated adsorption. Phys. Rev. B77, 115435.

Clarke, A., Imhoff, S., Gibbs, P., Cooley, J., Morris, C., Merrill, F., Hollander, B., Mariam, F., Ott, T., Barker, M., Tucker, T., Lee, W.-K., Fezzaa, K., Deriy, A., Patterson, B., Clarke, K., Montalvo, J., Field, R., Thoma, D., Smith, J., Teter, D., 2013. Proton radiography peers into metal solidification. Sci. Rep. 3. Article number: 2020.

Davydov, V.Yu., Klochikhin, A.A., Seisyan, R.P., Emtsev, V.V., Ivanov, S.V., Bechstedt, F., Furthmüller, J., Harima, H., Mudryi, A.V., Aderhold, J., Semchinova, O., Graul, J., 2002. Absorption and emission of hexagonal InN. Evidence of narrow fundamental band gap. Phys. Status Solidi B 229, r1.

Falson, J., Maryenko, D., Friess, B., Zhang, D., Kozuka, Y., Tsukazaki, A., Smet, J.H., Kawasaki, M., 2015. Even-denominator fractional quantum Hall physics in ZnO. Nat. Phys. 11, 347.

Feenstra, R.M., Chen, H., Ramachandran, V., Lee, C.D., Smith, A.R., Northrup, J.E., Zywietz, T., Neugebauer, J., Greve, D.W., 2000. Surface morphology of Gan surfaces during molecular beam epitaxy. Surf. Rev. Lett. 7, 601.

Gil, B. (Ed.), 1998. Group III Nitride Semiconductor Compounds. Clarendon Press, Oxford.

Hwang, E.S., Che, S.B., Saito, H., Wang, X.Q., Ishitani, Y., Yoshikawa, A., 2008. Cathodoluminescence study on spatial luminescence properties of InN/GaN multiple quantum wells consisting of 1-monolayer-thick InN wells/GaN matrix. J. Electron. Mater. 37, 597.

Hwang, E.S., Che, S.B., Ishitani, Y., Yoshikawa, A., 2009. Comparative cathodoluminescence characterization of ultrathin InN wells/GaN matrix MQWs grown on bulk-GaN and MOVPE-GaN. Phys. Status Solidi C 6, S405.

Inushima, T., Mamutin, V.V., Vekshin, V.A., Ivanov, S.V., Sakon, T., Motokawa, M., Ohoya, S., 2001. Physical properties of InN with the band gap energy of 1.1 eV. J. Cryst. Growth 481, 227–228.

Ishitani, Y., Masuyama, H., Terashima, W., Yoshitani, M., Hashimoto, N., Che, S.B., Yoshikawa, A., 2005. Bandgap energy of InN and its temperature dependence. Phys. Status Solidi C 2, 2276.

King, P.D.C., Veal, T.D., Kendrick, C.E., Bailey, L.R., Durbin, S.M., McConville, C.F., 2008. InN/GaN valence band offset: high-resolution X-ray photoemission spectroscopy measurements. Phys. Rev. B 78, 033308.

Koblmüller, G., Chu, R.M., Raman, A., Mishra, U.K., Speck, J.S., 2010. High-temperature molecular beam epitaxial growth of AlGaN/GaN on GaN templates with reduced interface impurity levels. J. Appl. Phys. 107, 043527.

Kusakabe, K., Yoshikawa, A., 2014. Proposal for realizing high-efficiency III-nitride semiconductor tandem solar cells with InN/GaN superstructure magic alloys fabricated at raised temperature (SMART). Proc. SPIE 8986 89861B, 1–10.

Kusakabe, K., Hashimoto, N., Itoi, T., Wang, K., Imai, D., Yoshikawa, A., 2016. Growth kinetics and structural perfection of $(InN)_1/(GaN)_{1-20}$ short-period superlattices on + c-GaN template in dynamic atomic layer epitaxy. Appl. Phys. Lett. 108, 152107.

Lim, D.H., Xu, K., Arima, S., Yoshikawa, A., Takahashi, K., 2002. Polarity inversion of GaN films by trimethyl–aluminum preflow in low-pressure metalorganic vapor phase epitaxy growth. J. Appl. Phys. 91, 6461.

Martin, R.W., Middleton, P.G., O'Donnell, K.P., Van der Stricht, W., 1999. Exciton localization and the Stokes' shift in InGaN epilayers. Appl. Phys. Lett. 74, 263.

Matsuoka, T., Okamoto, H., Nakao, M., Harima, H., Kurimoto, E., 2002. Optical bandgap energy of wurtzite InN. Appl. Phys. Lett. 81, 1246.

Morkoc, H., 1999. Nitride Semiconductors and Devices. Springer, Heidelberg.

Nakamura, S., Fasol, G., 1997. The Blue Laser Diode. Springer, Berlin.

Neugebauer, J., Zywietz, T.K., Scheffler, M., Northrup, J.E., Chen, H., Feenstra, R.M., 2003. Adatom kinetics on and below the surface: the existence of a new diffusion channel. Phys. Rev. Lett. 90, 056101.

Northrup, J.E., Van de Walls, C.G., 2004. Indium versus hydrogen-terminated GaN(0001) surfaces: surfactant effect of indium in a chemical vapor deposition environment. Appl. Phys. Lett. 84, 4322.

Northrup, J.E., Neugebauer, J., Feenstra, R.M., Smith, A.R., 2000. Structure of GaN (0001): the laterally contracted Ga bilayer model. Phys. Rev. B61, 9932.

Pankove, J.I., Moustakas, T.D. (Eds.), 1998. GaN, vol. 1. Academic Press, New York.

Pearton, S.J. (Ed.), 1997. GaN and Related Materials. Gordon and Breach, New York.

Terashima, W., Che, S.B., Ishitani, Y., Yoshikawa, A., 2006. Growth and characterization of AlInN ternary alloys in whole composition range and fabrication of InN/AlInN multiple quantum wells by RF molecular beam epitaxy. Jpn. J. Appl. Phys. 45 (Part 2), L539.

Toledo, N.G., Friedman, D.J., Farrell, R.M., Perl, E.E., Lin, C.-T., Bowers, J.E., Speck, J.S., Mishra, U.K., 2012. Design of integrated III-nitride/non-III-nitride tandem photovoltaic devices. J. Appl. Phys. 111, 054503.

Valdueza-Felip, S., Mukhtarova, A., Grenet, L., Bougerol, C., Durand, C., Eymery, J., Monroy, E., 2014. Improved conversion efficiency of as-grown InGaN/GaN quantum-well solar cells for hybrid integration. Appl. Phys. Express 7, 032301.

Vurgaftman, I., Meyer, J.R., 2003. Band parameters for nitrogen-containing semiconductors. J. Appl. Phys. 94, 3675.

Wang, X.Q., Yoshikawa, A., 2004. Molecular beam epitaxy growth of GaN, AlN and InN. Prog. Cryst. Growth Charact. Mater. 48–49, 42.

Wang, X.Q., Che, S.-B., Ishitani, Y., Yoshikawa, A., 2008. Hole mobility in Mg-doped p-type InN films. Appl. Phys. Lett. 92, 132108.
Wang, K., Imai, D., Kusakabe, K., Yoshikawa, A., 2016a. Leak path passivation by in situ Al-N for InGaN solar cells operating at wavelengths up to 570 nm. Appl. Phys. Lett. 108, 092105.
Wang, K., Imai, D., Kusakabe, K., Yoshikawa, A., 2016b. Proposal of leak path passivation for InGaN solar cells to reduce the leakage current. Appl. Phys. Lett. 108, 042108.
White, M.E., O'Donnell, K.P., Martin, R.W., Pereira, S., Deatcher, C.J., Watson, I.M., 2002. Photoluminescence excitation spectroscopy of InGaN epilayers. Mat. Sci. Eng. B93, 147–149.
Wu, J., Walukiewicz, W., Yu, K.M., Ager III, J.W., Haller, E.E., Lu, H., Schaff, W.J., Saito, Y., Nanishi, Y., 2002a. Unusual properties of the fundamental band gap of InN. Appl. Phys. Lett. 80, 3967.
Wu, J., Walukiewicz, W., Shan, W., Yu, K.M., Ager III, J.W., Haller, E.E., Lu, H., Schaff, W.J., 2002b. Effects of the narrow band gap on the properties of InN. Phys. Rev. B 66, 201403.
Wu, P.C., Losurdo, M., Kim, T.H., Choi, S.J., Bruno, G., Brown, A.S., 2007. In situ spectroscopic ellipsometry to monitor surface plasmon resonant group-III metals deposited by molecular beam epitaxy. J. Vac. Sci. Technol. B25, 1019.
Xu, K., Yano, N., Jia, A.W., Yoshikawa, A., Takahashi, K., 2001. Kinetic process of polarity selection in GaN growth by RF-MBE. Phys. Status Solidi B 228, 523.
Yoshikawa, A., Che, S.B., Yamaguchi, W., Saito, H., Wang, X.Q., Ishitani, Y., Hwang, E.S., 2007. Proposal and achievement of novel structure InN/GaN multiple quantum wells consisting of 1 ML and fractional monolayer InN wells inserted in GaN matrix. Appl. Phys. Lett. 90, 073101.
Yoshikawa, A., Che, S.B., Hashimoto, N., Saito, H., Ishitani, Y., Wang, X.Q., 2008. Fabrication and characterization of novel monolayer InN quantum wells in a GaN matrix. J. Vac. Sci. Technol. B 26, 1551.
Yoshikawa, A., Che, S.B., Ishitani, Y., Wang, X.Q., 2009. Advances in InN epitaxy and its material control by MBE towards novel InN-based QWs. J. Cryst. Growth 311, 2073.
Yoshikawa, A., Kusakabe, K., Hashimoto, N., Hwang, E.S., Itoi, T., 2016a. Dynamic atomic layer epitaxy of InN on/in + c-GaN matrix: effect of "In + N" coverage and capping timing by GaN layer on effective InN thickness. Appl. Phys. Lett. 108, 022108.
Yoshikawa, A., Kusakabe, K., Hashimoto, N., Hwang, E.S., Imai, D., Itoi, T., 2016b. Systematic study on dynamic atomic layer epitaxy of InN on/in + c-GaN matrix and fabrication of fine-structure InN/GaN quantum wells: Role of high growth temperature. J. Appl. Phys. http://dx.doi.org/10.1063/1.4967928.
A. Yoshikawa, K. Kusakabe, N. Hashimoto, D. Imai, and E. S. Hwang, Systematic study on dynamic atomic layer epitaxy of InN on/in + c-GaN matrix and fabrication of fine-structure InN/GaN quantum wells: Impact of excess In-atoms at high growth temperature, J. Appl. Phys., submitted.
Young, N.G., Farrell, R.M., Hu, Y.L., Terao, Y., Iza, M., Keller, S., DenBaars, S.P., Nakamura, S., Speck, J.S., 2013. High performance thin quantum barrier InGaN/GaN solar cells on sapphire and bulk (0001) GaN substrates. Appl. Phys. Lett. 103, 173903.
Yuki, A., Watanabe, H., Che, S.B., Ishitani, Y., Yoshikawa, A., 2009. 1–2 ML thick InN-based quantum wells with InGaN barriers for blue-green light emitters. Phys. Status Solidi C 6, S417.
Zhang, J.P., Wang, H.M., Gaevski, M.E., Chen, C.Q., Fareed, Q., Yang, J.W., Simin, G., Asif Khan, M., 2002. Crack-free thick AlGaN grown on sapphire using AlN/AlGaN superlattices for strain management. Appl. Phys. Lett. 80, 3542.

CHAPTER TEN

Nitride Semiconductor Nanorod Heterostructures for Full-Color and White-Light Applications

S. Gwo*,†,1, Y.J. Lu*, H.W. Lin*, C.T. Kuo*, C.L. Wu‡, M.Y. Lu*, L.J. Chen*
*National Tsing-Hua University, Hsinchu, Taiwan
†National Synchrotron Radiation Research Center (NSRRC), Hsinchu, Taiwan
‡National Cheng-Kung University, Tainan, Taiwan
[1]Corresponding author: e-mail address: gwo@phys.nthu.edu.tw

Contents

1. INTRODUCTION

Solid-state light sources based on full-color and white light-emitting diode (LED) and laser diode (LD) devices have gained much attention because of their tremendous potential for energy-efficient general illumination and advanced display applications (Nakamura et al., 2000; Pimputkar et al., 2009; Schubert, 2005, 2006). In particular, solid-state lighting technology has progressed rapidly in recent decades after the demonstration of *p*-type doing for gallium nitride (GaN; Amano et al., 1989) and high-brightness Indium gallium nitride (InGaN)/GaN heterostructure blue

Semiconductors and Semimetals, Volume 96
ISSN 0080-8784
http://dx.doi.org/10.1016/bs.semsem.2016.09.002
© 2017 Elsevier Inc. All rights reserved.

LED (Nakamura et al., 1994). For lighting, display, and solar energy harvesting applications, the InGaN alloy semiconductor system has been considered as the most promising material for full-visible-spectrum LEDs, LDs, and solar cells because of the direct band gap of $In_xGa_{1-x}N$, as shown in Fig. 1, can be continuously tuned from the NIR (InN: $x=1$, 0.65 eV) to the near-UV (GaN: $x=0$, 3.4 eV). Ideally, the luminous efficacy and color rendering capability of white LEDs can be optimized by light mixing of polychromatic (e.g., red, yellow, green, blue) InGaN emitters. However, at present, only short-wavelength (blue) InGaN/GaN planar quantum-well LEDs and LDs are efficient light-emitting devices due to deteriorated material properties (Nakamura, 1998), such as a large density of threading dislocations and a large thermodynamic miscibility gap for InN, as well as adverse strain-induced effects (Bernardini et al., 1997, 2001; Takeuchi et al., 1997; Waltereit et al., 2000; Yu et al., 1999) in the $In_xGa_{1-x}N$ active region.

Until now, the dramatic drop in the InGaN emission efficiency at long wavelengths has continued to hampers full-color and white-light applications based on InGaN/GaN heterostructures. Consequently, monolithic

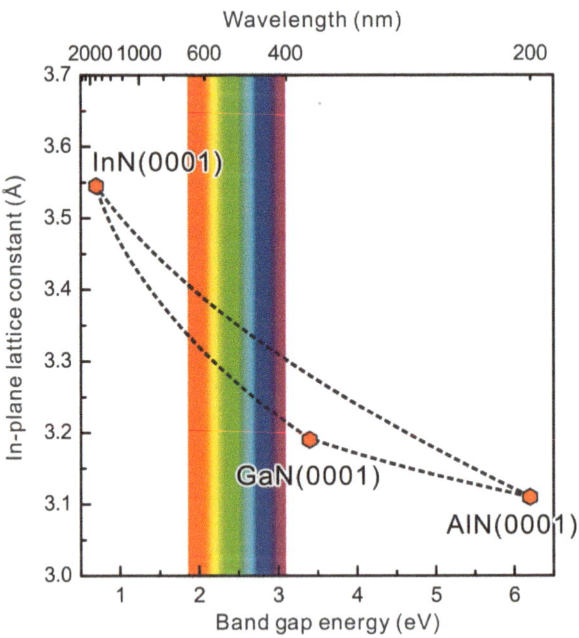

Fig. 1 Plot of band gap energies and in-plane lattice constant for the wurtzite III-nitride semiconductors (*hexagon symbols*) and their alloys (*dotted lines*) for the conventional thin films grown along the polar *c*-axis.

white LEDs are typically realized by the luminescence down-conversion technique using yellow phosphors, such as cerium-doped yttrium aluminum garnet (YAG: Ce), resulting in limited efficiency (efficacy) and color rendering. Recently, in order to reduce the glare and circadian disruption effects originating from a high level of short-wavelength blue light in the commercially available white LEDs for solid-state lighting, it has been suggested by the American Medical Association (AMA) that white LEDs emitting with less blue and more yellow and red wavelengths should be developed to replace current blue-LED-based devices. In fact, a major research direction in nitride semiconductors has been continuously devoted to improve the InGaN emission efficiency at longer wavelengths, including the spectral range between 550 and 590 nm is the well-known "green-yellow gap," where the highest spectral response region of the human eye resides in and none of the existing semiconductors can be used to make high-efficiency LEDs.

The origin of wavelength-dependent emission efficiency can be attributed mainly to the lattice mismatch between InN and GaN (InN is ~11% larger in lattice constant, see Fig. 1) and the polar nature of the wurtzite crystal structure. Conventional InGaN LEDs are based on planar InGaN/GaN quantum-well structures grown along the polar *c*-axis. Therefore, growth of high-In-content InGaN/GaN quantum wells would unavoidably result in a high density of defects and huge internal electrostatic fields (on the order of MV/cm). The internal fields in the InGaN wells spatially separate the electron and hole wave functions (so-called quantum-confined Stark effect, QCSE), making highly efficient longer-wavelength LEDs difficult to be achieved by using polar *c*-plane InGaN/GaN structures. For the low-In-content (blue) InGaN LEDs, the carrier localization phenomenon and ultrathin quantum-well structures (about 2–4 nm in well width) for nearly all commercial InGaN LEDs or LDs can alleviate the effects of high defect density and QCSE. Unfortunately, these are not applicable in the case of high-In-content (yellow and red) InGaN quantum wells because of the issue of increasingly large internal electrostatic fields.

Besides, there are other QCSE- and ultrathin-well-related detrimental features of *c*-plane InGaN LEDs, including spectral shift in the emission wavelength due to carrier screening of internal electrostatic fields at increasing drive current and large nonradiative Auger recombination loss. Therefore, avoiding QCSE in InGaN LEDs (Waltereit et al., 2000) by growing InGaN/GaN along the nonpolar direction has been considered as an important milestone to realize the ultimate solid-state light sources for general illumination and full-color applications. However, compared to their polar counterparts, the nonpolar InGaN/GaN heterostructure LEDs still suffer from the issue of a high density of defects, unless growing on native GaN substrates, which are not widely available and expensive.

2. ADVANTAGES OF NANOROD/NANOWIRE HETEROSTRUCTURES

Quasi 1D nitride semiconductor nanorods and nanowires (also known as nanocolumns or nanopillars) have attracted much research interest in the past two decades due to their unique and tunable properties for nanophotonic and nanoelectronic applications. Especially, 1D nitride semiconductor nanorod/nanowire heterostructures with well-defined axial/radial geometries and abruptly modulated compositions have become of particular interest to achieve unique device functionalities, such as LEDs, lasers, solar cells, sensors, and field-effect transistors, etc. (Kang et al., 2012; Kuykendall et al., 2015; Li and Waag, 2012; Qian et al., 2004; Ra et al., 2014; Zhao et al., 2015). In this chapter, we refer "nanorods" to 1D materials with moderate aspect ratios (length/diameter ∼ 10–100), in contrast to nanowires with larger aspect ratios.

To overcome the formidable challenges for structural imperfection and QCSE issues in InGaN/GaN semiconductor heterostructures, several studies reported in the past few years have suggested a promising solution for full-color and white solid-sate emitters using vertically self-aligned GaN nanorod arrays as strain-free or strain-reduced growth templates for the growth of InGaN/GaN nanorod or nanowire heterostructures (Guo et al., 2010, 2011; Hong et al., 2009; Kikuchi et al., 2004; Kim et al., 2004; Kishino et al., 2007, 2009, 2015; Lin et al., 2010; Lu et al., 2011; Nguyen et al., 2011, 2013). For example, without using any phosphor converters, it has been demonstrated that white InGaN/GaN nanorod heterostructure LEDs can be realized using ensemble emissions from an InGaN nanoemitter array embedded in GaN *p–n* nanorods (Guo et al., 2011; Lin et al., 2010; Nguyen et al., 2011, 2013). Furthermore, single InGaN nanodisks can be utilized for the fabrication of ultrasmall footprint, polychromatic nano-LEDs in the entire visible spectrum (Lu et al., 2011).

In general, the InGaN/GaN nanorod/nanowire heterostructure light emitters have the following unique advantages, in comparison to their planar heterostructure counterparts:

1. Improved material quality

   The nanorod/nanowire geometry offers a unique strain-relieving platform for heteroepitaxy. GaN nanorods/nanowires grown on silicon and sapphire are mate rials free of misfit dislocations and stacking faults (Chen et al., 2006; Colby et al., 2010; Kishino and Ishizara, 2015; Kuykendall et al., 2004; Lu et al., 2013), which enable the growth of high-crystalline-quality InGaN/GaN nanorod heterostructures with an excellent radiative efficiency (high quantum efficiency).
2. Full-color and white-light applications

   The nanorod/nanowire geometry also allows for the relaxation of lattice strain resulting from indium incorporation in the active regions of the InGaN/GaN nanorod heterostructures (Bardoux et al., 2009; Kawakami et al., 2006; Lin et al., 2010; Lu et al., 2011). The significantly reduced or completely eliminated strain in heterostructures facilitates the possibilities of efficient light emitters in the full-color range (red, orange, yellow, green, cyan, blue, violet), enabling color-tunable and white-light emissions by combining monolithic, polychromatic light emitters (Albert et al., 2013; Amitage and Tsubaki, 2010; Guo et al., 2011; Hong et al., 2011, 2015; Kishino et al., 2007, 2009, 2015; Kuykendall et al., 2007; Lin et al., 2010; Nguyen et al., 2011, 2013; Qian et al., 2005; Sekiguchi et al., 2010).
3. Versatile device architectures

   Nanorod heterostructures grown by plasma-assisted molecular beam epitaxy (PAMBE) and metal-organic vapor phase epitaxy (MOVPE) enable a complete control of nanorod orientation (aligned in polar or non-polar directions) and abrupt modulation of nanorod material composition in both axial (nanodisk and quantum disk; Kikuchi et al., 2004; Kim et al., 2004; Lin et al., 2010; Lu et al., 2011) or radial (coaxial core–shell or core–multishell; Dong et al., 2009; Koester et al., 2011; Lu et al., 2012; Qian et al., 2005; Yeh et al., 2012) heterostructure geometries.
4. Integrated waveguides and reflectors

   Utilizing the waveguiding modes of individual nitride semiconductor nanorods or nanowires with partially reflecting (backscattering) facets or monolithically integrated Bragg reflectors (Fu et al., 2015), as well as other photonic structures (e.g., plasmonic cavities; Lu et al., 2012, 2014; Wu et al., 2011), it is possible to realize nitride semiconductor nanorod/nanowire lasers. Moreover, the GaN nanorod arrays can act as tunable optical media by controlling their filling factor (rod ensembles; Chen et al., 2008a) and aspect ratio (individual rods; Chen et al., 2008b).

3. POLARIZATION EFFECTS

Group-III nitrides (i.e., III-nitrides, including AlN, GaN, InN, and their alloys) crystallize in the wurtzite form under typical growth conditions. The wurtzite crystal structure consists of a hexagonal close packed (hcp) lattice with a diatomic basis. In the case of III-nitride semiconductors, the diatomic basis involves group-III elements (Al, Ga, In; cation sublattice) and nitrogen (N; anion sublattice). Each atom in the sublattice is tetrahedrally coordinated with four nearest atoms in the sublattice of the opposite atom type. The cation and anion hcp sublattices are shifted with respect to each other along the *c*-axis of the wurtzite structure by the amount of u in fractional of the unit cell length (c). Here, u is defined as the dimensionless cell internal parameter and $u \cdot c$ is the cation–anion bond length along the *c*-axis. As shown in Fig. 2, the wurtzite structure is noncentrosymmetric (i.e., the lack of inversion symmetry), and the stacking of the hcp layers along the polar *c*-axis of the wurtzite structure has two inequivalent AaBbAaBb... (group-III polarity or + *c*-polarity) or aAbBaAbB... (N-polarity or −*c*-polarity) sequences. Due to this crystal symmetry property, group-III nitrides possess strong polarization properties, such as pyroelectricity (spontaneous polarization) and piezoelectricity (piezoelectric polarization), while centrosymmetric crystals do not exhibit such properties. According to

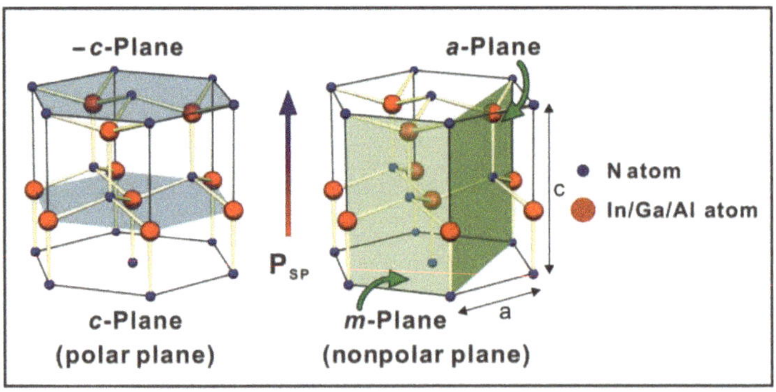

Fig. 2 Schematic diagram illustrating the hexagonal crystal structure of wurtzite III-nitrides (InN, GaN, AlN, and their alloys). The wurtzite structure has a tetrahedral coordination and lacks the mirror symmetry along the polar *c*-axis. The *light blue* and *green* shaded planes show polar +*c*-plane (Al/Ga/In-terminated (0001) face) and −*c*-plane (N-terminated $(000\overline{1})$ face), as well as nonpolar *m*-planes ($\{1\overline{1}00\}$) and *a*-planes ($\{11\overline{2}0\}$) of III-nitrides.

theoretical and experimental investigations, a large spontaneous polarization (P_{SP}) exists in bulk III-nitrides, which is directed toward the $-c$-axis ($[000\bar{1}]$) direction (Bernardini et al., 1997, 2001; Waltereit et al., 2000; Yu et al., 1999). There are a few structural parameters to characterize the wurtzite crystal structure, including the lattice parameters a and c, corresponding to the lattice periodicity in the hcp plane and the stacking periodicity along the polar c-axis (shown in Fig. 2), respectively, as well as the dimensionless cell internal parameter u (shown in Fig. 3).

In the ideal case of wurtzite crystal, each anion (cation) atom is tetragonally bonded with four nearest cation (anion) atoms, and the cell internal parameter u has the ideal value $u_0 = 3/8 = 0.375$. In group-III nitride crystals, the nonideality (spontaneous symmetry breaking) of the cell internal parameter ($u > u_0$) is the main cause that spontaneous polarizations are fairly large. According to theoretical and experimental investigations, the lattice constant ratio c/a in bulk III-nitrides is smaller than the ideal value $(1/u_0)^{1/2} = (8/3)^{1/2} \approx 1.633$. As a result, a large spontaneous polarization exists in a bulk III-nitride crystal, which is directed toward the $-c$-axis direction (shown in Fig. 2). According to a theoretical study by Bernardini et al. (2001), the III-nitride spontaneous polarizations (polarizations without any strain) have the following values: $-0.090\ C/m^2$ (AlN), $-0.034\ C/m^2$ (GaN), and $-0.042\ C/m^2$ (InN), where AlN has the largest value reported so far in comparison with known values of binary compound semiconductors.

In addition to spontaneous polarization, the absolute values of the piezoelectric constants in III-nitride compound semiconductors are up to one

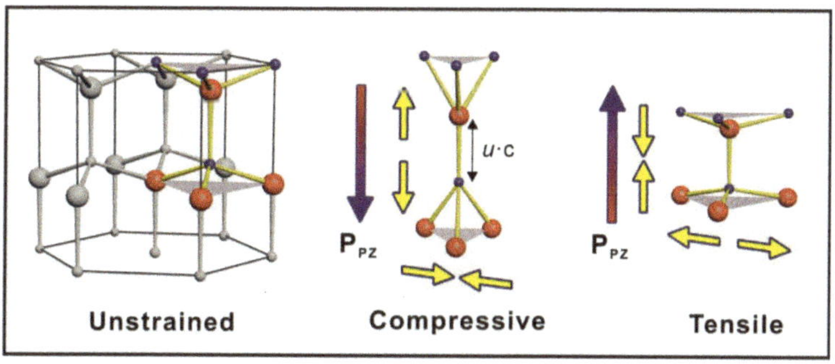

Fig. 3 Piezoelectric effects in wurtzite III-nitride semiconductors. In unstrained wurtzite crystal, the piezoelectric polarization ($\mathbf{P}_{PZ}$) is zero. Under the compressive and tensile strains, the resulting piezoelectric polarization fields are in the direction along the $+c$- and $-c$-axis, respectively.

order of magnitude larger than that in conventional III–V and II–VI compound semiconductors. Therefore, strains (compressive or tensile) in epitaxially grown III-nitride semiconductor films and heterostructures are significantly important because the spontaneous (P_{SP}; Fig. 2) and piezoelectric (P_{PZ}; Fig. 3) polarization fields have a well-defined orientation with respect to the crystal polarity (facets). Experimentally, heteroepitaxial III-nitride growth is performed preferably in the polar ([000]) or $\left([000\overline{1}]\right)$orientation, leading to the existence of both spontaneous and piezoelectric polarization fields and associated electrostatic charges at III-nitride semiconductor heterointerfaces that has been shown to influence carrier distributions, electric fields, and consequently a wide range of optical and electronic properties of nitride semiconductor devices (Waltereit et al., 2000; Yu et al., 1999).

In heterostructures or inhomogeneous alloy layers, variations in alloy composition are expected to create nonvanishing spontaneous and piezoelectric polarization charge densities. At an abrupt heterojunction interface between two III-nitride semiconductors with discontinuous polarizations ($\mathbf{P}=\mathbf{P}_{SP}+\mathbf{P}_{PZ}$), a polarization sheet charge density (σ_P) is developed at the heterointerface due to the difference in polarizations between two constituent media (1 and 2). According the Gauss's law, this sheet charge density can be obtained by using the following equation: $\sigma_P=-(\mathbf{P}_2-\mathbf{P}_1)\cdot\mathbf{n}_{12}=-\Delta\mathbf{P}\cdot\mathbf{n}_{12}$, where $\mathbf{P}_1$ and $\mathbf{P}_2$ are polarizations within media 1 and 2, respectively, and $\mathbf{n}_{12}$ is the unit vector of surface normal pointing from medium 1 to medium 2. In this definition, surface is a special case that a zero polarization appears on one side of the interface. For strained wurtzite III-nitride epitaxial layers grown along the polar direction, a piezoelectric polarization (along $+c$- or $-c$-axis) is developed in the epilayer, and it can be determined by the following equation for a biaxial stained epilayer:

$$P_{PZ}=\left(2e_{31}\epsilon_{\parallel}+e_{33}\epsilon_{\perp}\right)=(2e_{31}-\nu e_{33})\epsilon_{\parallel}=2\left(e_{31}-\frac{C_{31}}{C_{33}}e_{33}\right)\epsilon_{\parallel},$$

where e_{31} and e_{33} are piezoelectric coefficients, C_{31} and C_{33} are elastic constants, ν is the Poisson ratio, and $\epsilon_{\parallel}=(a-a_0)/a_0$ is the in-plane strain.

In the case of a coherently strained (pseudomorphic growth) InN/GaN quantum-well system, $\epsilon_{\parallel}$ in the ultrathin InN epilayer can be as large as -10% (compressively strained). Using the theoretical values reported by Bernardini et al. (2001) for InN, the piezoelectric polarization can be estimated to be as large as +0.150 C/m^2. In comparison to the small difference in spontaneous polarizations (−0.042 C/m^2 for InN and −0.034 C/m^2 for GaN), the polarization effects at the high-In-content InGaN/GaN are dominated by the piezoelectric polarization difference across the heterointerface. In the case of axial InGaN/GaN heterostructures grown on an array of GaN nanorods, the elastic relaxation of the strained InGaN/GaN heterostructures at the free nanorod sidewalls can alleviate or eliminate the issue of polarization-induced QCSE. Moreover, the GaN nanorod array as a growth template can block the dislocation lines originating at the interface with the lattice-mismatched substrates (Colby et al., 2010; Kishino and Ishizara, 2015). Both effects enable the realization of red, yellow, green, and blue InGaN nanoemitters monolithically with high crystalline quality on different types of substrates (Lin et al., 2010; Wölz et al., 2015; Zhao et al., 2016). In contrast to the InGaN system, the spontaneous polarization difference is quite large for InN/AlN or GaN/AlN heterojunction (−0.090 C/m^2 for AlN). Therefore, the spontaneous polarization effects become more dominant, especially for the cases that the lattice mismatch is smaller (AlGaN/GaN, 2.5% lattice mismatch between GaN and AlN), or the lattice strain at the heterojunction is negligibly small (e.g., at a near completely relaxed InN/AlN planar heterojunction; Kuo et al., 2011b).

There have been two kinds of nanorod light-emitting devices reported in the literature using PAMBE-grown radial or axial heterostructure geometries, such as axial InGaN/GaN disk-in-rod nanorod geometries and radial InGaN/GaN core–shell nanorod (see Fig. 4). In these nanorod heterostructures, the polarization fields and the associated QCSE effects can be significantly reduced or completely eliminated, thereby leading to improved recombination efficiency in the active InGaN regions, in comparison to the conventional *c*-plane planar InGaN/GaN quantum-well devices. However, in the axial nanorod heterostructures, although the piezoelectric polarization can be drastically reduced due to the efficient strain relaxation by free nanorod sidewall surfaces, the presence of spontaneous polarization field still need to be considered, particularly for the case of AlGaN/GaN nanorod heterostructures, as discussed earlier. For the radial core–shell nanorod heterostructures, the active regions are typically formed on the nonpolar sidewalls, which can be free of polarization effects. Moreover, the extremely large surface areas in radial nanorod/nanowire heterostructures significantly improved light extraction efficiency and spreading of injection current, promising a better emission quantum efficiency.

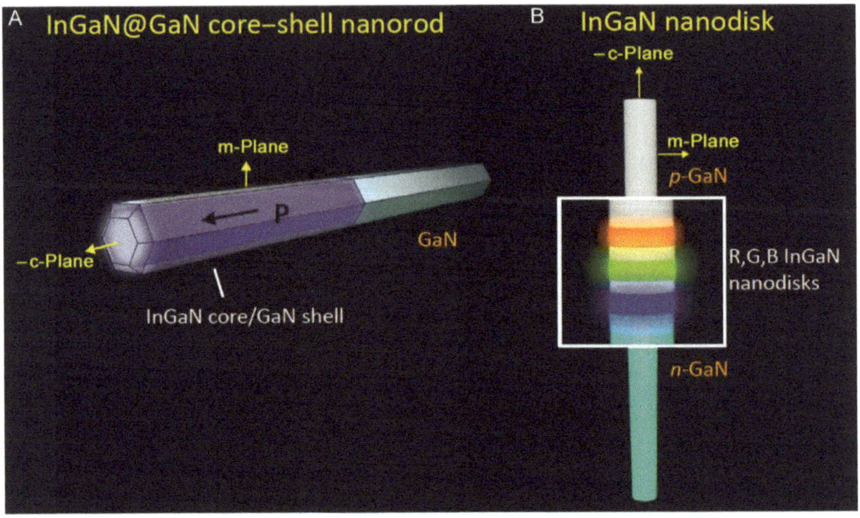

Fig. 4 Possible geometries of nitride semiconductor nanorod heterostructures grown by epitaxial techniques such as plasma-assisted molecular beam epitaxy (PAMBE). (A) Radial InGaN@GaN core–shell nanorod heterostructure. (B) Axial InGaN/GaN disk-in-rod nanorod heterostructure.

4. NANOROD/NANOWIRE GROWTH AND POLARITY CONTROL

The early growth studies of III-nitride nanowires were based on the catalyst-assisted vapor–liquid–solid growth mechanism (Chen et al, 2001). Typical growth catalysts used for GaN nanowire growth include nickel (Ni), iron (Fe), and gold (Au). In this growth technique, the fast growth rates and the capability of polarity control are important advantages for the growth III-nitride nanowires. In a study reported by Kuykendall et al. (2004), by exploiting the GaN and substrate epitaxial relation, highly oriented GaN nanowires were shown to preferentially grow into lithographically define, vertically aligned dense nanowire arrays along the nonpolar *m*-axis on γ-$LiAlO_2$(100) or along the polar *c*-axis on MgO(111). Recently, catalyst-directed orientation control of the growth axis has also been demonstrated for GaN nanowires grown along the nonpolar *m*- or *a*-axis using Au-rich or Ni-rich catalyst, respectively, on the same *r*-plane sapphire substrate and for GaN nanowires grown along the polar *c*-axis using Au-rich catalyst on *c*-plane GaN/sapphire (Kuykendall et al., 2015). The catalyst-assisted nanowire growth, however, involves the use of metal catalysts, which can induce electronic deep-trap states and structural imperfections. Moreover, controlling the indium composition in InGaN nanowires is very difficult. In the aspect of device applications, catalyst-free growth approaches for III-nitride nanorods and nanowires, such as MOVPE (also known as metal-organic chemical vapor deposition, MOCVD; Hersee et al., 2006) and PAMBE, are preferred choices for these purposes. In the catalyst-free grow methods, the inherent anisotropy of the wurtzite structure and the different growth rates along different crystallographic directions are utilized to grow nanorods and nanowires.

In the spontaneous growth process by PAMBE, vertically self-aligned nanorod array growth takes place at random sites on the N-polar surface (Bertness et al., 2011; Calleja et al., 2000, 2007; Chen et al., 2006; Fernández-Garrido et al., 2012; Shen et al., 2006; Tu et al., 2003; Wang et al., 2016; Yoshizawa et al., 1997). Furthermore, the sidewalls of well-formed III-nitride nanorod heterostructure have preferred nonpolar orientations. For example, the hexagonal nanorod heterostructure shown in Fig. 5 is enclosed by six *m*-plane sidewalls (Lu et al., 2012). Spatially ordered and uniform nanorod/nanowire array structures can also be obtained by using the selective-area growth (SAG) techniques in MOVPE (Hersee et al., 2006) and PAMBE (Bengoechea-Encabo et al., 2011; Brubaker et al., 2015; Kano et al., 2015; Kishino et al., 2009), where a patterned growth mask (e.g., Ti, TiN, SiO_2, Si_3N_4) is prepared on top of the grown surface. The opening areas of the growth mask control the nucleation, growth, and coalescence process for spatially ordered nanorod and nanowire formation. In the case of spontaneous nitride nanorod arrays grown by PAMBE, the nanorods are formed with the N-polarity, which could be confirmed by wet chemical etching and high-resolution scanning transmission electron microscopy (see Fig. 6). In the case of GaN nanorods grown by the PAMBE-based SAG technique on MOVPE GaN templates, nanorods generally are formed with the Ga-polarity, which can be controlled by the substrate polarity.

In comparison to PAMBE, MOVPE is commonly used to synthesize nitride semiconductors because of their scalability and fast growth rates. However, since low growth temperatures are required to achieve higher indium incorporation, MOVPE is unfavorable for producing single-crystalline, high-In-content InGaN alloys because of the thermal instability of InGaN, originating from the miscibility gap of InN. This can cause significant surface segregation of indium at high growth temperatures, which are required when using ammonia as the nitrogen precursor. Nevertheless,

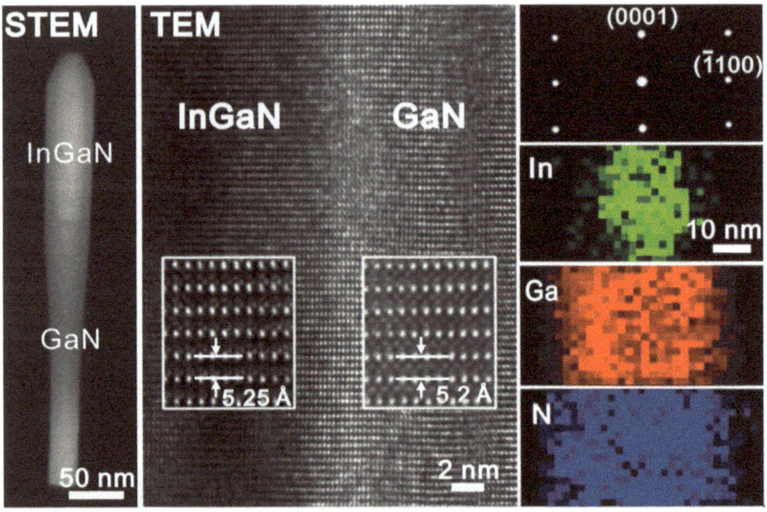

Fig. 5 Scanning transmission electron microscopy (STEM) and transmission electron microscopy (TEM) structural analyses of single-crystalline InGaN@GaN core–shell nanorod. The brighter area inside the nanorod in the high-angle angular dark field (HAADF) STEM image indicates the presence of InGaN core. Electron diffraction pattern and energy-dispersive X-ray spectroscopy (EDS) elemental mapping of In, Ga, and N shown on the right confirm the chemical composition of InGaN@GaN core–shell nanorods. *This figure is adapted with permission from Lu, Y.-J., Kim, J., Chen, H.-Y., Wu, C., Dabidian, N., Sanders, C.E., Wang, C.-Y., Lu, M.-Y., Li, B.-H., Qiu, X., Chang, W.-H., Chen, L.-J., Shvets, G., Shih, C.-K., Gwo, S., 2012. Plasmonic nanolaser using epitaxially grown silver film. Science 337, 450–453. Copyright AAAS 2012.*

the MOVPE method is the preferred approach for growing ordered, uniform radial core–multishell nanorod/nanowire arrays. In comparison, axial heterostructures grow along the length of the nanowires/nanorods, while the coaxial or core–shell heterostructures grow conformally around the nanorods or nanowires. The radial heterointerfaces in coaxial heterostructures surrounding the nanowire/nanorod increase the active area for current injection and carrier recombination. Furthermore, coaxial *c*-axis nanorod/nanowire heterostructure has the advantages of forming shells with nonpolar side facets.

On the other hand, PAMBE has the advantage to produce InGaN/GaN nanorods with high indium compositions because of the use of the radio frequency nitrogen plasma, which can produce reactive nitrogen radicals for growth of III-nitrides at lower temperatures (e.g., InN). However, in comparison to MOVPE, the PAMBE growth technique has a relatively slow deposition rate, limited by the flux of reactive nitrogen radicals for growth

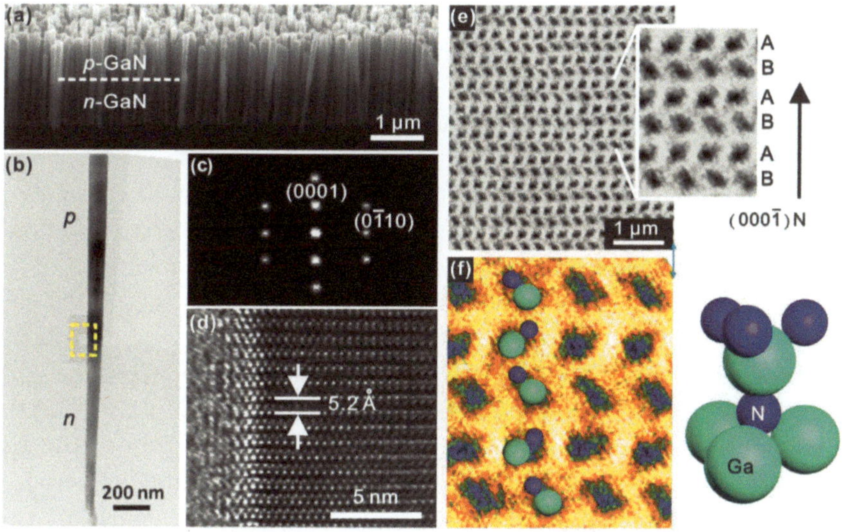

Fig. 6 Structural characterization of GaN nanorods with axial *p–n* junctions. (A) scanning electron microscopy (SEM) image of PAMBE-grown GaN nanorod arrays. The *p*- and *n*- segments of nanorods are clearly visible in the image with different image contrast. (B) Bright-field TEM image of a single GaN nanorod *p–n* junction. (C) Electron diffraction pattern at the nanorod *p–n* junction region. (D) High-resolution TEM image obtained at the *p–n* junction interface (marked area in (B)). The measured lattice parameter is in good agreement with the known value of strain-free wurtzite GaN crystal along the *c*-axis. In addition, no obvious defects are observed in the image, indicating that as-grown GaN nanorods exhibit a perfect wurtzite crystal structure. (E) High-resolution annular bright-field (ABF) STEM image of GaN nanorods showing the ABABAB stacking order of wurtzite structure. (F) Magnified false-color ABF-STEM image illustrates that N atoms are on the *top ends* of Ga–N dumbbells, demonstrating that the PAMBE-grown GaN nanorods have the nitrogen polarity (N-terminated top facets). *This figure is adapted with permission from Lu, Y.-J., Lu, M.-Y., Yang, Y.-C., Chen, H.-Y., Chen, L.-J., Gwo, S., 2013. Dynamic visualization of axial* p–n *junctions in single gallium nitride nanorods under electrical bias. ACS Nano 7, 7640–7647. Copyright ACS 2013.*

under nitrogen-rich conditions. Another important advantage of PAMBE is the precise interface control at the monolayer level, achievable using the technique of in situ reflection high-energy electron diffraction. It should be noted that PAMBE is a kinetics-controlled growth process. Besides the thermodynamically stable phases, metastable phases can also be prepared by controlling the growth parameters (group-III beam flux, flux ratio between group-III element and nitrogen plasma, and sample temperature). For example, in the case of nitride heterointerfaces (e.g., InN/GaN and InN/AlN; Kuo et al., 2011b; Wu et al., 2007) with large lattice mismatches

(>10% for coherently strained heterointerfaces), commensurate lattice matching condition (e.g., 8:9 for InN/AlN; Gwo et al., 2004; Wu et al., 2006) can be realized so that the PAMBE-grown InN/GaN and InN/AlN heterointerfaces exhibit greatly reduced strains (e.g., <0.5% for InN/AlN; Kuo et al., 2011b).

Interestingly, the spontaneous formation of "sausage"-like core–shell structure observed for the PAMBE growth process enables an additional strain relaxation mechanism in nanorod heterostructures. Examples of the sausage-like core–shell structure reported so far include an InGaN core embedded by a GaN shell (InGaN@GaN, see Fig. 5; Lu et al., 2012) and a GaN shell embedded by an AlGaN shell, where under the coherently strained conditions, the shell material has a uniaxial tensile strain, and the core material has a uniaxial compressive strain. The formation of sausage-like core–shell structures has been confirmed to be quite universal in PAMBE-grown nanorod heterostructures, including the axial InGaN/GaN disk-in-rod nanorod heterostructure grown by PAMBE (Woo et al., 2015a). The interplay between biaxial and axial strains allows for the formation of strain-free core–shell nanorod heterostructures without generation of a significant number of dislocations (see Fig. 5). Importantly, besides the nanorod sidewall surfaces, there are additional pathways to relax the large strains resulting from lattice mismatch in PAMBE-grown nanorod/nanowire heterostructures, which can significantly reduce or completely eliminate piezoelectric polarization effects.

5. DOPING AND SURFACE PROPERTIES

For most of these nanoscale device applications, it is indispensible to incorporate *n*- and *p*-type dopants into semiconductor nanostructures for the purpose of modulating their electrical properties and forming *p–n* junctions. In the literature, the formation of nanowire *p–n* junctions has been demonstrated by using cross-wire, axial (axial *p* and *n* segments), and coaxial (*p–n* or *n–p* core–shell) nanowire/nanorod techniques. In the case that amphoteric doping for the same type of semiconductor is not possible (e.g., ZnO), the heterostructure combinations of *n*-type (*p*-type) nanowires with *p*-type (*n*-type) nanowires or substrates have also been used to form *p–n* junction devices (Lupan et al., 2010; Ng et al., 2009; Tang et al., 2008). Recently, axially controlled modulation of doping and compositions along individual nanorods or nanowires has become a focused research area because of the easy implementation of vertically injected array devices

(Guo et al., 2010, 2011; Kikuchi et al., 2004; Kim et al., 2004; Lin et al., 2010; Nguyen et al., 2011).

Fig. 6 shows a GaN *p–n* junction nanorods array grown by PAMBE. Vertically self-aligned GaN nanorods arrays have been shown by transmission electron microscopy (TEM) and scanning transmission electron microscopy (STEM) to be dislocation-free single crystals. Most significantly for practical applications, contrary to other wide-band gap semiconductor materials, GaN nanorods can be doped as both *n*- and *p*-type materials, which is a prerequisite for electrically driven photonic devices. For nanorod and nanowire applications, the capabilities to delineate and quantitatively determine the electrostatic potentials and doping concentrations across the 3D semiconductor axial and radial *p–n* junctions on the nanoscale (Lu et al., 2013; Tchoulfian et al., 2014), as well as to correlate doping, structure, and carrier dynamics in single nanorods (Zhou et al., 2013) are crucial for understanding the fundamental properties and improving nanorods and nanowire device performance. Besides conventional techniques such as secondary ion mass spectroscopy and capacitance voltage measurements with limited spatial resolutions, a number of high-spatial-resolution dopant profiling techniques have recently been developed to fulfill the needs for next-generation semiconductor devices, which include electron holography, secondary electron (SE) contrast imaging (Lu et al., 2013; Tchoulfian et al., 2014), and electron-beam-induced current (Tchoulfian et al., 2014) in scanning electron microscopy (SEM), scanning probe microscopies (e.g., scanning Kelvin probe microscopy and light-modulated scanning tunneling spectroscopy), and synchrotron-radiation-based photoelectron spectromicroscopy (Kuo et al., 2009). For example, Fig. 7 shows an SE image of a single-axial GaN nanorod *p–n* junction under different electrical biasing conditions, which was carried out in an SEM chamber equipped with a multiple-probe manipulation and measurement system. The SE images of a single GaN nanorod *p–n* junction measured under different reverse biases show variable imaging contrast proportional to the magnitude of space charge in the depletion region, which can be used to delineate the electrostatic potential across the nanorod *p–n* junction under biasing conditions.

The PAMBE-grown, self vertically aligned III-nitride nanorods possess the nitrogen polarity such that the nanorod top facet is nitrogen-atom-terminated. Under optimized growth conditions, the sidewalls of as-grown III-nitride nanorods are composed of nonpolar facets (e.g., *m*-plane facets). The large surface-to-bulk ratio in the nanorod/nanowire material makes

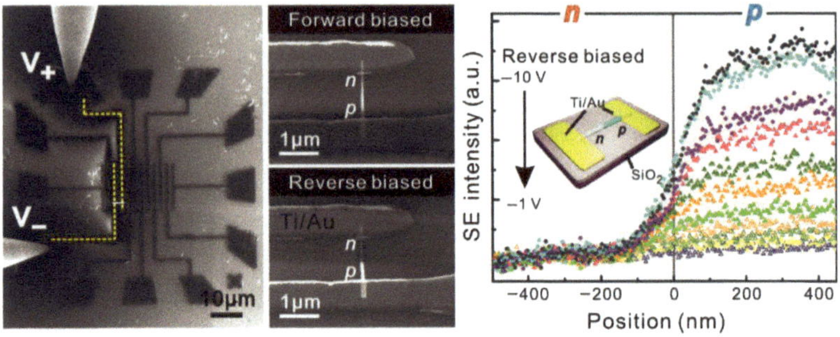

Fig. 7 Imaging of a single axial GaN nanorod *p–n* junction under reverse biasing conditions. This imaging experiment was carried out in an SEM chamber equipped with a multiple-probe manipulation and measurement system. The *yellow dash lines* in the SEM image shown on the left indicate the measuring electrodes with a single GaN nanorod *p–n* junction placed in between. The current–voltage measurements confirm that the reverse current is negligible at −10 V, under the measurement current resolution of picoampere. The secondary electron (SE) images of a single GaN nanorod *p–n* junction measured under different reverse biases show variable imaging contrast proportional to the magnitude of space charge in the depletion region. *This figure is adapted with permission from Lu, Y.-J., Lu, M.-Y., Yang, Y.-C., Chen, H.-Y., Chen, L.-J., Gwo, S., 2013. Dynamic visualization of axial* p–n *junctions in single gallium nitride nanorods under electrical bias. ACS Nano 7, 7640–7647. Copyright ACS 2013.*

the surface properties extremely important. For most of optoelectronic device applications, appropriate surface passivation schemes can greatly improve the device performance (Nguyen et al., 2013). In terms of surface properties, InN is an interesting III-nitride material to mention. In 2002, InN was discovered to be a narrow band gap material (Davydov et al., 2002), which has attracted much attention for applications in high-efficiency solar cells and high-speed electronic devices. One of the key issues for InN-based device applications is related to the unusual phenomenon of strong electron accumulation at as-grown InN surfaces except for a few special cases (Kuo et al., 2011a; Wu et al., 2008b; Zhao et al., 2012). For the realization of *p*-type doping in InN, surface electron accumulation leads to the formation of an *n*-type surface inversion layer, which prevents a direct electrical contact to the bulk InN. In contrast, this is a useful surface property for the purpose of sensor applications (Chang et al., 2010, 2011; Kao et al., 2012; Lu et al., 2007, 2008).

Recently, by using first principles calculations, the stable structures of InN growth surfaces have been shown to be In-adlayer terminated, which is independent of the film polarity. Moreover, the ubiquitous In-adlayers result in the surface states located above the conduction band minimum, and this is considered as one of the possible origins of surface electron accumulation. Based on their calculations, Van de Walle and Segev (2007) predicted the absence of electron accumulation at nonpolar surfaces, and it was experimentally confirmed at in situ cleaved *a*-plane InN surfaces (Ebert et al., 2011; Wu et al., 2008b). Recently, the pristine nonpolar sidewalls of PAMBE-grown InN nanowires have been confirmed to be free of surface electron accumulation and Fermi-level pinning (Zhao et al., 2012), demonstrating the importance of surface polarity control for nanorod/nanowire device applications.

6. III-NITRIDE NANOROD HETEROJUNCTION BAND ALIGNMENTS

Formation of semiconductor nanorod/nanowire heterostructures is essential for building functional nanodevices because that heterostructures allow for control of the flow and confinement of charge carriers, carrier dynamics, relative band position (band alignment), emission wavelength, and other important electronic and optical properties by modulating the comprising materials through incorporation of *p*- or *n*-type dopants, alloy composition, and layering of dissimilar materials.

In the past two decades (Martin et al., 1996; Moses et al., 2011), studies on heterojunction band alignments of III-nitrides have been a strong research focus because of their importance in the design and characterization of optoelectronic and electronic devices based on semiconductor heterostructures. However, there have been large uncertainties in the reported valence band offsets (VBOs) due to the strong influence of polarization effects resulting from spontaneous and piezoelectric polarizations in wurtzite III-nitrides. For example, Martin et al. (1996) have found that the discontinuities of piezoelectric polarizations in coherently strained heterojunctions (ultrathin III-nitride epitaxial layers grown on lattice-mismatched III-nitrides) would result in a strong forward–backward heterojunction asymmetry (i.e., A/B vs B/A growth sequence).

The polarization effects of nitride semiconductor heterostructures are strongly dependent on the crystal polarity. Wu et al. (2006, 2007, 2008a) and Kuo et al. (2011b, 2011c) showed that the polarization effects play an important role in determining the III-nitride semiconductor band lineup (specifically, VBOs) by using synchrotron-radiation-based photoelectron spectroscopy (PES). In the work reported by Kuo et al. (2011b), PES measurements show a dramatic difference for the valence band alignments of strain-relaxed commensurate InN (top-layer)/AlN (buried-layer) heterojunctions with opposite crystal polarities. The polarity control of InN/AlN heterojunction was realized by the growth of III-nitride buffer layers on top of nitridated *c*-plane sapphire substrates. The growth of an AlN buffer exhibits a cation-polar surface, while the additional growth of a low temperature GaN buffer layer before the AlN layer led to a nitrogen-polar surface. Under the commensurate growth conditions (Gwo et al., 2004; Wu et al., 2003, 2006), it has been found that the lattice strains in these films (including InN and AlN epilayers) are below 0.5%. Considering the reported piezoelectric coefficients, the piezoelectric polarizations of InN and AlN resulting from the epilayer strains are relatively small, which can be neglected in comparison with their large spontaneous polarizations ($-0.042\ \mathrm{C/m^2}$ for InN and $-0.090\ \mathrm{C/m^2}$ for AlN). Therefore, only the spontaneous polarizations have a significant influence on the measured VBOs at polar InN/AlN interfaces.

In the microscopic view, the atomic arrangements at InN/AlN polar interfaces are different, i.e., In–N–Al–N for In/Al-polarity, and N–In–N–Al for N-polarity. Since the direction of spontaneous polarization in the wurtzite structure is from the metal atom to the nearest neighbor nitrogen atom along the $-c$-axis, there are two opposite orientations of spontaneous polarizations across the heterojunction. As a result, a positive bound sheet charge is induced at the N-polar interface (as shown in Fig. 8A) and would cause an accumulation of mobile carriers (electrons) in unintentional *n*-type InN to form an interface dipole. In contrast, a negative polarization-induced sheet charge is induced at the In/Al-polar interface and would result in depletion of mobile electrons in InN to form an opposite interface dipole. The interface dipole at the polar heterojunction has been observed to have a remarkable influence on the measured core-level energy separation (ΔE_{CL}) of photoelectrons at the InN/GaN heterojuctions with opposite polarities (Wu et al., 2007). For the InN/AlN polar heterojunctions, the interface dipole would result in a shift of the Al 2*p* core-level emission energy from the buried AlN layer. As a consequence, the measured binding energy of Al 2*p* core-level emission can be smaller or larger, depending on the orientation of the interface dipole. Thus, the binding energy separation between Al 2*p* and In 4*d* core levels is $\Delta E_{\mathrm{CL}}^{(\mathrm{In/Al})} = \left(E_{\mathrm{B}}^{\mathrm{Al}} - eV_{\mathrm{int}}\right) - E_{\mathrm{B}}^{\mathrm{In}}$ and $\Delta E_{\mathrm{CL}}^{(\mathrm{N})} = \left(E_{\mathrm{B}}^{\mathrm{Al}} + e\widetilde{V}_{\mathrm{int}}\right) - E_{\mathrm{B}}^{\mathrm{In}}$, for the In/Al-polar and N-polar heterojunction, respectively, where eV_{int} and $e\widetilde{V}_{\mathrm{int}}$ are the photoelectron potential differences across the interface dipoles.

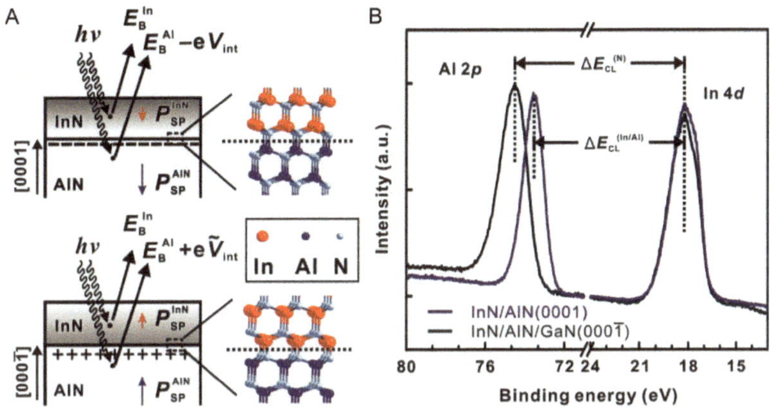

Fig. 8 (A) Schematic representation of the polar InN/AlN structures, including spontaneous polarizations, polarization-difference-induced interface sheet charges (bound charges) at the heterointerfaces with grown with opposite crystal polarities. Also shown are the interface-dipole-induced shift in the binding energy separation between Al 2*p* and In 4*d* core-levels and atomic arrangements of In/Al- and N-polar InN/AlN heterojunctions. (B) Core-level spectra of In/Al- and N-polar InN/AlN heterojunctions obtained by synchrotron radiation at 380 eV. *This figure is adapted with permission from Kuo, C.-T., Chang, K.-K., Shiu, H.-W., Lin, S.-C., Chen, C.-H., Gwo, S., 2011b. Spontaneous-polarization-induced heterojunction asymmetry in III-nitride semiconductors. Appl. Phys. Lett. 99, 022113. Copyright AIP 2011.*

Fig. 8B shows the corresponding photoelectron spectra for both types of polar heterojunctions. By aligning the In 4*d* core-level emission of thin InN/AlN heterojunctions with In/Al- and N-polarity, it can be clearly seen that there is a large energy difference of 1.0 eV for the Al 2*p* to the In 4*d*core-level separations across the In/Al-polar ($\Delta E_{CL}^{(In/Al)}$) and N-polar ($\Delta E_{CL}^{(N)}$) InN/AlN heterojunctions ($\Delta E_{CL}^{(N)} = 56.5$ eV and $\Delta E_{CL}^{(In/Al)} = 55.5$ eV). The large difference in ΔE_{CL} between In/Al- and N-polar InN/AlN heterojunctions is due to the internal electric fields induced by two opposite interface dipoles $\left[e\left(V_{int} + \widetilde{V}_{int}\right)\right]$, originating from the difference in spontaneous polarizations between InN and AlN. As a consequence, the measured VBO values at polar III-nitride heterojunctions are asymmetric in crystal polarity and depend on the difference in spontaneous polarizations (ΔP_{SP}). The ΔP_{SP} value between InN and AlN is the largest in III-nitrides, and as a result, the asymmetric band lineup is the most obvious one.

In the PES studies (Kuo et al., 2011b; Wu et al., 2007), it was confirmed that the measured VBO values are strongly affected by the piezoelectric and spontaneous polarization discontinuities in nearly strain-free III-nitride heterojunctions. Therefore, the determination of intrinsic band alignments of III-nitrides free from any polarization effects is very critical toward a full understanding of III-nitride heterojunctions. The introduction of the VBO measurement technique using synchrotron-based cross-sectional scanning photoelectron microscopy/spectroscopy (XSPEM/S) makes this task within reach of the present research capabilities. In comparison with the conventional PES technique for the measurement of VBOs using ultrathin heteroepitaxial films, the samples used for XSPEM/S are strain-free thick heteroepitaxial films free from the piezoelectric field fields. Additionally, the sample geometry for XSPEM/S is suitable for obtaining the intrinsic VBOs of a multiple heterojunction sample at the same time, which enable one to directly confirm the necessary transitivity rule of VBOs, rather than analyzing the results from several individual heterojunction samples.

In 2008 and 2011, we reported the intrinsic valence band alignments of wurtzite nitride semiconductor heterojunctions, which were investigated using XSPEM/S on a cleaved nonpolar III-nitride heterojunction surface (Fig. 9; Wu et al., 2008a) and nonpolar sidewalls of a vertically aligned axial triple-junction (InN/GaN/AlN/GaN) nanorod heterostructure array (Fig. 10; Kuo et al., 2011c), respectively. In these XSPEM/S studies, the cleavage surfaces were prepared by using heteroepitaxial samples grown on Si(111) wafers. The XSPEM/S measurements were performed on the cleaved sample surfaces, where the nonpolar nanorod side facets allow for the determination of "natural" (intrinsic) band alignments without the influence of spontaneous and piezoelectric polarization fields. For the triple-junction (InN/GaN/AlN/GaN) nanorod heterostructure sample, the VBOs of InN/GaN, GaN/AlN, and InN/AlN were experimentally confirmed to be 0.8 ± 0.1, 0.6 ± 0.1, and 1.4 ± 0.1 eV, respectively (Fig. 10). These results are in agreement with our previous data for heteroepitaxial films and obey the expected transitivity rule for the VBO measurements.

Table 1 summarizes the measured binding energy differences between core-level and valence band maximum ($E_V - E_{CL}$) for thick III-nitride epilayers (InN, GaN, and AlN), which show consistent results for different sets of samples and measurement geometries (crystal polarities). The experimental results confirm that these values are material constants, independent of film orientation and doping conditions. In contrast, core-level binding energy separations (ΔE_{CL}) measured for III-nitride heterojunction

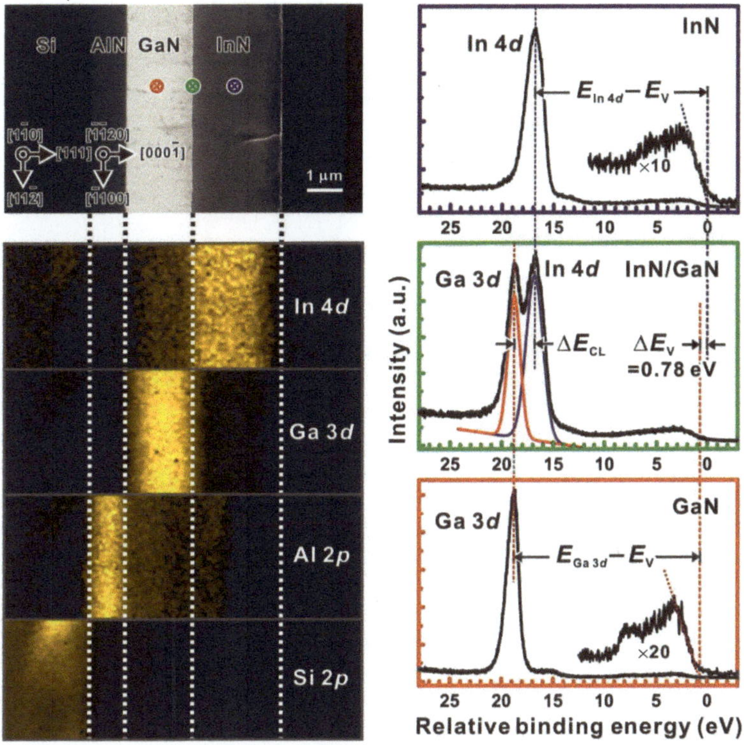

Fig. 9 Cross-section scanning photoelectron microscopy/spectroscopy (XSPEM/S) measurements performed on the in situ cleaved *a*-plane surface of wurtzite InN/GaN heterojunction. The directions shown in the SEM image (*upper left panel*) indicate the crystal directions in the heteroepitaxial III-nitrides-on Si(111) system grown by PAMBE. XSPEM images (*lower left panel*) acquired at different core-level emissions show consistency with the SEM image. The photoelectron spectra taken at selected regions across the InN/GaN heterojunction are displayed on the *right*. An InN/GaN valence band offset (VBO) of 0.8 eV can be determined from these measurements. *This figure is adapted with permission from Wu, C.-L., Lee, H.-M., Kuo, C.-T., Chen, C.-H., Gwo, S., 2008a. Cross-sectional photoelectron microscopy and spectroscopy of wurtzite InN/GaN heterojunction: measurement of "intrinsic" band lineup. Appl. Phys. Lett. 92, 162106. Copyright AIP 2008.*

samples are highly dependent on the crystal polarities. The VBOs (ΔE_V) of III-nitride A/B heterojunctions can be determined by the following equation: $\Delta E_V = (E_V - E_{CL})_A - (E_V - E_{CL})_B + (\Delta E_{CL})_{A/B}$, where A and B are two constituent heterojunction materials. Fig. 10 shows that, for the first time, III-nitride heterojunction band alignments can be measured with a high consistency, free from the polarization effects.

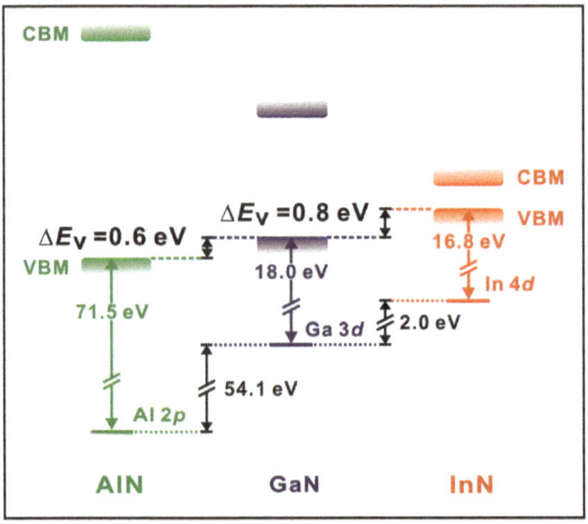

Fig. 10 Schematic diagram of measured intrinsic band alignments for InN, GaN, and AlN using a triple-junction (InN/GaN/AlN/GaN) nanorods heterostructure sample. The sample structure used for this study is an InN (200 nm)/GaN (1.2 μm)/AlN (0.9 μm)/GaN (100 nm) vertically aligned triple-junction nanorod array grown by PAMBE on a nitride Si_3N_4/Si(111) wafer. The measured values of ($E_{CL} - E_V$) and ΔE_{CL} are shown schematically in the band alignment diagram. The determined VBOs for InN/GaN, GaN/AlN, and InN/AlN heterojunctions are 0.8 ± 0.1, 0.6 ± 0.1, and 1.4 ± 0.1 eV, respectively. *This figure is adapted with permission from Kuo, C.-T., Chang, K.-K., Shiu, H.-W., Liu, C.-R., Chang, L.-Y., Chen, C.-H., Gwo, S., 2011c. Natural band alignments of InN/GaN/AlN nanorod heterojunctions. Appl. Phys. Lett. 99, 122101.*

7. DISK-IN-ROD NANOROD HETEROSTRUCTURES AS FULL-COLOR LIGHT EMITTERS

Both axial InGaN/GaN disk-in-rod and radial InGaN/GaN core–shell nanorod heterostructures grown by PAMBE-grown are ideal candidates for the realization of full-color and white-light emitters. In comparison, the disk-in-rod device geometry has the advantages of easy device implementation and ultrasmall footprint. However, it is important to note that the emissions from an InGaN/GaN nanodisk array highly nonuniform in emission colors of individual InGaN nanodisk. Fig. 11 shows the photoluminescence (PL) spectra of an InGaN nanodisk ensemble emitting over a wide spectral range, illustrating the important stochastic nature of InGaN nanodisks. In fact, each nanodisk array grown a specific

Table 1 Summary of PES and XSPEM/S Measured Binding Energy Differences Between Core-Level and Valence Band Maximum ($E_V - E_{CL}$) for Thick III-Nitride Epilayers, as Well as Core-Level Binding Energy Separations (ΔE_{CL}) and Valence Band Offsets (ΔE_V) for III-Nitride Heterojunctions With Different Crystal Polarities

Heterojunction	Polarity	Sample	$E_V - E_{CL}$ (eV)	ΔE_{CL} (eV)	ΔE_V (eV)
InN/GaN	In/Ga (0001)[1]	InN	16.77 (In 4*d*)	2.20	1.04
		GaN	17.93 (Ga 3*d*)		
	N $(000\overline{1})$[1]	InN	16.82 (In 4*d*)	1.69	0.54
		GaN	17.97 (Ga 3*d*)		
	Nonpolar[2]	Cleaved InN/GaN	16.84 (In 4*d*)	1.98	0.78
			18.04 (Ga 3*d*)		
	Nonpolar[3]	InN/GaN/AlN Nanorod	16.8 (In 4*d*)	2.0	0.8
			18.0 (Ga 3*d*)		
InN/AlN	In/Al (0001)[4]	InN	16.8 (In 4*d*)	55.5	0.8
		AlN	71.5 (Al 2*p*)		
	N $(000\overline{1})$[4]	InN	—	56.5	1.8
		AlN	—		
	Nonpolar[3]	InN/GaN/AlN Nanorod	16.8 (In 4*d*)	56.1	1.4
			71.5 (Al 2*p*)		
GaN/AlN	Nonpolar[3]	InN/GaN/AlN Nanorod	18.0 (Ga 3*d*)	54.1	0.6
			71.5 (Al 2*p*)		

These experimental data are compiled [1] From Wu et al. (2007), [2] Wu et al. (2008a), [3] Kuo et al. (2011c), and [4] Kuo et al. (2011b).

substrate temperature is composed of a broad and continuous distribution of polychromatic emitters due to the inherent alloy inhomogeneity among InGaN nanodisks. Nevertheless, we have found that the ensemble peak emission positions of InGaN nanodisk ensembles can be precisely controlled by the substrate temperature and In/Ga beam fluxes.

Using these InGaN ensembles, Lu et al. (2011) have been able to demonstrate that the dispersed single InGaN nanodisks embedded in GaN *p–n* nanorods can be utilized for the fabrication of ultrasmall footprint nano-LEDs in the entire visible spectrum. The InGaN/GaN disk-in-rod

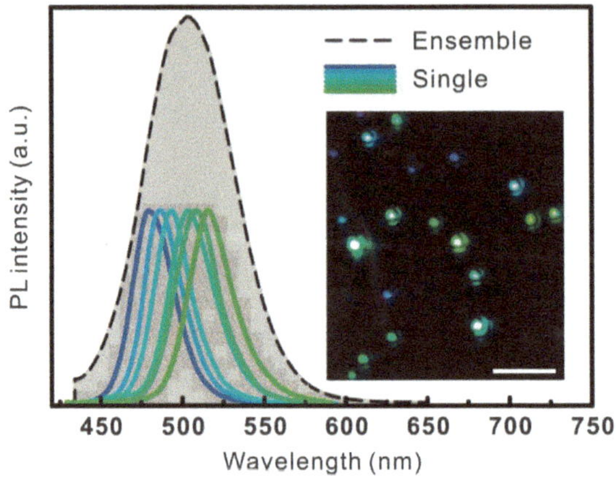

Fig. 11 Photoluminescene (PL) spectra measured for an InGaN nanodisk ensemble (*dash line*, collection area: ~50 μm in diameter) and dispersed InGaN nanodisks (*colored lines*). The *inset* shows an optical microscopy (OM) image of dispersed InGaN nanodisks dispersed on an oxidized Si substrate, which were optically excited by using an UV laser (scale bar in OM image: 5 μm). The ensemble PL spectrum (*dash curve*) centers at 504 nm with a wide spectral width. In comparison, the emissions from dispersed nanodisks are spectrally pure, ranging from *blue* to *green*. The axial InGaN/GaN nanodisk-in-nanorod heterostructure was grown by PAMBE within a vertically aligned GaN nanorod array. Each InGaN/GaN nanorod consists of a single InGaN nanodisk (nominal disk thickness: ~25 nm). *This figure is adapted with permission from Lin, H.-W., Lu, Y.-J., Chen, H.-Y., Lee, H.-M., Gwo, S., 2010. InGaN/GaN nanorod array white light-emitting diode. Appl. Phys. Lett. 97, 073101. Copyright AIP 2010.*

nano-LED structure shown in Fig. 12 was grown on an *n*-type Si(111) wafer by PAMBE. A catalyst-free 1.2 μm, *n*-type Si-doped GaN nanorod array was grown first on Si(111) along the wurtzite polar axis (−c direction), and 40-nm-thick InGaN nanodisks were grown on top of the *n*-type GaN nanorod array. In the final growth step, 800 nm, Mg-doped *p*-type GaN nanorods were grown on top of the InGaN nanodisks. Fig. 12A shows the fabricated InGaN/GaN nanodisk-in-nanorod LED using this structure, and Fig. 12B shows room-temperature current-voltage (*I–V*) characteristics measured for a single InGaN/GaN nanodisk–nanorod LED (nanodisk diameter: 90 nm; nanodisk thickness: 40 nm). The measured *I–V* characteristics indicate a good diode behavior. Especially, single nanodisk-in-nanorod diodes show no signs of leakage currents, allowing for the measurement of actual current density passing through a single InGaN/GaN nanodisk-in-nanorod

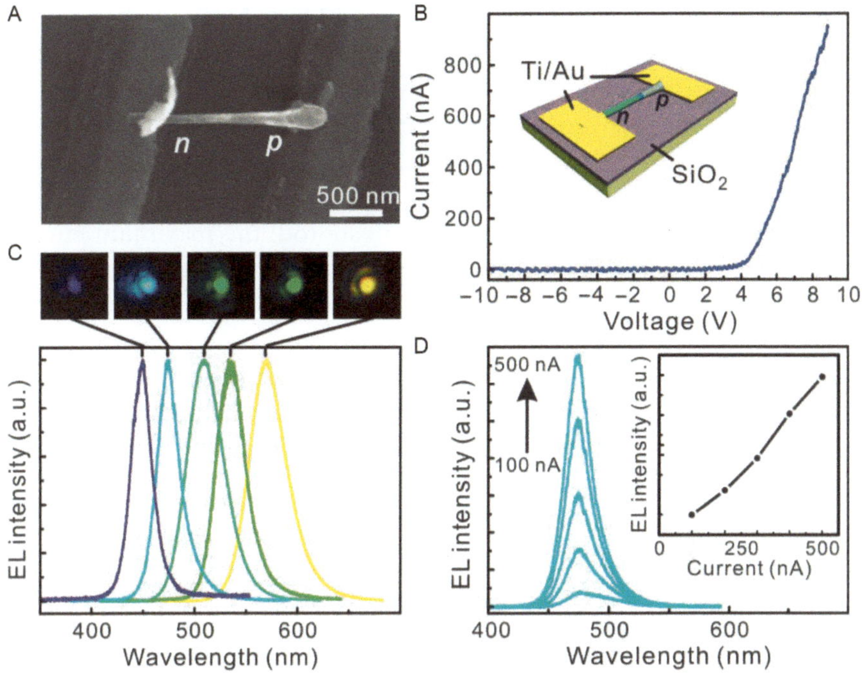

Fig. 12 (A) SEM image of a single InGaN/GaN nanodisk-in-nanorod LED. (B) Room-temperature *I–V* characteristics of the *blue* (475 nm) LED shown in (A). (C) Optical images of the full-color light emissions from single InGaN/GaN nanodisk-in-nanorod LEDs under forward electric bias (injection current: 500 nA). Normalized electroluminescence (EL) spectra show that emissions from single InGaN nanodisks are monochromatic with narrow spectral widths. (D) EL spectra of the InGaN/GaN nanodisk-in-nanorod LED emitting at 475 nm with injection currents varied from 100 to 500 nA. *Inset* shows the EL intensity exhibits a monotonic increase with increasing current for up to 500 nA, corresponding to a current density of 8000 A/cm^2. *This figure is adapted with permission from Lu, Y.-J., Lin, H.-W., Chen, H.-Y., Yang, Y.-C., Gwo, S., 2011. Single InGaN nanodisk light emitting diodes as full-color subwavelength light sources. Appl. Phys. Lett. 98, 233101. Copyright AIP 2011.*

structure. Fig. 12C illustrates examples of nanodisk-in-nanorod LEDs emitting from violet to yellow region at a constant injection current of 500 nA, demonstrating that the bright emissions from single nanodisk-in-nanorod LEDs is monochromatic with narrow spectral widths ranging from 25 to 40 nm. The Airy diffraction disks observed in the high-magnification optical microscopy (OM) images clearly illustrate the point source nature for single nanodisk-in-nanorod LEDs.

To compare with the conventional blue LED results, Lu et al. measured the electroluminescence (EL) spectra with increasing injection current 100–500 nA for the blue (475 nm) nanodisk-in-nanorod diode, as shown in Fig. 12D. The relation of integrated EL intensity with respect to the injection current is shown in the inset. The EL intensity exhibits a monotonic increase with increasing injection current up to 500 nA. Using the known diameter 90 nm of the nanorod, the maximum current density without showing efficiency degradation can be directly estimated to be 8000 A/cm^2. In comparison, this current density is extremely high in comparison to typical polar InGaN/GaN planar multiple-quantum-well LEDs with ultrathin well thicknesses. It was suggested that the demonstration of high current density in single nanodisk-in-nanorod LEDs without efficiency degradation (efficient droop at high injection current) is made possible by the thick (40 nm) and strain-free InGaN nanodisk structure.

8. TUNABLE WHITE LEDs BASED ON DISK-IN-ROD NANOROD HETEROSTRUCTURES

The disk-in-rod device geometry also has the advantage for the realization of tunable white LEDs (Guo et al., 2011; Li and Waag, 2012; Lin et al., 2010; Nguyen et al., 2011, 2013). Fig. 13A shows the key structure features in the white nanorod-array LED design by Lin et al. (2010). The white-light emission is realized by a stack of nanodisk emitter ensembles embedded in the GaN nanorods *p–n* junction for the desired and tunable light mixing effects. For InGaN/GaN planar multiple quantum-well structures, several groups have already demonstrated white LEDs. In these white LED designs, the InGaN active layer thicknesses have been limited to about 2–4 nm, and most of the color mixing was achieved with dichromatic emitters. In contrast, Lin et al. take full advantage of the strain-free nanodisk structure to stack nanodisk ensembles with varying thicknesses 10–25 nm. Furthermore, the numbers (thicknesses) and positions of red/green/blue (RGB) nanodisks with respect to the *p*-type GaN region of nanorod are important to obtain the desired light mixing conditions for the white emission shown in Fig. 13B. This device design consideration is due to a limited hole injection length. Moreover, the thick InGaN nanodisks employed here provide large and tunable active RGB region volumes as well as improved carrier capture with reduced electron overflow at high injection currents.

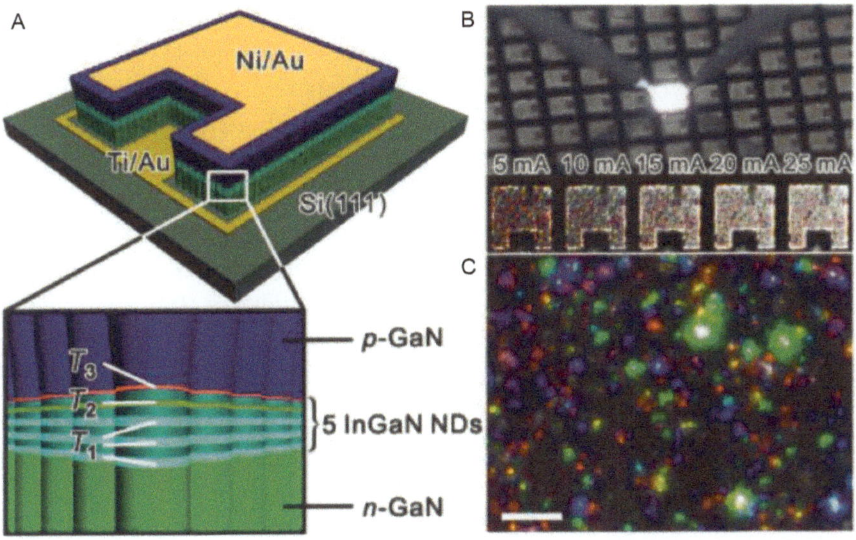

Fig. 13 Nanorods-array LED for white-light generation. (A) Schematic of device structure for white-light generation. The active regions contain multiple InGaN nanodisks. By controlling the number, thickness, and relative position of the constituent RGB nanodisks, white-light emission is realized by an ensemble of InGaN nanodisk stacks embedded in the GaN nanorod *p–n* junction. (B) OM image of nanorod-array LED emitting white light at 20 mA of injection current (chip size ~ 300 × 300 μm). Micro-EL images shown below were acquired under a 10 × objective lens. (C) μ-EL image at 20 mA under a 100 × objective lens reveal full-visible-spectrum emission from the white LED. *This figure is adapted with permission from Lin, H.-W., Lu, Y.-J., Chen, H.-Y., Lee, H.-M., Gwo, S., 2010. InGaN/GaN nanorod array white light-emitting diode. Appl. Phys. Lett. 97, 073101. Copyright AIP 2010.*

In the high-magnification OM image shown in Fig. 13C, spotty full-color emissions can be observed in detail. The low emission efficiency results from the fact that only a small number of nanorods are electrically injected by using a conventional *p*-type contact pad in this proof-of-concept experiment. There are several possible origins in low EL efficiency of InGaN nanodisks, including an imperfect *p*-type contact layer (both the *n*- and *p*-type ohmic contacts are Ti/Au metallic layers in this case) and the existence of stacking faults in thick InGaN nanodisks, as reported by Zhou et al. (2015). Recently, Musolino et al. (2014) have reported that indium tin oxide can be used as the *p*-type top contact with significantly lower turn-on voltage and a greatly improved extraction efficiency, about 10 times higher than the case using a conventional Ni/Au *p*-type contact. The full-color emissions of nanorod-array-based LEDs demonstrate a unique possibility to fabricate white LEDs with high color rendering capability. For the phosphor-based white LEDs, high color rendering would require a wide array of full-color phosphors, which has proven to be a difficult task because of the required temperature stability, quantum efficiencies, and chemical robustness. In the ensemble EL spectra measured for the InGaN/GaN nanorod-array LED, multiple ensemble peaks resulting from InGaN nanodisk arrays grown at different temperatures under constant III/N fluxes can be identified. From these EL spectra, it can be understood that the light mixing of emission bands causes the white-light emission. It is important to note that the observed emission bands (e.g., blue, green, yellow, red) correspond to the ensemble average behavior of $In_xGa_{1-x}N$ nanodisk emitters over a specific spectral range with the average alloy composition ($<x>$) controlled by the growth temperature under constant III/N fluxes, as shown in Fig. 11. For the nanorod-array LED shown in Fig. 13, EL emission bands show negligibly small spectral blue shifts with increasing current (5–25 mA). The small spectral blue shifts and linearly increased peak intensities with increasing current indicate insignificant polarization effects in the InGaN nanodisk emitters, which enable the drive-current-insensitive correlated color temperatures measured for the nanorod-array white LED. It was suggested that the residual small spectral shifts might result from a small spontaneous polarization mismatch between InGaN and GaN.

9. GREEN AND FULL-COLOR CORE–SHELL NANOROD PLASMONIC LASERS

The development of miniaturized nitride semiconductor lasers has continued to attract a great deal of attention in many scientific and technological research fields because of their transformative potentials for applications involving optical interconnects, ultrahigh-density data storage, advanced display technologies, biomedical imaging, molecule sensing, and so on (Hazari et al., 2015; Johnson et al., 2002; Li et al., 2015; Lu et al., 2012, 2014; Qian et al., 2008; Wu et al., 2011). Especially, a tremendous amount of research effort has been directed toward the development of new semiconductor lasers with ultrasmall size and low lasing threshold. Examples of these semiconductor lasers include vertical-cavity surface-emitting lasers, microcavity lasers, photonic-crystal laser, polariton lasers, semiconductor wire lasers, metal-cladded semiconductor microlasers, and plasmonic nanolasers, which exhibit unique and highly desirable laser features such as subwavelength, diffraction-unlimited, ultralow threshold, and ultrafast laser operation (Gwo and Shih, 2016).

Among these new types of semiconductor lasers, it has been proposed and demonstrated that metal-based nanoresonators (plasmonic nanocavities) can provide a path to break the diffraction limit of semiconductor laser size. This new type of diffraction-unlimited lasers was theoretically proposed by Bergman and Stockman (2003), and the physical mechanism was named as surface plasmon amplification by stimulated emission of radiation (SPASER or spaser for the device). Instead of relying on the feedback interaction between the photonic modes and the gain media, in a spaser the feedback occurs between the surface plasmons excited in a metal nanostructure and the adjacent optical gain media. The term "spaser" originally attributes to the feedback mechanism based on localized surface plasmons, but more recently, it has also been applied to plasmonic nanolasers based on propagating surface plasmon polaritons (SPPs). In both cases, the plasmon modes excited at metal surfaces in close proximity to semiconductor nanostructure gain media (e.g., semiconductor nanowires) can dramatically shrink the laser mode volume. Using noble metals such as silver (Ag) and gold (Au), the low-loss LPP and SPP modes can cover the visible and NIR spectral regions, allowing for schemes to confine the optical field into a deep subwavelength plasmonic nanocavity.

Furthermore, the realization of broadband tunable (in the visible wavelength range, all-color or full-color) semiconductor nanolasers with ultralow lasing thresholds can lead to transformative technological applications. This would require the availability of semiconductor gain media in a broad and tunable emission range. In the past two decades, tremendous research effort has been directed toward the development of novel semiconductor lasers with this capability. Without a doubt, the $In_xGa_{1-x}N$ alloy system is so far the most versatile and promising gain medium system for this purpose. Several full-color LED and laser devices have been demonstrated in the past few years based on the InGaN/GaN material system. In particular, this material system is very suitable for the development of electrically injected plasmonic nanolasers because of the well-developed growth and processing technologies for GaN-based nanorod *p–n* diodes.

Recently, Lu et al. (2012) achieved a milestone in InGaN-based plasmonic nanolaser. Especially, they succeeded in demonstrating the first continuous-wave (CW), low-threshold plasmonic nanolaser with a simulated 3D mode volume about three times the volume of the nanorod and exhibiting all of the necessary signatures for lasing. The basic device design is a metal–oxide–semiconductor (MOS) structure, using an atomically smooth epitaxial Ag film as the plasmonic material platform and a very short InGaN@GaN nanorod (total length of 470 nm and cross-section side length of 30 nm) as the gain medium (see Fig. 14), which enable lasing with an ultrashort gain length. The adoption of the new plasmonic material platform and the InGaN gain medium was the key to achieve this 3D diffraction-unlimited nanolaser.

The critical role of an atomically smooth, epitaxial Ag film can be qualitatively understood. The calculated energy-density distributions of the nanorod MOS structure for the lasing modes show that the SPP field is well confined in the low-permittivity oxide layer with an ultrasmall mode volume. Therefore, the SPP field is confined in thin dielectric (oxide) layer, which is typically around 5 nm in thickness. The conventional growth method using thermal evaporation typically yields granular films with surface roughness on the order of several nm, which can cause strong SPP scattering. In addition, grain boundaries also result in significant SPP scattering. Both of them contribute significantly to plasmonic losses. The usage of epitaxially grown atomically smooth Ag films minimizes these detrimental effects. A detailed quantitative analysis can be found in the online supporting information of Lu et al. (2012).

The usage of InGaN/GaN core–shell nanorods as the gain medium offers a large optical gain coefficient on the order of 10,000 cm^{-1} (which is about a factor of 5 larger than that of GaAs). Another important advantage of the core–shell nanorod gain medium is that it has a large modal gain because

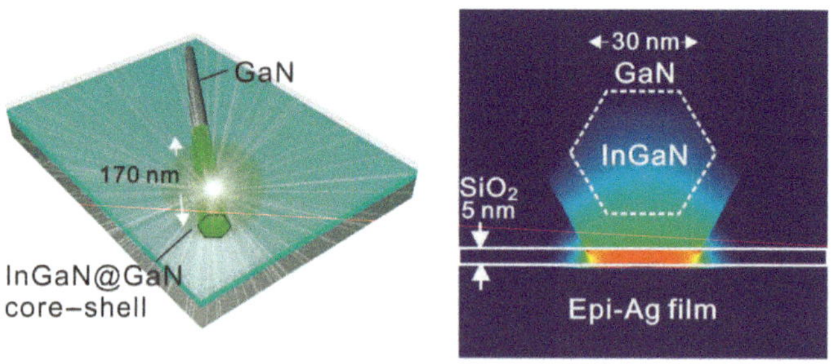

Fig. 14 A *green* plasmonic nanolaser structure, comprising of an epitaxial Ag film as the plasmonic material platform and a single InGaN@GaN core–shell nanorod as the gain medium. *This figure is adapted with permission from Lu, Y.-J., Kim, J., Chen, H.-Y., Wu, C., Dabidian, N., Sanders, C.E., Wang, C.-Y., Lu, M.-Y., Li, B.-H., Qiu, X., Chang, W.-H., Chen, L.-J., Shvets, G., Shih, C.-K., Gwo, S., 2012. Plasmonic nanolaser using epitaxially grown silver film. Science 337, 450–453. Copyright AAAS 2012.*

of the strong overlap between the gain medium and the lasing mode. Moreover, the core–shell nanorod geometry enables the incorporation of larger-In-content in the InGaN core despite the large lattice mismatch with the GaN shell. Therefore, the InGaN gain media can cover the whole visible spectral range for all-color laser emissions. As a laser device, this nanorod heterostructure geometry comes naturally with nonpolar side facets as the emission planes, which eliminate the adverse effects of the spatial separation of the electron and hole wave functions typically encountered in conventional polar quantum-well structures. The well-defined facets help maintain the spatial uniformity of the interface contact with the dielectric gap layer in the MIS structure, which confines the SPP field. As a consequence, the formation of smooth MIS interfaces can reduce the plasmonic loss and improve the cavity quality factor (Q).

Fig. 15 shows the InGaN/GaN core–shell nanorod gain media developed by Lu et al. (2014) for the development of semiconductor plasmonic nanolasers. This core–shell nanorod heterostructure comprises of an InGaN core, which is surrounding by a GaN shell (denoted by InGaN@GaN). These core–shell nanorods were grown on Si(111) wafers by PAMBE. Nanoscale structural and optical measurements using single nanorods confirmed both the presence of nonpolar (*m*-plane) side facets and a wide range of band gap tunability by controlling the In content. The PL spectra measured at room temperature show that emissions from individual nanorods continuously span from the blue to the red region of the visible spectrum (468–642 nm, see Fig. 15). The PL emission peaks are monochromatic and exhibit uniform intensity and linewidth, independent of the In content, across the visible spectrum range. The greatly improved PL properties in the long-wavelength (high-In-content) region are due to the absence of polar interfaces and surfaces. In the conventional wurtzite-type III-nitride heterostructures grown along the polar *c*-axis, the discontinuities of spontaneous and piezoelectric polarizations across the heterojunction induce large interface charges of opposite polarities, which result in a strong internal electric field and a greatly reduced carrier recombination efficiency. Thus, emissions from the nonpolar facets can have much larger gain coefficients, in comparison with that from the polar or semipolar III-nitride facets.

By using a CW excitation source at 405 nm (a commercial Blue-ray diode laser), the lasing characteristics were investigated by Lu and co-workers. As can be seen in Fig. 16, the emission spectra change from spontaneous emission at low excitation power (860 W/cm^2) to sharp lasing lines at high excitation power densities (4.7 kW/cm^2 and above). The lasing

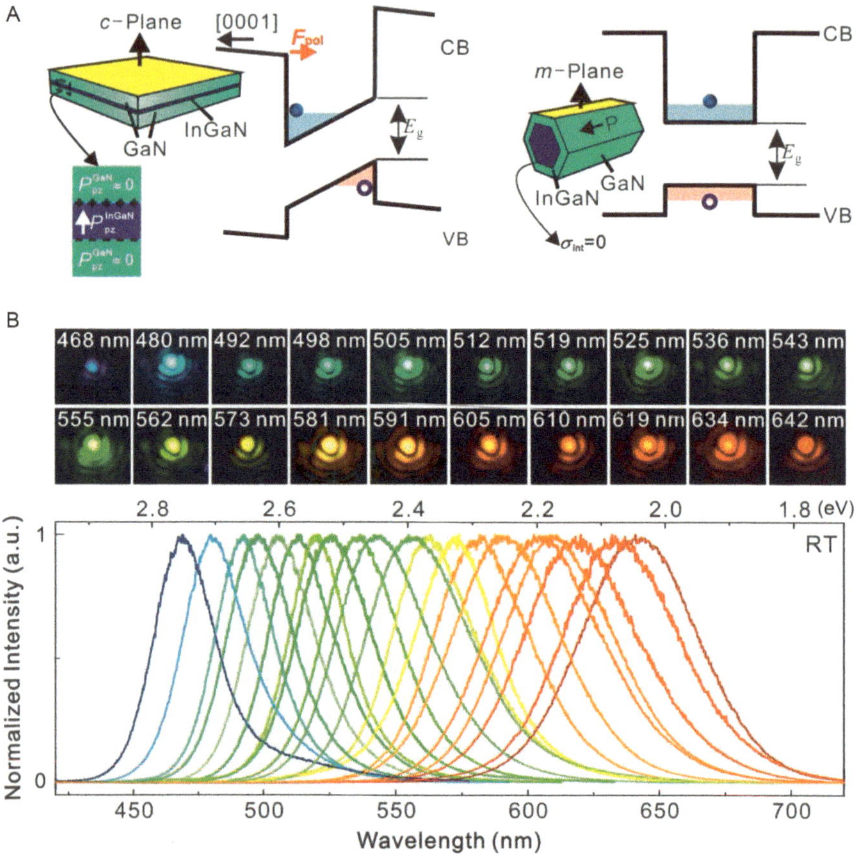

Fig. 15 (A) Schematic showing the adverse polarization effect on conventional polar InGaN/GaN quantum-well light-emitting devices, where F_{pol} is the electric field due to the presence of polarization-difference-induced interface charges. The polarization effect becomes more severe for high-In-content devices, resulting in poorer emission efficiency at long emission wavelengths. In contrast, the polarization effect is absent for the InGaN@GaN core–shell nanorod emitters with hexagonal nonpolar (*m*-plane) side facets. (B) Full-color PL images of the diffraction-limited light emissions from single InGaN@GaN core–shell nanorods dispersed on silicon substrates. These images were acquired at room temperature. Emission colors can be tuned from *blue* to *red* by incorporating different In contents in InGaN@GaN the nanorods. *This figure is adapted with permission from Lu, Y.-J., Wang, C.-Y., Kim, J., Chen, H.-Y., Lu, M.-Y., Chen, Y.-C., Chang, W.-H., Chen, L.-J., Stockman, M.I., Shih, C.-K., Gwo, S., 2014. All-color plasmonic nanolasers with ultralow thresholds: autotuning mechanism for single-mode lasing. Nano Lett. 14, 4381–4388. Copyright ACS 2014.*

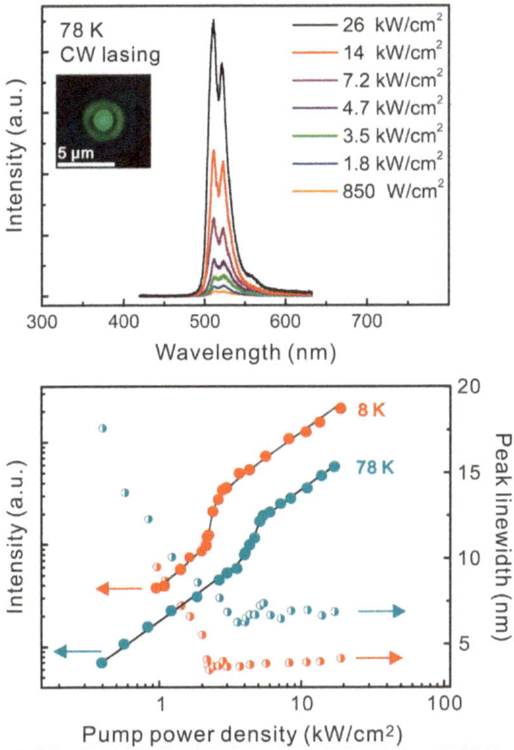

Fig. 16 Power-dependent lasing spectra of a single InGaN@GaN core–shell nanorod pumped by a CW 405 nm laser (*upper panel*). The *inset* shows the far-field laser spot with contrast fringes indicative of spatial coherence. Temperature-dependent lasing thresholds of the plasmonic cavity (*lower panel*). The log–log plots of output peak intensity vs the pump power at the main lasing peak (510 nm) are shown with the corresponding linewidth narrowing behavior when the plasmonic laser are measured at 8 K (*red*) and 78 K (*blue*). *This figure is adapted with permission from Lu, Y.-J., Kim, J., Chen, H.-Y., Wu, C., Dabidian, N., Sanders, C.E., Wang, C.-Y., Lu, M.-Y., Li, B.-H., Qiu, X., Chang, W.-H., Chen, L.-J., Shvets, G., Shih, C.-K., Gwo, S., 2012. Plasmonic nanolaser using epitaxially grown silver film. Science 337, 450–453. Copyright AAAS 2012.*

characteristics were identified by the clear "S"-shaped power dependence of the emitted intensity on the excitation power density on the log–log plot, accompanied by a clear signature of linewidth narrowing. The lasing threshold is identified as 3.7 kW/cm^2 at 78 K, and 2.1 kW/cm^2 at 8 K. Note that these threshold values are a few orders of magnitude lower than the previously achieved values using different gain media. Such a low threshold also allows for the first demonstration CW operation for plasmonic nanolasers.

In this study, the temporal coherence was also confirmed by measuring the second-order photon–photon correlation function (Lu et al., 2012).

Later on, Lu et al. (2014) demonstrated low-threshold, all-color plasmonic nanolasers on a single semiconductor platform using the MOS nanostructures fabricated by using Al_2O_3-covered epitaxial Ag films and InGaN@GaN semiconductor nanorods. It was shown that single-mode plasmonic nanolasers emitting across the full visible spectrum could be realized with ultralow lasing thresholds. The remarkable phenomenon of broadband tunable, single-mode lasing without the need of precise tuning of the resonator size was understood through theoretical simulations, which reveal a novel "autotuning" mechanism (Lu et al., 2014). This mechanism is consistent with the fact that propagating SPPs are not based on a resonant mechanism, which allow them to provide broadband confinement, and the emission wavelength is primarily determined by the gain medium.

As described earlier, the $In_xGa_{1-x}N$@GaN core–shell structure can accommodate a large-In composition range within the $In_xGa_{1-x}N$ core, spanning across the whole visible spectral range (Fig. 15). The lengths of nanorods measured for lasing characteristics range from 100 to 250 nm, and their diameter is about 30–50 nm. It is important to note that these nanorods dimensions are too small to sustain lasing conditions using these nanorods alone as the photonic resonators. Instead, lasing action is based on the presence of plasmonic resonators. The nanolaser device structure is exactly the same as the one used previously as shown schematically in Fig. 14. In comparison to the previous green plasmonic nanolaser device (Lu et al., 2012), two additional improvements were made. First of all, shorter nanorods were used, which lead to single-mode lasing. Second, instead of using SiO_2, an Al_2O_3 dielectric gap layer grown by atomic layer deposition was used. This yields a high-quality dielectric layer with precisely defined thickness, excellent conformity to the deposited surface, and high uniformity over a large area.

The individual nanorods to be measured were dispersed onto the substrate by drop casting diluted suspension solutions of nanorods. The SEM image showed spatially isolated single nanorods with an average spacing much larger than the optical detection area of localized laser emissions from single nanorods. Fig. 17 shows the emissions of single-mode plasmonic nanolasers in the full visible spectrum. OM images show single-mode blue, cyan, green, yellow, orange, and red laser emissions. The Airy disk diffraction patterns indicate the physical sizes of the emitters are on the

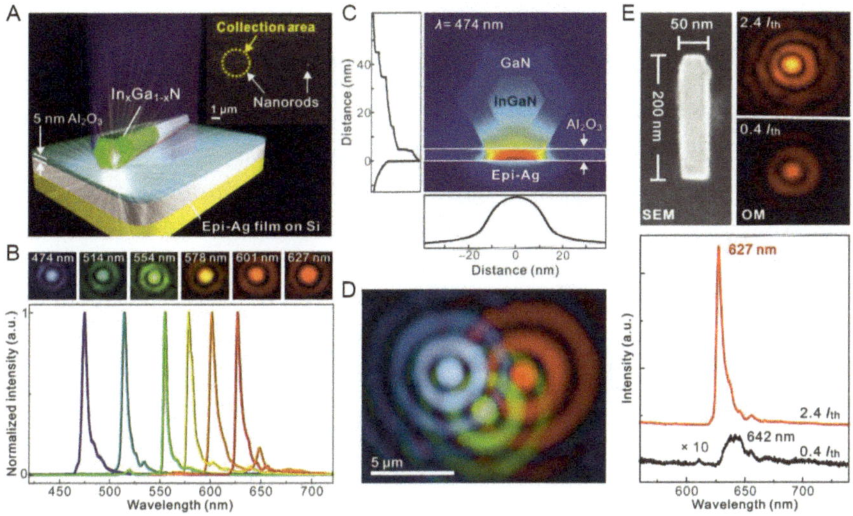

Fig. 17 Continuously tunable, all-color operation of nitride semiconductor plasmonic nanolasers using single InGaN@GaN core–shell nanorods with varied In compositions as the gain media. *This figure is adapted with permission from Lu, Y.-J., Wang, C.-Y., Kim, J., Chen, H.-Y., Lu, M.-Y., Chen, Y.-C., Chang, W.-H., Chen, L.-J., Stockman, M.I., Shih, C.-K., Gwo, S., 2014. All-color plasmonic nanolasers with ultralow thresholds: autotuning mechanism for single-mode lasing. Nano Lett. 14, 4381–4388. Copyright ACS 2014.*

subwavelength scale. These single-mode nanolasers have also been confirmed to have highly polarized (nearly 100%) emission along the nanorod axis. To prepare the sample that can emit RGB colors concurrently, three diluted suspension solutions of nanorods with different indium contents were sequentially applied to deposit RGB nanorods within close proximity of each other.

In this work, it has been clearly demonstrated that the single-mode emissions from these plasmonic nanolasers indeed cover the full visible spectral range without precisely tuning the cavity length. The lasing wavelength is primarily determined by the InGaN gain medium (In composition). This is a remarkable behavior, dramatically different from that of semiconductor lasers based on conventional photonic cavities. In semiconductor lasers such as vertical-cavity surface-emitting lasers, for each individual color, the emitting wavelength of gain medium needs to be specifically prepared and the cavity length needs to be fabricated individually for spectral matching with the gain medium and the desire emission wavelength. It is, therefore, important to have a more detailed theoretical study in the future to understand the detailed physical mechanism of plasmonic cavity-based nanolasers.

10. CONCLUSIONS AND OUTLOOK

In this chapter, we have discussed the advantages of nitride semiconductor nanorod/nanowire heterostructures as the important material platform for full-color and white-light sources. Some examples of full-color and white LEDs as well as nanorod lasers have been shown for both axial and radial InGaN/GaN nanorod heterostructures. In addition, the fundamental properties of nitride semiconductor nanorods/nanowires, including growth, polarity control, doping, surface properties, heterojunction band alignments, as well as some structural and optical properties have been presented. It is clear that this material platform is very promising for solid-state lighting and display applications. However, more studies are needed in order to facilitate both fundamental understanding and further device development.

For the fundamental studies, despite many studies on structural, electronic, and optical properties of nitride semiconductor nanorods and nanowires (Kioseoglou et al., 2015; Krause et al., 2016; Tourbot et al., 2011), more detailed studies for correlative structure–property–performance relationships on a single-nanostructure level (Lähnemann et al., 2011; Ra et al., 2014; Rigutti et al., 2014; Shahmohammadi et al., 2016) are required for a full understanding of their unique physical properties. For example, the presence or absence of strain in nanorod/nanowire heterostructures is extremely important for understanding their optical properties (Wölz et al., 2013). However, to directly compare with their optical properties, the exact strain state has to be confirmed on a single-nanostructure level. Recent advances in electron microscopy and synchrotron-based X-ray nanoprobe techniques can help shed important light on this issue (Martínez-Criado et al., 2014; Rigutti et al., 2014; Woo et al., 2015b). Moreover, microstructural defects such as stacking faults in nanowires have also been found to be important to understand their optical properties (Corfdir et al., 2014). Recently, the correlation of emission quenching (nonradiative carrier recombination) with stacking faults in thick InGaN nanodisks reported by Zhou et al. (2015) indicates that reducing this type of structural defect can lead to an improve emission efficiency for nanorod-based optoelectronic devices.

Regarding further device developments for improved performance, the emergence of a hybrid growth method using both PAMBE and MOVPE techniques (Albert et al., 2015) is a useful approach to take advantages of strengths in both approaches. Furthermore, flexible LEDs based on vertical nitride nanowires (Dai et al., 2015; Guan et al., 2016; Lee et al., 2011) using polymer-based transfer techniques or transferable epilayers grown on two-dimensional materials such as graphene (Chung et al., 2010) are interesting developments for nanorod/nanowire-based LEDs, which can broaden the application areas for lighting and display devices. Recently, monolithic integrated polychromatic (RGB) nanowire LED arrays (Wang et al., 2014) have been achieved by using a multistep selective-area PAMBE growth process. This is an important step for the realization of an advanced display device capable of controllable and tunable full-color light generation. We expect that multistep, SAG will become an enabling technique for nanorod/nanowire-array-based display devices in the future. Other device improvements to increase extraction efficiency (Musolino et al., 2014) and new injection schemes using alternating-current InGaN/GaN tunnel junctions (Sadaf et al., 2015, 2016) have also been devised. Moreover, integrated photonic systems using nitride semiconductor nanowires have been demonstrated (Tchernycheva et al., 2014). These are interesting directions to pursue further in the future.

In summary, the emergence of optoelectronic device applications based on nitride semiconductor nanorods and nanowires has brought a great deal of research interest in recent years because of the huge impact they can make in solid-state lighting and display technologies. The future in this exciting research area would rely on new device concepts, better material properties through efforts in improved growth techniques, innovative fabrication processes, and last but not least the fundamental understanding of their physical properties.

ACKNOWLEDGMENTS

The authors would like to acknowledge the financial supports from the Ministry of Science and Technology (MOST), National Science Council (NSC), and the Ministry of Education (MOE) in Taiwan for the nitride semiconductor research conducted in National Tsing-Hua University.

REFERENCES

Albert, S., Bengoechea-Encabo, A., Kong, X., Sanchez-Garcia, M.A., Calleja, E., Trampert, A., 2013. Monolithic integration of InGaN segments emitting in the blue, green, and red spectral range in single ordered nanocolumns. Appl. Phys. Lett. 102, 181103.

Albert, S., Bengoechea-Encabo, A., Ledig, J., Schimpke, T., Sánchez-Garcia, M.A., Strassburg, M., Waag, A., Calleja, E., 2015. Demonstration of (In, Ga)N/GaN core–shell micro light-emitting diodes grown by molecular beam epitaxy on ordered MOVPE GaN pillars. Cryst. Growth Des. 15, 3661–3665.

Amano, H., Kito, M., Hiramatsu, K., Akasaka, I., 1989. P-Type conduction in Mg-doped GaN treated with low-energy electron beam irradiation (LEEBI). Jpn. J. Appl. Phys. 28, L2112–L2114.

Amitage, R., Tsubaki, K., 2010. Multicolor luminescence from InGaN quantum wells grown over GaN nanowire arrays by molecular-beam epitaxy. Nanotechnology 21, 195202.

Bardoux, R., Kaneta, A., Funato, M., Kawakami, Y., Kikuchi, A., Kishino, K., 2009. Positive binding energy of a biexciton confined in a localized center formed in a single $In_xGa_{1-x}N/GaN$ quantum disk. Phys. Rev. B 79, 155307.

Bengoechea-Encabo, A., Barbagini, F., Fernandez-Garrido, S., Grandal, J., Ristic, J., Sanchez-Garcia, M.A., Caleeja, E., Jahn, J., Luna, E., Trampert, A., 2011. Understanding the selective area growth of GaN nanocolumns by MBE using Ti nanomasks. J. Cryst. Grow, 325–329.

Bergman, D.J., Stockman, M.I., 2003. Surface plasmon amplification by stimulated emission of radiation: quantum generation of coherent surface plasmons in nanosystems. Phys. Rev. Lett. 90, 027402.

Bernardini, F., Fiorentini, V., Vanderbilt, D., 1997. Spontaneous polarization and piezoelectric constants of III–V nitrides. Phys. Rev. B 56, R10024–R10027.

Bernardini, F., Fiorentini, V., Vanderbilt, D., 2001. Accurate calculation of polarization-related quantities in semiconductors. Phys. Rev. B 63, 193201.

Bertness, K.A., Sanford, N.A., Davydov, A.V., 2011. GaN nanowires grown by molecular beam epitaxy. IEEE. J. Sel. Top. Quantum Electron. 17, 847–858.

Brubaker, M.D., Duff, S.M., Harvey, T.E., Blanchard, P.T., Roshko, A., Sanders, A.W., Sanford, N.A., Bertness, K.A., 2015. Polarity-controlled GaN/AlN nucleation layers for selective-area growth of GaN nanowire arrays on Si(111) substrates by molecular-beam epitaxy. Cryst. Growth Des. 16, 596–604.

Calleja, E., Sánchez-García, M.A., Sánchez, F.J., Calle, F., Naranjo, F.B., Muñoz, E., Jahn, U., Ploog, K., 2000. Luminescence properties and defects in GaN nanocolumns grown by molecular beam epitaxy. Phys. Rev. B 62, 16826–16834.

Calleja, E., Ristić, J., Fernández-Garrido, S., Cerutti, L., Sánchez-García, M.A., Grandal, J., Trampert, A., Jahn, U., Sánchez, G., Griol, A., Sánchez, B., 2007. Growth, morphology, and structural properties of group-III-nitride nanocolumns and nanodisks. Phys. Stat. Sol. B 244, 2816–2837.

Chang, Y.-H., Chang, K.-K., Gwo, S., Yeh, J.A., 2010. Highly sensitive hydrogen detection using a Pt-catalyzed InN epilayer. Appl. Phys. Express 3, 114101.

Chang, Y.-H., Lu, Y.-S., Hong, Y.-L., Gwo, S., Yeh, J.A., 2011. Highly sensitive pH sensing using an indium nitride ion-sensitive field-effect transistor. IEEE Sens. J. 11, 1157–1161.

Chen, C.-C., Yeh, C.-C., Chen, C.-H., Yu, M.-Y., Liu, H.-L., Wu, J.-J., Chen, K.-H., Chen, L.-C., Peng, J.-Y., Chen, Y.-F., 2001. Catalytic growth and characterization of gallium nitride nanowires. J. Am. Chem. Soc. 123, 2791–2798.

Chen, H.-Y., Lin, H.-W., Shen, C.-H., Gwo, S., 2006. Structure and photoluminescence properties of epitaxially oriented GaN nanorods grown on Si(111) by plasma-assisted molecular-beam epitaxy. Appl. Phys. Lett. 89, 243105.

Chen, H.-Y., Lin, H.-W., Wu, C.-Y., Chen, W.-C., Chen, J.-S., Gwo, S., 2008a. Gallium nitride nanorod arrays as low-refractive-index transparent media in the entire visible spectral region. Opt. Express 16, 8106–8116.

Chen, H.-Y., Yang, Y.-C., Lin, H.-W., Chang, S.-C., Gwo, S., 2008b. Polarized photoluminescence from single GaN nanorods: effects of optical confinement. Opt. Express 6, 13465–13475.

Chung, K., Lee, C.-H., Yi, G.-C., 2010. Transferable GaN layers grown on ZnO-coated graphene layers for optoelectronic devices. Science 330, 655–657.

Colby, R., Liang, Z., Wildeson, I.H., Ewoldt, D.A., Sands, T.D., Garcia, R.E., Stach, E.A., 2010. Dislocation filtering in GaN nanostructures. Nano Lett. 10, 1568–1573.

Corfdir, P., Hauswald, C., Zettler, J.K., Flissikowski, T., Lähnemann, J., Fernández-Garrido, S., Geelhaar, L., Grahn, H.T., Brandt, O., 2014. Stacking faults as quantum wells in nanowires: density of states, oscillator strength, and radiative efficiency. Phys. Rev. B 90, 195309.

Dai, X., Messanvi, A., Zhang, H., Durand, C., Eymery, J., Bougerol, C., Julien, F.H., Tchernycheva, M., 2015. Flexible light-emitting diodes based on vertical nitride nanowires. Nano Lett. 15, 6958–6964.

Davydov, V.Yu., Klochikhin, A.A., Seisyan, R.P., Emtsev, V.V., Ivanov, S.V., Bechstedt, F., Furthmüller, J., Harima, H., Mudryi, A.V., Aderhold, J., Semchinova, O., Graul, J., 2002. Absorption and emission of hexagonal InN. Evidence of narrow fundamental band gap. Phys. Stat. Sol. B 229, r1–r3.

Dong, Y., Tian, B., Kempa, T.J., Lieber, C.M., 2009. Coaxial group III-nitride nanowire photovoltaics. Nano Lett. 9, 2183–2187.

Ebert, Ph., Schaafhausen, S., Lenz, A., Sabitova, A., Ivanova, L., Dähne, M., Hong, Y.-L., Gwo, S., Eisele, H., 2011. Direct measurement of the band gap and Fermi level position at InN(1120). Appl. Phys. Lett. 98, 062103.

Fernández-Garrido, S., Kong, X., Gotschke, T., Calarco, R., Geelhaar, L., Trampert, A., Brandt, O., 2012. Spontaneous nucleation and growth of GaN nanowires: the fundamental role of crystal polarity. Nano Lett. 12, 6119–6125.

Fu, A., Gao, H., Petrov, P., Yang, P., 2015. Widely tunable distributed Bragg reflectors integrated into nanowire waveguide. Nano Lett. 15, 6909–6913.

Guan, N., Dai, X., Messanvi, A., Zhang, H., Yan, J., Gautier, E., Bougerol, C., Julien, F.H., Durand, C., Eymery, J., Tchernycheva, M., 2016. Flexible white light emitting diodes based on nitride nanowires and nanophosphors. ACS Photonics 3, 597–603.

Guo, W., Zhang, M., Banerjee, A., Bhattacharya, P., 2010. Catalyst-free InGaN/GaN nanowire light emitting diodes grown on (001) silicon by molecular beam epitaxy. Nano Lett. 10, 3355–3359.

Guo, W., Banerjee, A., Bhattacharya, P., Ooi, B.S., 2011. InGaN/GaN disk-in-nanowire white light emitting diodes on (001) silicon. Appl. Phys. Lett. 98, 193102.

Gwo, S., Shih, C.-K., 2016. Semiconductor plasmonic nanolasers: current status and perspectives. Rep. Prog. Phys. 79, 086501.

Gwo, S., Wu, C.-L., Shen, C.-H., Chang, W.-H., Hsu, T.M., Wang, J.-S., Hsu, J.-T., 2004. Heteroepitaxial growth of wurtzite InN films on Si(111) exhibiting strong near-infrared photoluminescence at room temperature. Appl. Phys. Lett. 84, 3765–3767.

Hazari, A., Aiello, A., Ng, T.K., Ooi, B.S., Bhattacharya, P., 2015. III-Nitride disk-in-nanowire 1.2 μm monolithic diode laser on (001) silicon. Appl. Phys. Lett. 107, 191107.

Hersee, S.D., Sun, X., Wang, X., 2006. The controlled growth of GaN nanowires. Nano Lett. 6, 1808–1811.

Hong, C.-C., Ahn, H., Wu, C.-Y., Gwo, S., 2009. Strong green photoluminescence from $In_xGa_{1-x}N$/GaN nanorod arrays. Opt. Express 17, 17227–17233.

Hong, Y.J., Lee, C.-H., Yoon, A., Kim, M., Seong, H.-K., Chung, H.J., Sone, C., Park, Y.J., Yi, G.-C., 2011. Visible-color-tunable light-emitting diodes. Adv. Mater. 23, 3284–3288.

Hong, Y.J., Lee, C.-H., Yoo, J., Kim, Y.-J., Jeong, J., Kim, M., Yi, G.-C., 2015. Emission color-tuned light-emitting diode microarrays of nonpolar $In_xGa_{1-x}N$/GaN multishell nanotube heterostructures. Sci. Rep. 5, 18020.

Johnson, J.C., Choi, H.-J., Knutsen, K.P., Schaller, R.D., Yang, P., Saykally, R.J., 2002. Single gallium nitride nanowire lasers. Nat. Mater. 1, 106–110.

Kang, M.S., Lee, C.-H., Park, J.B., 2012. Gallium nitride nanostructures for light-emitting diode applications. Nano Energy 44, 391–400.

Kano, T., Yoshida, R., Miyagawa, R., Mizuno, Y., Oto, T., Kishino, K., 2015. GaN nanocolumn arrays with diameter <30 nm prepared by two-step selective area growth. Electron. Lett. 51, 2125–2126.

Kao, K.-W., Hsu, M.-C., Chang, Y.-H., Gwo, S., Yeh, J.A., 2012. A sub-ppm acetone gas sensor for diabetes detection using 10 nm thick ultrathin InN FETs. Sensors 12, 7157–7168.

Kawakami, Y., Suzuki, S., Kaneta, A., Funato, M., Kikuchi, A., Kishino, K., 2006. Origin of high oscillator strength in green-emitting InGaN/GaN nanocolumns. Appl. Phys. Lett. 89, 163124.

Kikuchi, A., Kawai, M., Tada, M., Kishino, K., 2004. InGaN/GaN multiple quantum disk nanocolumn light-emitting diodes grown on (111) Si substrate. Jap. J. Appl. Phys. 43, L1524–L1526.

Kim, H.-M., Cho, Y.-H., Lee, H., Kim, S.I., Ryu, S.R., Kim, D.Y., Kang, T.W., Chung, K.S., 2004. High-brightness light emitting diodes using dislocation-free indium gallium nitride/gallium nitride multiquantum-well nanorods arrays. Nano Lett. 4, 1059–1062.

Kioseoglou, J., Pavloudis, Th., Kehagias, Th., Komninou, Ph., Karakostas, Th., Latham, C.D., Rayson, M.J., Briddon, P.R., Eickhoff, M., 2015. Structural and electronic properties of GaN nanowires with embedded $In_xGa_{1-x}N$ nanodisks. J. Appl. Phys. 118, 034301.

Kishino, K., Ishizara, S., 2015. Selective-area growth of GaN nanocolumns on Si(111) substrates for application to nanocolumn emitters with systematic analysis of dislocation filtering effect of nanocolumns. Nanotechnology 26, 225602.

Kishino, K., Kikuchi, A., Sekiguchi, H., Ishizawa, S., 2007. InGaN/GaN nanocolumn LEDs emitting from blue to red. In: Proc. SPIE 6473, 64730T.

Kishino, K., Sekiguchi, H., Kikuchi, A., 2009. Improved Ti-mask selective-area growth (SAG) by rf-plasma-assisted molecular beam epitaxy demonstrating extremely uniform GaN nanocolumn arrays. J. Cryst. Grow. 311, 2063–2068.

Kishino, K., Yanagihara, A., Ikeda, K., Yamano, K., 2015. Monolithic integration of four-colour InGaN-based nanocolumn LEDs. Electron. Lett. 51, 852–854.

Koester, R., Hwang, J.-S., Salomon, D., Chen, X., Bougerol, C., Barnes, J.-P., Dang, D.L.S., Rigutti, L., de Luna Bugallo, A., Jacopin, G., Tchernycheva, M., Durand, C., Eymery, J., 2011. M-plane core–shell InGaN/GaN multiple-quantum-wells on GaN wires for electroluminescent devices. Nano Lett. 11, 4839–4845.

Krause, T., Hanke, M., Brandt, O., Trampert, A., 2016. Counterintuitive strain distribution in axial (In, Ga)N/GaN nanowires. Appl. Phys. Lett. 108, 032103.

Kuo, C.-T., Lee, H.-M., Shiu, H.-W., Chen, C.-H., Gwo, S., 2009. Direct imaging of GaN *p–n* junction by cross-sectional scanning photoelectron microscopy and spectroscopy. Appl. Phys. Lett. 94, 122110.

Kuo, C.-T., Lin, S.-C., Chang, K.-K., Shiu, H.-W., Chang, L.-Y., Chen, C.-H., Tang, S.-J., Gwo, S., 2011a. Is electron accumulation universal at InN polar surfaces? Appl. Phys. Lett. 98, 052101.

Kuo, C.-T., Chang, K.-K., Shiu, H.-W., Lin, S.-C., Chen, C.-H., Gwo, S., 2011b. Spontaneous-polarization-induced heterojunction asymmetry in III-nitride semiconductors. Appl. Phys. Lett. 99, 022113.

Kuo, C.-T., Chang, K.-K., Shiu, H.-W., Liu, C.-R., Chang, L.-Y., Chen, C.-H., Gwo, S., 2011c. Natural band alignments of InN/GaN/AlN nanorod heterojunctions. Appl. Phys. Lett. 99, 122101.

Kuykendall, T., Pauzauskie, P.J., Zhang, Y., Goldberger, J., Sirbuly, D., Denlinger, J., Yang, P., 2004. Crystallographic alignment of high-density gallium nitride nanowire arrays. Nat. Mater. 3, 524–528.

Kuykendall, T., Ulrich, P., Aloni, S., Yang, P., 2007. Complete composition tunability of InGaN nanowires using a combinatorial approach. Nat. Mater. 6, 951–956.

Kuykendall, T.R., Schwartzberg, A.M., Aloni, S., 2015. Gallium nitride nanowires and heterostructures: toward color-tunable and white-light sources. Adv. Mater. 27, 5805–5812.

Lähnemann, J., Brandt, O., Pfüller, C., Flissikowski, T., Jahn, U., Luna, E., Hanke, M., Knelangen, M., Trampert, A., Grahn, H.T., 2011. Coexistence of quantum-confined Stark effect and localized states in an (In, Ga)N/GaN nanowire heterostructure. Phys. Rev. B 84, 155303.

Lee, C.-H., Kim, Y.-J., Hong, Y.J., Jeon, S.-R., Bae, S., Hong, B.H., Yi, G.-C., 2011. Flexible inorganic nanostructure light-emitting diodes fabricated on graphene films. Adv. Mater. 23, 4614–4619.

Li, S., Waag, A., 2012. GaN based nanorods for solid state lighting. J. Appl. Phys. 111, 071101.

Li, K.H., Liu, X., Wang, Q., Zhao, S., Mi, Z., 2015. Ultralow-threshold electrically injected AlGaN nanowire ultraviolet lasers on Si operating at low temperature. Nat. Nanotechnol. 10, 140–144.

Lin, H.-W., Lu, Y.-J., Chen, H.-Y., Lee, H.-M., Gwo, S., 2010. InGaN/GaN nanorod array white light-emitting diode. Appl. Phys. Lett. 97, 073101.

Lu, Y.-S., Huang, C.-C., Yeh, J.A., Chen, C.-F., Gwo, S., 2007. InN-based anion selective sensors in aqueous solutions. Appl. Phys. Lett. 91, 202109.

Lu, Y.-S., Ho, C.-L., Yeh, J.A., Lin, H.-W., Gwo, S., 2008. Anion detection using ultrathin InN ion selective field effect transistors. Appl. Phys. Lett. 92, 212102.

Lu, Y.-J., Lin, H.-W., Chen, H.-Y., Yang, Y.-C., Gwo, S., 2011. Single InGaN nanodisk light emitting diodes as full-color subwavelength light sources. Appl. Phys. Lett. 98, 233101.

Lu, Y.-J., Kim, J., Chen, H.-Y., Wu, C., Dabidian, N., Sanders, C.E., Wang, C.-Y., Lu, M.-Y., Li, B.-H., Qiu, X., Chang, W.-H., Chen, L.-J., Shvets, G., Shih, C.-K., Gwo, S., 2012. Plasmonic nanolaser using epitaxially grown silver film. Science 337, 450–453.

Lu, Y.-J., Lu, M.-Y., Yang, Y.-C., Chen, H.-Y., Chen, L.-J., Gwo, S., 2013. Dynamic visualization of axial *p–n* junctions in single gallium nitride nanorods under electrical bias. ACS Nano 7, 7640–7647.

Lu, Y.-J., Wang, C.-Y., Kim, J., Chen, H.-Y., Lu, M.-Y., Chen, Y.-C., Chang, W.-H., Chen, L.-J., Stockman, M.I., Shih, C.-K., Gwo, S., 2014. All-color plasmonic nanolasers with ultralow thresholds: autotuning mechanism for single-mode lasing. Nano Lett. 14, 4381–4388.

Lupan, O., Pauporté, T., Viana, B., 2010. Low-voltage UV-electroluminescence from ZnO-nanowire array/p-GaN light-emitting diodes. Adv. Mater. 22, 3298–3302.

Martin, G., Botchkarev, A., Rockett, A., Morkoç, H., 1996. Valence-band discontinuities of wurtzite GaN, AlN, and InN heterojunctions measured by x-ray photoemission spectroscopy. Appl. Phys. Lett. 68, 2541–2543.

Martínez-Criado, C., Segura-Ruiz, J., Alén, B., Eymery, J., Rogalev, A., Tucoulou, R., Homs, A., 2014. Exploring single semiconductor nanowires with a multimodal hard X-ray nanoprobe. Adv. Mater. 26, 7873–7879.

Moses, P.G., Miao, M., Yan, Q., Van de Walle, C.G., 2011. Hybrid functional investigations of band gaps and band alignments for AlN, GaN, InN, and InGaN. J. Chem. Phys. 134, 084703.

Musolino, M., Tahraoui, A., Limbach, F., Lähnemann, J., Jahn, U., Brandt, O., Geelhaar, L., Riechert, H., 2014. Understanding peculiarities in the optoelectronic characteristics of light emitting diodes based on (In, Ga)N/GaN nanowires. Appl. Phys. Lett. 105, 083505.

Nakamura, S., 1998. The roles of structural imperfections in InGaN-based blue light-emitting diodes and laser diodes. Science 281, 956–961.

Nakamura, S., Mukai, T., Senoh, M., 1994. Candela-class high-brightness InGaN/AlGaN double-heterostructure blue-light-emitting diodes. Appl. Phys. Lett. 64, 1687–1689.

Nakamura, S.S., Pearton, S.J., Fasol, G., 2000. The Blue Laser Diode: The Complete Story, second ed. Springer-Verlag, Berlin.

Ng, A.M.C., Xi, Y.Y., Hsu, Y.F., Djurišić, A.B., Chan, W.K., Gwo, S., Tam, H.L., Cheah, K.W., Fong, P.W.K., Lui, H.F., Surya, C., 2009. GaN/ZnO nanorod light emitting diodes with different emission spectra. Nanotechnology 20, 445201.

Nguyen, H.P.T., Zhang, S., Cui, K., Han, X., Fathololoumi, S., Couillard, M., Botton, G.A., Mi, Z., 2011. p-Type modulation doped InGaN/GaN dot-in-a-wire white-light-emitting diodes monolithically grown on Si(111). Nano Lett. 11, 1919–1924.

Nguyen, H.P.T., Zhang, S., Connie, A.T., Kibria, M.D., Wang, Q., Shih, I., Mi, Z., 2013. Breaking the carrier injection bottleneck of phosphor-free nanowire white light-emitting diodes. Nano Lett. 13, 5437–5442.

Pimputkar, S., Speck, J.S., Denbaars, S.P., Nakamura, S., 2009. Prospects for LED lighting. Nat. Photonics 3, 180–182.

Qian, F., Li, Y., Gradečak, S., Wang, D., Barrelet, C.J., Lieber, C.M., 2004. Gallium nitride-based nanowire radial heterostructures for nanophotonics. Nano Lett. 4, 1975–1979.

Qian, F., Gradečak, S., Li, Y., Wen, C.-Y., Lieber, C.M., 2005. Core/multishell nanowire heterostructures as multicolor, high-efficiency light-emitting diodes. Nano Lett. 5, 2287–2291.

Qian, F., Li, Y., Gradečak, S., Park, H.-G., Dong, Y., Ding, Y., Wang, Z.L., Lieber, C.M., 2008. Multi-quantum-well nanowire heterostructures for wavelength-controlled lasers. Nat. Mater. 7, 701–706.

Ra, Y.-H., Navamathavan, R., Yoo, H.-I., Lee, C.-R., 2014. Single nanowire light-emitting diodes using uniaxial and coaxial InGaN/GaN multiple quantum wells synthesized by metalorganic chemical vapor deposition. Nano Lett. 14, 1537–1545.

Rigutti, L., Blum, I., Shinde, D., Hernández-Maldonado, D., Lefebvre, W., Houard, J., Vurpillot, F., Vella, A., Tchernycheva, M., Durand, C., Eymery, J., Deconihout, B., 2014. Correlation of microphotoluminescence spectroscopy, scanning transmission electron microscopy, and atom probe tomography on a single nano-object containing an InGaN/GaN multiquantum well system. Nano Lett. 14, 107–114.

Sadaf, S.M., Ra, Y.-H., Nguyen, H.P.T., Djavid, M., Mi, Z., 2015. Alternating-current InGaN/GaN tunnel junction nanowire white-light emitting diodes. Nano Lett. 15, 6696–6701.

Sadaf, S.M., Ra, Y.H., Szkopek, T., Mi, Z., 2016. Monolithically integrated metal/semiconductor tunnel junction nanowire light-emitting diodes. Nano Lett. 16, 1076–1080.

Schubert, E.F., 2005. Solid-state light sources getting smart. Science 308, 1274–1278.

Schubert, E.F., 2006. Light-Emitting Diodes, second ed. Cambridge University Press, Cambridge.

Sekiguchi, H., Kishino, K., Kikuchi, A., 2010. Emission color control from blue to red with nanocolumn diameter of InGaN/GaN nanocolumn arrays. Appl. Phys. Lett. 96, 231104.

Shahmohammadi, M., Ganiere, J.-D., Zhang, H., Ciechonski, R., Vescovi, G., Kryliouk, O., Tchernycheva, M., Jacopin, G., 2016. Excitonic diffusion in InGaN/GaN core-shell nanowires. Nano Lett. 16, 243–249.

Shen, C.-H., Chen, H.-Y., Lin, H.-W., Gwo, S., Klochikhin, A.A., Davydov, V.Yu., 2006. Near-infrared photoluminescence from vertical InN nanorod arrays grown on silicon: effects of surface electron accumulation layer. Appl. Phys. Lett. 88, 253104.

Takeuchi, T., Sota, S., Katsuragawa, S., Komori, M., Takeuchi, H., Amano, H., Akasaki, I., 1997. Quantum-confined start effect due to piezoelectric fields in GaInN strained quantum wells. Jpn. J. Appl. Phys. 36, L382–L385.

Tang, Y.B., Chen, Z.H., Song, H.S., Lee, C.S., Cong, H.T., Cheng, H.M., Zhang, W.J., Bello, I., Lee, S.T., 2008. Vertically aligned p-type single-crystalline GaN nanorod arrays on n-type Si for heterojunction photovoltaic cells. Nano Lett. 8, 4191–4195.

Tchernycheva, M., Messanvi, A., de Luna Bugallo, A., Jacopin, G., Lavenus, P., Rigutti, L., Zhang, H., Halioua, Y., Julien, F.H., Eymery, J., 2014. Integrated photonic platform based on InGaN/GaN nanowire emitters and detectors. Nano Lett. 14, 3515–3520.

Tchoulfian, P., Donatini, F., Levy, F., Dussaigne, A., Ferret, P., Pernot, J., 2014. Direct imaging of p–n junctions in core–shell GaN wires. Nano Lett. 14, 3491–3498.

Tourbot, G., Bougerol, C., Grenier, A., Den Hertog, M., Sam-Giao, D., Cooper, D., Gilet, P., Gayral, B., Daudin, B., 2011. Structural and optical properties of InGaN/GaN nanowire heterostructures grown by PA-MBE. Nanotechnology 22, 075601.

Tu, L.W., Hsiao, C.L., Chi, T.W., Lo, I., Hsieh, K.Y., 2003. Self-assembled vertical GaN nanorods grown by molecular-beam epitaxy. Appl. Phys. Lett. 82, 1601–1603.

Van de Walle, C.G., Segev, D., 2007. Microscopic origins of surface states on nitride surfaces. J. Appl. Phys. 101, 081704.

Waltereit, P., Brandt, O., Trampert, A., Grahn, H.T., Menniger, J., Ramsteiner, M., Reiche, M., Ploog, K.H., 2000. Nitride semiconductors free of electrostatic fields for efficient white light-emitting diodes. Nature 406, 865–868.

Wang, R., Nguyen, H.P.T., Connie, A.T., Lee, J., Shih, I., Mi, Z., 2014. Color-tunable, phosphor-free InGaN nanowire light-emitting diode arrays monolithically integrated on silicon. Opt. Express 22, A1768–A1775.

Wang, P., Yuan, Y., Zhao, C., Wang, X., Zheng, X., Rong, X., Wang, T., Sheng, B., Wang, Q., Zhang, Y., Bian, L., Yang, X., Xu, F., Qin, Z., Li, X., Zhang, X., Shen, B., 2016. Lattice-polarity-driven epitaxy of hexagonal semiconductor. Nano Lett. 16, 1328–1334.

Wölz, M., Ramsteiner, M., Kaganer, V.M., Brandt, O., Geelhaar, L., Riechert, H., 2013. Strain engineering of nanowire multi-quantum well demonstrated by Raman spectroscopy. Nano Lett. 13, 4053–4059.

Wölz, M., Hauswald, C., Flissikowski, T., Gotschke, T., Fernández-Garrido, S., Brandt, O., Grahn, H.T., Geelhaar, L., Riechert, H., 2015. Epitaxial growth of GaN nanowires with hight structural perfection on a metallic TiN film. Nano Lett. 15, 3743–3747.

Woo, S.Y., Bugnet, M., Nguyen, H.P.T., Mi, Z., Botton, G.A., 2015a. Atomic ordering in InGaN alloys within nanowire heterostructures. Nano Lett. 15, 6413–6418.

Woo, S.Y., Gauquelin, N., Nguyen, H.P.T., Mi, Z., Botton, G.A., 2015b. Interplay of strain and indium incorporation in InGaN/GaN dot-in-a-wire nanostructures by scanning transmission electron microscopy. Nanotechnology 26, 344002.

Wu, C.-L., Wang, J.-C., Chan, M.-H., Chen, T.T., Gwo, S., 2003. Heteroepitaxy of GaN on Si(111) realized with a coincident-interface AlN/β-Si_3N_4(0001) double-buffer structure. Appl. Phys. Lett. 83, 4530–4532.

Wu, C.-L., Shen, C.-H., Gwo, S., 2006. Valence band offset of wurtzite InN/AlN commensurate heterojunction determined by photoelectron spectroscopy. Appl. Phys. Lett. 88, 032105.

Wu, C.-L., Lee, H.-M., Kuo, C.-T., Gwo, S., Hsu, C.-H., 2007. Polarization-induced valence-band alignments at cation- and anion-polar InN/GaN heterojunctions. Appl. Phys. Lett. 91, 042112.

Wu, C.-L., Lee, H.-M., Kuo, C.-T., Chen, C.-H., Gwo, S., 2008a. Cross-sectional photoelectron microscopy and spectroscopy of wurtzite InN/GaN heterojunction: measurement of "intrinsic" band lineup. Appl. Phys. Lett. 92, 162106.

Wu, C.-L., Lee, H.-M., Kuo, C.-T., Chen, C.-H., Gwo, S., 2008b. Absence of Fermi level pinning at cleaved nonpolar InN surfaces. Phys. Rev. Lett. 101, 106803.

Wu, C.-Y., Kuo, C.-T., Wang, C.-Y., He, C.-L., Lin, M.-H., Ahn, H., Gwo, S., 2011. Plasmonic green nanolaser based on a metal–oxide–semiconductor structure. Nano Lett. 11, 4256–4260.

Yeh, T.-W., Lin, Y.-T., Stewart, L.S., Dapkus, P.D., Sarkissian, R., O'Brien, J.D., Ahn, B., Nutt, S.R., 2012. InGaN/GaN multiple quantum wells grown on nonpolar facets of vertical GaN nanorods arrays. Nano Lett. 12, 3257–3262.

Yoshizawa, M., Kikuchi, A., Mori, N., Kishino, K., 1997. Growth of self-organized GaN nanostructures on Al_2O_3(0001) by RF-radical Source molecular beam epitaxy. Jap. J. Appl. Phys. 36, L459–L462.

Yu, E.T., Dang, X.Z., Asbeck, P.M., Lau, S.S., Sullivan, G.J., 1999. Spontaneous and piezoelectric polarization effects in III-V nitride heterostructures. J. Vac. Sci. Technol. B 17, 1742–1749.

Zhao, S., Fathololoumi, S., Bevan, K.H., Liu, D.P., Kibria, M.G., Li, Q., Wang, G.T., Guo, H., Mi, Z., 2012. Tuning the surface charge properties of epitaxial InN nanowires. Nano Lett. 12, 2877–2882.

Zhao, S., Nguyen, H.P.T., Kibra, M.G., Mi, Z., 2015. III-Nitride nanowire optoelectronics. Prog. Quantum Electron. 44, 14–68.

Zhao, C., Ng, T.K., Wei, N., Prabaswara, A., Alias, M.S., Janjua, B., Shen, C., Ooi, B.S., 2016. Facile formation of high-quality InGaN/GaN quantum disks-in-nanowires on bulk-metal substrates for high-power light emitters. Nano Lett. 16, 1056–1063.

Zhou, M.-Y., Lu, Y.-J., Gwo, S., Gradečak, S., 2013. Correlation of doping, structure, and carrier dynamics in a single GaN nanorod. Appl. Phys. Lett. 102, 253104.

Zhou, X., Lu, M.-Y., Lu, Y.-J., Jones, E.J., Gwo, S., Gradečak, S., 2015. Nanoscale optical properties of indium gallium nitride/gallium nitride nanodisk-in-rod heterostructures. ACS Nano 9, 2868–2875.

CHAPTER ELEVEN

III-Nitride Electrically Pumped Visible and Near-Infrared Nanowire Lasers on (001) Silicon

P. Bhattacharya*,[1], A. Hazari*, S. Jahangir*, W. Guo†, T. Frost*
*University of Michigan, Ann Arbor, MI, United States
†University of Massachusetts, Lowell, MA, United States
[1]Corresponding author: e-mail address: pkb@umich.edu

Contents

1. INTRODUCTION

Crystalline semiconductor nanowires epitaxially grown on a variety of substrates have been a subject of interest for some time (Fernández-Garrido et al., 2013; Huang et al., 2010; Li et al., 2003; Lu & Lieber, 2006; Persson et al., 2004; Wallentin et al., 2013; Wang et al., 2006, 2011; Wu et al., 2004). In addition to the investigation of their growth phenomena, there has been intense research to elucidate their electronic, optical, structural, and magnetic properties (Banerjee et al., 2011; Bhattacharya et al., 2004; Chang et al., 2003; Deshpande et al., 2015; Grossklaus et al., 2013; Jahangir et al., 2013; Law et al., 2004; Yan et al., 2015; Yang et al., 2002; Zhao et al., 2004). In addition, nanowires have been exploited for device applications (Arafin et al., 2013; Frost et al., 2014; Greytak et al., 2005; Jahangir et al., 2014; Li et al., 2012). The more recently demonstrated III-nitride nanowires, that can be epitaxially grown on silicon, have

Semiconductors and Semimetals, Volume 96
ISSN 0080-8784
http://dx.doi.org/10.1016/bs.semsem.2016.07.002
© 2017 Elsevier Inc.
All rights reserved.

generated more interest because of their favorable material properties and the wide spectral range of the observed luminescence from InGaN/GaN nanowire heterostructures, which includes the visible range (Guo et al., 2010; Jahangir et al., 2014; Qian et al., 2004). While quantum wells (QWs) that are incorporated in visible light sources cannot emit at longer wavelengths beyond green (540 nm), III-nitride nanowire heterostructures have demonstrated emission in the red (610 nm) and beyond, up to 1.6 μm (Hazari et al., 2015a; Jahangir et al., 2014). At the other end of the spectrum, these nanowire heterostructures can also emit at extremely short wavelengths (Arafin et al., 2013; Connie et al., 2015). Since the nanowire heterostructures can be grown on (001) silicon substrates, they provide the possibility of placing important technologies such as solid-state lighting, mobile projectors and displays, infrared detectors and sensors, single photon sources, and silicon photonics on the preferred and more economical silicon platform. It therefore does not come as a surprise that light sources with the III-nitride nanowires have emerged as important devices for research and development. Whereas most of the work is focused on the development of visible light emitting diodes (LEDs) (Arafin et al., 2013; Hersee et al., 2009; Jahangir et al., 2014; Mi & Chang, 2009; Pan et al., 2013), considerable research is under way in the development of lasers (Arafin et al., 2013; Greytak et al., 2005; Hazari et al., 2015a; Jahangir et al., 2015; Li et al., 2012; Mi & Chang, 2009). III-nitride nanowires and nanowire heterostructures for these device applications can be grown in two ways: by self-organization of the seed crystal in which the nanowires of ~60–80 nm diameter are randomly positioned and their diameters can usually be varied in the 20–80 nm range (Chang et al., 2009; Guo et al., 2010, 2011a); they can also be grown in a regular array through a suitable mask, usually SiO_2 on Si, formed by patterning and nanolithography (Kishino et al., 2008; Wierer et al., 2012). The nanowire diameters in this case tend to be larger and are ~100–500 nm.

In(Ga,Al)N nanowires are grown catalyst-free on silicon substrates with aerial density in the range of 10^8–10^{11} cm^{-2} and lengths up to a few micrometers. The nanowires grow vertically in the wurtzite crystalline form with the *c*-axis aligned to the direction of growth. The In alloy composition in InGaN nanowires can be varied to produce emission in the 366–1400 nm range (Guo et al., 2010; Hazari et al., 2015a). InGaN disks, 2–3 nm in thickness, can be inserted in Ga(Al)N nanowires to take advantage of quantum sized effects in the design of light emitters. The nanowires grown on Si are relatively free of extended defects, compared to planar GaN layers grown on lattice-mismatched substrates, because of the large surface-to-volume

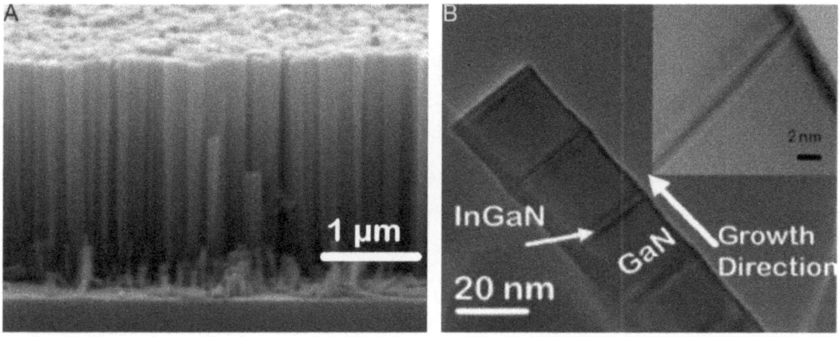

Fig. 1 (A) Scanning electron microscope (SEM) image of GaN nanowire array and (B) transmission electron microscope (TEM) image of InGaN disks in GaN nanowire. *Inset* shows magnified image of a single InGaN disk. *Reprinted from Guo, W., Banerjee, A., Bhattacharya, P., Ooi, B.S., 2011a. InGaN/GaN disk-in-nanowire white light emitting diodes on (001) silicon. Appl. Phys. Lett. 98. 193102-1–193102-3 with the permission of AIP Publishing.*

ratio (Guo et al., 2010; Nam et al., 2005; Songmuang et al., 2007). Relevant scanning electron microscope (SEM) and transmission electron microscope (TEM) images of an array of nanowires and a single GaN nanowire with InGaN disks are shown in Fig. 1A and B, respectively. The nanowires and disk-in-nanowires have reduced polarization field due to the radial relaxation of strain during epitaxy, thereby exhibiting smaller radiative lifetimes (Jahangir et al., 2013). This is the principle factor which enables emission at longer wavelengths with the incorporation of large amounts of In in the InGaN disks, compared to that in QWs. The blue shift of the emission peak with bias due to the quantum confined Stark effect is also reduced (Frost et al., 2014; Guo et al., 2010; Jho et al., 2002). The measured surface recombination velocity in GaN nanowires is very small and $\sim 10^3$ cm/s (Schlager et al., 2008). These favorable materials characteristics make the nanowires very suitable for device applications. Single nanowires or arrays have been used for fabrication of LEDs, lasers, polariton lasers, and single photon sources.

One of the challenges in the development of practical nanowire light sources has been the realization of a monolithic electrically pumped nanowire laser. This chapter is devoted to the epitaxy, design, fabrication, and characteristics of edge-emitting electrically injected III-nitride nanowire lasers that have peak emission ranging from 533 nm to 1.3 μm.

2. MOLECULAR BEAM EPITAXY (MBE) OF III-NITRIDE NANOWIRES ON (001) SILICON

A good understanding of the epitaxial growth of GaN nanowires on silicon has emerged from a substantial body of groundbreaking work

(Deshpande et al., 2015; Fernández-Garrido et al., 2013; Grossklaus et al., 2013; Guo et al., 2010; Nguyen et al., 2012; Wang et al., 2006; Yan et al., 2015; Yoshizawa et al., 1997). However, for completeness, the essential features are reiterated here. The growth being described is undertaken in a Veeco Gen II MBE system equipped with a N_2 plasma source. The typical Ga flux for growth of GaN nanowires at a substrate temperature of 819°C is 1.17×10^{-7} Torr and the N_2 plasma flow rate is 1 sccm. The In and Al fluxes are adjusted, depending on the desired composition of InGaN or AlGaN and are typically in the range of 5×10^{-8} Torr and 1.8×10^{-8} Torr, respectively. For growth of GaN nanowires, as the source species arrive on the heated substrate there is initially a preferential bonding of N with Si over that of N with Ga and the Ga incorporation rate is very low. The initial formation of the ~3 nm thick SiN_x layer, shown in Fig. 2, largely removes the epitaxial constraint of the substrate. The time-dependent stress field generated on the SiN_x layer helps in the formation of the initial GaN clusters (or seed crystals) and nanowire growth then proceeds parallel to the SiN_x layer with large Ga incorporation. Nitrogen-rich conditions are maintained, which slows down the random movement of the adatoms on Si and helps to accumulate them at the nucleation centers. The nitrogen-rich condition has a similar effect at the top of the nanowire as growth of the nanowire proceeds. SEM images show that the GaN nanowires do not grow on preferred nucleation sites but grow randomly with varying diameters. The nanowire diameter and aerial density are controlled by varying the V/III ratio. The nanowire density is also controlled by the thickness of the SiN_x layer. At the other extreme, a thick SiN_x layer inhibits nanowire nucleation and the density of the nanowires falls sharply. Due to the formation of the amorphous SiN_x layer, nucleation

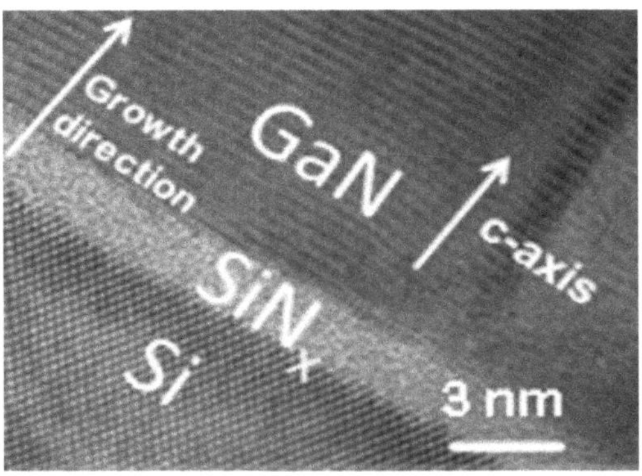

Fig. 2 High-resolution TEM image of Si/SiN_x/GaN nanowire interface.

and growth of the nanowires proceed in a similar way on both (001)- and (111)-oriented Si substrates (Guo et al., 2010; Paek et al., 2009). The SiN_x layer also reduces the ~13.5% stress at the interface, allowing nearly stress-free growth of nanowires. This also reduces the defect density at the interface, unlike planar heteroepitaxy of GaN films on sapphire or silicon. As shown in Fig. 3, there are three adatom paths that contribute to the growth of the nanowires. These are: (i) the adatoms that impinge on the top surface of the nanowires; (ii) adatoms that impinge on the Si substrate surface, diffuse toward the base of the nanowires and ultimately cross the energy barrier between Si and GaN; and (iii) adatoms that impinge on the nanowire sidewalls. In the latter two cases, adatoms diffuse on the nanowire sidewalls toward the top (0001) surface, which has the lowest chemical potential. At the same time, there is desorption of adatoms from all the surfaces.

Some of the nanowires in the array can grow tilted due to thickness non-uniformity of the SiN_x layer (Grossklaus et al., 2013). Coalescence can result between two or more slightly misaligned nanowires, or between two adjacent nanowires of increasing diameter during epitaxy. For example, the temperature is lowered for growth of InGaN nanowires or disks and for growth of p-type regions. The crystallographic (*c*-axis) misorientation

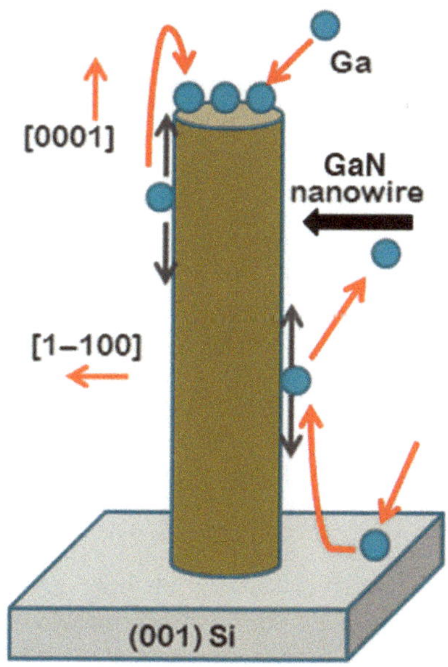

Fig. 3 Illustration of adatom migration, incorporation, and desorption during epitaxial growth of GaN nanowires.

between coalescing nanowires is accommodated by the formation of arrays of dislocations and stacking faults, which can behave electronically as non-radiative deep level defects. We have performed deep level transient spectroscopy measurements on GaN nanowire/Au Schottky diodes and have identified electron and hole traps whose density in the nanowires increases with their aerial density (Jahangir et al., 2014). These traps directly affect the luminescent output from nanowire LEDs (Calleja et al., 1996; Jahangir et al., 2014). Arrhenius plots of these traps are shown in Fig. 4.

Another nonradiative recombination process that adversely affects the performance of luminescent devices is Auger recombination, which is expected to be small in wide-bandgap semiconductors. We have determined

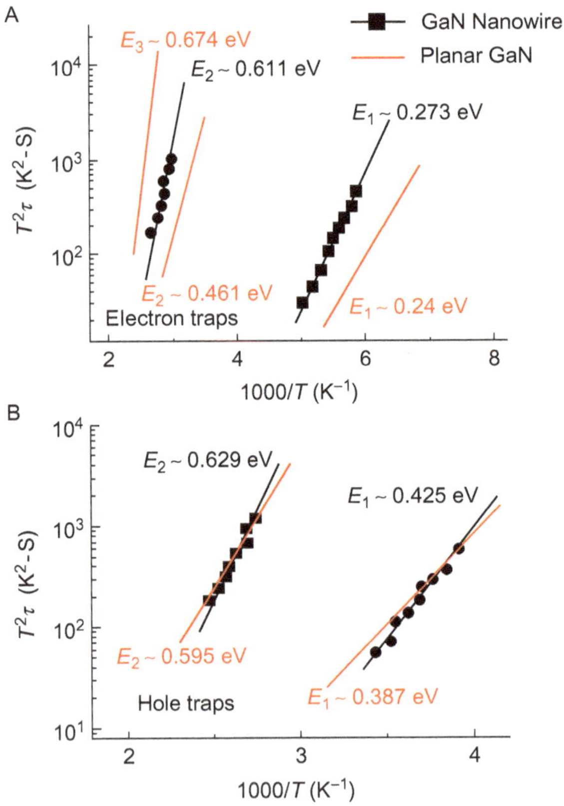

Fig. 4 Arrhenius plots of deep level (A) electron traps and (B) hole traps identified in GaN. *Reprinted, with permission, from Jahangir, S., Schimpke, T., Strassburg, M., Grossklaus, K.A., Millunchick, J.M., Bhattacharya, P., 2014. Red-emitting (λ = 610 nm) $In_{0.51}Ga_{0.49}N$/GaN disk-in-nanowire light emitting diodes on silicon. IEEE J. Quantum Electron. 50 (7), 530–537. © 2014 IEEE.*

the Auger recombination coefficients in InGaN nanowires and InGaN disks in GaN nanowires from excitation-dependent and time-resolved photoluminescence measurements (Guo et al., 2011b). The In content in the InGaN alloys in both types of samples varied around 35–40%. The values of the Auger coefficient varied in the range 10^{-33}–10^{-32} cm^6 s^{-1}, which are in reasonably good agreement with theoretical predictions for InGaN alloy semiconductors (Piprek, 2010). The low values of the Auger coefficient implies that the threshold current of lasers made with these materials will not be seriously affected by Auger recombination.

Finally, it is important to discuss the nature and dimensionality of quantum confinement in the 2-nm thick InGaN disk regions of the nanowires. High-resolution transmission electron microscopy done on InGaN/GaN disk-in-nanowire samples clearly indicates the formation of a single island in the disk region, without any noticeable wetting layer (Deshpande et al., 2015). We therefore believe that the islands are Volmer–Weber quantum dots, whose lateral dimensions depend on the nanowire diameter. Energy minimization calculations confirm the formation of such islands (Nakajima et al., 2001). Furthermore, single photon emission is observed by optical excitation of a single InGaN disk in a single nanowire (Deshpande et al., 2013, 2014). Therefore, we believe that the nanowire devices being described in this chapter are, in reality, InGaN/GaN quantum dot lasers.

3. FABRICATION OF NANOWIRE WAVEGUIDES AND ELECTRICALLY PUMPED EDGE-EMITTING VISIBLE LASERS ON (001) SILICON

A schematic of an edge-emitting nanowire laser and the nanowire laser heterostructure are shown in Fig. 5A. Multiple $In_xGa_{1-x}N$ disks ($0.35 < x < 1.0$) with GaN or $In_yGa_{1-y}N$ barriers are placed in the center of the $In_{0.06}Ga_{0.94}N$ waveguide. The cladding layers consist of lattice-matched $In_{0.18}Al_{0.82}N$ and the p- and n-GaN contact layers are doped with Mg and Si, respectively. The heterostructure for waveguide propagation loss measurements is similar, except that the $In_xGa_{1-x}N$ disks are not inserted and there is no doping in the GaN contact layers. The nanowires in the arrays have an average diameter of 60 nm and the typical nanowire density is $\sim 2 \times 10^{10}$ cm^{-2}. The nanowire fill factor is 0.35. A typical room temperature photoluminescence spectrum from a laser heterostructure is shown in Fig. 6. The internal quantum efficiency of the nanowire heterostructure arrays is measured by excitation-dependent and time-resolved luminescence

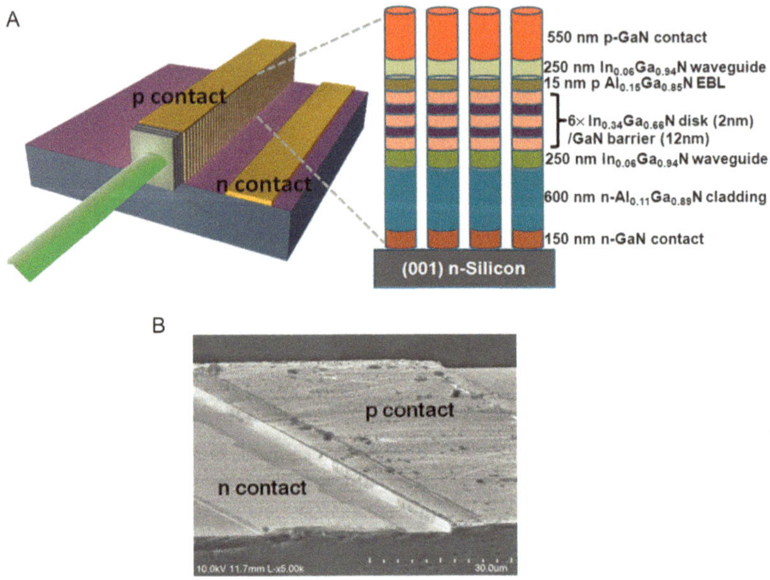

Fig. 5 (A) Schematic of an edge-emitting nanowire array laser and the nanowire laser heterostructure and (B) SEM image of fabricated device. *Reprinted (adapted) with permission from Frost, T., Jahangir, S., Stark, E., Deshpande, S., Hazari, A., Zhao, C., Ooi, B.S., Bhattacharya, P., 2014. Monolithic electrically injected nanowire array edge-emitting laser on (001) silicon. Nano Lett. 14, 4535–4541. Copyright (2014) American Chemical Society.*

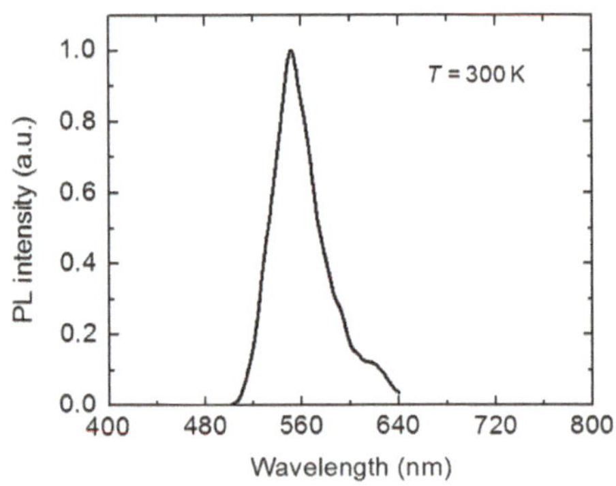

Fig. 6 Room temperature photoluminescence from an InGaN/GaN nanowire laser heterostructure emitting in green (560 nm).

measurements at room temperature. The radiative efficiency of as-grown $In_xGa_{1-x}N/GaN$ disks emitting in the visible range (430–650 nm) is found to be ~35–40% (Guo et al., 2010; Jahangir et al., 2014). We have also investigated the passivation of III-nitride nanowires with parylene, which is deposited at room temperature using a physical vapor deposition (PVD) process

(Jahangir et al., 2014). It is found that such passivation results in an increase of the radiative efficiency by ~10% due to the reduction of surface recombination on the nanowire sidewalls.

To fabricate waveguides for propagation loss measurements, the undoped nanowire samples are planarized and passivated with parylene. Ridge waveguides of 50 μm width and lengths ranging from 0.5 to 2 mm are defined by standard photolithography and dry etching. Waveguide propagation loss, including scattering loss and substrate leakage, in the nanowire waveguides is obtained from the measurement of transmission of subbandgap excitation at one end as a function of waveguide length (Fig. 7). Typically measured loss coefficients in the nanowire waveguides are in the range of 9–10 dB/cm (Frost et al., 2014; Hazari et al., 2015a). To fabricate edge-emitting lasers, the processing steps are very similar. After passivation with parylene, a slight etch back is done to expose the p-GaN nanowire tips and these are treated in $(NH_4)_2S$ to reduce the p-contact resistance. Ridges of 4–10 μm width and 50 μm width (broad area lasers) are defined by photolithography and dry etching. Ni/Au/ITO and Al contacts to the p-GaN nanowires and to the top surface of the Si substrate are formed by PVD. The substrate is cleaved and the nanowire/parylene edges are further planarized by focused ion beam etching, followed by SiO_2/TiO_2 or $ZnSe/MgF_2$ distributed Bragg reflector (DBR) deposition (depending on the emission wavelength). Typical facet reflectives are

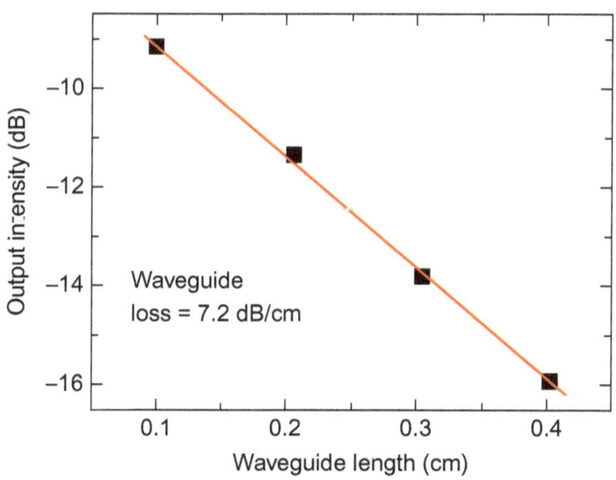

Fig. 7 Output intensity of nanowire heterostructure ridge waveguides as a function of waveguide length. *Reprinted (adapted) with permission from Frost, T., Jahangir, S., Stark, E., Deshpande, S., Hazari, A., Zhao, C., Ooi, B.S., Bhattacharya, P., 2014. Monolithic electrically injected nanowire array edge-emitting laser on (001) silicon. Nano Lett. 14, 4535–4541. Copyright (2014) American Chemical Society.*

0.72 and 0.95 and measurements are made of the coherent output from the low reflectivity facet. A SEM image of a fabricated device is shown in Fig. 5B.

4. MODE CONFINEMENT AND PROPAGATION IN NANOWIRE LASERS

An inherent problem encountered in the design of III-nitride heterostructure nanowire lasers is the small index difference between GaN and the AlGaN alloys, even for large Al contents. For example, the difference between the refractive index of GaN (Ejder, 1971) and that of $Al_{0.15}Ga_{0.85}N$ (Laws et al., 2000), commonly incorporated as the waveguide cladding layer is 0.052. Alloys with larger Al content are generally avoided due to larger lattice mismatch and associated defect density. On the other hand, the index difference between GaN and lattice-matched $In_{0.18}Al_{0.82}N$ is 0.22. Therefore, thicker cladding layers can be grown without creating defects and the mode confinement in a GaN or dilute InGaN waveguide layer is expected to be better than that with AlGaN cladding. Fig. 8 shows the X-ray diffraction spectrum of lattice-matched planar $In_{0.18}Al_{0.82}N$ layers grown on GaN at a substrate temperature of 497°C. The confined near-Gaussian mode calculated by the transfer matrix method in an InGaN/GaN/InAlN disk-in-nanowire laser heterostructure designed to emit at $\lambda = 630$ nm (red) is shown in Fig. 9. The mode confinement factor is calculated to be 0.018.

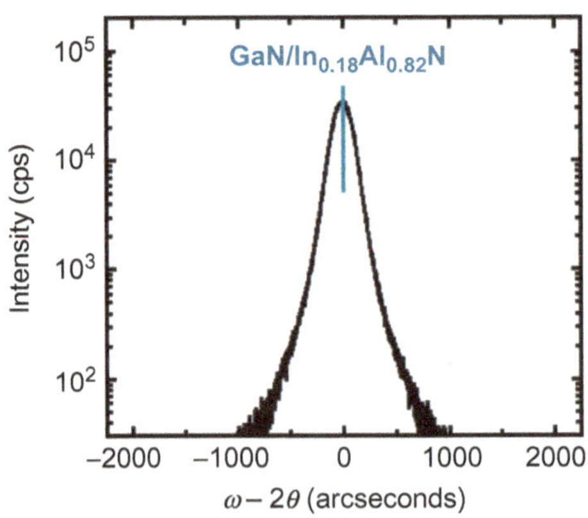

Fig. 8 X-ray diffraction spectrum of lattice-matched planar $In_{0.18}Al_{0.82}N$ molecular beam epitaxial layer grown on (0001) GaN substrate. *Reprinted, with permission, from Frost, T., Banerjee, A., Sun, K., Chuang, S.-L., Bhattacharya, P., 2013. InGaN/GaN quantum dot red laser. IEEE J. Quantum Electron. 49 (11), 923–931. © 2013 IEEE.*

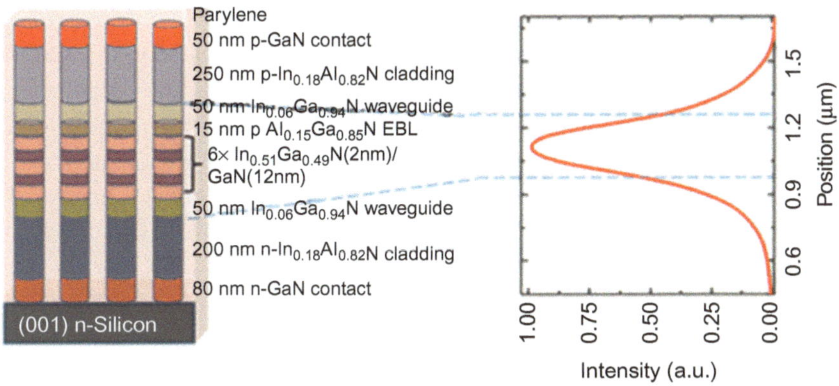

Fig. 9 InGaN/GaN disk-in-nanowire red laser heterostructure with the calculated mode profile. *Reprinted from Jahangir, S., Frost, T., Hazari, A., Yan, L., Stark, E., LaMountain, T., Mirecki-Millunchick, J., Ooi, B.S., Bhattacharya, P., 2015. Small signal modulation characteristics of red-emitting (λ = 610 nm) III-nitride nanowire array lasers on (001) silicon. Appl. Phys. Lett. 106, 071108 with the permission of AIP Publishing.*

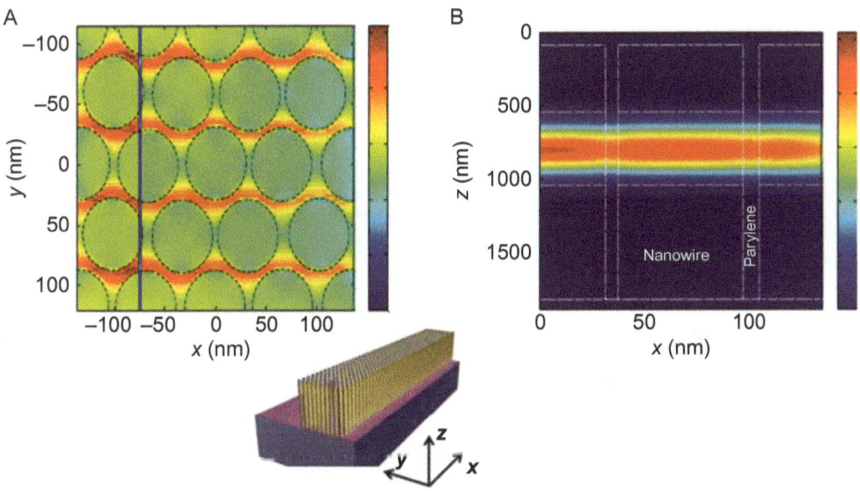

Fig. 10 Calculated optical field distribution in the (A) *x*–*y* and (B) *x*–*z* plane. *Reprinted (adapted) with permission from Frost, T., Jahangir, S., Stark, E., Deshpande, S., Hazari, A., Zhao, C., Ooi, B.S., Bhattacharya, P., 2014. Monolithic electrically injected nanowire array edge-emitting laser on (001) silicon. Nano Lett. 14, 4535–4541. Copyright (2014) American Chemical Society.*

We have performed detailed calculations of the characteristics of the photon field and propagation of photon modes in the in-plane direction along the waveguides using the three-dimensional finite difference time-domain technique (Frost et al., 2014; Heo et al., 2013). The nanowire–parylene composite guiding medium is modeled as a hexagonal close pack with nanowire diameter of 60 nm and average spacing of 67 nm. A random offset in the spacing is incorporated to avoid the formation of a photonic crystal. It is found that the optical field is continuous over the nanowires and the light generated by electron–hole recombination can freely propagate in the in-plane direction in the waveguide (Fig. 10). Plane wave expansion calculations of the cavity dispersion confirm that the waveguide region can be treated as a nanowire–parylene composite with an average index of ~2.0. Size (diameter) variation of the nanowire and the nanowire density will vary this average index by a small amount, but will not affect the optical field distribution adversely.

5. CHARACTERISTICS OF VISIBLE NANOWIRE LASERS

We have characterized green- and red-emitting disk-in-nanowire electrically pumped lasers (Frost et al., 2014; Mi & Chang, 2009). Here we briefly describe the steady-state and transient (small-signal modulation) characteristics of red-emitting ($\lambda = 610$ nm) lasers. Room temperature light–current–voltage (L–I–V) characteristics of a 1 mm × 10 μm ridge waveguide laser, measured with continuous wave (CW) current injection, are shown in Fig. 11. The facet reflectivities of this device are 0.35 and 0.95 and the output is measured from the low reflectivity facet. The threshold

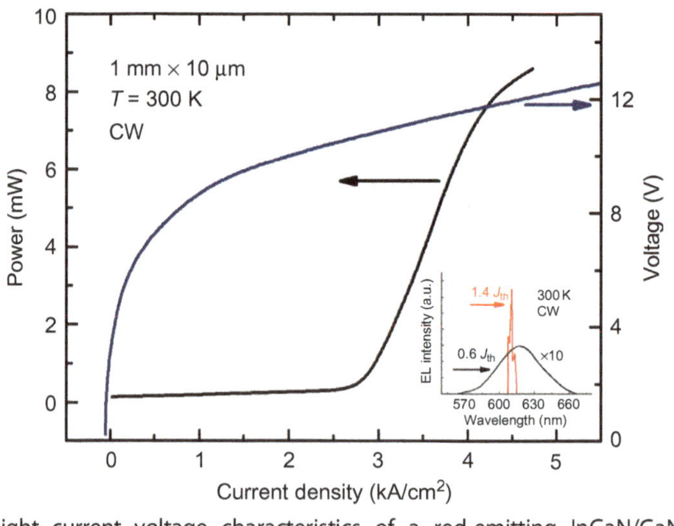

Fig. 11 Light–current–voltage characteristics of a red-emitting InGaN/GaN disk-in-nanowire laser at room temperature. *Reprinted from Jahangir, S., Frost, T., Hazari, A., Yan, L., Stark, E., LaMountain, T., Mirecki-Millunchick, J., Ooi, B.S., Bhattacharya, P., 2015. Small signal modulation characteristics of red emitting ($\lambda = 610$ nm) III-nitride nanowire array lasers on (001) silicon. Appl. Phys. Lett. 106, 071108 with the permission of AIP Publishing.*

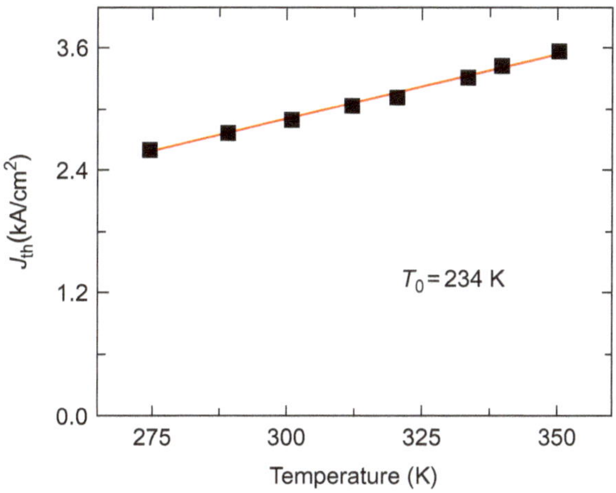

Fig. 12 Measured temperature dependence of J_{th} of red-emitting laser.

current density $J_{th}=2.85$ kA/cm^2 and the output slope efficiency and wall-plug efficiency are 0.088 W/A and 0.57%, respectively. The output spectrum above and below threshold is shown in the inset to the figure. The spectral linewidth of the dominant mode is ~9 Å. It should be noted that the threshold current densities measured for this laser and for similar green-emitting disk-in-nanowire lasers ($J_{th}=1.76$ A/cm^2) are significantly lower than green-emitting InGaN/GaN QW lasers. The only other semiconductor laser that emits in the red is currently realized with GaAs/InGaAlP QWs, which have $J_{th}\sim 5$–10 kA/cm^2 (Ikeda et al., 1985; Rennie et al., 1993). Therefore, the nanowire laser on Si has the promise of being a very useful device. Perhaps an even more encouraging aspect of the nanowire red laser characteristics is its temperature stability. Fig. 12 shows the measured temperature dependence of J_{th} for the laser of Fig. 9. A value of the characteristic temperature $T_0=234$ K is derived based on the relation $J_{th}(T)=J_{th}(0)\exp(T/T_0)$. In comparison, values of $T_0\sim 40$–50 K are reported for red-emitting InGaAlP/GaAs lasers (Ikeda et al., 1985; Rennie et al., 1993). We believe the consistently high values of T_0 that we have measured in all nanowire lasers is a result of the small rate of Auger recombination and low leakage of injected carriers due to large band offsets in the InGaN/GaN heterostructure. The measured blue shift of the emission peak in the range of 1.71–4 kA/cm^2 is ~49.31 meV (14.8 nm) for the red-emitting laser. This corresponds to a polarization field of 1098 kV/cm, which is significantly smaller than those measured and reported for green-emitting InGaN/GaN QW LEDs (Jho et al., 2002). Measurements of the steady-state characteristics have also been made on lasers with varying cavity lengths. Values of the internal quantum

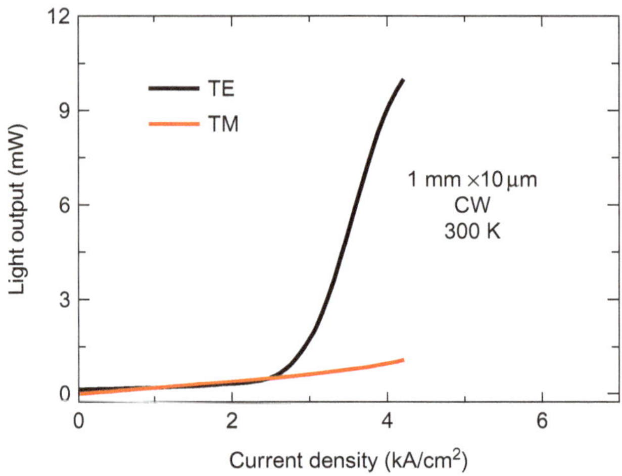

Fig. 13 Polarization-dependent output from a single facet of red-emitting nanowire array laser.

efficiency of the quantum dots and the differential gain can be derived from these measurements and these are 0.54 and 3.0×10^{-17} cm^2 for the red-emitting lasers. Fig. 13 shows the polarization-dependent L–I characteristics of the lasers. It is evident that the transverse electric mode is dominant, which is because it has a higher gain than the TM mode and is better confined in the waveguide.

The differential gain and other useful performance characteristics of a laser can also be derived from the response to high frequency small-signal modulation of a laser. These measurements are made at room temperature on 4 μm × 400 μm ridge waveguide red- and green-emitting lasers. The measurements are made with the help of a bias $-T$, a high-speed detector and a spectrum analyzer. The measured damping-limited modulation response of a red-emitting InGaN/GaN disk-in-nanowire laser as a function of injection is shown in Fig. 14. Values of differential gain $dg/dn = 3.1 \times 10^{-17}$ cm^2 and gain compression factor $\in \sim 2 \times 10^{-17}$ cm^3 are derived from analysis of these data. It may be noted that the value of dg/dn is in good agreement with that obtained from the steady-state measurements. This is generally the case for all the nanowire lasers we have fabricated and characterized.

The spectral gain near threshold, measured by the Hakki–Paoli technique, is shown in Fig. 15. The peak modal gain is 27 cm^{-1} and is therefore 4.5 cm^{-1} per disk layer. The linewidth enhancement factor α was also determined from

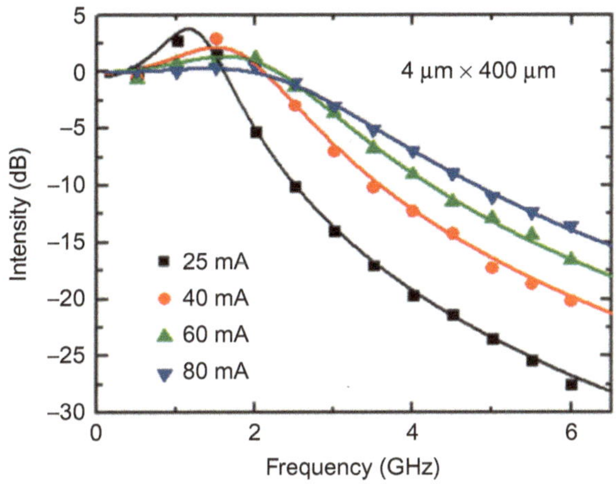

Fig. 14 Measured small-signal modulation response of a ridge waveguide red-emitting InGaN/GaN disk-in-nanowire array laser at 300 K for varying dc injection current values. *Reprinted from Jahangir, S., Frost, T., Hazari, A., Yan, L., Stark, E., LaMountain, T., Mirecki-Millunchick, J., Ooi, B.S., Bhattacharya, P., 2015. Small signal modulation characteristics of red emitting (λ = 610 nm) III-nitride nanowire array lasers on (001) silicon. Appl. Phys. Lett. 106, 071108 with the permission of AIP Publishing.*

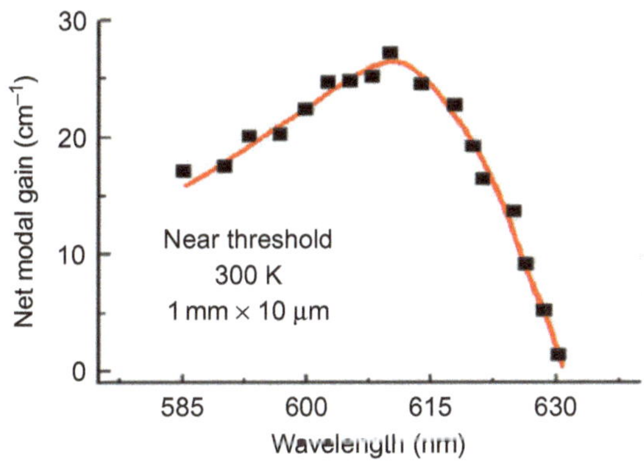

Fig. 15 Red NW laser spectral gain plot from Hakki–Paoli measurements.

the Hakki–Paoli measurements. The variation of the α with λ of a 10 μm × 1 mm ridge waveguide red-emitting laser is shown in Fig. 16. The linewidth enhancement factor exhibits a minimum value of ~0.2 at 610 nm. This experimentally observed trend of α has been predicted from theoretical calculations of quantum dot lasers by Miyake and Asada (1989). The minimum in α is related to the strong peaking of the gain spectra of quantum dots. This result is further evidence of the formation of InGaN/GaN quantum dots in the disk regions of the nanowires.

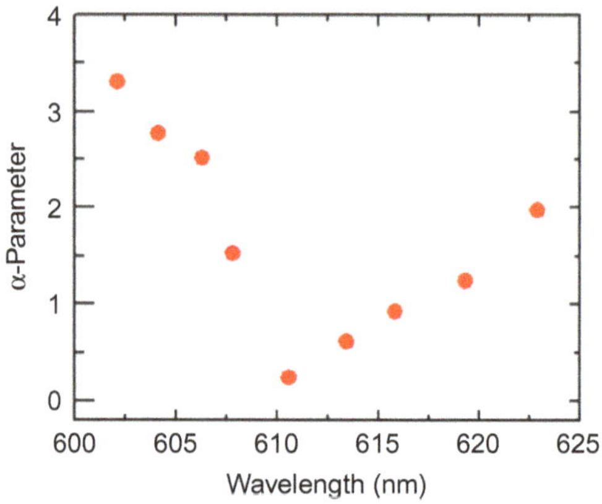

Fig. 16 Measured α-parameter as a function of emission wavelength. *Reprinted from Jahangir, S., Frost, T., Hazari, A., Yan, L., Stark, E., LaMountain, T., Mirecki-Millunchick, J., Ooi, B.S., Bhattacharya, P., 2015. Small signal modulation characteristics of red emitting (λ = 610 nm) III-nitride nanowire array lasers on (001) silicon. Appl. Phys. Lett. 106, 071108 with the permission of AIP Publishing.*

6. ELECTRICALLY PUMPED 1.3 μM DISK-IN-NANOWIRE LASERS ON SILICON

A monolithic electrically injected laser on (001)Si is useful and important for silicon photonics. Current devices that are being investigated for this application are directly grown In(Ga)As/GaAs quantum dot lasers, which have to be grown on misoriented substrates and are therefore incompatible with Si complementary metal-oxide semiconductor (CMOS) technology, and wafer bonded III–V lasers on Si. With the excellent performance characteristics of visible InGaN/GaN disk-in-nanowire lasers, it is natural to consider the realization of near-infrared lasers. For application in silicon photonics, it is desirable to have an emission wavelength of 1.2 μm or higher. There are two problems associated with this. The In content in the InGaN disk regions of the red-emitting nanowire lasers is ~50%. To extend the emission wavelength to ~1.3 μm, the In content in the disk has to be increased substantially, which might lead to composition inhomogeneity and a large density of defects. Second, the optical constants of InGaN alloys which are needed for the design of long wavelength III-nitride lasers are not readily available. We have therefore recently measured the optical constants of epitaxial wurtzite $In_xGa_{1-x}N$ $(0<x<0.73)$ in the $400<x<1687$ nm wavelength range (Hazari et al., 2015b). The measurements were made on ~600 nm thick planar epitaxial layers grown on (001)Si substrates by

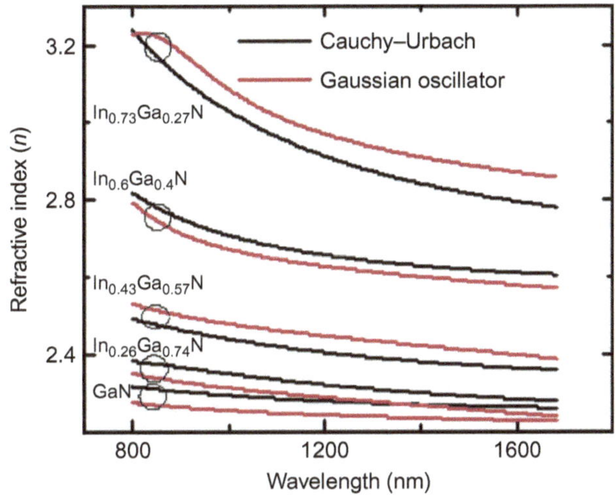

Fig. 17 Refractive index of InGaN as a function of wavelength obtained from reflectance measurements. *Reprinted from Hazari, A., Bhattacharya, A., Frost, T., Zhao, S., Baten, M.Z., Mi, Z., Bhattacharya, P., 2015b. Optical constants of InxGa$_{1-x}$N ($0 \leq x \leq 0.73$) in the visible and near-infrared wavelength regimes. Opt. Lett. 40 (14), 3304–3307 with the permission of OSA Publishing.*

MBE. The refractive indices and extinction coefficients were derived from analysis of the measured data using the Cauchy–Urbach and Gaussian oscillator models. The refractive indices are shown in Fig. 17.

We have incorporated $In_{0.85}Ga_{0.15}N$ disks in GaN nanowires as the gain media and have reported the characteristics of 1.2 μm electrically pumped lasers fabricated with the nanowire arrays (Hazari et al., 2015a). The radiative efficiency of the disks is ~17%, probably due to a large density of defects formed during epitaxy. Nonetheless, the lasers are characterized by $J_{th}=1.24$ kA/cm^2, $T_0=242$ K in the temperature range of 2–77°C and $dg/dn=5.6\times10^{-17}$ cm^2. These results motivated us to investigate the incorporation of binary InN disks with $In_{0.4}Ga_{0.6}N$ barriers in between for the realization of lasers with peak emission at 1.3 μm. For better optical confinement of the guided mode, the nanowire array laser was designed as a graded refractive index separate confinement heterostructure (GRIN-SCH). The nanowire heterostructure grown on (001)Si and the calculated optical mode is shown in Fig. 18. The 150 nm graded cladding layer from GaN to $In_{0.4}Ga_{0.6}N$ was grown in 10 equal steps of 15 nm thickness. Low temperature photoluminescence spectrum from the nanowire laser heterostructure shown in Fig. 19 indicates emission from the InN disks with peak at 1.6 μm. Edge-emitting lasers suitable for electrical injection were fabricated in the same way as the visible lasers, described earlier.

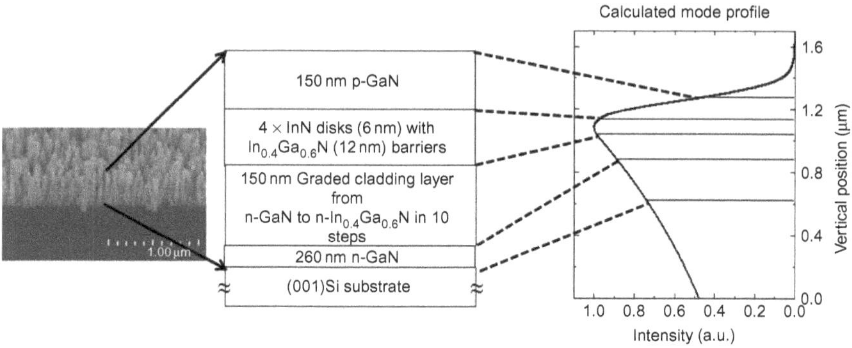

Fig. 18 Schematic illustration of the near-IR ($\lambda = 1.3$ μm) disk-in-nanowire array laser heterostructure with a SEM image of the nanowire array. The calculated mode profile is shown alongside.

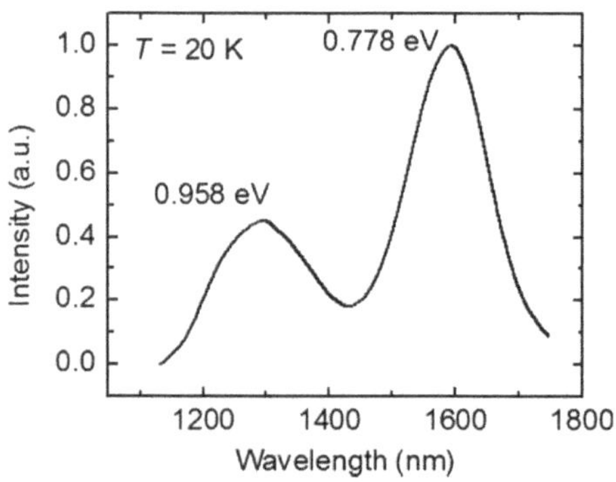

Fig. 19 Low temperature photoluminescence spectrum from the near-IR nanowire laser heterostructure.

The $MgF_2/ZnSe$ DBR facets reflectivity are 0.68 and 0.88 and measurements were made of emission from the low reflectivity facet.

The steady-state L–I characteristics at room temperature of a 50 μm × 2 mm ridge waveguide laser under CW and pulsed (5% duty cycle) bias conditions are shown in Fig. 20. Output powers up to ~10 mW are measured at room temperature without any heat sinking or facet cooling. The output spectrum shown in the inset confirms 1.3 μm peak emission. The slope efficiency is 0.14 W/A. A low value of $I_{th} = 673$ mA is measured. This value can be further lowered with better design of the laser heterostructure. The variation of J_{th} with temperature was measured and is plotted in Fig. 21, from which a high value of $T_0 = 241$ K is derived in the temperature range $275 < T < 350$ K.

For on-chip data transmission with a monolithic electrically pumped laser, it is important to have knowledge of the small-signal modulation

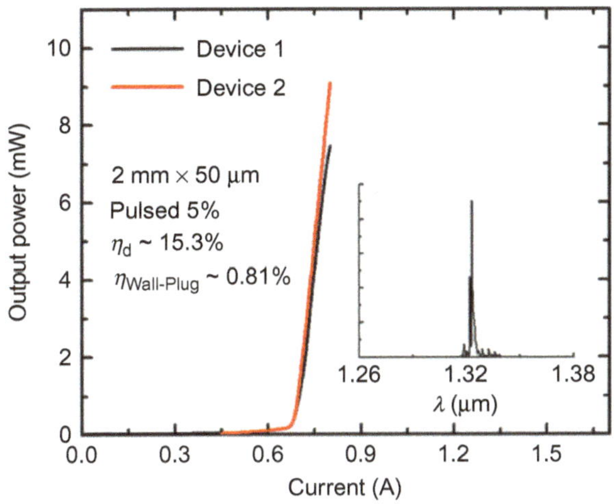

Fig. 20 Light–current characteristics of broad area near-IR lasers. The *inset* shows the output characteristics for an injection current of 810 mA.

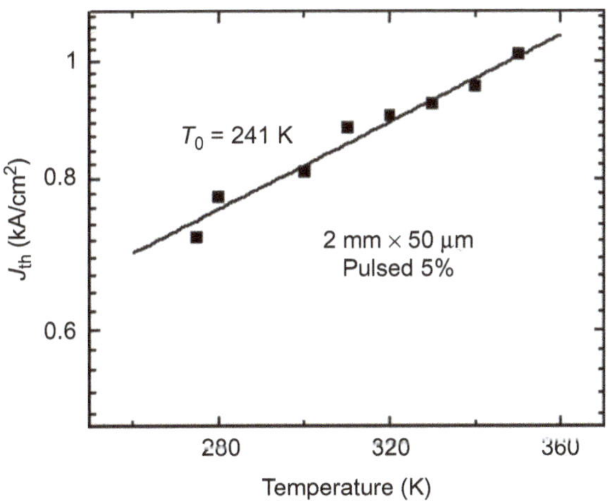

Fig. 21 Temperature dependence of the threshold current density.

bandwidth of the device. The measured modulation response for different values of dc injection current is shown in Fig. 22. A −3 dB bandwidth of 2.4 GHz is measured for the highest injection current. The damping factor γ_d and the resonance frequency f_r of the modulation response are derived from the plots for each injection current. f_r is plotted against $(I - I_{th})^{1/2}$ in Fig. 23A where I is the injection current and I_{th} is the threshold current to obtain the differential gain and γ_d is plotted against f_r^2 in Fig. 23B to obtain

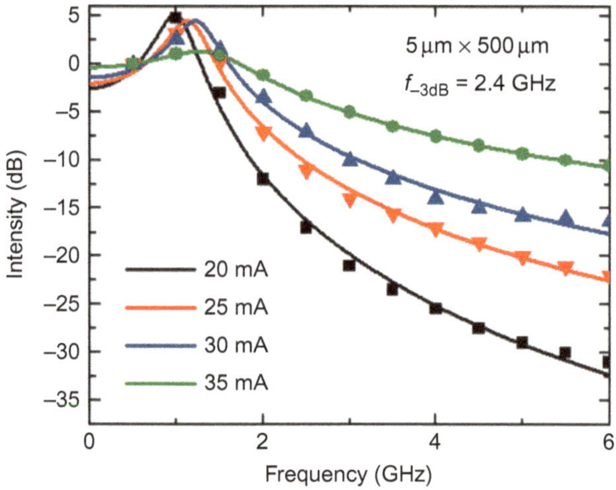

Fig. 22 Measured small-signal modulation response of a ridge waveguide near-IR nanowire laser at 300 K for varying dc injection current values.

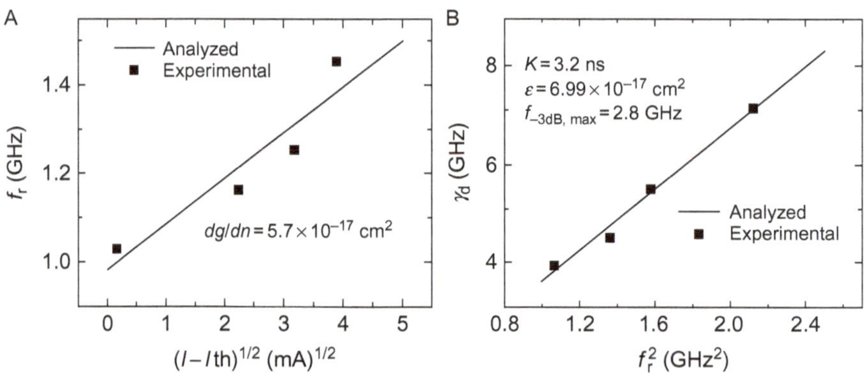

Fig. 23 (A) Variation of resonance frequency f_r as a function of $(I - I_{th})^{1/2}$ and (B) dependence of damping factor γ_d on square of resonance frequency f_r.

the gain compression factor. From analysis of these data, a value of the differential gain $dg/dn = 5.7 \times 10^{-17}$ cm^2 and gain compression factor $\epsilon = 6.99 \times 10^{-17}$ cm^3 are derived. This value is very low and comparable to those reported for high performance semiconductor QW and quantum dot lasers.

A preliminary measurement of the variation of laser output power with time for a constant CW injection was made without any heat sinking or

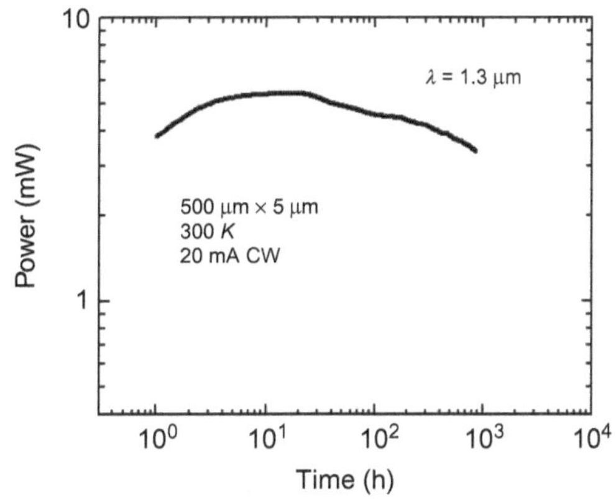

Fig. 24 Measured output power of 1.3 μm disk-in-nanowire laser as a function of time with constant CW current injection.

active cooling. The data are shown in Fig. 24, which indicate a lifetime of ~1000 h. Much higher lifetimes will be measured in lasers with better design and fabrication and with adequate mounting and heat sinking.

7. CONCLUSION

A random array of III-nitride nanowires epitaxially grown on (001)Si substrate enables the realization of monolithic electrically injected edge-emitting lasers. The gain region of these lasers are 2-nm thick $In_xGa_{1-x}N$ disks placed in the waveguide, whose composition is varied in the range $0.34 < x < 1.0$ to tune the peak wavelength of the luminescence. It is evident that quantum dots are formed in the InGaN disk region due to strain relaxation during growth of the nanowires. We have described here the measured characteristics of nanowire lasers with emission in both the visible (green and red) and the near-infrared (1.3 μm) spectral range. The nanowire waveguides exhibit propagation loss ~10 dB/cm. The lasers are characterized by low threshold, compared to visible III-nitride QW lasers on GaN, excellent temperature stability of the threshold current density ($T_0 > 200$ K) and large differential gain ($\sim 5 \times 10^{-17}$ cm^2). The small-signal modulation characteristics of the lasers have also been measured and the lasers exhibit modulation bandwidths ~2–5 GHz and very low gain compression factor, $\epsilon \sim 4 \times 10^{-17}$ cm^3. With optimization of heterostructure design, fabrication techniques and adequate heat sinking during operation, the visible and 1.3 μm lasers will be useful for displays, plastic fiber communication, and silicon photonics.

REFERENCES

Arafin, S., Liu, X., Mi, Z., 2013. Review of recent progress of III-nitride nanowire lasers. J. Nanophoton. 7, 074599.

Banerjee, A., Dogan, F., Heo, J., Manchon, A., Guo, W., Bhattacharya, P., 2011. Spin relaxation in InGaN quantum disks in GaN nanowires. Nano Lett. 11, 5396–5400.

Bhattacharya, S., Banerjee, D., Adu, K.W., Samui, S., Bhattacharya, S., 2004. Confinement in silicon nanowires: optical properties. Appl. Phys. Lett. 85 (11), 2008–2010.

Calleja, E., Sanchez, F.J., Basak, D., Sanchez-Garcia, M.A., Munoz, E., Izpura, I., Calle, F., Tijero, J.M.G., Sanchez-Rojas, J.L., Beaumont, B., Lorenzini, P., Gibart, P., 1996. Yellow luminescence and related deep states in undoped GaN. Phys. Rev. B 55 (7), 4689–4694.

Chang, Y.Q., Wang, D.B., Luo, X.H., Xu, X.Y., Chen, X.H., Li, L., Chen, C.P., Wang, R.M., Xu, J., Yu, D.P., 2003. Synthesis, optical, and magnetic properties of diluted magnetic semiconductor $Zn_{1-x}Mn_xO$ nanowires via vapor phase growth. Appl. Phys. Lett. 83 (19), 4020–4022.

Chang, Y.-L., Li, F., Fatehi, A., Mi, Z., 2009. Molecular beam epitaxial growth and characterization of non-tapered InN nanowires on Si(111). Nanotechnology 20, 345203.

Connie, A.T., Zhao, S., Sadaf, S.M., Shih, I., Mi, Z., Du, X., Lin, J., Jiang, H., 2015. Optical and electrical properties of Mg-doped AlN nanowires grown by molecular beam epitaxy. Appl. Phys. Lett. 106. 213105-1–213105-5.

Deshpande, S., Heo, J., Das, A., Bhattacharya, P., 2013. Electrically driven polarized single-photon emission from an InGaN quantum dot in a GaN nanowire. Nat. Commun. 4, 1675.

Deshpande, S., Frost, T., Hazari, A., Bhattacharya, P., 2014. Electrically pumped single-photon emission at room temperature from a single InGaN/GaN quantum dot. Appl. Phys. Lett. 105. 141109-1–141109-5.

Deshpande, S., Frost, T., Yan, L., Jahangir, S., Hazari, A., Liu, X., Mirecki-Millunchick, J., Mi, Z., Bhattacharya, P., 2015. Formation and nature of InGaN quantum dots in GaN nanowires. Nano Lett. 15, 1647–1653.

Ejder, E., 1971. Refractive index of GaN. Phys. Status Solidi A 6, 445–448.

Fernández-Garrido, S., Kaganer, V.M., Sabelfeld, K.K., Gotschke, T., Grandal, J., Calleja, E., Geelhaar, L., Brandt, O., 2013. Self-regulated radius of spontaneously formed GaN nanowires in molecular beam epitaxy. Nano Lett. 13 (7), 3274–3280.

Frost, T., Jahangir, S., Stark, E., Deshpande, S., Hazari, A., Zhao, C., Ooi, B.S., Bhattacharya, P., 2014. Monolithic electrically injected nanowire array edge-emitting laser on (001) silicon. Nano Lett. 14, 4535–4541.

Greytak, A.B., Barrelet, C.J., Li, Y., Lieber, C.M., 2005. Semiconductor nanowire laser and nanowire waveguide electro-optic modulators. Appl. Phys. Lett. 87. 151103-1–151103-3.

Grossklaus, K.A., Banerjee, A., Jahangir, S., Bhattacharya, P., Millunchick, J.M., 2013. Misorientation defects in coalesced self-catalyzed GaN nanowires. J. Cryst. Growth 371, 142–147.

Guo, W., Zhang, M., Banerjee, A., Bhattacharya, P., 2010. Catalyst-free InGaN/GaN nanowire light emitting diodes grown on (001) silicon by molecular beam epitaxy. Nano Lett. 10, 3355–3359.

Guo, W., Banerjee, A., Bhattacharya, P., Ooi, B.S., 2011a. InGaN/GaN disk-in-nanowire white light emitting diodes on (001) silicon. Appl. Phys. Lett. 98. 193102-1–193102-3.

Guo, W., Zhang, M., Bhattacharya, P., Heo, J., 2011b. Auger recombination in III-nitride nanowires and its effect on nanowire light-emitting diode characteristics. Nano Lett. 11, 1434–1438.

Hazari, A., Aiello, A., Ng, T.-K., Ooi, B.S., Bhattacharya, P., 2015a. III-nitride disk-in-nanowire 1.2 μm monolithic diode laser on (001)silicon. Appl. Phys. Lett. 107. 191107-1–191107-5.

Hazari, A., Bhattacharya, A., Frost, T., Zhao, S., Baten, M.Z., Mi, Z., Bhattacharya, P., 2015b. Optical constants of $In_xGa_{1-x}N$ ($0 \leq x \leq 0.73$) in the visible and near-infrared wavelength regimes. Opt. Lett. 40 (14), 3304–3307.

Heo, J., Jahangir, S., Xiao, B., Bhattacharya, P., 2013. Room-temperature polariton lasing from GaN nanowire array clad by dielectric microcavity. Nano Lett. 13, 2376–2380.

Hersee, S.D., Fairchild, M., Rishinaramangalam, A.K., Ferdous, M.S., Zhang, L., Varangis, P.M., Swartzentruber, B.S., Talin, A.A., 2009. GaN nanowire light emitting diodes based on templated and scalable nanowire growth process. Electron. Lett. 45 (1), 75–76.

Huang, J.Y., Zhong, L., Wang, C.M., Sullivan, J.P., Xu, W., Zhang, L.Q., Mao, S.X., Hudak, N.S., Liu, X.H., Subramanian, A., Fan, H., Qi, L., Kushima, A., Li, J., 2010. In situ observation of the electrochemical lithiation of a single SnO_2 nanowire electrode. Science 330 (6010), 1515–1520.

Ikeda, M., Mori, Y., Sato, H., Kaneko, K., Watanabe, N., 1985. Room-temperature continuous-wave operation of an AlGaInP double heterostructure laser grown by atmospheric pressure metalorganic chemical vapor deposition. Appl. Phys. Lett. 47 (1027), 1027–1028.

Jahangir, S., Banerjee, A., Bhattacharya, P., 2013. Carrier lifetimes in green emitting InGaN/GaN disks-in-nanowire and characteristics of green light emitting diodes. Phys. Status Solidi C 10 (5), 812–815.

Jahangir, S., Schimpke, T., Strassburg, M., Grossklaus, K.A., Millunchick, J.M., Bhattacharya, P., 2014. Red-emitting ($\lambda = 610$ nm) $In_{0.51}Ga_{0.49}N$/GaN disk-in-nanowire light emitting diodes on silicon. IEEE J. Quantum Electron. 50 (7), 530–537.

Jahangir, S., Frost, T., Hazari, A., Yan, L., Stark, E., LaMountain, T., Mirecki-Millunchick, J., Ooi, B.S., Bhattacharya, P., 2015. Small signal modulation characteristics of red-emitting ($\lambda = 610$ nm) III-nitride nanowire array lasers on (001) silicon. Appl. Phys. Lett. 106, 071108.

Jho, Y.D., Yahng, J.S., Oh, E., Kim, D.S., 2002. Field-dependent carrier decay dynamics in strained $In_xGa_{1-x}N$/GaN quantum wells. Phys. Rev. B 66. 035334-1–035334-11.

Kishino, K., Sekiguchi, H., Kikuchi, A., 2008. Improved Ti-mask selective-area growth (SAG) by rf-plasma-assisted molecular beam epitaxy demonstrating extremely uniform GaN nanocolumn arrays. J. Cryst. Growth 311, 2063–2068.

Law, M., Goldberger, J., Yang, P., 2004. Semiconductor nanowires and nanotubes. Annu. Rev. Mater. Res. 34, 83–122.

Laws, G.M., Larkins, E.C., Harrison, I., Molloy, C., Somerford, D., 2000. Improved refractive index formulas for the $Al_xGa_{1-x}N$ and $In_yGa_{1-y}N$ alloys. J. Appl. Phys. 89 (2), 1108–1115.

Li, C., Zhang, D., Han, S., Liu, X., Tang, T., Zhou, C., 2003. Diameter-controlled growth of single-crystalline In_2O_3 nanowires and their electronic properties. Adv. Mater. 15 (2), 143–146.

Li, Q., Wright, J.B., Chow, W.W., Luk, T.S., Brener, I., Lester, L.F., Wang, G.T., 2012. Single-mode GaN nanowire lasers. Opt. Express 20 (16), 17873–17879.

Lu, W., Lieber, C.M., 2006. Semiconductor nanowires. J. Phys. D: Appl. Phys. 39 (21), R387–R406.

Mi, Z., Chang, Y.-L., 2009. III-V compound semiconductor nanostructures on silicon: epitaxial growth, properties, and applications in light emitting diodes and lasers. J. Nanophoton. 3. 031602–031602-19.

Miyake, Y., Asada, M., 1989. Spectral characteristics of linewidth enhancement factor α of multidimensional quantum wells. Jpn. J. Appl. Phys. 28 (7), 1280–1281.

Nakajima, K., Ujihara, T., Miyashita, S., Sazaki, G., 2001. Effects of misfit dislocations and AlN buffer layer on the GaInN/GaN phase diagram of the growth mode. J. Appl. Phys. 89 (1), 146–153.

Nam, C.-Y., Jaroenapibal, P., Tham, D., Luzzi, D.E., Evoy, S., Fischer, J.E., 2005. Diameter-dependent electromechanical properties of GaN nanowires. Nano Lett. 6 (2), 153–158.

Nguyen, H.P.T., Cui, K., Zhang, S., Djavid, M., Korinek, A., Botton, G.A., Mi, Z., 2012. Controlling electron overflow in phosphor-free InGaN/GaN nanowire white light-emitting diodes. Nano Lett. 12, 1317–1323.

Paek, J.H., Nishiwaki, T., Yamaguchi, M., Sawaki, N., 2009. Catalyst free MBE-VLS growth of GaAs nanowires on (111)Si substrate. Phys. Status Solidi C 6 (6), 1436–1440.

Pan, C., Dong, L., Zhu, G., Niu, S., Yu, R., Yang, Q., Liu, Y., Wang, Z.L., 2013. High-resolution electroluminescent imaging of pressure distribution using a piezoelectric nanowire LED array. Nat. Photon. 7, 752–758.

Persson, A.I., Larsson, M.W., Stenström, S., Ohlsson, B.J., Samuelson, L., Wallenberg, L.R., 2004. Solid-phase diffusion mechanism for GaAs nanowire growth. Nat. Mater. 3 (10), 677–681.

Piprek, J., 2010. Efficiency droop in nitride-based light-emitting diodes. Phys. Status Solidi A 207 (10), 2217–2225.

Qian, F., Li, Y., Gradečak, S., Wang, D., Barrelet, C.J., Lieber, C.M., 2004. Gallium nitride-based nanowire radial heterostructures for nanophotonics. Nano Lett. 4 (10), 1975–1979.

Rennie, J., Okajima, M., Watanabe, M., Hatakoshi, G., 1993. High temperature (74°C) CW operation of 634 nm InGaAlP laser diodes utilizing a multiple quantum barrier. IEEE J. Quantum Electron. 29 (6), 1857–1862.

Schlager, J.B., Bertness, K.A., Blanchard, P.T., Robins, L.H., Roshko, A., Sanford, N.A., 2008. Steady-state and time-resolved photoluminescence from relaxed and strained GaN nanowires grown by catalyst-free molecular-beam epitaxy. J. Appl. Phys. 103. 124309-1–124309-6.

Songmuang, R., Landré, O., Daudin, B., 2007. From nucleation to growth of catalyst-free GaN nanowires on thin AlN buffer layer. Appl. Phys. Lett. 91. 251902-1–251902-3.

Wallentin, J., Anttu, N., Asoli, D., Huffman, M., Åberg, I., Magnusson, M.H., Siefer, G., Fuss-Kailuweit, P., Dimroth, F., Witzigmann, B., Xu, H.Q., Samuelson, L., Deppert, K., Borgström, M.T., 2013. InP nanowire array solar cells achieving 13.8% efficiency by exceeding the ray optics limit. Science 339 (6123), 1057–1060.

Wang, G.T., Talin, A.A., Werder, D.J., Creighton, J.R., Lai, E., Anderson, R.J., Arslan, I., 2006. Highly aligned, template-free growth and characterization of vertical GaN nanowires on sapphire by metal-organic chemical vapour deposition. Nanotechnology 17, 5773–5780.

Wang, D., Pierre, A., Kibria, M.G., Cui, K., Han, X., Bevan, K.H., Guo, H., Paradis, S., Hakima, A.-R., Mi, Z., 2011. Wafer-level photocatalytic water splitting on GaN nanowire arrays grown by molecular beam epitaxy. Nano Lett. 11, 2353–2357.

Wierer Jr., J.J., Li, Q., Koleske, D.D., Lee, S.R., Wang, G.T., 2012. III-nitride core–shell nanowire arrayed solar cells. Nanotechnology 23, 194007.

Wu, Y., Cui, Y., Huynh, L., Barrelet, C.J., Bell, D.C., Lieber, C.M., 2004. Controlled growth and structures of molecular-scale silicon nanowires. Nano Lett. 4 (3), 433–436.

Yan, L., Jahangir, S., Wight, S.A., Nikoobakht, B., Bhattacharya, P., Millunchick, J.M., 2015. Structural and optical properties of disc-in-wire InGaN/GaN LEDs. Nano Lett. 15, 1535–1539.

Yang, P., Yan, H., Mao, S., Russo, R., Johnson, J., Saykally, R., Morris, N., Pham, J., He, R., Choi, H.-J., 2002. Controlled growth of ZnO nanowires and their optical properties. Adv. Funct. Mater. 12 (5), 323–331.
Yoshizawa, M., Kikuchi, A., Mori, M., Fujita, N., Kishino, K., 1997. Growth of self-organized GaN nanostructures on Al_2O_3 (0001) by RF-radical source molecular beam epitaxy. Jpn. J. Appl. Phys. 36 (2), L459–L462.
Zhao, X., Wei, C.M., Yang, L., Chou, M.Y., 2004. Quantum confinement and electronic properties of silicon nanowires. Phys. Rev. Lett. 92 (23). 236805-1–236805-4.

CHAPTER TWELVE

Exploring the Next Phase in Gallium Nitride Photonics: Cubic Phase Light Emitters Heterointegrated on Silicon

C. Bayram[1], R. Liu
University of Illinois at Urbana–Champaign, Champaign, IL, United States
[1]Corresponding author: e-mail address: cbayram@illinois.edu

Contents

1. INTRODUCTION TO PHOTONICS

Photonics is the science, engineering, and technology of generating, controlling, and detecting light waves and photons, which are particles of light, via electronics (i.e., electrically charged particles). It explores a wide variety of wavelengths, from gamma rays to radio, including X-rays, ultraviolet (UV), visible, and infrared light. Photonics, propelled by breakthroughs in optics and electronics, has evolved into a useful tool in consumer devices, space exploration, and medicine, becoming an integral part of our daily lives.

Semiconductors and Semimetals, Volume 96
ISSN 0080-8784
http://dx.doi.org/10.1016/bs.semsem.2016.07.001
2017 Published by Elsevier Inc.

Nanophotonics, the union of light science, quantum device engineering, and nanotechnology conducted at the submicron scale, is further poised to revolutionize how we utilize light, and integration will be the next key to fully benefit from and scale up impacts of this new paradigm. From conventional (e.g., lighting) to emerging (e.g., internet-of-things) applications, technological and economic needs have driven the push to develop photonic-based solutions and the enabling technologies. The aim is to facilitate fast, efficient, and low-power technology solutions with less footprint and at an ever more affordable device price point (e.g., consumer electronics (barcode scanners, blue-ray players), energy (lighting, renewable technologies), communications and information technology (internet), health and medicine (eye surgery, medical instruments, cancer detection), advanced manufacturing (laser cutting and machining), defense and national security (infrared camera, remote sensing), entertainment (holography, laser shows)). Thus, it is safe to predict that *the 21st century will depend as much on photonics as the 20th century depended on electronics*. Advances in the field of nanophotonics require innovation in both integration and enhancement capabilities of traditional as well as novel photonic products.

2. BACKGROUND

2.1 Lighting Technologies, Solid-State Lighting, and Light-Emitting Diodes

Lighting technologies have been developed to substitute for sunlight and have been pivotal in the development of mankind. Over the past 200 years, lighting has evolved from being a by-product of burning a medium, such as a candle or an oil-based lamp, to the first incandescent lamp (that generates light by heating a filament), and then the fluorescent lamp (that generates light by heating mercury or a halogen source). All such lighting forms have limited efficiencies in the <25% efficiency range at best, which means 75% of the input power is mostly lost as heat. During the past 50 years, new lighting technologies have been investigated to break the 50% efficiency barrier for *generating more light than heat*. The resulting technology, *solid-state lighting (SSL)*, is based on *light-emitting diodes* (LEDs) and can drastically reduce the energy consumption to help lower greenhouse gas emissions (Weisbuch et al., 2015). With a 9% market penetration of SSL in general lighting alone, lighting electricity consumption is to be cut by 50% and greenhouse gas emissions by >260 million tons, which is equivalent to eliminating all car gas emissions in the world. *The ultimate goal is to solve the issues preventing global SSL adoption and, thus, help prevent global warming.*

LEDs are semiconductor devices with stacks of p-type, i-type (intrinsic), and n-type layers. When a voltage is applied between p- and n-layers, holes and electrons passing through the i-layer recombine and emit photons (i.e., light). The light (i.e., photon) wavelength (energy), power (i.e., photon count), and efficiency (i.e., photon emission per injected electron–hole pair) depend strongly on the bandgap of the i-layer, current through, and light output power. Interestingly, blue LEDs, under low current density (<10 A/cm^2), have high efficiency ($\sim$75%) but low light output power (<1 W), whereas under high current density ($\sim$100 A/cm^2), they have high light output power (>10 W required for SSL) but low efficiency (<50%) (Fig. 1). This phenomenon is called *droop* and is a major bottleneck to the world-wide adoption of white LEDs (currently composed of phosphor-converted blue LEDs) for SSL. Another approach to generating white light necessary for SSL is via color mixing, where primary color blue, green, and red LEDs are mixed. Color mixing doubles energy savings with respect to the phosphor-down conversion approach because our eye responds to green light the most (Fig. 1). Although blue and red LEDs with high output power and high efficiencies are available, green LEDs with such characteristics are not, creating the so-called *green gap* problem (Fig. 1) (Auf der Maur et al., 2016).

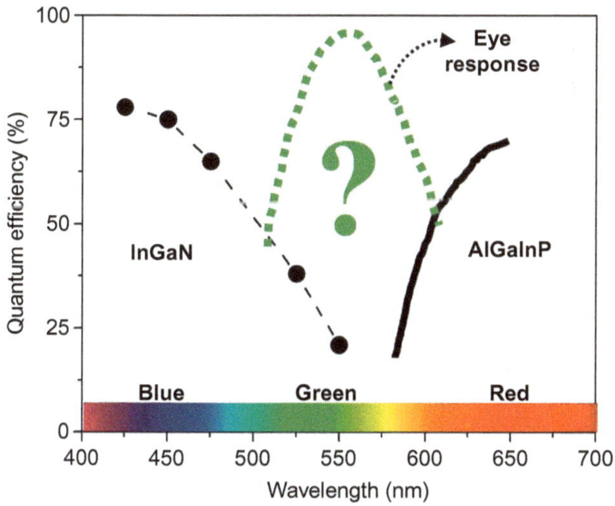

Fig. 1 Existing LED research has not yet addressed the limited LED performance in the green spectrum, where our eye has the peak response.

2.2 Current Barriers in Green LEDs

The awarding of the 2014 Nobel Prize in Physics to GaN-based "blue" LEDs has pushed fundamental research and development of the GaN-based LEDs forward. Today, LEDs based on $In_xGa_{(1-x)}N$ alloy are the most promising candidates for fulfilling the green gap. InGaN is a direct wide bandgap semiconductor with an emission that can span the useful visible spectrum from 400 nm up to 700 nm via increasing the indium content (x) of the $In_xGa_{(1-x)}N$ alloy. Basically, the alloy is precision engineered for light emission at a target wavelength and incorporated into the LED structure as the active layer. Although phosphor-converted blue LEDs are ready available for green spectrum emission, curling and delamination of phosphor prevents employment of such emitters for extended periods of time and for high-power lighting applications. Comparing the existing (i.e., color-mixing vs phosphor-converted) white color generation approaches, color-mixing approach promises: (1) more natural white light spectra and (2) higher efficiency lighting technology than any other white lighting means. The color-mixed approach will only realize its potential advantage over the phosphor-converted alternatives when green LEDs can be achieved with power conversion efficiencies in the 60–70% range (whereas today their efficiencies are ~20% as shown in Fig. 1). Many scientists have been working in this area more than a decade, and has contributed significantly to this body of research which has led directly to technological breakthroughs that have transformed LEDs into a lighting solution. *Based on this available body of research, six fundamental barriers in LEDs have been identified* and summarized in Fig. 2.

Almost without exception, green emitting LEDs employ hexagonal phase GaN materials (Fig. 3) and suffer from *high indium content* (due to required high indium content for green emission and miscibility gap of InGaN in these high-In content layers), *defectivity* (due to employment of lattice- and thermal-mismatched growth on nonnative substrates), *low p-doping* (due to large (e.g., >200 meV) activation energy of Mg impurity), *asymmetry* between electron and hole density/mobility (due to difference between p-dopant Mg and n-dopant Si activation energies (i.e., 200 meV vs 30 meV)), *polarization* (due to noncentrosymmetry and inherit piezoelectricity of the hexagonal lattice), and *Auger* nonradiative recombination (under high (>10 A/cm^2) injection densities) (Avrutin et al., 2013; Chang et al., 2012; Cho et al., 2013; Karpov, 2015; Lin et al., 2014; Meyaard et al., 2013; Tsao et al., 2014; Verzellesi et al., 2013; Weisbuch et al., 2015; Wierer et al., 2013). Hence, addressing the "green gap" problem requires a technological breakthrough.

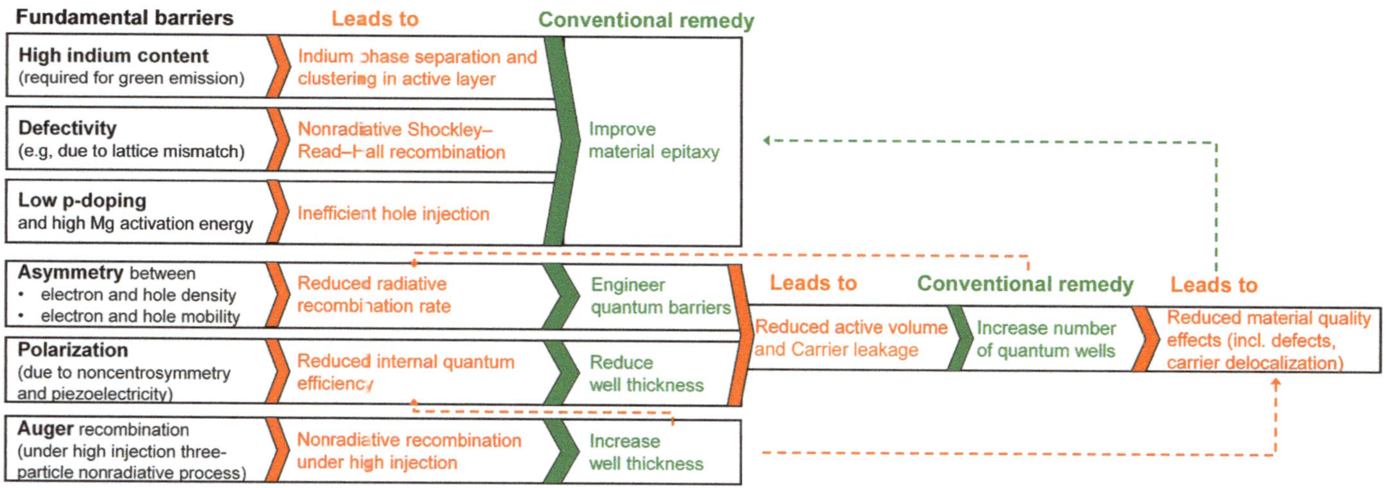

Fig. 2 Fundamental barriers (i.e., high indium content, defectivity, low p-doping, asymmetry, polarization, Auger) in today's III-nitride green light-emitting diodes and conventional remedies. Despite the best efforts, all visible LEDs are bottlenecked by these six barriers. Across the visible spectrum, green LEDs are the most affected and have the (1) lowest efficiency ($<20\%$) and (2) lowest power (<10 mW) under 35 A/cm^2 injection at 25°C, far from theoretical potential.

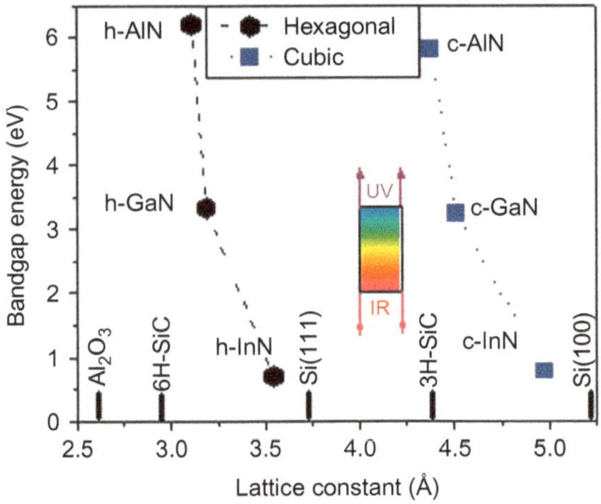

Fig. 3 Cubic phase GaN has a smaller bandgap than hexagonal one and can provide an innovative solution that will address droop and green gap issues of SSL.

Of the 10,000+ journal articles that have been written on GaN-based materials, less than 2% reported on the cubic material phase of GaN, and a yet fewer experimentally studied light-emitting structures of cubic phase (As et al., 2000; Brandt et al., 1998; Chichibu et al., 2001, 2003a; DeCuir et al., 2007; Goldys et al., 1999; Kohler et al., 2002; Li et al., 2005, 2007; Nakadaira and Tanaka, 1997; Radosavljevic et al., 2014; Tanaka and Nakadaira, 2000; Taniyasu et al., 2000; Wu et al., 1999; Yang et al., 1999). As such, there exists a need for a better approach for generating cubic phase GaN materials that could not only close the "green gap" but also open up cubic phase for applications ranging from polarization-free photonics (As, 2009) and room-temperature ferromagnetism (Chitta et al., 2004) to high-temperature spintronics (Bub et al., 2014), normally-off transistors (Abe et al., 2006; Ghosh et al., 2014; Granzner et al., 2011), and single-photon emitters (Kako et al., 2014). *This chapter aims filling this knowledge gap in an effort to break these six fundamental barriers limiting SSL.*

3. CUBIC GALLIUM NITRIDE

In this chapter, we review our novel approach capable of enabling single-crystalline, single-phase cubic GaN. We start our discussion with a new crystallographic modeling of hexagonal-to-cubic phase transition in a U-shaped groove and identify critical nanosubstrate pattern designs in doing so. Then, we will correlate light emission and material phase using photonic structures with various hexagonal-to-cubic phase ratio. Finally, we will discuss the implications of this body of research in photonics. Overall, here we establish a fundamental scientific understanding of nanosubstrate patterning on the uniformity, phase stability, and phase mitigation of GaN, and propose a systematic methodology for generating a stable and pure cubic phase material for photonic structures.

3.1 Problem Statement

The majority of the binary semiconductors have either cubic (zinc-blende) or hexagonal (wurtzite) phase. Although cubic and hexagonal structures have different symmetry (cubic and hexagonal have fourfold ($T_d^2\overline{4}3m$) and sixfold ($P6_3mc$) symmetry, respectively (Dalpian and Wei, 2004)), the two structures are in fact very similar. Both structures have the same local tetrahedral bonding and differ only in their third-nearest-neighbor atomic arrangement. Thus, the total energy difference $E_{\mathrm{Total}}^{\mathrm{hexagonal}} - E_{\mathrm{Total}}^{\mathrm{cubic}}$ is usually very small, on the order of a few meV/atom (Dalpian and Wei, 2004). An ionic compound such as GaN prefers the hexagonal phase, whereas covalent compounds such as GaAs or CdTe prefer the cubic phase thermodynamically. Thus, conventional GaN materials are in hexagonal phase and have been grown on similar (three- or sixfold) symmetry substrates including sapphire (Al_2O_3), 4H-SiC, and Si(111). However, *cubic phase has potential superiority over hexagonal phase* for certain device applications and integrity of GaN devices. For example, cubic GaN is *polarization-free* along the growth direction, shows a *higher mobility* resulting from its *lower phonon scattering* in the higher crystallographic symmetry (Dhara et al., 2004), has a *high p-type conductivity* (Dhara et al., 2004), *higher electron drift velocity* (Caerano et al., 1999), requires *less indium* content for green emission thanks to its lower bandgap, and has a *cleavage plane* for laser fabrication. It is also predicted theoretically that c-GaN quantum wells have *higher optical gain* (Ohtoshi et al., 1998) and *smaller Auger losses* (Delaney et al., 2009) than that of hexagonal GaN wells. Overall, cubic phase GaN has the *potential to overcome four fundamental barriers* in LEDs. *Nonetheless, cubic phase devices have almost been unexplored due to phase instability.* Thermodynamical instability of cubic phase GaN, epilayer-substrate chemical incompatibility (i.e., III-N vs III-As), and large lattice mismatch (i.e., ~20%) are some of the key issues bottlenecking epitaxial deposition schemes limiting studies in cubic phase materials.

3.2 State of the Art

In the literature, cubic phase GaN has first been realized via direct deposition on cubic phase substrates such GaAs (As et al., 2000; Taniyasu et al., 2000), 3C-SiC (Abe et al., 2006; Chichibu et al., 2003b), Si(100) (Wei et al., 2000), and MgO (Garcia et al., 2015; Sanorpim et al., 2011). However, such direct deposition on nonnative substrates suffers from high defectivity, structural metastability (i.e., phase mixing), chemical metastability (i.e., tendency toward spinodal decomposition), and in some instances (such as GaAs) chemical incompatibility (Kemper et al., 2011; Scolfaro, 2002; Waheeda et al., 2014; Xu et al., 2002). Another approach to form cubic phase GaN is to mitigate hexagonal phase into a cubic one by impurity incorporation (such as Ga (Dhara et al., 2004) and Mn (Cui et al., 2003)). This technique relies on p-d orbital repulsion between the 3d impurity levels requiring a large density of impurity incorporation, thus reducing the resulting material quality; it is thus unusable in device applications (Dalpian and Wei, 2004). Recently, substrate patterning (e.g., with a V-shaped [Si{111}–Si{111}] groove structures) (Bayram et al., 2014; Lee et al., 2005; Stark et al., 2013) was introduced. This idea is based on the crystallographic fact that hexagonal crystal $\langle 0001 \rangle$ direction and cubic crystal $\langle 111 \rangle$ direction are equivalent (Bayram et al., 2015a). It follows from this fact that if two hexagonal phase growth fronts merge, a cubic phase would form. However, GaN material grown in such a V-groove is reported to possess a mixed phase of cubic and hexagonal phase GaN material rather than a pure cubic phase material (Lee et al., 2005; Stark et al., 2013). Overall, *there are no methods to synthesize a pure cubic phase material* without hexagonal phase incursions, let alone a cubic phase material that is stable above 500 °C for use in devices. The first step in addressing this important need is to understand hexagonal-to-cubic phase transition mechanisms.

3.3 Novel Approach: Hexagonal-to-Cubic Transition

Recently, we proposed *a new method to enable cubic III-nitrides via U-shaped [Si{111}–Si{100}–Si{$1\bar{1}\bar{1}$}] groove structures* (Bayram et al., 2014, 2015a). Based on the crystallographic fact that h-crystal $\langle 0001 \rangle$ direction and c-crystal $\langle 111 \rangle$ direction are equivalent, when two h-phase growth fronts merge with an angle of ~110 degree (i.e., the angle between the two Ga–N bonds in the hexagonal tetrahedral bonding), a c-phase would form after seam (Fig. 4). To realize, here we employ anisotropic nanopatterning of on-axis

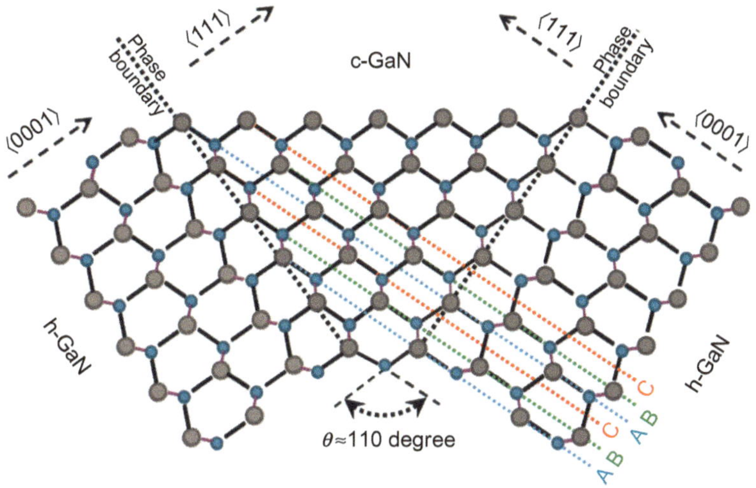

Fig. 4 Hexagonal-to-cubic phase transition. Sketch shows h-phase ⟨0001⟩ direction and c-phase ⟨111⟩ direction equivalency. When two h-phase growth fronts merge with an angle of ~110 degree (i.e., the angle between the two Ga–N bonds in the hexagonal tetrahedral bonding), a c-phase would form after seam.

CMOS-compatible Si(100) substrates and create a novel U-shaped nanopattern with a crystallographic angle of 54.74 degree between the (100) and (111)Si surfaces (Figs. 5–7). Thanks to our proprietary growth conditions (Bayram et al., 2015b), we nucleate h-GaN selectively on the Si(111) surfaces. Thus, continuous growth leads to two h-GaN growth fronts meet in an apex angle (θ) of 54.74 degree $\times$ 2 = 109.48 degree, that is exactly the angle between the two Ga-N bonds in hexagonal tetrahedral bonding. Our novel technique enables transformation of the hexagonal (h-) GaN into a cubic (c-) GaN after seam (Fig. 7).

3.4 Crystallographic Modeling

Crystallographic geometry modeling based on the schematic of the growth pattern of GaN in a U-groove is carried out in this section. Fig. 5 shows the model sketch of GaN in a U-groove. The V-shaped red dashed lines refer to the phase boundaries where hexagonal-to-cubic phase transitions occurs, and the top (red) dashed line indicates the c-GaN top surface. Under the selective growth conditions, GaN nucleates on the Si(111) facets only and these h-GaN growth fronts meet at a 109.48 degree angle in the middle of the groove (see Fig. 7 for the experimental result). After the middle of

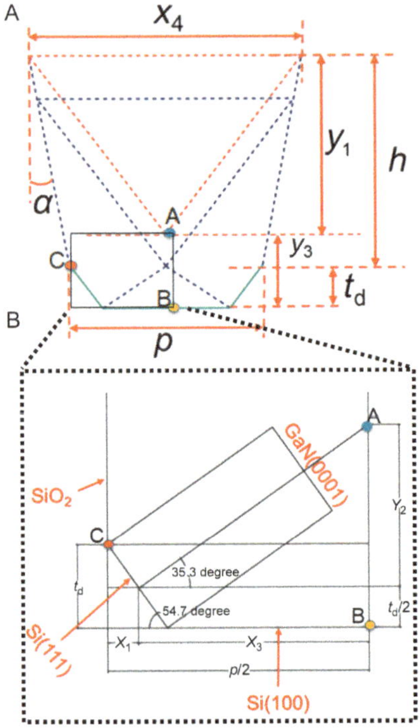

Fig. 5 Sketch of GaN deposition in U-groove model (under selective growth conditions) is shown with dimensions labeled. (A) The *red dashed triangle* indicates the central region where GaN has phase transition from hexagonal to cubic. *Light blue lines* indicate the silicon surfaces, whereas the *dark blue lines* are for illustrative purposes only. (B) Zoomed-in schematic of the *boxed region* indicated with continuous *black box* in (A). Points labeled as A, B, C in (B) correspond to the same points A, B, C in (A) (Liu and Bayram, 2016a).

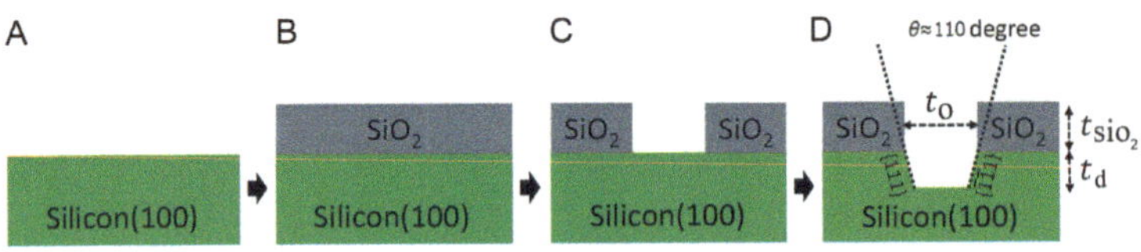

Fig. 6 Sketch of U-nanogroove fabrication steps. (A) RCA cleaning, (B) oxide (SiO_2) growth, (C) lithography and reactive ion etching of oxide (SiO_2), and (D) KOH dip (10%). Thanks to selective KOH etching between Si{100} and Si{111} planes, an ideal apex angle of 110° is achieved. *Adapted from Bayram, C., Ott, J., Shiu, K.-T., Cheng, Z. Y., Kim, J., Razeghi, M., Sadana, D.K., 2014. Cubic phase GaN on nano-grooved Si (100) via maskless selective area epitaxy.* Adv. Funct. Mater. ***24*** *(28) 4491.*

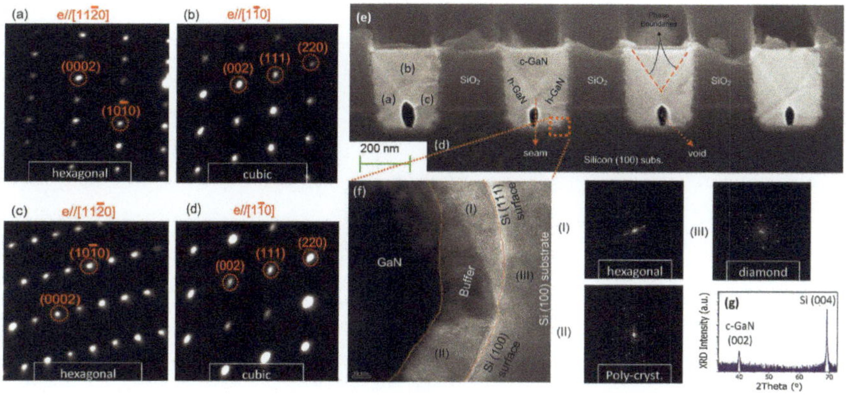

Fig. 7 A new method of cubic phase synthesis: hexagonal-to-cubic transformation. (A–E) Fast Fourier transform (FFT) diffraction patterns in (A–D) belong to (A, C), (B), (D) regions shown in (E) corresponding to h-GaN, c-GaN, and Si(100), respectively. Electron beam is parallel to $[11\bar{2}0]$ and $[1\bar{1}0]$ zones of wurtzite and cubic phase GaN, respectively. (F) HR-TEM image of the GaN {0001}–Si{111} interface shows crystalline AlN buffer on Si{111} surfaces, whereas poly AlN on Si{100}. FFT diffraction patterns from selected area regions (I), (II), and (III) are observed to be hexagonal single crystalline, polycrystalline, and diamond single crystalline. (G) Omega/2Theta XRD scan of GaN on a U-shaped nanopatterned Si(100) shows a clear cubic phase diffraction peak.

the growth fronts meet (point A in Fig. 5), GaN grown on top will phase transition to cubic phase. Under these experimental/crystallographic observations, the geometrical modeling is carried out as follows (Liu and Bayram, 2016a).

In the following derivation x_i and y_i are various dimensions as shown in Fig. 5, h_c is the critical GaN deposition thickness (defined as the GaN deposition height above Si(100) that maximizes cubic phase GaN coverage on the U-groove surface), t_d is the etch depth, p is the opening width, and α is the oxide sidewall angle.

Fig. 5B shows the zoomed in region indicated by the dotted square in Fig. 5A. Points A, B, C in Fig. 5A and B correspond to the same exact locations. The crystallographic angles (54.74 degree between the (100) and (111) Si surfaces) are shown accordingly.

From Fig. 5B, we see that:

$$\tan 35.3° = \frac{y_2}{x_3} \tag{1}$$

$$\tan 54.7° = \frac{t_d}{2x_1} \tag{2}$$

$$x_1 + x_3 = \frac{p}{2} \tag{3}$$

$$y_3 = y_2 + \frac{t_d}{2} \tag{4}$$

Rearranging (1), (2), and (3), we get:

$$y_2 = x_3 \tan 35.3° \tag{5}$$

$$x_1 = \frac{t_d}{2 \tan 54.7°} \tag{6}$$

$$x_3 = \frac{p}{2} - x_1 \tag{7}$$

Substituting (6) into (7):

$$x_3 = \frac{p}{2} - x_1 = \frac{p}{2} - \frac{t_d}{2 \tan 54.7°} \tag{8}$$

Then using (8) in (5):

$$y_2 = x_3 \tan 35.3° = \left(\frac{p}{2} - \frac{t_d}{2 \tan 54.7°}\right) \tan 35.3° \tag{9}$$

From Fig. 5A,

$$x_4 = p + 2\ h_C \tan \alpha \tag{10}$$

$$y_1 + y_3 = h_C + t_d \tag{11}$$

$$\tan 35.3° = \frac{x_4}{2y_1} \tag{12}$$

Rearranging (12) using (10), and (11) using (4):

$$y_1 = \frac{x_4}{2 \tan 35.3°} = \frac{p + 2\ h_C \tan \alpha}{2 \tan 35.3°} \tag{13}$$

$$h = y_1 + y_3 - t_d = y_1 + y_2 - \frac{t_d}{2} \tag{14}$$

Substituting (13) and (9) in (14), we get:

$$h_C = \frac{p + 2\ h_C \tan \alpha}{2 \tan 35.3°} + \left(\frac{p}{2} - \frac{t_d}{2 \tan 54.7°}\right) \tan 35.3° - \frac{t_d}{2} \tag{15}$$

Solving for h_c

$$h_C = \left(1 - \frac{\tan\alpha}{\tan 35.3°}\right)^{-1} \left[\left(\frac{p}{2} - \frac{t_d}{2\tan 54.7°}\right)\tan 35.3° + \frac{p}{2\tan 35.3°} - \frac{t_d}{2}\right] \tag{16}$$

Simplify and plugging in values for the tangents, we have the relationship between critical thickness and the patterning parameters:

$$h_C = \frac{[1.06p - 0.75t_d]}{\left(1 - \frac{\tan\alpha}{0.71}\right)} \tag{17}$$

If α is negligible ($\approx 0°$), the relationship simplifies to (Liu and Bayram, 2016a):

$$h_C = [1.06p - 0.75t_d] \tag{18}$$

3.5 Substrate Preparation

Our approach moves away from the conventional V-shaped [Si{111}–Si {111}] groove structures which have generally resulted in highly defective and mixed phase GaN (Lee et al., 2005; Stark et al., 2013). To increase the control over cubic phase purity, we consider a U-shaped [SiO_2–Si{111}–Si {100}–Si{111}–SiO_2] groove for generating the cubic phase (Bayram et al., 2014).

Fig. 6 shows the schematic process flow for silicon(100) substrate preparation prior to MOCVD growth. First, CMOS-compatible on-axis Si(100) substrate is cleaned via standard RCA process. Then, 500 nm-thick thermal oxide (SiO_2) is grown, followed by deep UV (193 nm) lithography. Subsequently, reactive ion etching of oxide (SiO_2) is realized to expose Si(100) surface on parts of the wafer. After another RCA-like cleaning process, the wafer is submersed into KOH (10%) solution for ~1 min to selectively etch Si and expose {111} facet family. By this approach, two different groove structures [SiO_2–Si{111}–Si{100}] and [SiO_2–Si{111}–Si {100}–Si{111}–SiO_2] are formed on on-axis Si(100) parent substrate, as shown in Fig. 6D. After desired groove structures are formed and surface native oxide is removed via dilute HF treatment, wafers are loaded inside the MOCVD chamber for growth.

3.6 Material Growth and Characterization

Our selective MOCVD scheme (Bayram et al., 2015b) consists of four steps: (1) ammonia-free heat up, (2) prealuminization, (3) AlN buffer deposition, and (4) GaN layer deposition. Under selective MOCVD growth conditions, GaN regrowth is nucleated on the Si{111} family of planes rather than Si{100} ones, as shown earlier. Thus, [SiO_2–Si{111}–Si{100}–Si{111}–SiO_2] groove structures have only two nucleation sites for GaN regrowth: Si(111) and Si$(\bar{1}\,\bar{1}1)$ surfaces. When GaN growth fronts initiated from facing directions meet, a void is formed in the middle where the seam forms (Fig. 7). The size of the void depends on the groove period as well as the growth parameters.

Fig. 6D provides a representative example of a U-shaped groove with corresponding substrate-nanopattern parameters to be inquired. Crystallographic relations between cubic and hexagonal phase of GaN and Si(100) are used to investigate optimum hexagonal-to-cubic phase transition parameters of opening width (t_O), oxide thickness (t_{SiO_2}), and etch depth (t_d) that minimizes hexagonal incursions in cubic phase materials.

Based on our proposed crystallographic geometry modeling (see Section 3.4), a schematic of the growth pattern of GaN in a U-groove seen from the side is shown in Fig. 8 (Liu and Bayram, 2016a).

For complete c-GaN surface coverage in these U-grooves, the critical GaN deposition thickness (h_c) is shown obeying $h_c = (1.06p - 0.75t_d)/(1 - \tan\alpha/0.71)$, where p is the opening width of the U-groove at the dielectric/substrate interface, t_d is the silicon etch depth, and α is the dielectric sidewall angle (Figs. 5A and 8) (Liu and Bayram, 2016a). Additionally,

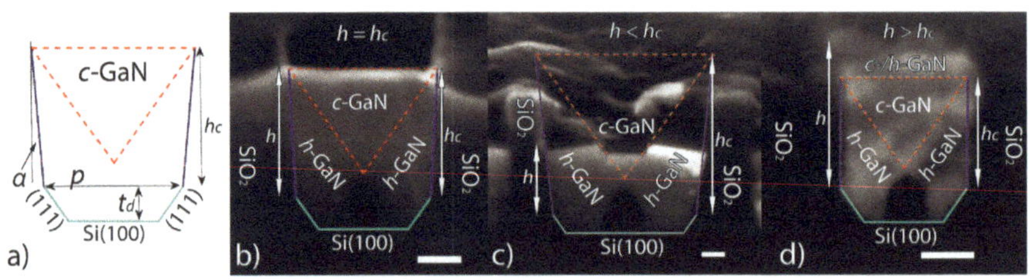

Fig. 8 (A) Schematic defining critical deposition thickness (h_c), etch depth (t_d), opening width (p), and sidewall angle (α). Cross-sectional SEM views of GaN deposited in a U-groove when: (B) ($h = h_c$) complete c-GaN surface coverage is achieved ($ff \approx 100\%$), (C) ($h < h_c$) the U-groove is under-filled with GaN ($ff \approx 40\%$), and (D) ($h > h_c$) the U-groove is over-filled with GaN ($ff \approx 120\%$). This leads to phase mixing. The *white scale bars* in the *bottom right* of each figure correspond to 100 nm (Liu and Bayram, 2016a).

we define a quantity, fill factor (ff) as $ff = h/h_c$, to quantify the ratio between actual and ideal GaN deposition thickness for complete c-GaN surface coverage. Here h is defined as the GaN deposition thickness above Si(100) substrate level (Fig. 8B–D).

Fig. 8B–D shows cross-sectional scanning electron microscope (SEM) images of GaN in three different U-grooves. Fig. 8B is an ideal case ($ff \approx 100\%$) where GaN has a perfect fill ($h = h_c$) for the patterning parameters of $t_d = 70$ nm, $p = 320$ nm; and $\alpha \approx 4°$. The top c-GaN(100) surface is seen very flat. Fig. 8C shows an under-filled case ($h < h_c$, $ff \sim 40\%$) with $t_d = 85$ nm, $p = 670$ nm; and $\alpha \approx 4°$. On the surface, a small amount of the flat c-GaN(100) can be seen in the center flanked by two h-GaN $(1\bar{1}01)$ planes, which make a 7-degree angle to the (100) planes of both the Si substrate and c-GaN. For c-GaN to cover the entire surface, more deposition is needed. Fig. 8D shows an over-filled case ($h > h_c$, $ff \sim 130\%$) in which too much GaN has been deposited for $= t_d = 65$ nm, $p = 230$ nm; and $\alpha \approx 0°$. There is no identifiable crystal plane on the surface, implying overdeposition of GaN ($h > h_c$) causes the over-filled GaN to phase mix (Bayram et al., 2014).

Various U-grooves (with p ranging from 200 to 1110 nm, and t_d from 65 to 85 nm) are fabricated, and GaN is deposited in these various U-grooves to create a variety of ff ranging from 30% to 210%. Cathodoluminescence is used to study c- and h-GaN emission properties (Liu and Bayram, 2016b); it provides insights into the composition of the surface by properly adjusting the electron acceleration voltage and deliberately creating electron–hole pairs at specific depths.

Fig. 9 shows the room (Fig. 9A at 280 K) and low (Fig. 9B at 5.7 K) temperature CL spectra of 12 different types of GaN on U-groove structures. Our control sample is the $p = \infty$ nm, where there is no opposing wall or h-GaN growth front, and therefore no phase transition, leading to pure h-GaN. The ff, p, and t_d of U-grooves are listed beside to their respective curves in Fig. 9A.

For the two U-grooves with $ff \approx 100\%$, which corresponds to the ideal case for complete cubic phase surface coverage (see SEM image in Fig. 8B), the CL emission spectra show strong c-GaN presence with band-to-band photon energies of 3.22 eV at 280 K (Fig. 9A), and 3.29 eV at 5.7 K (Fig. 9B), marked with the dotted vertical lines. These two U-grooves exhibit additional luminescence mechanisms such as the two donor–acceptor pairs (DAP_1 and DAP_2). We believe DAP_1 is an impurity level

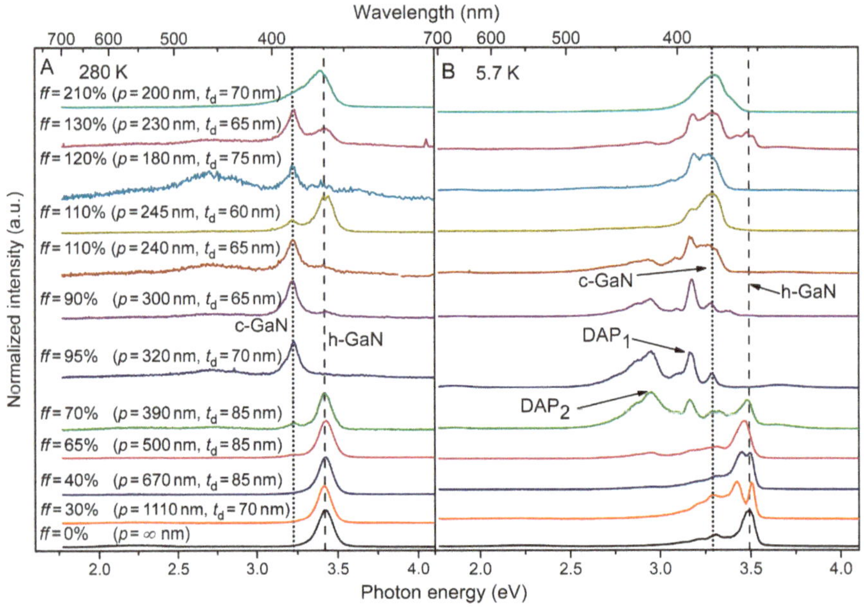

Fig. 9 CL spectra of GaN on various U-grooves at (A) 280 K and (B) 5.7 K. When *ff*≪100%, mostly h-GaN emission (*dashed lines*) is observed, suggesting incomplete c-GaN surface coverage at 280 K and 5.7 K. When *ff*≈100%, strong cubic band-to-band transition is observed (*dotted lines*) at 280 K and cubic DAPs are observed at 5.7 K. When *ff*≫100%, both c- and h-GaN emissions are recorded at 280 K, which become dominantly c-GaN at 5.7 K. This study used an injection current of 2 nA and electron acceleration voltage of 2 kV, so the interactive depth of electrons is restricted to the top ~50 nm (Liu and Bayram, 2016a).

(e.g., Si, C, V_{Ga}), which is present in both phases of GaN that has shown to be closer to the conduction band of c-GaN (~100 meV) than that of h-GaN (~180 meV). DAP_2 shows involvement of another deeper defect level (~230 meV) that is not seen in h-GaN. LO-phonons of both DAPs, especially DAP_2, are evident at ~80 meV intervals less than the main transitions.

For *ff* < 100% (the bottom five U-grooves in Fig. 2), CL emissions are dominated by the h-GaN, as seen in the *ff* = 0 ($p = \infty$ nm) U-groove at 3.42 and 3.49 eV (dashed lines) at 280 and 5.7 K, respectively. These U-grooves correspond to the cross-sectional SEM image in Fig. 8C, with a varying portion of c-GaN in the center, flanked by two $(1\bar{1}01)$ planes of h-GaN. As the *ff* approaches 100% from 30%, c-GaN emission becomes observable and intensifies, resulting in a superposition of both h- and c-GaN, indicating the amount of c-GaN in the center relative to the entire groove opening is increasing.

For *ff* > 100% (the top five U-grooves in Fig. 9), CL spectra show a mixture of h- and c-GaN emissions. These U-grooves correspond to the cross-sectional SEM image in Fig. 1D, where the U-groove is overfilled, and no clear crystal plane is distinguishable. At 280 K, h- and c-GaN emissions of various amplitudes can be seen without any particular pattern. The CL spectra suggest that once h_c is passed, the subsequently deposited GaN is of mixed phase. Interestingly, despite the various intensities of h-GaN emission at 280 K across these U-grooves, all of them show predominantly the band-to-band emission of c-GaN at 5.7 K. DAP_1 can also be seen in four of the five U-grooves, and its intensity seems to correlate to the intensity of c-GaN emission at 280 K. We believe that such dominance of c-GaN emission at 5.7 K even in mixed-phase GaN materials is due to carrier confinement in c-GaN as its bandgap (3.29 eV) is smaller than that of h-GaN (3.49 eV).

To study the GaN phase along the vertical growth direction $\langle 100 \rangle$, we vary the CL electron acceleration voltage and record the emission spectra. The relationship between maximum electron penetration depth and primary electron acceleration voltage is described by Kanaya–Okayama formula: $D = 0.0267 \times A \times E_0^{5/3} / \rho \times Z^{8/9}$ where D is the maximum penetration depth, A the atomic weight, E_0 the beam energy, ρ the density, and Z the weighted average atomic number (Kanaya and Okayama, 1972). Fig. 10 (inset) shows the maximum electron penetration depth as a function of electron acceleration voltage for GaN materials, assuming the similar behavior in both h- and c-GaN. Fig. 10 shows the CL spectra of the optimized c-GaN ($ff \approx 100\%$) U-groove recorded for various electron acceleration voltages (from 1 to 8 kV) under a constant current of 2 nA at room temperature. At voltages up to 2 kV ($D = 42$ nm), no h-GaN emission is observed. This indicates that the GaN is composed almost entirely of c-GaN, in good agreement with our modeling. Beginning at 3 kV ($d = 84$ nm) h-GaN emission in the CL spectrum is observed. As voltage increases, more EHPs are created in the deeper parts of the U-grooves, and as a result, more EHPs recombine in the h-GaN region, emitting at the bandgap of h-GaN, as shown in Fig. 10.

Further electron backscatter diffraction (EBSD) measurements are carried out in order to study the distribution and location of the different GaN phases on the U-groove surfaces. Fig. 11 shows the plane-view EBSD images overlaid on top of SEM images of the (a) $ff = 40\%$, (b) $ff = 65\%$, and

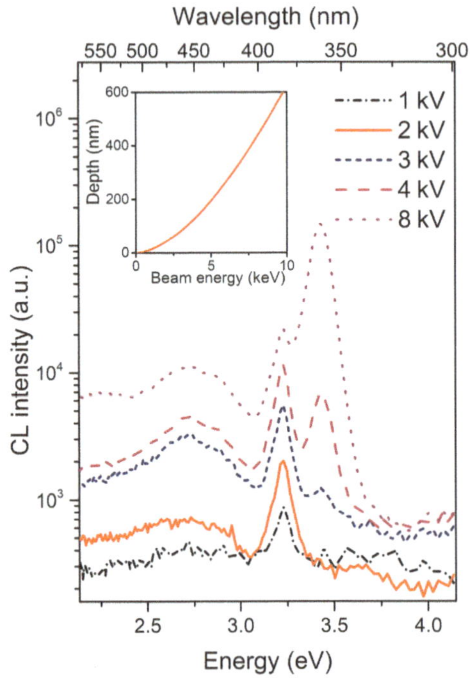

Fig. 10 CL spectra of GaN on U-grooves with *ff* ≈ 100% are shown as a function of electron acceleration voltage. *Inset* shows calculated maximum electron penetration depth for GaN material. At 2 kV (range = 44 nm) and below, electrons are not able to penetrate the c-GaN to reach h-GaN. At 3 kV (range = 84 nm) and above, EHPs are being created in the h-GaN that corresponds to increase in h-GaN emission in the CL spectra (Liu and Bayram, 2016a).

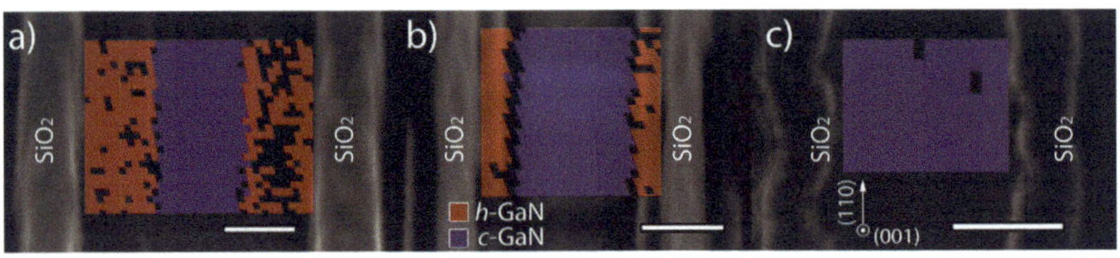

Fig. 11 Top-view SEM images with EBSD overlay of GaN on U-grooves with (A) *ff* = 40%, (B) *ff* = 65%, and (C) *ff* ≈ 100%. Each pixel has resolution of 20 × 20 nm, and *red, blue, gray* colors correspond to h-GaN, c-GaN, and indistinguishable phases, respectively. The crystal orientation of c-GaN is identified and labeled in (C). The *white scale bars* in the *bottom right* of each figure correspond to 200 nm (Liu and Bayram, 2016a).

(c) *ff* ≅ 100% U-grooves. The red, blue, and gray pixels indicate hexagonal, cubic, and indistinguishable crystallography structures, respectively, and have the spatial resolution dimension of 20 nm × 20 nm. EBSD identifies and confirms the GaN in the cubic phase, with its (001) plane on the top surface and its (110), $(1\overline{1}0)$ planes parallel and perpendicular to the U-grooves, respectively. As the *ff* of the U-groove approaches 100%, the surface coverage of c-GaN increases, until the entire surface is composed of only c-GaN, as shown in Fig. 11C. EBSD shows that the c-GaN formed through phase transition in a U-groove is of single phase and proves U-groove nanopatterning is a feasible and reliable method of forming single-phase c-GaN surface.

4. FUTURE PROSPECTS OF CUBIC GaN MATERIALS

Almost without exception, GaN LEDs (or any kind of GaN devices) are grown on three- or sixfold symmetry surfaces because naturally occurring GaN hexagonal (wurtzite) material phase has sixfold symmetry. GaN LEDs grown on such surfaces are thus hexagonal which is *noncentrosymmetric* (meaning that Ga and N atoms cannot be interchanged in the lattice), leading to residual *spontaneous polarization* fields (Wood and Jena, 2008). Moreover, hexagonal GaN materials are *piezoelectric*—meaning that under stress additional piezoelectric fields are generated (Wood and Jena, 2008). Both of these polarization fields, spontaneous and piezoelectric, are along $\langle 0001 \rangle$ growth direction, which is also the carrier injection direction in vertical transport devices such as LEDs, lasers, and detectors, and thus detrimental to recombination dynamics and device efficiency (Avrutin et al., 2013; Chang et al., 2012; Cho et al., 2013; Karpov, 2015; Lin et al., 2014; Meyaard et al., 2013; Tsao et al., 2014; Verzellesi et al., 2013). On the other hand, cubic phase materials do not possess any polarization fields along the $\langle 001 \rangle$ direction, which *opens up polarization-free photonics research.* Other advantages of cubic phase material in photonic devices include having a cleavage plane and higher optical gain (Caerano et al., 1999; Dhara et al., 2004; Ohtoshi et al., 1998). Yet, cubic phase GaN is one of the least studied materials due to the phase instability. Even today, there are debates on the exact bandgap of the GaN binary compound, and lack of knowledge on bandgap of the $In_xGa_{(1-x)}N$ ternary compounds as well as their bowing parameters as a function of indium content (x). As such, the investigation of the emission characteristics of cubic phase GaN and its dependence on the In-content will be a valuable contribution to the field and may solve the long-standing debate on the emission fundamentals of cubic phase InGaN alloys.

Polarization-free GaN photonic research has primarily focused on switching the growth plane from inherently polar ones to inherently polarization-free (a.k.a. nonpolar) ones. Practically, the most common non-polar substrate is the m-plane freestanding GaN one (i.e., making 90 degree with respect to ⟨0001⟩ direction) that is sliced along {1100} planes out of GaN boules grown along ⟨0001⟩ direction. Hence, m-plane substrates are small in area (~cm^2) and are expensive. Moreover, (1) high-quality epitaxial regrowth of GaN on nonpolar surfaces is challenging due to low surface energy of nonpolar planes (Sun et al., 2011; Zhao et al., 2012), and (2) alloying (such as $In_xGa_{(1-x)}N$) is significantly degraded for nonpolar structures (Zhao et al., 2012), limiting the promise of such substrates in LEDs and laser diodes applications. It is important to note that such research is still conducted on the hexagonal material phase of GaN that has piezoelectricity, leading to electric fields (kV/cm). An alternative method of enabling polarization-free GaN devices is switching to the cubic phase GaN materials that are completely free of spontaneous and piezoelectric polarization along the common ⟨001⟩ direction. Moreover, the optical gain of a cubic phase GaN/InGaN quantum well is larger than that of hexagonal one because of its smaller electron and hole effective masses as well as its weaker TE-polarization on account of the small spin–orbit splitting energy (Kim and Han, 1998). The understanding of fundamental cubic phase material behavior in alloying and investigating alloyed cubic phase InGaN material will enable a more accurate prediction of polarization-free photonic devices, a necessary requirement for enabling high-efficiency, high-power vertical devices. Moreover, due to the cubic GaN possessing a lower bandgap, cubic InGaN active layers are expected to have ~10% less indium content (Kitamura et al., 2001; Park and Lee, 2010) and possess no miscibility gap for green emission (Caetano et al., 2006)—making cubic phase InGaN active layers ideal for green emitters.

Radiative recombination rate is proportional to the square of the overlap integral of the electron and hole wave functions as the radiative recombination rate, according to Fermi Golden Rule, is proportional to the square matrix element of the electron–phonon coupling Hamiltonian between electron and hole wave functions, which reduces to square of the overlap integral for spontaneous recombination. Our simulation results suggest significant improvement (more than two orders) in the overlap integral when a cubic phase material is used rather than a hexagonal material phase (see Fig. 12A). This means that cubic phase active layers will have more than four orders of magnitude larger radiative recombination rates. Experimentally, we have recently demonstrated room-temperature emission from cubic phase GaN materials integrated on nanopatterned Si and reported on the

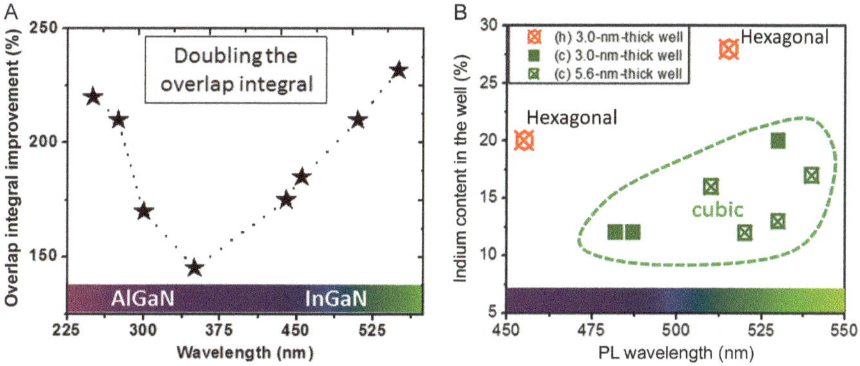

Fig. 12 Our preliminary investigation shows cubic phase quantum well (QW) structures being ideally suited for green LEDs thanks to improved electron–hole overlap and reduced indium content in the active layer. (A) Our preliminary calculations of the electron–hole wave function overlap integral in h- and c-QWs show more than two orders of magnitude improvement in the latter. (B) Comparison of c- and h-InGaN/GaN MQWs shows (i.e., for 3-nm-thick InGaN QW) cubic active layers needing (~10%) less indium in the InGaN well than hexagonal ones.

cubic GaN emission characteristics (Bayram et al., 2014; Liu and Bayram, 2016a,b). Overall, cubic GaN materials are very promising for polarization-free photonics, and integrated approach enables a scalable solution.

5. CONCLUSION

As stated previously, nearly all of the 10,000+ articles in the GaN and related area utilize an inherently polar material referencing hexagonal (wurtzite) phase of GaN. This approach precludes the ability of the LED to have high radiative recombination rates, leading to degraded efficiency performance. Particularly, larger the indium content for green emission, larger is the lattice mismatch and piezoelectric fields in a green LED. Thus, further In-content in the active layer for green emission lowers the LED efficiency. In addition, all GaN-based LEDs—without exception—have reduced light emission efficiency as injection current increases—a detrimental phenomenon known as "droop"—limiting the power output of GaN LEDs. The "droop" effect alone is shadowing the theoretical performance of LEDs, preventing the world-wide adoption of LEDs, and exists in every spectrum. The biggest contributors to "droop" effect in green LEDs are believed to be the high indium content, defectivity, low p-doping, asymmetry, polarization, and Auger.

To address these current limitations, *we propose a cubic phase LED approach that has intrinsically more isotropic properties* (including smaller effective masses, higher drift velocity, higher carrier mobility, higher doping efficiency, higher optical gain, smaller Auger losses) than hexagonal counter parts, has no polarization, and requires at least $\sim 10\%$ less indium in the active layer (with respect to hexagonal ones) for the green emission.

Despite the environmental, technological, and sociological advantages of LEDs over existing lighting sources, world-wide adoption of SSL is lagging due to both technical and financial hurdles. *LEDs of cubic phase provide innovative solutions that can address droop and green gap issues enabling global SSL adoption, bringing substantial energy savings, helping prevent global warming, and generating novel research fronts.*

ACKNOWLEDGMENTS

This work was carried out in the Micro and Nanotechnology Laboratory and Frederick Seitz Materials Research Laboratory Central Research Facilities, University of Illinois at Urbana–Champaign, IL, USA.

REFERENCES

Abe, M., Nagasawa, H., Potthast, S., Fernandez, J., Schormann, J., Josef, D., Lischka, K., 2006. Cubic GaN/AlGaN HEMTs on 3C-SiC substrate for normally-off operation. IEICE Trans. Electron. E89-C (7), 1057–1063.

As, D.J., 2009. Cubic group-III nitride-based nanostructures—basics and applications in optoelectronics. Microelectron. J. 40, 204–209.

As, D.J., Richter, A., Bush, J., Lubbers, M., Mimkes, J., Lischka, K., 2000. Electroluminescence of a cubic GaN/GaAs (001) p-n junction. Appl. Phys. Lett. 76, 13–15.

Auf der Maur, M., Pecchia, A., Penazzi, G., Rodrigues, W., Di Carlo, A., 2016. Efficiency drop in green InGaN/GaN light emitting diodes: the role of random alloy fluctuations. Phys. Rev. Lett. 116, 027401.

Avrutin, V., Hafiz, S.A., Zhang, F., Ozgur, U., Morkoc, H., 2013. InGaN light-emitting diodes: efficiency-limiting processes at high injection. J. Vac. Sci. Technol. A 31 (5), 050809.

Bayram, C., Ott, J., Shiu, K.-T., Cheng, C.-W., Zhu, Y., Kim, J., Razeghi, M., Sadana, D.K., 2014. Cubic phase GaN on nano-grooved Si (100) via maskless selective area epitaxy. Adv. Funct. Mater. 24 (28), 4492.

Bayram, C., Cheng, C.-W., Sadana, D.K., Shiu, K.-T., 2015a. Dual phase gallium nitride material formation on (100) silicon. U.S. Patent 9,048,173 (issued June 2, 2015).

Bayram, C., Cheng, C.-W., Sadana, D.K., Shiu, K.-T., 2015b. Selective gallium nitride regrowth on (100) silicon. U.S. Patent 9,059,075 (issued June 16, 2015).

Brandt, O., Mullhauser, J.R., Yang, B., Yang, H., Ploog, K.H., 1998. Optical properties of cubic GaN and (In, Ga)N. Phys. E 2, 532–538.

Bub, J.H., Schaefer, A., Schupp, T., As, D.J., Hagele, D., Rudolp, J., 2014. High temperature electron spin dynamics in bulk cubic GaN: nanosecond spin lifetimes far above room-temperature. Appl. Phys. Lett. 105, 182404.

Caerano, E.W.S., Filho, R.N.C., Freire, V.N., da Costa, J.A.P., 1999. Velocity overshoot in zincblende and wurtzite GaN. Solid State Commun. 110, 469–472.

Caetano, C., Teles, L.K., Marques, M., Dal Pino, A., Ferreira, L.G., 2006. Phase stability, chemical bonds, and gap bowing of InxGa1-xN alloys: comparison between cubic and wurtzite structures. Phys. Rev. B 74, 045215.

Chang, M.-H., Das, D., Varde, P.V., Pecht, M., 2012. Light emitting diodes reliability review. Microelectron. Reliab. 52, 762–782.

Chichibu, S.F., Sugiyama, M., Onuma, T., Kitamura, T., Nakanishi, H., Kuroda, T., Tacheuchi, A., Sota, T., Ishida, Y., Okumura, H., 2001. Localized exciton dynamics in strained cubic In0.1Ga0.9N/GaN multiple quantum wells. Appl. Phys. Lett. 79, 4319.

Chichibu, S.F., Onuma, T., Sota, T., DenBaars, S.P., Nakamura, S., Kitamura, T., Ishida, Y., Okumura, H., 2003a. Influence of InN mole fraction on the recombination processes of localized excitons in strained cubic InxGa1-xN/GaN multiple quantum wells. J. Appl. Phys. 93, 2051–2054.

Chichibu, S.F., Onuma, T., Aoyama, T., Nakajima, K., Ahmet, P., Chikyow, T., Sota, T., DenBaars, S.P., Nakamura, S., Kitamura, T., Ishida, Y., Okumura, H., 2003b. Recombination dynamics of localized excitons in cubic InxGa1-xN/GaN multiple quantum wells grown by radio frequency molecular beam epitaxy on 3C-SiC. J. Vac. Sci. Technol. B 21 (4), 1856–1862.

Chitta, V.A., Coaquira, J.A.H., Fernandez, J.R.L., Duarte, C.A., Leite, J.R., Schikora, D., As, D.J., Lischka, K., Abramof, E., 2004. Room temperature ferromagnetism in cubic GaN epilayers implanted with Mn+ ions. Appl. Phys. Lett. 85, 3777–3779.

Cho, J., Schubert, E.F., Kim, J.K., 2013. Efficiency droop in light emitting diodes: challenges and countermeasures. Laser Photon. Rev. 7 (3), 408–421.

Cui, Y., Lazorov, V.K., Goetz, M.M., Liu, H., Robertson, D.O., Gajdardziska-Josifovska, M., Li, L., 2003. Cubic GaN formation in Mn/GaN multilayer films grown on 6H-SiC (0001). Appl. Phys. Lett. 82, 4666–4668.

Dalpian, G.M., Wei, S.-H., 2004. Hole-mediated stabilization of cubic GaN. Phys. Rev. Lett. 93, 216401.

DeCuir, E.A., Fred, E., Manasreh, M.O., Schormann, J., As, D.J., Lischka, K., 2007. Near-infrared intersubband absorption in nonpolar cubic GaN/AlN superlattices. Appl. Phys. Lett. 91, 041911.

Delaney, K.T., Rinke, P., Van de Walle, C.G., 2009. Auger recombination rates in nitrides from first principles. Appl. Phys. Lett. 94, 191109.

Dhara, S., Datta, A., Wu, C.T., Lan, Z.H., Chen, K.H., Wang, Y.L., Hsu, C.W., Shen, C.H., Chen, L.C., Chen, C.C., 2004. Hexagonal-to-cubic phase transformation in GaN nanowires by Ga+ implantation. Appl. Phys. Lett. 84, 5473–5475.

Garcia, V.D.C., Hinostroza, I.E.O., Echavarria, A.E., Luna, E.L., Rodriguez, A.G., Vidal, M.A., 2015. Bulk lattice parameter and band gap of cubic InxGa(1-x)N (001) alloys on MgO (100) substrates. J. Cryst. Growth 418, 120–125.

Ghosh, S., Bag, A., Jana, S.K., Mukhopadhyay, P., Dinara, S.M., Kabi, S., Biswas, D., 2014. An unified analytical model for design consideration of doped cubic and undoped hexagonal AlGaN/GaN MIS gate HEMTs. Solid State Electron. 96, 1–8.

Goldys, E.M., Godlewski, M., Langer, R., Barski, A., Bergman, P., Monemar, B., 1999. Analysis of the red optical emission in cubic GaN grown by molecular-beam epitaxy. Phys. Rev. B 60, 5464–5469.

Granzner, R., Tschumak, E., Kittler, M., Tonisch, K., Jatal, W., Pezoldt, J., As, D., Schwierz, F., 2011. Vertical design of cubic GaN-based high electron mobility transistors. J. Appl. Phys. 110, 114501.

Kako, S., Holmes, M., Sergent, S., Burger, M., As, D.J., Arakawa, Y., 2014. Single photon emission from cubic GaN quantum dots. Appl. Phys. Lett. 104, 011101.

Kanaya, K., Okayama, S., 1972. Penetration and energy-loss theory of electrons in solid targets. J. Phys. D Appl. Phys. 5, 43.

Karpov, S., 2015. ABC-model for interpretation of internal quantum efficiency and its droop in III-nitride LEDs: a review. Opt. Quant. Electron. 47 (6), 1293–1303.

Kemper, R.M., Schupp, T., Haberlen, M., Niendorf, T., Maier, H.-J., Dempewolf, A., Bertram, F., Christen, J., Kirste, R., Hoffman, A., Lindner, J., As, D.J., 2011. Anti-phase domains in cubic GaN. J. Appl. Phys. 110, 123512.

Kim, C.-H., Han, B.-H., 1998. Valence subbands and optical gain in wurtzite and zinc-blende strained GaN/AlGaN quantum wells. Solid State Commun. 106 (3), 127–132.

Kitamura, T., Suzuki, Y., Ishida, Y., Shen, X.Q., Nakanishi, H., Chichibu, S.F., Shimizu, M., Okumura, H., 2001. Optical properties of cubic InGaN/GaN multiple quantum wells on 3C-SiC substrates by radio-frequency plasma-assisted molecular beam epitaxy. Phys. Status Solidi A 188 (2), 705–709.

Kohler, U., As, D.J., Potthast, S., Khartchenko, A., Lischka, K., Noriega, O.C., Meneses, E.A., Tabata, A., Rodrigues, S.C.P., Scolfaro, L.M.R., Sipahi, G.M., Leite, J.R., 2002. Optical characterization of cubic AlGaN/GaN quantum wells. Phys. Status Solidi A 192, 129–134.

Lee, S.C., Pattada, B., Hersee, S.D., Jiang, Y.-B., Brueck, S.R.J., 2005. Nanoscale spatial phase modulation of GaN on V-grooved Si substrate—cubic phase GaN on Si(100) for monolithic integration. IEEE J. Quantum Electron. 41 (4), 596–605.

Li, S., Schormann, J., Pawlis, A., As, S.J., Lischka, K., 2005. Cubic InGaN/GaN multi-quantum wells and AlGaN/GaN distributed Bragg reflectors for application in resonant cavity LEDs. Microelectron. J. 36, 963–968.

Li, S., Schormann, J., As, D.J., Lischka, K., 2007. Room temperature green light emission from nonpolar cubic InGaN/GaN multi-quantum-wells. Appl. Phys. Lett. 90, 071903.

Lin, G.-B., Shan, Q., Wang, Y., Li, T., Schubert, E.F., 2014. "U-turn" feature in the efficiency-versus-current curve of GaInN/GaN light-emitting diodes. Appl. Phys. Lett. 105, 221116.

Liu R., Bayram C., 2016a. Maximizing cubic phase gallium nitride surface coverage on nano-patterned silicon (100). Appl. Phys. Lett. 109, 042103.

Liu, R., Bayram, C., 2016b. Cathodoluminescence study of luminescence centers in hexagonal and cubic phase GaN hetero-integrated on Si(100). J. Appl. Phys. 120, 025106.

Meyaard, D.S., Lin, G.-B., Cho, J., Schubert, E.F., Shim, H., Han, S.-H., Kim, M.-H., Sone, C., Kim, Y.S., 2013. Identifying the cause of the efficiency droop in GaInN light-emitting diodes by correlating the onset of high injection with the onset of the efficiency droop. Appl. Phys. Lett. 102, 251114.

Nakadaira, A., Tanaka, H., 1997. Stimulated emission at 34 K from an optically pumped cubic GaN/AlGaN heterostructure grown by metalorganic vapor-phase epitaxy. Appl. Phys. Lett. 71, 812.

Ohtoshi, T., Niwa, A., Kuroda, T., 1998. Orientation dependence of optical gain in zincblende-GaN strained-quantum-well lasers. IEEE J. Sel. Top. Quantum Electron. 4 (3), 527–530.

Park, S.-H., Lee, Y.-T., 2010. Optical properties of zinc-blende InGaN/GaN quantum well structures and comparison with experiment. Chin. Phys. Lett. 27 (4), 044208.

Radosavljevic, A., Radovanovic, J., Milanovic, V., 2014. Optimization of cubic GaN/AlGaN quantum well-based structures for intersubband absorption in the infrared spectral range. Solid State Commun. 182, 38–42.

Sanorpim, S., Kuntharin, S., Parinyataramas, J., Yaguchi, H., Iwahashi, Y., Orihara, M., Hijikata, Y., Yoshida, S., 2011. High cubic-phase purity InN on MgO (001) using cubic-phase GaN as a buffer layer. Phys. Semicond. 1399, 131–132.

Scolfaro, L.M.R., 2002. Phase separation in cubic group-III nitride alloys. Phys. Status Solidi A 190 (1), 15–22.

Stark, C.J.M., Detchprohm, T., Lee, S.C., Jiang, Y.-B., Brueck, S.R.J., Wetzel, C., 2013. Green cubic GaInN/GaN light-emitting diode on microstructured silicon (100). Appl. Phys. Lett. 103, 232107.

Sun, Q., Yerino, C.D., Leung, B., Han, J., Coltrin, M.E., 2011. Understanding and controlling heteroepitaxy with the kinetic Wulff plot: a case study with GaN. J. Appl. Phys. 110, 053517.

Tanaka, H., Nakadaira, A., 2000. Cubic GaN light emitting diode grown by metalorganic vapor-phase epitaxy. IEICE Trans. Electron. E83-C, 585–590.

Taniyasu, Y., Suzuki, K., Lim, D.H., Jia, A.W., Shimotomai, M., Kato, Y., Kobayashi, M., Yoshikawa, A., Takahashi, K., 2000. Cubic InGaN/GaN double-heterostructure light emitting diodes grown on GaAs (001) substrates by MOVPE. Phys. Status Solidi A 180, 241–246.

Tsao, J.Y., Crawford, M.H., Coltrin, M.E., Fischer, A.J., Koleske, D.D., Subramania, G.S., Wang, G.T., Wierer, J.J., Karlicek, R.F., 2014. Toward smart and ultra-efficient solid-state lighting. Adv. Opt. Mater. 2, 809–836.

Verzellesi, G., Saguatti, D., Meneghini, M., Bertazzi, F., Goano, M., Meneghesso, G., Zanoni, E., 2013. Efficiency droop in InGaN/GaN blue light-emitting diodes: physical mechanisms and remedies. J. Appl. Phys. 114, 071101.

Waheeda, S.N., Zainal, N., Hassan, Z., Novikov, S.V., Akimov, A.V., Kent, A.J., 2014. Low fraction of hexagonal inclusions in thick and bulk cubic GaN layers. Appl. Surf. Sci. 317, 1010–1014.

Wei, C.H., Xie, Z.Y., Li, L.Y., Yu, Q.M., Edgar, J.H., 2000. MOCVD growth of cubic GaN on 3C-SiC deposited on Si (100) substrates. J. Electron. Mater. 29 (3), 317–321.

Weisbuch, C., Piccardo, M., Martinelli, L., Iveland, J., Peretti, J., Speck, J.S., 2015. The efficiency challenge of nitride light-emitting diodes for lighting. Phys. Status Solidi A 212 (5), 899–913.

Wierer, J.J., Tsao, J.Y., Sizov, D.S., 2013. Comparison between blue lasers and light-emitting diodes for future solid-state lighting. Laser Photon. Rev. 7 (6), 963–993.

Wood, C., Jena, D., 2008. Polarization Effects in Semiconductors: From Ab Initio Theory to Device Applications. Springer, New York.

Wu, J., Yaguchi, H., Onabe, K., Shiraki, Y., 1999. Stimulated emission from optically pumped cubic GaN/AlGaN double heterostructures. J. Cryst. Growth 197, 73–77.

Xu, M., Liu, C.X., Liu, H.F., Luo, G.M., Chen, X.M., Yu, W.X., Cui, S.F., Li, J.H., Chen, H., Mai, Z.H., Zhou, J.M., Jia, Q.J., Zheng, W.L., Jiang, X.M., 2002. Thermal stability of cubic GaN film grown by molecular-beam epitaxy on GaAs(001). Phys. Lett. A 299, 79–84.

Yang, H., Zheng, L.X., Li, J.B., Wang, X.J., Xu, D.P., Wang, Y.T., Hu, X.W., Han, P.D., 1999. Cubic-phase GaN light-emitting diodes. Appl. Phys. Lett. 74, 2498.

Zhao, Y., Yan, Q., Huang, C.-Y., Huang, S.-C., Hsu, P.S., Tanaka, S., Pan, C.-C., Kawaguchi, Y., Fujito, K., Van de Walle, C.G., Speck, J.S., DenBaars, S.P., Nakamura, S., Feezell, D., 2012. Indium incorporation and emission properties of nonpolar and semipolar InGaN quantum wells. Appl. Phys. Lett. 100, 201108.

INDEX

Note: Page numbers followed by "*f*" indicate figures, and "*t*" indicate tables.

B

C

D

K

L

M

N

Q

R

S

CONTENTS OF VOLUMES IN THIS SERIES

Volume 4 **Physics of III–V Compounds**

Volume 5 **Infrared Detectors**

Volume 6 **Injection Phenomena**

Volume 11 **Solar Cells**

H. J. Hovel, Introduction; Carrier Collection, Spectral Response, and Photocurrent; Solar Cell Electrical Characteristics; Efficiency; Thickness; Other Solar Cell Devices; Radiation Effects; Temperature and Intensity; Solar Cell Technology

Volume 12 **Infrared Detectors (II)**

W. L. Eiseman, J. D. Merriam, and R. F. Potter, Operational Characteristics of Infrared Photodetectors
P. R. Bratt, Impurity Germanium and Silicon Infrared Detectors
E. H. Putley, InSb Submillimeter Photoconductive Detectors
G. E. Stillman, C. M. Wolfe, and J. O. Dimmock, Far-Infrared Photoconductivity in High Purity GaAs
G. E. Stillman and C. M. Wolfe, Avalanche Photodiodes
P. L. Richards, The Josephson Junction as a Detector of Microwave and Far-Infrared Radiation
E. H. Putley, The Pyroelectric Detector – An Update

Volume 13 **Cadmium Telluride**

K. Zanio, Materials Preparations; Physics; Defects; Applications

Volume 14 **Lasers, Junctions, Transport**

N. Holonyak, Jr., and M. H. Lee, Photopumped III–V Semiconductor Lasers
H. Kressel and J. K. Butler, Heterojunction Laser Diodes
A. Van der Ziel, Space-Charge-Limited Solid-State Diodes
P. J. Price, Monte Carlo Calculation of Electron Transport in Solids

Volume 15 **Contacts, Junctions, Emitters**

B. L. Sharma, Ohmic Contacts to III–V Compounds Semiconductors
A. Nussbaum, The Theory of Semiconducting Junctions
J. S. Escher, NEA Semiconductor Photoemitters

Volume 16 **Defects, (HgCd)Se, (HgCd)Te**

H. Kressel, The Effect of Crystal Defects on Optoelectronic Devices
C. R. Whitsett, J. G. Broerman, and C. J. Summers, Crystal Growth and Properties of $Hg_{1-x}Cd_xSe$ Alloys
M. H. Weiler, Magnetooptical Properties of $Hg_{1-x}Cd_xTe$ Alloys
P. W. Kruse and J. G. Ready, Nonlinear Optical Effects in $Hg_{1-x}Cd_xTe$

Volume 17 **CW Processing of Silicon and Other Semiconductors**

J. F. Gibbons, Beam Processing of Silicon
A. Lietoila, R. B. Gold, J. F. Gibbons, and L. A. Christel, Temperature Distributions and Solid Phase Reaction Rates Produced by Scanning CW Beams

Volume 18 **Mercury Cadmium Telluride**

Volume 19 **Deep Levels, GaAs, Alloys, Photochemistry**

Volume 20 **Semi-Insulating GaAs**

Volume 21 **Hydrogenated Amorphous Silicon**

Part A

Volume 22 **Lightwave Communications Technology**

Part A

Part B

Part C

Part D

Part E

Volume 23 **Pulsed Laser Processing of Semiconductors**

Volume 24 **Applications of Multiquantum Wells, Selective Doping, and Superlattices**

Volume 25 **Diluted Magnetic Semiconductors**

Volume 26 **III–V Compound Semiconductors and Semiconductor Properties of Superionic Materials**

Volume 27 **High Conducting Quasi-One-Dimensional Organic Crystals**

Volume 28 **Measurement of High-Speed Signals in Solid State Devices**

Volume 29 Very High Speed Integrated Circuits: Gallium Arsenide LSI

Volume 30 Very High Speed Integrated Circuits: Heterostructure

Volume 31 Indium Phosphide: Crystal Growth and Characterization

Volume 32 Strained-Layer Superlattices: Physics

Volume 33 Strained-Layer Superlattices: Material Science and Technology

Volume 34 **Hydrogen in Semiconductors**

Volume 35 **Nanostructured Systems**

Volume 36 **The Spectroscopy of Semiconductors**

Volume 37 **The Mechanical Properties of Semiconductors**

Volume 38 **Imperfections in III/V Materials**

Volume 39 **Minority Carriers in III–V Semiconductors: Physics and Applications**

Volume 40 **Epitaxial Microstructures**

Volume 41 **High Speed Heterostructure Devices**

Volume 42 **Oxygen in Silicon**

Volume 43 **Semiconductors for Room Temperature Nuclear Detector Applications**

Volume 44 **II–IV Blue/Green Light Emitters: Device Physics and Epitaxial Growth**

Volume 45 **Effect of Disorder and Defects in Ion-Implanted Semiconductors: Electrical and Physiochemical Characterization**

Volume 46 **Effect of Disorder and Defects in Ion-Implanted Semiconductors: Optical and Photothermal Characterization**

A. Othonos and C. Christofides, Photoluminescence and Raman Scattering of Ion Implanted Semiconductors. Influence of Annealing

C. Christofides, Photomodulated Thermoreflectance Investigation of Implanted Wafers. Annealing Kinetics of Defects

U. Zammit, Photothermal Deflection Spectroscopy Characterization of Ion-Implanted and Annealed Silicon Films

A. Mandelis, A. Budiman, and M. Vargas, Photothermal Deep-Level Transient Spectroscopy of Impurities and Defects in Semiconductors

R. Kalish and S. Charbonneau, Ion Implantation into Quantum-Well Structures

A. M. Myasnikov and N. N. Gerasimenko, Ion Implantation and Thermal Annealing of III–V Compound Semiconducting Systems: Some Problems of III–V Narrow Gap Semiconductors

Volume 47 **Uncooled Infrared Imaging Arrays and Systems**

R. G. Buser and M. P. Tompsett, Historical Overview

P. W. Kruse, Principles of Uncooled Infrared Focal Plane Arrays

R. A. Wood, Monolithic Silicon Microbolometer Arrays

C. M. Hanson, Hybrid Pyroelectric-Ferroelectric Bolometer Arrays

D. L. Polla and J. R. Choi, Monolithic Pyroelectric Bolometer Arrays

N. Teranishi, Thermoelectric Uncooled Infrared Focal Plane Arrays

M. F. Tompsett, Pyroelectric Vidicon

T. W. Kenny, Tunneling Infrared Sensors

J. R. Vig, R. L Filler, and Y. Kim, Application of Quartz Microresonators to Uncooled Infrared Imaging Arrays

P. W. Kruse, Application of Uncooled Monolithic Thermoelectric Linear Arrays to Imaging Radiometers

Volume 48 **High Brightness Light Emitting Diodes**

G. B. Stringfellow, Materials Issues in High-Brightness Light-Emitting Diodes

M. G. Craford, Overview of Device Issues in High-Brightness Light-Emitting Diodes

F. M. Steranka, AlGaAs Red Light Emitting Diodes

C. H. Chen, S. A. Stockman, M. J. Peanasky, and C. P. Kuo, OMVPE Growth of AlGaInP for High Efficiency Visible Light-Emitting Diodes

F. A. Kish and R. M. Fletcher, AlGaInP Light-Emitting Diodes

M. W. Hodapp, Applications for High Brightness Light-Emitting Diodes

J. Akasaki and H. Amano, Organometallic Vapor Epitaxy of GaN for High Brightness Blue Light Emitting Diodes

S. Nakamura, Group III–V Nitride Based Ultraviolet-Blue-Green-Yellow Light-Emitting Diodes and Laser Diodes

Volume 49 **Light Emission in Silicon: from Physics to Devices**

D. J. Lockwood, Light Emission in Silicon

G. Abstreiter, Band Gaps and Light Emission in Si/SiGe Atomic Layer Structures

T. G. Brown and D. G. Hall, Radiative Isoelectronic Impurities in Silicon and Silicon-Germanium Alloys and Superlattices
J. Michel, L. V. C. Assali, M. T. Morse, and L. C. Kimerling, Erbium in Silicon
Y. Kanemitsu, Silicon and Germanium Nanoparticles
P. M. Fauchet, Porous Silicon: Photoluminescence and Electroluminescent Devices
C. Delerue, G. Allan, and M. Lannoo, Theory of Radiative and Nonradiative Processes in Silicon Nanocrystallites
L. Brus, Silicon Polymers and Nanocrystals

Volume 50 **Gallium Nitride (GaN)**

J. I. Pankove and T. D. Moustakas, Introduction
S. P. DenBaars and S. Keller, Metalorganic Chemical Vapor Deposition (MOCVD) of Group III Nitrides
W. A. Bryden and T. J. Kistenmacher, Growth of Group III–A Nitrides by Reactive Sputtering
N. Newman, Thermochemistry of III–N Semiconductors
S. J. Pearton and R. J. Shul, Etching of III Nitrides
S. M. Bedair, Indium-based Nitride Compounds
A. Trampert, O. Brandt, and K. H. Ploog, Crystal Structure of Group III Nitrides
H. Morkoç, F. Hamdani, and A. Salvador, Electronic and Optical Properties of III–V Nitride based Quantum Wells and Superlattices
K. Doverspike and J. I. Pankove, Doping in the III-Nitrides
T. Suski and P. Perlin, High Pressure Studies of Defects and Impurities in Gallium Nitride
B. Monemar, Optical Properties of GaN
W. R. L. Lambrecht, Band Structure of the Group III Nitrides
N. E. Christensen and P. Perlin, Phonons and Phase Transitions in GaN
S. Nakamura, Applications of LEDs and LDs
I. Akasaki and H. Amano, Lasers
J. A. Cooper, Jr., Nonvolatile Random Access Memories in Wide Bandgap Semiconductors

Volume 51A **Identification of Defects in Semiconductors**

G. D. Watkins, EPR and ENDOR Studies of Defects in Semiconductors
J.-M. Spaeth, Magneto-Optical and Electrical Detection of Paramagnetic Resonance in Semiconductors
T. A. Kennedy and E. R. Claser, Magnetic Resonance of Epitaxial Layers Detected by Photoluminescence
K. H. Chow, B. Hitti, and R. F. Kiefl, μSR on Muonium in Semiconductors and Its Relation to Hydrogen
K. Saarinen, P. Hautojärvi, and C. Corbel, Positron Annihilation Spectroscopy of Defects in Semiconductors
R. Jones and P. R. Briddon, The *Ab Initio* Cluster Method and the Dynamics of Defects in Semiconductors

Volume 51B **Identification Defects in Semiconductors**

G. Davies, Optical Measurements of Point Defects
P. M. Mooney, Defect Identification Using Capacitance Spectroscopy

Volume 52 **SiC Materials and Devices**

Volume 53 **Cumulative Subjects and Author Index Including Tables of Contents for Volumes 1–50**

Volume 54 **High Pressure in Semiconductor Physics I**

Volume 55 **High Pressure in Semiconductor Physics II**

Volume 56 **Germanium Silicon: Physics and Materials**

Volume 57 **Gallium Nitride (GaN) II**

Volume 58 **Nonlinear Optics in Semiconductors I**

Volume 63 **Chemical Mechanical Polishing in Si Processing**

Volume 64 **Electroluminescence I**

Volume 65 **Electroluminescence II**

Volume 66 **Intersubband Transitions in Quantum Wells: Physics and Device Applications II**

Volume 67 **Ultrafast Physical Processes in Semiconductors**

Volume 68 **Isotope Effects in Solid State Physics**

Volume 69 **Recent Trends in Thermoelectric Materials Research I**

Volume 70 **Recent Trends in Thermoelectric Materials Research II**

Volume 71 **Recent Trends in Thermoelectric Materials Research III**

Volume 72 **Silicon Epitaxy**

Volume 73 **Processing and Properties of Compound Semiconductors**

Volume 79 **Semiconducting Chalcogenide Glass II**

Volume 80 **Semiconducting Chalcogenide Glass III**

Volume 81 **Conducting Organic Materials and Devices**

Volume 82 **Semiconductors and Semimetals**

Volume 83 **Semiconductors and Semimetals**

Volume 84 **Semiconductors and Semimetals**

Volume 85 **Semiconductors and Semimetals**

Volume 86 **Advances in Semiconductor Lasers**

Volume 87 **Advances in Photovoltaics: Volume 1**

Volume 88 **Oxide Semiconductors**

Volume 89 **Advances in Photovoltaics: Part 2**

Volume 90 **Advances in Photovoltaics: Part 3**

Volume 91 **Defects in Semiconductors**

FSC
www.fsc.org
MIX
Papier | Fördert gute Waldnutzung
FSC® C083411

Kolibri 360 GmbH
Ferdinand-Jühlke-Straße 7
99095 Erfurt, Deutschland
produktsicherheit@kolibri360.de